Tinbergen's Legacy in *Behaviour*: Sixty Years of Landmark Stickleback Papers

Tinbergen's Legacy in *Behaviour*:

ty Years of Landmark Stickleback Papers

Edited by

Frank von Hippel

BRILL

LEIDEN • BOSTON

2010

This book is printed on acid-free paper.

ISBN 978 90 04 17029 2

PRINTED IN THE NETHERLANDS

Contents

Acknowledgements

In 1992, I returned to Berkeley from three months of observational field work on colobus monkeys in Kenya, under the mentoring of Thelma Rowell. The work was meant to be the basis of my PhD. Although productive, the experience convinced me to become an experimental scientist, not an observational one. I asked George Barlow to tutor me in the classical ethological literature in order to generate ideas for a new dissertation. For one semester, George and I met weekly in his office to discuss the works of Niko Tinbergen, Konrad Lorenz and other important figures in the development of ethology.

George had a profound understanding of classical ethology, having worked as a post-doctoral fellow with Lorenz at the Max Planck Institute (1958–1960) and as a visiting scientist during his sabbatical with Tinbergen at Oxford (1973–1974, coincident with Tinbergen's receipt of the Nobel Prize). I hoped to study sexual selection and parental care in an experimental framework focused on a specific set of questions on how parental ability may be assessed during courtship, and I asked George what study animal I should use. George's response was immediate: "Stickleback." Since then I have used threespine stickleback to study a spectrum of topics, including sexual selection, parental care, rapid evolution and speciation, biogeography, invasive species biology, and trophic ecology. More recently, I have found stickleback to be a powerful tool in ecotoxicology.

George realized that I had broad interests, and that I needed a model organism to explore them. He was a cichlid man, and his first inclination may have been to suggest cichlids. But he saw that stickleback would be a perfect fit for me. I started my first stickleback study with a handful of spare aquaria that George had in storage. I repaired the cracks and set them up on make-shift racks in Thelma's lab on the third floor of Hilgard Hall. I bought kiddie pools for holding my fish, and put them on the Hilgard balconies, which overlooked the Berkeley Hills. I happened upon a local stickleback population whose males had black nuptial coloration, and the females in this population even developed some red (unfortunately, this special population apparently no longer exists). I then ran the experiments for my Ph.D. with a pocket change budget and George's guidance.

Nearly twenty years later, I have the opportunity to compile special stickleback publications from sixty years of the journal *Behaviour*. I am reminded of reading these papers the first time, back in George's office, and I can hear George gently prodding me to think about how the concepts can inform my work.

George passed away on July 14, 2007, just one year after he attended his last conference, the Fifth International Conference on Stickleback Behaviour and Evolution at the University of Alaska Anchorage, where he was honoured for his lifetime of scientific achievements. He was delighted by all the new ideas at the conference, and quick to point out how presenters' research could be improved by adopting approaches from classical ethology. I give George my deep thanks for starting my research career with these wonderful little fish, and for providing me with a grounding in a field that by the early 1990s was nearly forgotten.

Several people helped me with this book. My wife Cathy edited my essays, and Michael Bell provided useful comments on the Introduction. My commentary on Physiology and Behaviour grew out of several years of discussions and grant writing with Richard Bernhardt, John Postlethwait, Todd O'Hara, William Cresko and Loren Buck. Michiel Thijseen, the senior acquisitions editor in science at Brill, provided guidance on the structure of the book, and Sabine Steenbeek of Brill provided contractual support for contributing authors. Theo Bakker and Frietson Galis helped me to contact some of the Dutch authors of the original stickleback papers. Brian Wisenden helped with the book's title, as it "ties together the man Tinbergen, his role in establishing the field, his role in establishing the journal, and his choice of study organism." The scans of all articles originally published before 2000 were generously provided by JSTOR and are reproduced here with their permission.

This book is dedicated to my children, Max and Sophia (who know more about stickleback than is reasonable) and Sam (who, at age two, cannot *yet* be expected to have mastered the subject).

Frank A. von Hippel
Department of Biological Sciences
University of Alaska Anchorage
3211 Providence Dr.
Anchorage, AK 99508-4614
USA
frank@uaa.alaska.edu

Introduction: The Stickleback Model

Frank A. von Hippel

"With respect to reptiles and fishes," wrote Charles Darwin (1871) in *The Descent of Man, and Selection in Relation to Sex*, "too little is known of their habits to enable us to speak of their marriage arrangements. The stickleback (Gasterosteus), however, is said to be a polygamist; and the male during the breeding-season differs conspicuously from the female." With this introduction to the stickleback, Darwin set the stage for its development as a model organism in behavioural studies. And just one year after the publication of *The Descent of Man*, Darwin (1872) published *The Expression of the Emotions in Man and Animals*, which, along with his chapter on instinct in *The Origin of Species* (Darwin, 1859) launched ethology as a science by applying the ideas of natural and sexual selection to the study of behaviour. He concluded his chapter on instinct with a summary of his theory of natural selection, as applied to behaviour: "Finally, it may not be a logical deduction, but to my imagination it is far more satisfactory to look at such instincts as the young cuckoo ejecting its fosters-brothers, — ants making slaves, — the larvæ of ichneumonidæ feeding within the live bodies of caterpillars, — not as specially endowed or created instincts, but as small consequences of one general law leading to the advancement of all organic beings, — namely, multiply, vary, let the strongest live and the weakest die."

Darwin used the comparative approach with both his animal and human subjects. The co-discoverer of natural selection, Alfred Russel Wallace, wrote in his review of Darwin's book on the expression of emotions (Wallace, 1873):

> By means of a series of questions sent to correspondents in various parts of the world, Mr. Darwin has ascertained that many well-known modes of expression are almost universal. Even such an apparently conventional action as the shrug of helplessness or apologetic refusal has been observed among various savage races. Being thus proved to be natural, not

> an acquired, expression, it becomes necessary to account for it, and this is done on the principle of antithesis; every part of the expression being the opposite of that which implies determination and action. Comparatively few human expressions, on the other hand, can be distinctly recognized in animals, that of sneering by raising the upper lip on one side, and thus showing the canine teeth, being one of the most curious.

The technique of comparative biology, as established by Darwin, was taken up by Niko Tinbergen. It was Tinbergen who built upon what was already a good body of work on threespine stickleback (first described by Linnaeus in 1758 as *Gasterosteus aculeatus*, which means 'belly bone spined') and made them a model in ethology. Tinbergen's book (1951), *The Study of Instinct*, marked a critical achievement in the maturation of ethology as a scientific enterprise. In it, Tinbergen illustrated his vision of ethology with comparative examples from throughout the animal kingdom, but with special attention paid to threespine stickleback. Of the 210 pages of text, threespine stickleback appears on 42 of them, as well as the photo plate opposite the title page. The highly ritualized behaviour of the stickleback appealed to Tinbergen's penchant for experimentation. The stickleback responded well to models, and this discovery gave rise to a flurry of experiments using dummy males and females. The ease of maintaining stickleback in the lab also appealed to Tinbergen, for here was an animal that could be studied in natural and artificial environments. Tinbergen's reliance on the stickleback and his other favourite species led one reviewer to write, "In its current stage, ethology resembles a patchwork quilt, pieced together with bits of stickleback, wasp, cichlid, duck, herring gull, etc., the whole seeming to cover the reality better than any theory thus far advanced" (Greenberg, quoted in Kruuk, 2003).

Three years before the publication of *The Study of Instinct*, in 1948, Tinbergen founded the journal *Behaviour* to serve the burgeoning field of ethology. The first issue of the journal included Tinbergen and van Iersel's (1948) classic paper, "'Displacement reactions' in the three-spined stickleback." By the time Tinbergen received the Nobel Prize in 1973, his former students, such as Jan van Iersel, Piet Sevenster and Desmond Morris, had greatly expanded the utility of the stickleback model through their discoveries in the behaviour of the reproductive cycle.

The threespine stickleback's role as a model organism expanded beyond the confines of ethology in the 1960s and 1970s to include morphological evolution as well. A major appeal of the threespine stickleback was its widespread distribution in marine, brackish and fresh waters of the northern hemisphere. This not only provided local availability of wild-caught fish for investigators in Europe, Asia and North America, but also allowed evolutionary biologists to study processes such as selection, adaptation, gene flow and speciation with an ideal "natural experiment." Ubiquitous, diverse threespine stickleback in recently deglaciated fresh waters on offshore islands led McPhail and Lindsey (1970) to conclude that resident freshwater populations have evolved repeatedly from oceanic populations. The fossil record of threespine stickleback demonstrates that this repetitive diversification has been taking place for at least ten million years (Bell, 1994). Although the marine phenotype has apparently remained nearly unchanged for at least 13 million years (Bell et al., 2009), colonizing populations diverge rapidly in fresh water (Klepaker, 1993; Bell et al., 2004; Kristjánsson, 2005), leading to reproductive isolation from their marine ancestral populations in a matter of decades (Furin, 2006). Marine colonists have given rise to hundreds of thousands of freshwater isolates, each of which has evolved morphological and behavioural traits in fairly predictable patterns depending on such features as predation regime, prey availability, and clarity and ionic strength of the water (Bell & Foster, 1994a). Therefore, the threespine stickleback is among the best model organisms to study iterative and rapid evolution in nature.

The diversification of threespine stickleback led to an additional feature that makes them a valuable study system in evolutionary biology: that is, the formation of species pairs, which may be grouped into six varieties (McKinnon & Rundle, 2002). These species pairs make it possible to distinguish historically contingent and deterministic aspects of speciation, including both its geographic context (allopatric, parapatric, or sympatric) and the processes driving reproductive isolation (prezygotic and postzygotic) (e.g., Hagen, 1967; McPhail, 1969; von Hippel & Weigner, 2004; Karve et al., 2008).

The stickleback model became sufficiently important to a broad range of scientists that in 1976 Robert Wootton published the first book devoted to it, *The Biology of the Sticklebacks*. In 1981, Coad published a 142 page bibliography of the stickleback literature with over 2,000 entries. This was followed in 1983 by *Die Stichlinge* (Paepke, 1983), in 1984 by *A Functional Biology of Sticklebacks* (Wootton, 1984), in 1991 by *The Family Gasterosteidae of World Fish Fauna* (Ziuganov, 1991), in 1994 by *The Evolutionary Biology of the Threespine Stickleback* (Bell & Foster, 1994a), and in 2007 by *Biology of the Three-Spined Stickleback* (Östlund-Nilsson et al., 2007). Along the way, the threespine stickleback also began to serve an important role in the fields of ecotoxicology and developmental genetics.

The success of the threespine stickleback model in ecotoxicology arises in part from its widespread distribution. Contaminated sites throughout much of the northern hemisphere, including near-shore marine habitats and brackish and fresh waters, contain native stickleback populations. Stickleback appear to be a tough fish in the face of chemical onslaught (e.g., Katsiadaki et al., 2002b; Bernhardt et al., 2006), and so they may persist in contaminated environments in which other fish species perish. Furthermore, males of stickleback species are unique in producing a glue protein called spiggin (named after the Swedish word for stickleback, "spigg") that they use in nest construction. This protein is produced under the influence of androgens, hence its presence in females indicates that androgenic endocrine disruptors are present in the environment, and it can also be used as a biomarker for chemicals with anti-androgenic properties (Katsiadaki et al., 2002a). This unique biomarker of androgen-disrupting contaminants is coupled in the stickleback with an oestrogen biomarker, the presence of vitellogenin (egg yolk protein precursor) in males (Katsiadaki et al., 2002b; Hahlbeck et al., 2004). Males only produce vitellogenin when they are exposed to oestrogenic contaminants. These and other advantages of the stickleback model in ecotoxicology (Katsiadaki, 2007; Sanchez et al., 2008) are rendered far more powerful by the rapid development of molecular tools for stickleback, which came about due to advances in developmental genetics.

The threespine stickleback became a model in developmental genetic studies because its model status as a field organism in evolutionary studies could be combined with laboratory genetic techniques adapted from other model organisms, such as zebrafish (*Danio rerio*). This coupling of the ability to perform sophisticated genetic techniques in the laboratory with the extraordinary knowledge of stickleback ecology and evolution gleaned from over a century of study in the field motivated the scientific community to sequence the stickleback genome, with the initial assembly released in 2007. This major advance led to the rapid development of sophisticated genetic tools in stickleback, such as high-throughput genome sequencing (Baird et al., 2008). These tools, in turn, have been employed in the full array of fields that use the stickleback model, from its original use in ethology, to its subsequent utility in evolutionary biology (reviewed by Kingsley et al., 2004; Kingsley & Peichel, 2007), to its applied employment in ecotoxicology (e.g., Geoghegan et al., 2008), and of course in its use in developmental genetics (e.g., Shapiro et al., 2004; Kimmel et al., 2005). Because the stickleback is now an excellent model organism for a variety of laboratory and field studies to which molecular genetic methods can be applied, it has been dubbed a "supermodel" (Gibson, 2005).

The stickleback supermodel began its modelling career in ethology due to the pioneering work of Tinbergen and his Dutch colleagues. Therefore, this book begins with Russell and Russell's (1985) history of the stickleback's development into an ethological model. Many of the critical papers in the development of the stickleback model, from 1948 until today, were published in the pages of *Behaviour*. Most of these papers focused on threespine stickleback, but important contributions were made via the study of other stickleback species as well. This book comprises selected papers grouped into five conceptual areas, beginning with (1) the stages in the stickleback reproductive cycle (establishment of a territory and nest building, courtship and parental care); because the reproductive cycle has received the most attention since Tinbergen and his colleagues began investigating it, this book devotes the most space to its treatment. This is followed by discussion of (2) homosexuality, cannibalism and sexual strategies, (3) predators and parasites, (4) physiology and behaviour, and (5) behavioural genetics, phylogenetics & speciation. Many of the chosen articles are too long to republish in their entirety, in which case excerpts are included with the pertinent parts; partial pages that are not part of the intended text are obscured with an overlay in order to draw the reader's attention to the selected text. Nine retrospectives accompany the original articles, as authors look back upon their contributions and comment on the development of their research and the discipline.

Notes on terminology

While European authors typically use the common name "three-spined stickleback," the official common name in North America is "threespine stickleback" (Nelson et al., 2004).

The word stickleback is pluralized the same way the word fish is. Fish and stickleback should be used as the plural in reference to multiple individuals of a single species. However, reference to more than one species should use fishes and sticklebacks with a terminal "s" as the plural. Despite this convention, many authors use "sticklebacks" to refer to multiple individuals of the same species.

Ever since the first printed accounts of sticklebacks, tremendous taxonomic confusion has accompanied the publication of various Latin names. This has been particularly prevalent in the older threespine stickleback literature due to the lack of understanding of how similar resident freshwater isolates were founded repeatedly by marine colonists and evolved in parallel, creating a "phylogenetic raceme" (Bell & Foster, 1994b). For example, the terms trachurus, semiarmatus and leiurus were frequently used in the literature to refer to different lateral plate morphs of threespine stickleback, and they were often employed as separate nominal species or subspecies (Wootton, 1976), although it is now known that they are not. Furthermore, the terms were inconsistently applied (Bell & Foster, 1994b). Hence, Hagen and Gilbertson (1972) coined the terms "complete," "partial," and "low" for the three lateral plate morphs described by trachurus, semiarmatus and leiurus in the older literature.

The terminology used in the articles included in this book has been left as originally published. Because the articles are republished here as they originally appeared in print, they include both their beauty and their blemishes; errors are not corrected not matter how blatant.

The cover of this book is a special illustration of stickleback species drawn by Alexander Francis Lydon (1836–1917) for the book *British Fresh-Water Fishes*, which was published in 1879. The illustration includes the rough-tailed stickleback, half-armed stickleback, smooth-tailed stickleback, short-spined stickleback, four-spined stickleback and ten-spined stickleback. The first four are all threespine stickleback (*Gasterosteus aculeatus*), demonstrating the taxonomic confusion that accompanied the stickleback family at that time. The four-spined stickleback is most likely a threespine stickleback with four spines (a rare trait found in disparate populations), though it could be what is now known as *Apeltes quadracus*, which would be curious because it is known as an eastern North American endemic (Wootton, 1976). The latter possibility would suggest that the species was extirpated from the British Isles. The author, the Reverend W. Houghton (1879), notes that, "The four-spined stickleback (*G. spinulosus*) appears to be very rare. It was first discovered by Dr. J. Stark in a pond near Edinburgh in 1830." The ten-spined stickleback, then *Gasterosteus pungitius*, is now the ninespine stickleback, *Pungitius pungitius*.

A note on culture

One final note on the illustrious stickleback: it has its place in human culture as well as in science. Houghton (1879) wrote of the threespine stickleback, "these fish are caught in incredible numbers in the Baltic about the middle of November, when they assemble on the coasts of that sea, and are taken by fishermen in boat loads. They are boiled, and the oil they contain is skimmed off the water: a bushel of fish is said to yield two gallons of oil. The refuse is spread over the ground for manure." In my hometown of Anchorage, Alaska, one of our local waterways, Ship Creek, is known in the indigenous Dena'ina language as Dgheyay Leht, 'Where Stickleback Run.' The Dena'ina relied on the anadromous threespine stickleback ('needlefish'), boiled for soup, when salmon were scarce. "Though generally of a small size," wrote Houghton (1879), "the Sticklebacks yield to none in point of interest."

References

Baird, N.A., Etter, P.D., Atwood, T.S., Currey, M.C., Shiver, A.L., Lewis, Z.A., Selker, E.U., Cresko, W.A. & Johnson, E.A. (2008). Rapid SNP discovery and genetic mapping using sequenced RAD markers. — PLoS ONE 3: e3376.

Bell, M.A. (1994). Paleobiology and evolution of threespine stickleback. — In: The evolutionary biology of the threespine stickleback (Bell, M.A. & Foster, S.A., eds). Oxford University Press, Oxford, p. 439-471.

Bell, M.A., Aguirre, W.E. & Buck, N.J. (2004). Twelve years of contemporary armor evolution in a threespine stickleback population. — Evolution 48: 814-824.

Bell, M.A. & Foster, S.A. (eds) (1994a). The evolutionary biology of the threespine stickleback. — Oxford University Press, Oxford.

Bell, M.A. & Foster, S.A. (1994b). Introduction to the evolutionary biology of the threespine stickleback. — In: The evolutionary biology of the threespine stickleback (Bell, M.A. & Foster, S.A., eds). Oxford University Press, Oxford, p. 1-27.

Bell, M.A., Stewart, J.D. & Park, P.J. (2009). The world's oldest fossil threespine stickleback fish. — Copeia 2009: 256-265.

Bernhardt, R.R., von Hippel, F.A. & Cresko, W.A. (2006). Perchlorate induces hermaphroditism in threespine sticklebacks. — Environ. Toxicol. Chem. 25: 2087-2096.

Coad, B.W. (1981). A bibliography of the sticklebacks (Gasterosteidae: Osteichthyes). — National Museum of Natural Sciences, National Museums of Canada, Ottawa.

Darwin, C. (1859). On the origin of species by means of natural selection. — Murray, London.

Darwin, C. (1871). The descent of man, and selection in relation to sex. — Murray, London.

Darwin, C. (1872). The expression of the emotions in man and animals. — Murray, London.

Furin, C.G. (2006). The role of assortative mating between a recently derived resident-freshwater population of threespine stickleback (*Gasterosteus aculeatus*) and its putative anadromous ancestor. — MS thesis, University of Alaska Anchorage.

Geoghegan, F., Katsiadaki, I., Williams, T.D. & Chipman, J.K. (2008). A cDNA microarray for the three-spined stickleback, *Gasterosteus aculeatus* L., and analysis of the interactive effects of oestradiol and dibenzanthracene exposures. — J. Fish Biol. 72: 2133-2153.

Gibson, G. (2005). The synthesis and evolution of a supermodel. — Science 307: 1890-1891.

Hagen, D.W. (1967). Isolating mechanisms in threespine sticklebacks (*Gasterosteus*). — J. Fish. Res. Board Can. 24: 1637-1692.

Hagen, D.W. & Gilbertson, L.G. (1972). Geographic variation and environmental selection in *Gasterosteus aculeatus* L. in the Pacific northwest, America. — Evolution 26: 32-51.

Hahlbeck, E., Katsiadaki, I., Mayer, I., Adolfsson-Erici, M., James, J. & Bengtsson, B.E. (2004). The juvenile three-spined stickleback (*Gasterosteus aculeatus* L.) as a model organism for endocrine disruption — II — Kidney hypertrophy, vitellogenin and spiggin induction. — Aquat. Toxicol. 70: 311-326.

von Hippel, F.A. & Weigner, H. (2004). Sympatric anadromous-resident pairs of threespine stickleback species in young lakes and streams at Bering Glacier, Alaska. — Behaviour 141: 1441-1464.

Houghton, W. (1879). British fresh-water fishes. — W. Mackenzie, London.

Karve, A.D., von Hippel, F.A. & Bell, M.A. (2008). Isolation between sympatric anadromous and resident threespine stickleback species in Mud Lake, Alaska. — Environ. Biol. Fishes 81: 287-296.

Katsiadaki, I. (2007). The use of the stickleback as a sentinel and model species in ecotoxicology. — In: Biology of the three-spined stickleback (Östlund-Nilsson, S., Mayer, I. & Huntingford, F.A., eds). CRC Press, Boca Raton, p. 319-351.

Katsiadaki, I., Scott, A.P., Hurst, M.R., Matthiessen, P. & Mayer, I. (2002a). Detection of environmental androgens: a novel method based on enzyme-linked immunosorbent assay of spiggin, the stickleback (*Gasterosteus aculeatus*) glue protein. — Environ. Toxicol. Chem. 21: 1946-1954.

Katsiadaki, I., Scott, A.P. & Mayer, I. (2002b). The potential of the three-spined stickleback (*Gasterosteus aculeatus* L.) as a combined biomarker for oestrogens and androgens in European waters. — Mar. Environ. Res. 54: 725-728.

Kimmel, C.B., Ullmann, B., Walker, C., Wilson, C., Currey, M., Phillips, P.C., Bell, M.A., Postlethwait, J.H. & Cresko, W.A. (2005). Evolution and development of facial bone morphology in threespine sticklebacks. — P. Natl. Acad. Sci. USA 102: 5791-5796.

Kingsley, D.M. & Peichel, C.L. (2007). The molecular genetics of evolutionary change in sticklebacks. — In: Biology of the three-spined stickleback (Ostlund-Nilsson, S., Mayer, I. & Huntingford, F.A., eds). CRC Press, Boca Raton.

Kingsley, D.M., Zhu, B.L., Osoegawa, K., De Jong, P.J., Schein, J., Marra, M., Peichel, C., Amamiya, C., Schluter, D., Balabhadra, S., Friedlander, B., Cha, Y.M., Dickson, M., Grimwood, J., Schmutz, J., Talbot, W.S. & Myers, R. (2004). New genomic tools for molecular studies of evolutionary change in threespine sticklebacks. — Behaviour 141: 1331-1344.

Klepaker, T. (1993). Morphological changes in a marine population of threespine stickleback, *Gasterosteus aculeatus*, recently isolated in fresh water. — Can. J. Zool. 71: 1251-1258.
Kristjánsson, B.K. (2005). Rapid morphological changes in threespine stickleback, *Gasterosteus aculeatus*, in freshwater. — Environ. Biol. Fishes 74: 357-363.
Kruuk, H. (2003). Niko's nature; a life of Niko Tinbergen and his science of animal behaviour. — Oxford University Press, Oxford.
McKinnon, J.S. & Rundle, H.D. (2002). Speciation in nature: the threespine stickleback model systems. — Trends Ecol. Evol. 17: 480-488.
McPhail, J.D. (1969). Predation and the evolution of a stickleback (*Gasterosteus*). — J. Fish. Res. Board Can. 26: 3183-3208.
McPhail, J.D. & Lindsey, C.C. (1970). Freshwater fishes of northwestern Canada and Alaska. — Fisheries Research Board of Canada, Ottawa.
Nelson, J.S., Crossman, E.J., Espinosa-Pérez, H., Findley, L.T., Gilbert, C.R., Lea, R.N. & Williams, J.D. (2004). Common and scientific names of fishes from the United States, Canada and Mexico, 6th edn. — American Fisheries Society, Herndon, VA.
Östlund-Nilsson, S., Mayer, I. & Huntingford, F.A. (eds) (2007). Biology of the three-spined stickleback. — CRC Press, London.
Paepke, H.-J. (1983). Die stichlinge. — Ziemsen Verlag, Wittenberg.
Russell, W.M.S. & Russell, C. (1985). Sticklebacks and ethology. — Behaviour 93: 184-193.
Sanchez, W., Katsiadaki, I., Piccini, B., Ditche, J.M. & Porcher, J.M. (2008). Biomarker responses in wild three-spined stickleback (*Gasterosteus aculeatus* L.) as a useful tool for freshwater biomonitoring: a multiparametric approach. — Environ. Int. 34: 490-498.
Shapiro, M.D., Marks, M.E., Peichel, C.L., Blackman, B.K., Nereng, K.S., Jonsson, B., Schluter, D. & Kingsley, D.M. (2004). Genetic and developmental basis of evolutionary pelvic reduction in threespine sticklebacks. — Nature 428: 717-723.
Tinbergen, N. (1951). The study of instinct. — Oxford University Press, Oxford.
Tinbergen, N. & van Iersel, J.J.A. (1948). "Displacement reactions" in the three-spined stickleback. — Behaviour 1: 56-63.
Wallace, A.R. (1873). [Review of] The expression of the emotions in man and animals by Charles Darwin, M.A., F.R.S., & c. London: Murray. 1872. — Quarterly Journal of Science 3: 113-118.
Wootton, R.J. (1976). The biology of the sticklebacks. — Academic Press, London.
Wootton, R.J. (1984). A functional biology of sticklebacks. — University of California Press, Berkeley.
Ziuganov, V. (1991). The family Gasterosteidae of world fish fauna. — Nauka Publishing House, Leningrad.

STICKLEBACKS AND ETHOLOGY

by

W. M. S. RUSSELL and CLAIRE RUSSELL

(Dept. of Sociology, University of Reading, Berks. RG1 6NH, England)

(Acc. 7-IX-1984)

In a paper published in 1972 (p. 232), WOOTTON wrote that 'the reproductive behaviour of the male three-spined stickleback has become one of the classical subjects for detailed analysis'. This is no overstatement. The three-spined stickleback occupies a unique place in the history of experimental ethology. It is the main species discussed in TINBERGEN's 'Objectivistic Study' paper (1942), and it takes pride of place in "The Study of Instinct" (1951), with 42 entries in the index, compared with 25 for the herring gull, which comes second. It has a similar position in classical comparative papers by other ethologists, such as that of BAERENDS on fish behaviour in 1957. Of the other four or more stickleback species (NORMAN, 1937, p. 63; MORRIS, 1958, pp. 1-3), only the ten-spined early received considerable attention: three-spined and ten-spined figure largely in MORRIS's classical paper on courtship ceremonies (1956). One or both of these species were discussed at every International Ethology Conference from 1952 to 1971. A paper on the three-spined by TINBERGEN & VAN IERSEL appeared in the first issue of "Behaviour" (1947). Of the first nine "Behaviour" Supplements, three were on sticklebacks: those of VAN IERSEL and SEVENSTER concerned the three-spined and came from Leiden; that of MORRIS concerned the ten-spined and came from Oxford, which in this particular context may be considered a daughter foundation of Leiden. Our present Symposium bears eloquent witness to the flood of stickleback work since then. In experimental ethology, sticklebacks have played a part comparable to that of the genus *Drosophila* in genetics, and the three-spined to that of *Drosophila melanogaster*.

Why is this? The choice of species for experimental purposes is an important aspect of animal experiment, as one of us discussed some years ago in a book on humane experimental technique (RUSSELL & BURCH, 1959, pp. 141 ff.). The use of the stickleback in ethology is a success story. Why? If a species is suitable, the more that is learned about it, the more refined experiment and analysis become possible, and the more interesting questions are opened up. This kind of momentum develops, provided the early work is of high quality, ideally performed at a major centre of the science concerned. By the 1930s, there was a considerable literature of observations and simple experiments on sticklebacks, notably their reproductive behaviour (VAN IERSEL, 1953; MORRIS, 1958). The great Victorian naturalist Frank BUCKLAND, for instance, describes a three-spined stickleback shaking up the nest materials, 'and dragging out the eggs, and then pushing them into their receptacles again, and tucking them up with his snout, and arranging the whole to his mind, and again and again adjusting it till he was satisfied' (cited in NORMAN, 1947, p. 307). The 'detailed analysis' began in the 1930s in the Netherlands, already one guarantee of success. In ethology, when we are impressed by a piece of meticulous and elegant experiment and analysis, we say it is of Dutch quality. Specifically, as we celebrate this week, it began in Leiden. Konrad LORENZ launched the discipline of ethology; Niko TINBERGEN made it a sophisticated experimental science. The first decisive event, therefore, was TINBERGEN's choice of the three-spined stickleback as a favourite experimental object. A second critical moment came when TINBERGEN had to hand over the direction of the work at Leiden to a successor: Jan VAN IERSEL took over (VAN IERSEL, 1953, p. 1) after 1947, and from that moment on the continuing progress of stickleback work, and its huge contribution to ethology, was assured. The present conference testifies to the world-wide expansion of this work, and its venue to the continuing importance of the primary base and centre in Leiden.

All this implies that, as we put it, the species is suitable. Why? The stickleback is a common little fish, for generations the prey of man, and especially of boy. Some 200 years ago, for instance, the great radical and pioneer of birth control, Francis PLACE, was fishing the ditches of London with his school-mates for 'pricklebacks', 'called by the boys "little bats"' (THALE, 1972, pp. 59-60). Common-ness no doubt helps. But commonness is neither sufficient nor even necessary for success as an experimental animal. The relatively common European hamster is rarely used for experiment; the almost world-wide laboratory stocks of the much-used golden hamster were all laboratory-bred from a single female with 12 young found near Aleppo—only two other adults have ever been recorded in the wild (HINDLE, 1947; HINDLE & MAGALHAES, 1957). So what is the background of the stickleback's success?

The major vertebrate groups most extensively studied in ethology are the higher mammals, the birds, and the teleost fishes. The forebrains of higher mammals involve sheets of tissue thrown into folds and wrinkles, which permit an exceptional wealth of interconnections, the basis for learning and integrated intelligence. As C. J. Herrick showed in 1924, the front parts of the brains of birds and teleost fishes are quite different from this. By convergent use of different structures, both birds and teleosts have evolved solid masses of tissue, with scattered islands of cells. As we observed in our book "Human Behaviour" (1961, especially pp. 56-7, 460-62), such brains are well suited for the development through conditioning of extremely complex systems of automatic behaviour, in part programmed in advance as innate. In his study of conditioning in pigeons (1941), Emil Diebschlag showed that conditioned reactions in these birds involve all the characteristics also shown in innate automatic behaviour, including key stimuli, vacuum activity, the formation of reaction chains, and compartmentalisation (cf. Russell & Russell, 1961, pp. 58-75, 466-71); no doubt this applies generally to birds and teleosts. As a result, birds and teleosts are well suited for the refined analysis of complex automatic systems, whether innate or conditioned, and of the conditioning process itself, which still affects behaviour in higher mammals.

Birds and teleosts were therefore the main vertebrate groups studied in the early days of ethology, and only later were ethological methods applied to higher mammals, including monkeys, apes and man. The study of birds and teleosts remains, however, an indispensable base, to which ethology must continually return, for two good reasons. In the Introduction to "The Expression of the Emotions", Darwin tells us he attended to 'the expression of the several passions' in animals, to correct the danger of biassed interpretation, more liable to occur in observing man. But this bias may also affect our observations of monkeys and apes. Their displacement sexual activities, for instance, only interpreted as such in the 1950s by Reynolds and Claire Russell (1970), have always been supposed autochthonous, even by scientists of the stature of Buffon. This has created problems for zoo directors with prudish patrons or superiors. In 1968, in a park in Attard, Malta, we saw a colony of hamadryas baboons. Presented by a British Army unit, they had to be shown to the public, but their cage was boarded up so that only their feet were visible, lest their supposedly sexual antics give offence. Bias of this kind is less likely to affect observations of birds and teleosts.

The second reason applies especially to the fishes, whose aquatic environment makes them in some ways more open to fine experimental manipulations than birds. The finest published ethological studies of monkeys and apes do not begin to compare, in refinement of measurement and intervention, or in intensity of analysis, with the finest experiments on fishes. The "Behaviour" Supplement of VAN IERSEL in 1953 was a revelation of what could be achieved here. It is not too much to say that a centre like Leiden is a base-camp for the whole ethological enterprise.

In 1952, BAERENDS published a classification of the forms of social life in fishes. He distinguished two kinds of society of fishes of roughly the same age, the school and the reproductive territorial colony. Because of their shared signals, he regarded the territory-holders of a reproductive colony as making up a true society, and the later stickleback work of VAN DEN ASSEM (1967) has fully justified this. Some pelagic species with no parental care school for life. Other species migrate in season to a breeding site, where one or both sexes cease to school and set up breeding territories. A possible third kind of society shows various kinds of rank order. But BAERENDS interpreted this, at least in some species, as a transitional arrangement during the passage from schooling to territorial life. Here again, his suggestion has been largely confirmed by the later stickleback work of VAN DEN ASSEM (1967), and more recently BAKKER & SEVENSTER (1983), though VAN DEN ASSEM found traces of rank order lingering into the stable territorial phase.

The teleost species with the most complex behaviour are evidently those which show both schooling and a territorial reproductive phase: they are also the ones that show more or less complex forms of parental behaviour. The transition to the reproductive phase means that its onset and development can be observed, and also retarded or hastened for experimental purposes (see the paper by BAGGERMAN in this issue). The sticklebacks belong in this category, and are therefore highly suitable subjects for experimental ethology.

BAERENDS noted that some territorial species build nests on the bottom, and others at or near the surface. Among sticklebacks, the ten-spined builds a nest in dense water-weeds off the bottom; the three-spined builds a nest on the bottom in a relatively open situation (MORRIS, 1958; see also the paper by KETELE in this issue). HOOGLAND, MORRIS & TINBERGEN (1957) showed that the three-spine armament offers better protection from predators than the softer ten spines, and that this probably determines the different nest-sites of the two species. MORRIS (1958, p. 143) notes that 'the less favourable position for the eggs of [the three-spined] is correlated with more heavily protected eggs, and a more intense fanning cycle', hence more complex and interesting parental behaviour. Moreover, the ten-spined remains shy long after being brought to the laboratory, and often builds in corners difficult to observe, whereas the better protected and therefore bolder three-spined 'can be observed without special precautions within a day, even in tanks with very little cover'. According to HOOGLAND, MORRIS & TINBERGEN, this is one reason why the three-spined 'is so well suited for laboratory studies of its behaviour'.

In his survey of social life in fishes, BAERENDS concluded (1952, p. 217): 'we can now affirm that fishes permit us to study essential problems concerning the whole vertebrate stock'. The three-spined stickleback, so complex in its behaviour and so well suited for the laboratory, is an outstanding example of this. Apart from contributions of technique, stickleback studies have thrown light on many fundamental processes in vertebrate behaviour. SEVENSTER's "Behaviour" Supplement on displacement fanning, for instance (1961), is the most detailed analysis of a single displacement act. More recently (1968), SEVENSTER studied conditioning in the three-spined. He showed that, if biting a rod is used as the conditioned reaction, and the opportunity to court a ripe female as the reward, this reward actually interferes with the reaction, by promoting a sexual mood incompatible with biting. Hence in conditioning experiments we must consider the place of both reaction and reinforcement in the behaviour of the species. Another application of this principle was learned by friends of ours the hard way. They had just had expensive new fabric covers put on their furniture. The very first evening, their cat sharpened his claws on the new covers. Full of indignation, our friends picked up the offending animal, and put him out in the garden. This one-shot training was completely effective. After this, every time the cat wanted to go out in the garden, he sharpened his claws on the new covers.

SEVENSTER has continued his work on conditioning, and more and more intriguing complications emerge (see his paper in this issue), which all studies of vertebrate conditioning will have to reckon with.

The stickleback's repertoire includes many different categories of behaviour of general significance for vertebrates. Hence there have been fundamentally valuable stickleback studies on feeding, schooling, migration and the transition to the reproductive phase, and especially reproductive behaviour, including territory-holding, nest-building, courtship and parental care. Among feeding studies, we may mention that of Beatrice TUGENDHAT (*e.g.* 1960), then working with Niko TINBERGEN, for it illustrates another point: the value of work on sticklebacks for acquiring a mastery of ethological principles and techniques. For this author, as Beatrice GARDNER with her husband R. A. GARDNER, was later to make the wonderful achievement of teaching true language to a chimpanzee (*e.g.* 1969; for the criteria for true language, see RUSSELL & RUSSELL, 1971).

Most published stickleback studies have, however, concerned reproductive behaviour. In 1972, WOOTTON reported a detailed study of three-spined reproductive behaviour under natural conditions in the River Wear. He found close agreement, even quantitatively, with the laboratory results, a striking vindication of the laboratory work.

Two of the behaviour categories will illustrate the general vertebrate relevance of the stickleback work. VAN DEN ASSEM (1967, p. 2) noted what a favourable subject the three-spined is 'for quantitative work on territory problems' of general interest. Hence the importance of his own and VAN IERSEL's (1958) studies of territory in the three-spined. The special relations between neighbours, noted by BAERENDS (1952) and studied in detail by VAN DEN ASSEM, alone and later with VAN DER MOLEN (1969), have a general significance, as Claire RUSSELL has discussed (1975), for mammals and even man (cf. also RUSSELL & RUSSELL, 1984, p. 31).

The study of parental behaviour is of supreme importance, because the whole achievement of our species is based on the development of parental behaviour and its off-shoot, which Claire RUSSELL (1981) has called social parental behaviour. This can be defined as parent-like care for any young or even adult individuals, in a species that can recognize its own young. The qualification is needed, for the three-spined and also the brook stickleback will retrieve strange young (even of other species), presumably through lack of discrimination (VAN IERSEL, 1953, p. 20; WINN, 1960, p. 434). In monkeys and dolphins, social parental behaviour is extended to strangers, and even to individuals of other species (KUMMER, 1968, pp. 116-18; RUSSELL & RUSSELL, 1971, pp. 174-5). Highly-developed parental and social parental behaviour promote individual exploration, intelligence and creativity (RUSSELL & RUSSELL, 1982). To take one expression of this in man, renaissance societies have tended in the direction of the welfare state (RUSSELL & RUSSELL, in press).

These developments are only possible in social vertebrates, for, as Caryl P. HASKINS has emphasized (1952, *passim*), social insects can be sociable only with close relatives. The evolution of parental care in vertebrates is therefore of great interest. It has evolved repeatedly in different groups (RUSSELL, 1965). In some bird species, there is division of labour between two parents: thus among ostriches, both sexes incubate eggs, but the male alone rears the hatched young (BENOIT, 1956). In fishes, whose eggs hatch more rapidly, the same parent or parents care for both eggs and young. Care may be uniparental (by one parent) or biparental (by both). In uniparental fishes, though the female provides care in some cichlid species (BAERENDS, 1952, p. 211), care by the male is more common, perhaps because, as CALLAN has pointed out in general (1970), males are demographically more expendable, for parental care entails risks from predation. In uniparental mammals, for obvious reasons, the rearing parent is always female. Ignorance of the parental arrangements in a species can have practical consequences. NICHOLSON (1957) told the true story of a (human) couple who bought their children a pair of hamsters, 'in the hope that they would breed and thus admirably inform the children about reproduction and family life in mammals'. Not knowing that hamsters are uniparental, they omitted to remove the male after mating. In due course, the children learned some startling facts of life, for 'Father hamster killed Mother hamster and ate the offspring'.

Care of eggs and young by one or both parents has evolved in a number of teleost families (BAERENDS, 1952; RUSSELL, 1965), presumably by convergence and sometimes perhaps parallelism. There is remarkably detailed convergence in some reproductive activities between sticklebacks and the anabantids studied by FORSELIUS (1957), though they belong to different orders (NORMAN, 1937, pp. 3-4). The evolution of parental care is intriguing in a fish like the three-spined, which has been described as 'especially destructive to the spawn and young fry of other fishes' (NORMAN, 1947, p. 132), and indeed will sometimes eat its own, as VAN IERSEL has described (1953, p. 20; for the brook stickleback, see the paper by MOODIE in this issue). This problem has been explored recently by FEUTH-DE BRUIJN & SEVENSTER (1983).

The sticklebacks are of course uniparental, only the male caring for eggs and young; but SEVENSTER has told us (personal communication, 1983) the female fifteen-spined (sea) stickleback shows an incipient tendency to stay near the nest without eating the eggs for as long as ten minutes, the male permitting this. It would be of great interest to know if this incipient tendency to biparental care could be enhanced by artificial selection. In this context, BAKKER (this issue) has found yet another experimental use for the three-spined, which turns out to be highly suitable for work on the genetics and artificial selection of behaviour traits.

Thanks to sticklebacks and their observers, we have come a long way in the couple of centuries since Oliver GOLDSMITH described the bony fishes as a 'fierce unmindful tribe... that leave their spawn without any protection' (in "An History of the Earth, and Animated Nature", published posthumously in 1774; cf. RUSSELL, 1979, p. 62). VAN IERSEL's "Behaviour" Supplement of 1953 on the three-spined was an unprecedentedly detailed analysis of parental behaviour, and this may serve as our final illustration of the importance of sticklebacks in ethology.

References

ASSEM, J. VAN DEN (1967). Territory in the three-spined stickleback, *Gasterosteus aculeatus* L. An experimental study in intra-specific competition. — Behaviour Suppl. 16, p. 1-164.

—— & MOLEN, J. N. VAN DER (1969). Waning of the aggressive response in the three-spined stickleback upon constant exposure to a conspecific. 1. A preliminary analysis of the phemomenon. — Behaviour 34, p. 286-324.

BAERENDS, G. P. (1952). Les sociétés et les familles de poissons. — Colloques Internat. d. Centre Nat. de la Recherche Scient., Paris 34, p. 207-18.

—— (1957). The ethological analysis of fish behavior. — In: The physiology of fishes. 2. Behavior (M. E. BROWN, ed.), p. 229-269. Academic Press, New York.

BAGGERMAN, B. The role of external factors and biological rhythms in the regulation of the reproductive season of the stickleback. — Behaviour, this issue.

BAKKER, Th. C. M. Selection on aggression in sticklebacks. — Behaviour, this issue.

—— & SEVENSTER, P. (1983). Determinants of dominance in male sticklebacks (*Gasterosteus aculeatus* L.). — Behaviour 86, p. 55-71.

BENOIT, J. (1956). États physiologiques et instinct de reproduction chez les oiseaux. — In: L'instinct dans le comportement des animaux et de l'homme (various authors), p. 177-260. Masson, Paris.

CALLAN, H. (1970). Ethology and society. — Clarendon Press, Oxford.

DIEBSCHLAG, E. (1941). Über den Lernvorgang bei der Haustaube. — Z. vergl. Physiol. 28, p. 67-104.

FEUTH-DE BRUIJN, E. & SEVENSTER, P. (1983). Parental reactions to young in sticklebacks (*Gasterosteus aculeatus* L.). — Behaviour 83, p. 186-203.

FORSELIUS, S. (1957). Studies of anabantid fishes, 1. A qualitative description of the reproductive behaviour in territorial species investigated under laboratory conditions, with special regard to the genus *Colisa*. An introduction. 2. Anatomy and histology of some endocrine organs. Quantitative description of growth and nest-building activity. 3. A theoretical discussion of the differentiation, function, causation and regulation of reproductive behaviour. — Zoologiska Bidrag fran Uppsala 32, p. 93-597.

GARDNER, R. A. & GARDNER, B. T. (1969). Teaching sign language to a chimpanzee. — Science 165, p. 664-672.

HASKINS, C. P. (1952). Of societies and men. — Allen and Unwin, London.

HERRICK, C. J. (1924). Neurological foundations of animal behavior. — Henry Holt, New York.

HINDLE, E. (1947). The golden hamster. — In: The UFAW handbook on the care and management of laboratory animals (1st edn, A. N. WORDEN, ed.), p. 196-202. Baillière, Tindall and Cox, London.

—— & MAGALHAES, H. (1957). The golden hamster. — In: *Idem* (2nd edn, A. N. WORDEN & W. LANE-PETTER, eds), p. 324-335. UFAW, London.

HOOGLAND, R., MORRIS, D. & TINBERGEN, N. (1957). The spines of sticklebacks (*Gasterosteus* and *Pygosteus*) as means of defence against predators (*Perca* and *Esox*). — Behaviour 10, p. 205-236.

IERSEL, J. J. A. VAN (1953). An analysis of the parental behaviour of the male three-spined stickleback (*Gasterosteus aculeatus* L.). — Behaviour Suppl. 3, p. 1-159.
—— (1958). Some aspects of territorial behaviour of the male three-spined stickleback. — Archs néerlandaises de Zoologie 13, Suppl. 1, p. 381-400.
KETELE, A. Interspecific competition between *G. aculeatus* f. *leiura* and *Pungitius pungitius*. — Behaviour, this issue.
KUMMER, H. (1968). Social organization of hamadryas baboons. A field study. — S. Karger, Basel.
MOODIE, G. E. E. Filial egg-cannibalism in the brook stickleback *Culaea inconstans*. — Behaviour, this issue.
MORRIS, D. (1956). The function and causation of courtship ceremonies. — In: L'instinct dans le comportement des animaux et de l'homme (various authors), p. 261-84. Masson, Paris.
—— (1958). The reproductive behaviour of the ten-spined stickleback (*Pygosteus pungitius* L.). — Behaviour Suppl. 6, p. 1-154.
NICHOLSON, C. S. S. (1957). Children and animals. — New Era in Home and School 38, p. 116-119.
NORMAN, J. R. (1937). British Museum (Natural History): Illustrated guide to the fish gallery. — British Museum, London.
—— (1947). A history of fishes (3rd edn). — Ernest Benn, London.
RUSSELL, C. (1970). The concept of pseudosex. — Guy's Hospital Gazette 84, p. 241-245.
—— (1975). The chimpanzee carnival. Food, space and social behaviour. — Biology and Human Affairs 40, p. 77-100.
—— (1981). Kinship symbols and their evolution. — Social Biology and Human Affairs 45, p. 119-144.
—— & RUSSELL, W. M. S. (1961). Human behaviour: a new approach. — Deutsch, London.
—— & —— (1971). Language and animal signals. — In: Linguistics at large (N. MINNIS ed.), p. 159-194. Gollancz, London.
—— & —— (1982). Cultural evolution. 1. Culture, culture transmission, and culture change. 2. Cultural selection. — Social Biology and Human Affairs 47, (No. 1) p. 17-38; (No. 2) p. 1-17.
—— & —— (1984). Overpopulation crisis. — Social Biology and Human Affairs 49, p. 23-42.
—— & —— (in press). Orwell, apocalypse and overpopulation. — In: Storm Warnings: the imagination of future societies (C. GREENLAND, E. S. RABKIN & G. SLUSSER eds). Southern Illinois University Press.
RUSSELL, W. M. S. (1965). Les troubles des relations parents-enfants chez les vertébrés autres que l'homme. — In: Psychiatrie Animale (A. BRION & H. EY, eds), p. 245-255 (transl. J. M. Favreu). Desclée de Brouwer, Paris.
—— (1979). Biology and literature in Britain 1500-1900. 1. From the Renaissance to the Romantics. — Biology and Human Affairs 44, p. 50-72.
—— & BURCH, R. L. (1959). The principles of humane experimental technique. — Methuen, London.
SEVENSTER, P. (1961). A causal analysis of a displacement activity (fanning in *Gasterosteus aculeatus* L.). — Behaviour Suppl. 9, p. 1-170.
—— (1968). Motivation and learning in sticklebacks. — In: The central nervous system and fish behavior (D. INGLE ed.), p. 233-245. University of Chicago Press, London.
—— (1983). — Personal communication.
——. Cognition in sticklebacks: an experiment on operant conditioning. — Behaviour, this issue.
THALE, M. (ed.) (1972). The autobiography of Francis Place (1771-1854). — University Press, Cambridge.

TINBERGEN, N. (1942). An objectivistic study of the innate behaviour of animals. — Bibliotheca biotheoretica D, i, 2, p. 39-98.
—— (1951). The study of instinct. — Clarendon Press, Oxford.
—— & IERSEL, J. J. A. VAN (1947). 'Displacement reactions' in the three-spined stickleback. — Behaviour 1, p. 56-64.
TUGENDHAT, B. (1960). The normal feeding behaviour of the three-spined stickleback (*Gasterosteus aculeatus* L.). — Behaviour 15, p. 284-318.
WINN, H. E. (1960). Biology of the brook stickleback *Eucalia inconstans* (Kirtland). — Amer. Midl. Nat. 63, p. 424-438.
WOOTTON, R. J. (1972). The behaviour of the male three-spined stickleback in a natural situation: a quantitative description. — Behaviour 41, p. 232-241.

The Reproductive Cycle

Frank A. von Hippel

> Owing to the element which fishes inhabit, little is known about their courtship, and not much about their battles. The male stickleback (*Gasterosteus leiurus*) has been described as 'mad with delight' when the female comes out of her hiding-place and surveys the nest which he has made for her. 'He darts round her in every direction, then to his accumulated materials for the nest, then back again in an instant; and as she does not advance he endeavours to push her with his snout, and then tries to pull her by the tail and side-spine to the nest.' The males are said to be polygamists; they are extraordinarily bold and pugnacious, whilst 'the females are quite pacific.' Their battles are at times desperate; 'for these puny combatants fasten tight on each other for several seconds, tumbling over and over again, until their strength appears completely exhausted.' With the rough-tailed stickleback (*G. trachurus*) the males whilst fighting swim round and round each other, biting and endeavouring to pierce each other with their raised lateral spines. The same writer adds, 'the bite of these little furies is very severe. They also use their lateral spines with such fatal effect, that I have seen one during a battle absolutely rip his opponent quite open, so that he sank to the bottom and died.' When a fish is conquered, 'his gallant bearing forsakes him; his gay colours fade away; and he hides his disgrace among his peaceable companions, but is for some time the constant object of his conqueror's persecution.' — (Darwin, 1871)

Niko Tinbergen finished his PhD on the 12th of April 1932, married Lies Rutten two days later, and then celebrated his 25th birthday. In July, Niko and Lies launched on an expedition to Greenland as part of the International Polar Year 1932–1933, an experience that deeply affected his perspective on animal behaviour (Kruuk, 2003). Two years later, in 1934, Tinbergen began his academic career in Leiden, and quickly established a productive laboratory focused on birds, insects and fish, with a unique line of research on threespine stickleback. The stickleback was a natural fit for Tinbergen; it had stereotyped and well characterized behaviour that could be tested and disrupted with simple experiments.

Among Tinbergen's students was a standout who took a leadership role in the lab, a young man named Jan Joost ter Pelkwijk ('Pelk'). Pelk was a friend of Niko's younger brother Luuk, who himself would later become an important ecologist, though tragically Luuk committed suicide in 1955 at the age of 39. Like Niko, Luuk suffered from terrible depression (Kruuk, 2003).

Niko Tinbergen supervised Pelk and the other students in a series of classic experiments, which included dummy experiments where nesting male stickleback were induced to attack model fish painted red and to lead female models to the nest. They even reported that male stickleback tried to attack the red postal van as it drove past the laboratory window! Tinbergen's group compartmentalized the stickleback's courtship sequence into a series of stimulus-response events (Figs. 1, 2), thereby giving rise and evidence to the concept of sign-stimuli, releasers and the innate releasing mechanism (IRM), originally formulated by Konrad Lorenz (1932, 1935, 1937). It was at that point in his career that Tinbergen met Lorenz, with whom he later shared important intellectual exchanges and at the end of his career, in 1973, the Nobel Prize. Niko received his Nobel Prize four years after his older brother Jan received a Nobel for Economics; it was the first time that two Nobels were handed out to siblings (Kruuk, 2003).

Six years after Niko started his lab in Leiden, on the 10th of May, 1940, the German army invaded Holland. On the 9th of September, 1942, Tinbergen was arrested, and like over 1,200 other leading intellectuals in Holland, he was held as a prisoner of war in the Beekvliet hostage camp; his internment lasted for two years, until the 11th of September, 1944, when the Germans abandoned the facility (Kruuk, 2003). During his captivity, Tinbergen wrote and illustrated a number of books for his children, including *Het Verhaal van Jan Stekel* [*The Tale of John Stickle*], which covers stickleback life history, including the subject of his future paper with Hoogland and Morris (Hoogland et al., 1957) on spines as a mean of defence against pike (see the section, "Predators & Parasites"). Meanwhile, various people, including Niko's brother Luuk and his half-Jewish wife Tilde and her family, hid from the Nazis in the Tinbergen home in Leiden (Kruuk, 2003). After his release from Beekvliet, Tinbergen helped the resistance by typing messages, and

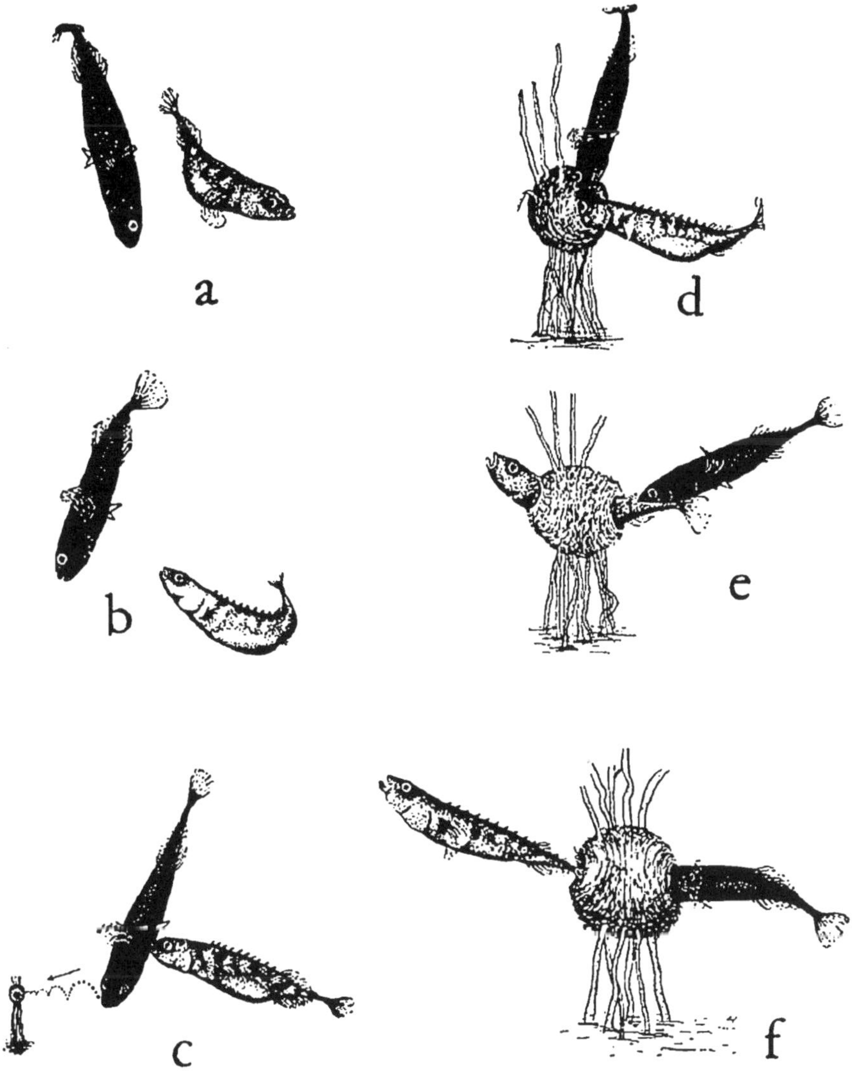

Figure 1. "The courtship pattern of *Pygosteus*. (a) The male dances to the female. (b) The female turns to the male. (c) The male dances to the nest with the female following. (d) The male shows the nest entrance to the female. (e) The male shivers on the female's tail. (f) The female leaves the nest and the male enters to fertilise the eggs." Drawing by Desmond Morris (1958). Desmond Morris was a Ph.D. student of Tinbergen's at Oxford, and subsequently a best-selling author (he wrote *The Naked Ape* and many other books), successful painter and art historian (he wrote *The Biology of Art* and a book on ancient Cypriot pottery), prolific film maker, curator of mammals at the London Zoo, and director of the Institute of Contemporary Art in London (Kruuk, 2003).

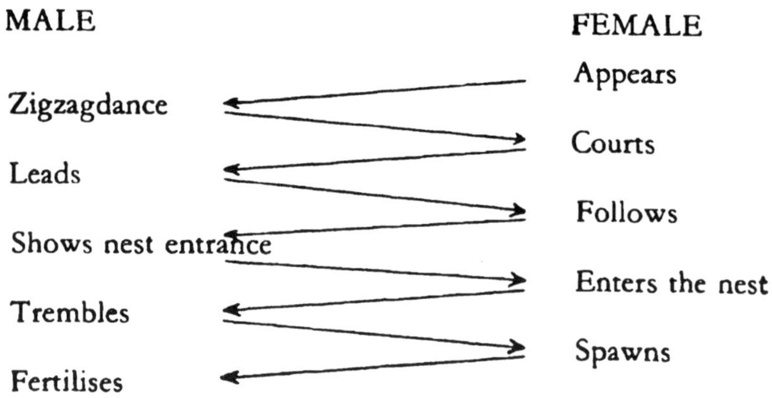

Figure 2. "The reaction chain for *Gasterosteus* given by Tinbergen (1942)." From Morris (1958).

eventually he was found out. Fortunately, the discovery was made just days before Holland was liberated and the Nazis did not have time to arrest and execute him.

After the liberation, Tinbergen took stock of his situation and began to rebuild his research programme in Leiden. Much of the news was tragic. Tinbergen's prized student Pelk had been killed in 1942 at the age of 27 by the Japanese while on a Dutch mine-sweeper, and Tinbergen's mentor had died of hunger towards the end of the war (Kruuk, 2003). And while Tinbergen had suffered a great deal from the Nazis, Lorenz had joined the Nazi party and lent his research to their ideology, even writing papers about the dangers of racial interbreeding based on ideas from animal domestication. Lorenz himself was held as a prisoner of war by the Russians from 1944 until 1948. Later, Tinbergen helped Lorenz to rejoin the scientific fold (Kruuk, 2003).

In a flurry of post-war productivity, Tinbergen re-established his lab in Leiden, wrote landmark papers and his famous book *The Study of Instinct* (completed in 1948, but not published until 1951), and founded the journal *Behaviour* in 1948 to serve the burgeoning field of ethology (Kruuk, 2003). *Behaviour* was the first ethological journal published in English. Tinbergen moved to Oxford one year after founding the journal, at which point his former student Gerard Baerends took on the reins, which he held as managing editor for nearly four decades (Kruuk, 2003). One of Tinbergen's landmark papers was published in the first issue; it took on a paradox in stickleback courtship behaviour. In order to understand the paradox, one must first understand the stickleback reproductive cycle.

Male threespine stickleback compete for territories (territorial phase), build nests (nest-building phase), compete for females with courtship displays (sexual phase), and care for the embryos and fry (parental phase). The transition between the nest-building phase and the sexual phase is marked by the male creeping through his nest and thereby creating a tunnel with an entrance and exit (Sevenster, 1961). During courtship, the male performs a zigzag dance toward the female, swimming back and forth in rapid movements. The male then attempts to lead the female to his nest. If the female follows the male to the nest, the male points to the nest entrance, and when the female enters the nest, the male quivers on her tail to stimulate her to spawn (Fig. 2). The male then follows the female through the nest and fertilizes her eggs, marking the transition from the sexual to the parental phase. She then leaves or he chases her away (Sevenster, 1961; see Fig. 3 for a

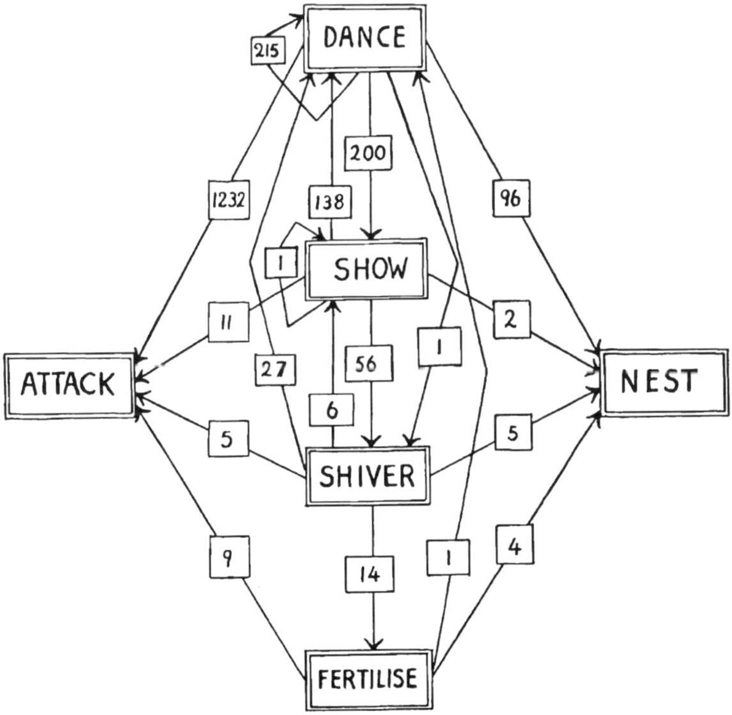

Figure 3. "Male courtship diagram, showing the number of times each action was followed by another action." From Morris (1958).

male courtship diagram). During the parental phase, the male cares for the embryos by fanning water through the nest and by guarding them. After the embryos hatch into fry, the male continues to guard them and he retrieves them when they swim away so they can be kept together in a defendable cluster. As the Reverend W. Houghton (1879) put it,

> For the first few days the little fry are seen close to the nest, lying for the most part on their sides inactively; but as they grow they become more vigorous, and anxious to see something of the world. But the father-fish is slow to encourage such juvenile desires, for is he not well aware that danger lurks on every side? And so, as I myself have seen, should some little occupant of the nursery, moved by piscine curiosity, stray away a little too far from the paternal abode, "does his father know he is out?" Yes, indeed, very soon, and after him he hies, seizes the young truant in his mouth, as a cat would her kitten, and shoots him out right upon the nest.

During nest building, courtship, and parental care, males display nest-directed activities (Sevenster, 1961). These activities include fanning, boring and gluing. Fanning moves water through the nest which, during the parental phase, provides embryos with oxygen and removes carbon dioxide (van Iersel, 1953; Sevenster, 1961). Boring (or pushing) shapes the nest. Gluing is accomplished with a mucus secretion produced by the kidneys; the secretion holds nest materials together. Fanning is crucial to embryo survivorship, and boring and gluing are crucial to nest construction and maintenance. Yet, despite no obvious adaptive significance ('survival value' in Tinbergen's parlance), all three activities show a significant increase in frequency during courtship, even when the male has no embryos in the nest and hence no offspring to benefit.

This curious puzzle was just the sort that set Tinbergen's lab into brainstorming mode during his years in Leiden following the conclusion of World War II. It was a problem that Tinbergen considered sufficiently important that he published his solution to it with Jan van Iersel in the first issue of his new journal *Behaviour* in 1948. This paper is now considered a classic, and it opens this section of the book. At the time, van Iersel was Tinbergen's senior assistant, and in 1949, when Tinbergen moved to Oxford, van Iersel became his successor in Leiden. While Tinbergen conducted relatively few studies on stickleback following the move to Oxford, van Iersel made a career studying stickleback behaviour, in collaboration with his future Leiden successor and Tinbergen student, Piet Sevenster (Kruuk, 2003).

The earliest attempts to explain the cause of nest-directed activities during the sexual phase related to nest condition. Warington (1852), Evers (1878), and Steche (1914) proposed that fanning during the sexual phase removed loose material from the nest. Wrede (1909) suggested that fanning during the sexual phase allowed the male to judge the strength of his nest. Leiner (1929) proposed that the fanning stream cleaned the nest. Morris (1952) observed that sexual fanning prevented the nest from filling with silt. Tinbergen and van Iersel (1948), however, saw nest-directed activities, particularly courtship fanning, as behaviours occurring out of context, or so-called "displacement activities."

Tinbergen (1952) defined a displacement activity as "an activity, belonging to the executive motor pattern of an instinct other than the instinct(s) activated." Displacement activities occur when "an activated drive is denied discharge though its own consummatory act(s)" and the displacement activities are "outlets through which the thwarted drives can express themselves in motion." An example of a displacement activity would be when an animal encounters another at a territorial border and experiences the conflicting motivations to fight and to flee, but instead engages in a seemingly irrelevant behaviour such as preening or feeding.

The obvious factors that activate parental activities, such as eggs or fry in the nest, are absent during the sexual phase. Hence nest-directed activities during the sexual phase appear to be activated by a nonparental drive. Adriaan Kortlandt (1940) coined such activation by a seemingly inappropriate drive as allochthonous activation. Activation in the normal situation is autochthonous. In this same important 1940 paper, Kortlandt also theorized on displacement activities and the hierarchical organization of behaviour, but his views did not receive the attention of those of Tinbergen and Lorenz (Kruuk, 2003). Later, Kortlandt (1955) wrote a scathing review of the work of Tinbergen and Lorenz, in which he focused much of his criticism on *The Study of Instinct*.

Because nest-directed activities have no obvious adaptive significance during the sexual phase, and yet they increase in frequency when the male is exposed to a gravid female, Tinbergen and van Iersel (1948) hypothesized that these behaviours, especially fanning, are displacement activities activated allochthonously by the sexual drive. Tinbergen and van Iersel suggested that nest-directed activities seem irrelevant to the task of inducing the female to lay her eggs in the nest because the male must leave the female and therefore cease courtship in order to return to the nest. They hypothesized that nest-directed activities occur when the female is reluctant to follow the male to the nest, thereby thwarting the male's sexual tendency and causing a surplus of sexual drive (Tinbergen & van Iersel, 1948; van Iersel, 1953). They hypothesized that this surplus of drive finds an outlet though courtship fanning, so Tinbergen and van Iersel's hypothesis is known as the Surplus Hypothesis (Sevenster, 1961). More generally, Tinbergen and van Iersel demonstrated that displacement activities that had evolved a signal function were ritualized; they differed from those that were not employed as signals (and from the behaviour in context) by being more stereotyped and conspicuous.

The Surplus Hypothesis assumes that the parental centre is activated by the sexual drive and that parental activating factors are absent. Yet, Piet Sevenster (1961) presented evidence that the sexual drive inhibits the parental centre, at least during the parental phase, and that the parental system is activated whenever fanning occurs. Sevenster suggested that even during the sexual phase the parental system is activated by the presence of the nest, and he pointed out that fanning occurs almost exclusively at the nest. Sevenster therefore saw no reason to invoke an allochthonous activation of fanning during the sexual phase. Sevenster also showed that a high number and older age of eggs in the nest, water with a high carbon dioxide content, and internal parental factors all facilitate courtship fanning. For these reasons, Sevenster rejected the Surplus Hypothesis. Desmond Morris (1958) also questioned the Surplus Hypothesis because male ninespine stickleback do not show an increase in nest-directed activities such as fanning (Fig. 4), or other displacement activities, when thwarted by the female. Yet their behaviour in many other respects is similar to the behaviour of the threespine stickleback.

Sevenster (1961) presented results suggesting that the sexual, aggressive, and parental drives are somewhat mutually inhibitory, although the inhibition between the aggressive and parental drives is not as strong as that between the sexual and parental drives. These mutual inhibitions do not mean that during one phase the behaviour associated with the other two phases is not expressed. Indeed, during the first few days of the parental phase the male continues to court females (van Iersel, 1953), and the male must remain

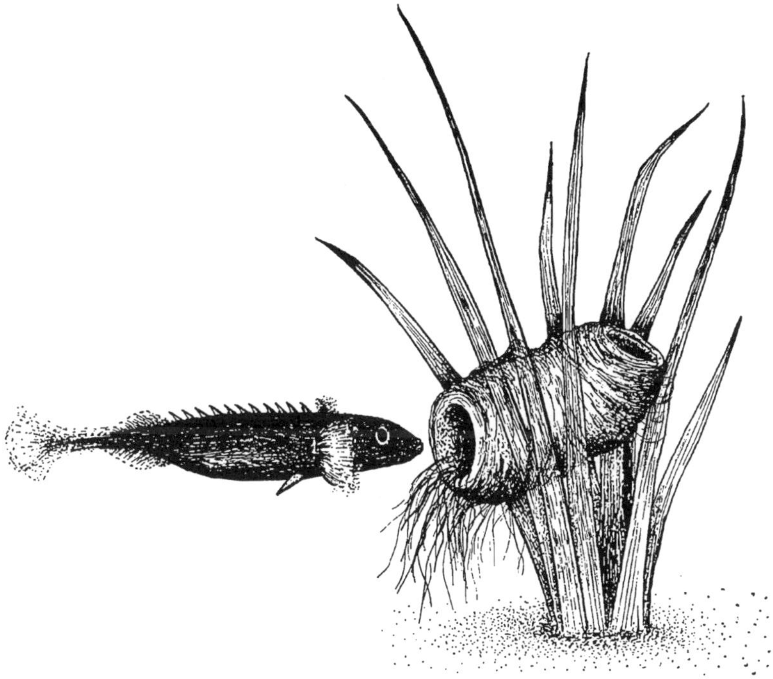

Figure 4. "The nest of *Pygosteus*, showing a male fanning at the entrance." Drawing by Desmond Morris (1958).

aggressive while parental to chase away potential predators of eggs and fry. Mutual inhibition implies that when one drive is stimulated, the others are inhibited, but the relations between the drives may shift continuously.

Sevenster observed that fanning during the sexual phase occurs when neither the sexual nor the aggressive drive are maximally stimulated, but rather when both are at some intermediate level and therefore in conflict for expression. Given his results that fanning during the sexual phase is activated autochthonously and is facilitated by parental factors, Sevenster proposed that when the sexual and aggressive drives are in conflict, their inhibition on the parental drive may be reduced. If this level of inhibition is reduced enough that parental factors, such as the presence of the nest, acquire a level of stimulation above the threshold for activation of parental behaviour, then fanning occurs. This is known as the Disinhibition Hypothesis because the parental drive is disinhibited due to a conflict between two competing drives. Because a gravid female entering a territory presents stimuli for both aggressive and sexual behaviour, a conflict between the two drives could disinhibit parental behaviour, leading to courtship fanning (Sevenster, 1961).

The Surplus Hypothesis and the Disinhibition Hypothesis were later challenged by other researchers. David McFarland, who was Tinbergen's successor upon his retirement at Oxford (Kruuk, 2003), extended the early nest-condition explanations by suggesting that nest-directed activities during courtship are done to keep the nest in proper condition to receive the female (McFarland, 1974; Cohen & McFarland, 1979). Wilz (1970a, 1970b) proposed a causal role for nest-directed activities, including creeping through, in the regulation of aggressive and sexual tendencies by linking these activities with a behavioural pattern called dorsal pricking. In dorsal pricking a male moves forward on the underside of a female with his spines erect as if he were trying to prick the female and push her toward the water's surface. Wilz hypothesized that a male performs dorsal pricking to keep the female from approaching his nest so that he is not subject to further aggressive stimuli and so that he may then go to the nest and display nest-directed activities. These activities, Wilz suggested, enable a male to switch from an aggressive state to a sexual state. Dorsal pricking, in this view, evolved as a mechanism to allow a male to perform nest-directed activities by stalling the female during courtship. Alternatively, Sargent (1982) proposed that dorsal pricking allows a courting male to judge when he can lead the female to his nest with minimal risk of courtship intrusion or nest raiding by neighbouring males.

Rohwer (1978) and Ridley and Rechten (1981; see section Homosexuality, Cannibalism & Sexual Strategies) proposed a signal function for the origin of fanning during courtship: because females prefer to spawn in nests already containing eggs but the eggs are hidden in the nest, males would be selected to show females that they have eggs by fanning the eggs during courtship. Rohwer (1978) further proposed that males construct canopied nests because the nests prevent females from assessing presence or absence of eggs from a distance. Females would be able to select males with eggs by observing which males fanned. Males without eggs would be selected to deceptively signal to females that they already have eggs by also fanning during courtship, "and displacement fanning would become established as a deceptive social releaser" (Ridley & Rechten, 1981). von Hippel (1996) extended these ideas by suggesting that nest-directed activities are part of courtship itself, and that they serve to advertise the quality of the male's parental care.

The transition from thinking about stickleback courtship in terms of competing drives to thinking of it in terms of competing individuals coincided with the fading away of ethology and the emergence of behavioural ecology. That transition can be seen in the time series of papers presented in this section, which begins with Niko Tinbergen and Jan van Iersel's (1948) classic paper on displacement activities. This is followed by two papers by Desmond Morris (1957, 1958 excerpts), who provides a retrospective of his work in this section. The intellectual development of the ideas continues with excerpts from Piet Sevenster's (1961) classic monograph on displacement fanning.

The next part of this section examines other aspects of the stickleback reproductive cycle, beginning with excerpts of Harman V.S. Peeke's (1969) analysis of habituation of aggression, and his retrospective of this work. Excerpts from Robert Wootton's (1971) paper on aggression in parental male stickleback continues the development of our understanding of aggression, and is accompanied by a retrospective from Wootton.

Aggression is one of the more obvious aspects of a male stickleback's behaviour (Fig. 5), and something that Tinbergen (1968) even related to human warfare in his paper, "War and peace in animals and man." Tinbergen's application of ethological principles to human behaviour led to one of the first riots in the "sociobiology debate" that began in the late 1960s and raged on for two decades. The demonstrators described Tinbergen (quoted in Kruuk, 2003) as "a notorious biologist who has been presenting the theories of innate aggressiveness, instinctive urges for territory, need for 'living space,' etc. which have been popularized in the writing of Konrad Lorenz, Robert Ardrey, Desmond Morris and other so-called scientists." The demonstrators wrote, "During the discussion period it was pointed out to Tinbergen how his ideas were fascist, the same ideas the Nazis encouraged and developed and that they had no scientific base whatsoever... When his flimsy rhetoric failed to divert the discussion, the chairman of the meeting broke up the meeting and protected this Nazi from being fully exposed to the people." Another quote from the demonstrators (Kruuk, 2003) put it this way: "During his talk on sticklebacks and seagulls he emphasized all the fascist ideas such as innate aggression, instinctive urge for territory (private property), need for living space (lebensraum) etc., the same ideas which can be found in Hitler's *Mein Kampf*. Despite the fact that Tinbergen repeatedly cautioned against

Figure 5. "A dominant male dragging away a rival." Drawing by Desmond Morris (1958).

extrapolating the results of animal experiments to man, he ignored his warning and proceeded to do just that." These were criticisms of a man who had been a prisoner of war of the Nazis for two years, and who advocated for social justice.

This section also includes excerpts of Theo Bakker and Piet Sevenster's (1983) analysis of dominance in male stickleback, along with Bakker's retrospective of this work. Dominance, it turns out, relates to the brightness of the male's nuptial colouration. And it was the male's nuptial colouration that first attracted the interest of naturalists and scientists in this little fish. Darwin (1871) wrote,

> The male stickleback (*Gasterosteus leiurus*)... is described by Mr. Warington, as being then 'beautiful beyond description.' The back and eyes of the female are simply brown, and the belly white. The eyes of the male, on the other hand, are 'of the most splendid green, having a metallic luster like the green feathers of some humming-birds. The throat and

belly are of a bright crimson, the back of an ashy-green, and the whole fish appears as though it were somewhat translucent and glowed with an internal incandescence.' After the breeding-season these colours all change, the throat and belly become of a paler red, the back more green, and the glowing tints subside.

That with fishes there exists some close relation between their colours and their sexual functions we can clearly see; — firstly, from the adult males of certain species being differently coloured from the females, and often much more brilliantly; — secondly, from these same males, whilst immature, resembling the mature females; — and, lastly, from the males, even of those species which at all other times of the year are identical in colour with the females, often acquiring brilliant tints during the spawning-season. We know that the males are ardent in their courtship and sometimes fight desperately together. If we may assume that the females have the power of exerting a choice and of selecting the more highly-ornamented males, all the above facts become intelligible through the principle of sexual selection... I cannot, from the several previous considerations, give up the belief that the males which are the most attractive to the females, from their brighter colours or other ornaments, are commonly preferred by them; and that the males have thus been rendered more beautiful in the course of ages.

This beauty of the males, of course, was a focal point of studies using Pelk's wax dummies in Tinbergen's lab in Leiden to test the concept of Lorenz's innate releasing mechanism (ter Pelkwijk & Tinbergen, 1937). Gerard Baerends, who was Tinbergen's first Ph.D. student and became the most prominent ethologist in Holland after Tinbergen moved to Oxford (Kruuk, 2003), followed up on Pelk and Tinbergen's work with a re-analysis of the dummy experiments and their relation to the innate releasing mechanism concept (Baerends, 1985); excerpts of that paper are reproduced here.

A fascinating development in the study of stickleback colour was the discovery of the "white" stickleback of Nova Scotia. Excerpts of two studies on the white stickleback are included here. In the first paper, MacDonald et al. (1995a) compared subtidal versus intertidal breeding strategies, including over different substrates, in a site with the greatest known diversity of reproductive behaviour in *Gasterosteus*. In the second paper, MacDonald et al. (1995b) focused on the behaviour of white stickleback breeding in the intertidal zone. MacDonald et al. (1995a) hypothesized that this novel reproductive strategy shown by some white stickleback males of dispersing their embryos over rocks in the intertidal zone evolved after the "emancipation from parental care." Only in the white stickleback have researchers discovered the evolutionary loss of paternal care. Clearly, these studies fall within the realm of frequency-dependent selection and the science of behavioural ecology that replaced classical ethology.

On the other side of the colour spectrum from white stickleback are those with black nuptial colouration, which have been discovered in California, Washington and British Columbia (see Kynard, 1978 in section Homosexuality, Cannibalism & Sexual Strategies; von Hippel, 1999). McDonald (not to be confused with MacDonald), Reimchen and Hawryshyn (1995; excerpts included here) demonstrated that black nuptial colouration evolves in bog habitats where the spectral irradiance in the water is shifted toward longer wavelengths, and that sexual selection for nuptial colouration is operating from visual contrast rather than hue (red or black). Sarah Kraak, Theo Bakker and Sandra Hočevar (2000) linked male colouration to nest site preference based on habitat characteristics in wild stickleback. This study (excerpts included here) has the experimental simplicity and elegance of a Tinbergen study — Tinbergen essentially created the experimental approach of manipulating one aspect of an animal's environment in the wild, rather than controlling all but one aspect of an animal's environment in the lab (Kruuk, 2003). Kraak et al. (2000) performed the simple manipulation of planting macrophytes in a stream channel at two depths and then observed which males built nests under the macrophytes as opposed to in control sites. In the final study of this section, Victor Rush, Jeffrey McKinnon, Michael Abney and Craig Sargent (2003; excerpts) examined changes in the contrast of the different elements of the stickleback's nuptial colouration pattern in free-swimming males exposed to other males or females.

These studies round out our exploration of nuptial colouration by demonstrating that the innate releasing mechanism associated with red colouration as demonstrated by Pelkwijk and Tinbergen (1937) and re-analyzed by Baerends (1985) depends entirely on the environmental and social context. Tinbergen seldom applied ecological principles to his ethological studies, despite the influence of his ecologist brother Luuk (Kruuk, 2003). But here we can see how rich the story unfolds when grounded in ecology and frequency-dependent selection of different behavioural strategies. Behavioural ecology, so ably developed by theoreticians in the 1970s and later, found in the stickleback an ideal model to test the new ideas, just as the ethologists had made such prolific use of the stickleback before.

From the articles of Tinbergen and Morris and Sevenster to those printed on the most recent pages of *Behaviour*, one can appreciate how much the style and substance of science has changed. The anthropomorphic wording is gone for sound intellectual reasons, though the change has meant the loss of a wonderful tenor of prose. Rather than 100–200 page articles on a collection of topics, papers now focus on a single narrow subject with no superfluous text. A glance at a typical reference section shows how the literature to draw upon for any particular idea has snowballed. For example, Piet Sevenster's (1961) classic analysis of displacement fanning, at 170 pages of text, includes only 17 references, six of which are by Tinbergen. The last article of this section, by Rush et al. (2003), exactly one-tenth as long, includes 44 references, and remarkably, one of them is *The Study of Instinct* — still relevant six decades after its publication.

References

Baerends, G.P. (1985). Do the dummy experiments with sticklebacks support the IRM-concept? — Behaviour 93: 258-277.

Bakker, T.C.M. & Sevenster, P. (1983). Determinants of dominance in male sticklebacks (*Gasterosteus aculeatus* L.). — Behaviour 86: 55-71.

Cohen, S. & McFarland, D.J. (1979). Time-sharing as a mechanism for the control of behaviour sequences during the courtship of the three-spined stickleback (*Gasterosteus aculeatus*). — Anim. Behav. 27: 270-283.

Darwin, C. (1871). The descent of man, and selection in relation to sex. — Murray, London.

Evers, M. (1878). Zur charakteristik des stichlings (*Gasterosteus aculeatus*). — Jahresber. Naturw. Ver. Elberfeld 5: 26-46.

von Hippel, F.A. (1996). Sexual selection and parental care in the threespine stickleback (*Gasterosteus aculeatus*). — PhD thesis, University of California, Berkeley.

von Hippel, F.A. (1999). Black male bellies and red female throats: color changes with breeding status in a threespine stickleback. — Environ. Biol. Fishes 55: 237-244.

Hoogland, R., Morris, D. & Tinbergen, N. (1957). The spines of sticklebacks (*Gasterosteus* and *Pygosteus*) as means of defence against predators (*Perca* and *Esox*). — Behaviour 10: 205-236.

Houghton, W. (1879). British fresh-water fishes. — W. Mackenzie, London.

van Iersel, J.J.A. (1953). An analysis of the parental behaviour of the male three-spined stickleback (*Gasterosteus aculeatus* L.). — Behaviour, Supplement III: 1-159.

Kortlandt, A. (1940). Wechselwirkung zwischen instinkten. — Arch. Neerl. Zool. 4: 443-520.

Kortlandt, A. (1955). Aspects and prospects of the concept of instinct. — Arch. Neerl. Zool. 11: 155-284.

Kraak, S.B.M., Bakker, T.C.M. & Hocevar, S. (2000). Stickleback males, especially large and red ones, are more likely to nest concealed in macrophytes. — Behaviour 137: 907-919.

Kruuk, H. (2003). Niko's nature; a life of Niko Tinbergen and his science of animal behaviour. — Oxford University Press, Oxford.

Kynard, B.E. (1978). Breeding behavior of a lacustrine population of threespine sticklebacks (*Gasterosteus aculeatus* L.). — Behaviour 67: 178-206.

Leiner, M. (1929). Okologische studien an *Gasterosteus aculeatus*. — Zeitschr. Morph. u. Okol. Tiere 14: 360-399.

Lorenz, K. (1932). Betrachtungen über das erkennen der arteigenen triebhandlungen der vögel. — J. f. Ornithol. 80: 50-98.

Lorenz, K. (1935). Der kumpan in der umwelt des vogels. — J. f. Ornithol. 83: 137-215.

Lorenz, K. (1937). Ueber die bildung des instinktsbegriffs. — Naturw. 25: 289-300.

MacDonald, J.F., Bekkers, J., Macisaac, S.M. & Blouw, D.M. (1995a). Intertidal breeding and aerial development of embryos of a stickleback fish (*Gasterosteus*). — Behaviour 132: 1183-1206.

MacDonald, J.F., MacIsaac, S.M., Bekkers, H. & Blouw, D.M. (1995b). Experiments on embryo survivorship, habitat selection, and competitive ability of a stickleback fish (*Gasterosteus*) which nests in the rocky intertidal zone. — Behaviour 132: 1207-1221.

McDonald, C.G., Reimchen, T.E. & Hawryshyn, C.W. (1995). Nuptial colour loss and signal masking in *Gasterosteus*: an analysis using video imaging. — Behaviour 132: 963-977.

McFarland, D.J. (1974). Time-sharing as a behavioural phenomenon. — In: Advances in the study of behavior (Lehrman, D.S., Rosenblatt, J.S., Hinde, R.A. & Shaw, E., eds). Academic Press, New York, p. 201-225.

Morris, D. (1952). Homosexuality in the ten-spined stickleback (*Pygosteus pungitius* L.). — Behaviour 4: 233-261.

Morris, D. (1957). "Typical Intensity" and its relation to the problem of ritualisation. — Behaviour 11: 1-12.

Morris, D. (1958). The reproductive behaviour of the ten-spined stickleback (*Pygosteus pungitius* L.). — Behaviour, Supplement VI: 1-154.

Peeke, H.V.S. (1969). Habituation of conspecific aggression in the three-spined stickleback (*Gasterosteus aculeatus* L.). — Behaviour 35: 137-156.

ter Pelkwijk, J.J. & Tinbergen, N. (1937). Eine reizbiologische analyse einiger verhaltensweisen von *Gasterosteus aculeatus* L. — Z. f. Tierpsychol. 1: 193-200.

Ridley, M. & Rechten, C. (1981). Female sticklebacks prefer to spawn with males whose nests contain eggs. — Behaviour 76: 152-161.

Rohwer, S. (1978). Parent cannibalism of offspring and egg raiding as a courtship strategy. — Am. Nat. 112: 429-440.

Rush, V.N., McKinnon, J.S., Abney, M.A. & Sargent, R.C. (2003). Reflectance spectra from free-swimming sticklebacks (*Gasterosteus*): social context and eye-jaw contrast. — Behaviour 140: 1003-1019.

Sargent, R.C. (1982). Territory quality, male quality, courtship intrusions, and female nest-choice in the threespine stickleback, *Gasterosteus aculeatus*. — Anim. Behav. 30: 364-374.

Sevenster, P. (1961). A causal study of a displacement activity (fanning in *Gasterosteus aculeatus* L.). — Behaviour, Supplement IX: 1-170.

Steche, O. (1914). Brehms tierleben bd. 3: die fische (neu bearbeitet von O. Steche). — Leipzig: 347.

Tinbergen, N. (1951). The study of instinct. — Oxford University Press, Oxford.
Tinbergen, N. (1952). "Derived" activities; their causation, biological significance, origin and emancipation during evolution. — Q. Rev. Biol. 27: 1-32.
Tinbergen, N. (1968). On war and peace in animals and man. — Science 160: 1411-1418.
Tinbergen, N. & van Iersel, J.J.A. (1948). "Displacement reactions" in the three-spined stickleback. — Behaviour 1: 56-63.
Warington, R. (1852). Observations on the natural history of the watersnail and fish (*Gasterosteus leiurus*) kept in confined and limited portion of water. — Ann. Mag. Nat. Hist. 2: 273-280.
Wilz, K.J. (1970a). Causal and functional analysis of dorsal pricking and nest activity in the courtship of the three-spined stickleback *Gasterosteus aculeatus*. — Anim. Behav. 18: 115-124.
Wilz, K.J. (1970b). Self-regulation of motivation in the three-spined stickleback (*Gasterosteus aculeatus* L.). — Nature 226: 465-466.
Wootton, R.J. (1971). Measures of the aggression of parental male three-spined sticklebacks. — Behaviour 40: 228-261.
Wrede, T. (1909). Die zucht des dreistachligen stichlings (*Gasterosteus aculeatus*) im zimmeraquarium. — Blatt. Aquar. Terrer. Kunde 20: 736-740.

"DISPLACEMENT REACTIONS" IN THE THREE-SPINED STICKLEBACK

by

N. TINBERGEN and J. J. A. VAN IERSEL
(Zoological Laboratory, Leiden, Netherlands)
(With 4 textfigures)
(Received 4. VI. 1946)

Several observers of animal behaviour have called attention to the fact that an animal may, under certain conditions, perform quite "irrelevant" movements. For instance, fighting domestic cocks go through the motions of picking up some food; fighting starlings preen their feathers, courting male birds of paradise wipe their bills, fighting avocets occasionally assume the sleeping attitude, *etc.* This phenomenon has been subjected to an analysis by KORTLANDT (1940) and by TINBERGEN (1939, 1940). They showed that such irrelevant movements occur when the animal is under the influence of a powerful urge, but at the same time is in some way prevented from expressing this urge in the appropriate way. The animal, under such conditions, does not suspend action, but vents its urge in movements belonging to another drive. The behaviour "sparks over" (MAKKINK, 1936) from one drive to another. Whereas a good name has been given to it in German which has been generally accepted ("Übersprunghandlung" KORTLANDT, "Übersprungbewegung" TINBERGEN), no suitable English term has been found HUXLEY's term "habit preening" for displacement preening in the great crested grebe (HUXLEY, 1914) is to be rejected because the phenomenon has probably nothing to do with habit. Neither "sparking-over activity", nor "sham activity" (CARPENTER, 1934), nor "substitute activity" (KIRKMAN, 1937; TINBERGEN, 1939) nor "irrelevant movement" (RAND, 1943) seems to be entirely satisfactory. The Rev. E. A. ARMSTRONG suggested to one of us the name "displaced reaction", which was changed by Prof. HUXLEY during a discussion on the subject into "displacement reaction". This term is being adopted by Mr. ARMSTRONG in the forthcoming revised edition of his "Bird display" and will be used here. Displacement reactions may appear under a variety of conditions, most of which are of one of the following three types:

1. The fighting drive cannot express itself by actual fighting because a conflicting drive (the escape drive) is aroused at the same time. This situation is of general occurrence during boundary disputes between two territorial fighters.

2. The mating drive cannot express itself by actual mating because the required signal movement or releaser is not given by the mate. This is the cause of the occurrence of displacement reactions during courtship.

3. When an external stimulus, after having activated a drive, suddenly stops, the drive may use displacement reactions as an outlet. This situation seems to give rise to phenomena like post coition ceremonies.

Certain movements, therefore, can be aroused in two ways, *viz.*, either by activation of its "own" drive, or by activation of another drive, in which latter case we speak of a displacement reaction. The displacement reaction gets its impulse from a strange drive, its impulse is "allochthonous"; when activated by its own drive a movement or rather its impulse is "autochthonous". Often there are differences between the allochthonous movement and its autochthonous "example". For instance, the displacement reaction may be very vigorous, in case its allochthonous urge is very strong. The displacement preening or the displacement wiping-of-the-bill may be very intensive, very hasty, or/and very much emphasised. Further, the displacement reaction is often less complete than its example: it may consist of only part of the chain of movements composing the example. Thus a fighting cock will rarely really eat something; its displacement reaction consists of the pecking movement only, actual seizing and swallowing of food is not observed.

Comparative study of displacement reactions has further led to the following, partly tentative, conclusions:

1. The form of the displacement reaction is characteristic of each species. For instance, every male starling preens when its fighting drive is obstructed; a cock or a male skylark always makes pecking movements under the same conditions; a pigeon always preens when the mating urge is frustrated, and so on. This is based on the fact that the displacement reactions are innate. A species may have a number of different displacement reactions, each belonging to a special situation.

2. In some cases, displacement reactions have a secondary function, serving as a social signal for releasing special responses in other individuals of the same species. Displacement reactions initiated by the fighting drive

often have threat function, those by the mating drive may have the power to release sexual responses in the partner.

3. Those displacement reactions which do have releaser function seem to differ from such as have no communicative function by what has been called ritualisation. Instead of being identical, or very nearly so, with the autochthonous example, they differ from it by showing superposed movements which actually serve to make the movement more conspicuous. Thus a displacement reaction with threat function may display special weapons or bright colours.

The above is a sketchy review of the work thus far done. For particulars we refer to KORTLANDT (1940) and to TINBERGEN (1940).

During the spring of 1946 we made some more observations on displacement reactions in the three-spined stickleback, on which we shall now report. We have decided to publish them because we are able to supplement and to correct former statements made by the senior author and because we are able to substantiate some of his tentative general conclusions put forward in 1940.

Obstruction of the fighting urge

When a male stickleback in his territory meets another male, he chases it. Outside his territory, however, he will not attack but avoids a fight, and, if attacked, he flees. Where two territories meet, fierce fighting often ensues on the boundary. In these combats, actual fighting is interrupted by threatening. A threatening male stands in a vertical position, head pointing downward, and making the movements of picking something up from the bottom (fig. 1). This movement, which experiments with dummies have shown to function as a threat, intimidating other males, is, in origin, a displacement activity, and can be experimentally aroused by obstructing the fighting drive (TINBERGEN, 1940). In previous papers, this movement was interpreted as displacement feeding; new observations have shown that we have to do with digging, *viz.*, digging a hole for the nest, an activity which, when autochthonously aroused, precedes the actual nestbuilding. In most cases the incompleteness of the movement does not allow us to see whether the animal is going to pick up food or to remove sand. Occasionally, however, especially when the fighting drive is very strongly activated, the movement becomes more complete and

Fig. 1. Male *Gasterosteus aculeatus* L. threatening.

develops into perfect digging. This is very rarely, if ever, seen under "normal" conditions, but we discovered that by putting several males together into one aquarium (for instance, five males into a tank measuring 200 × 50

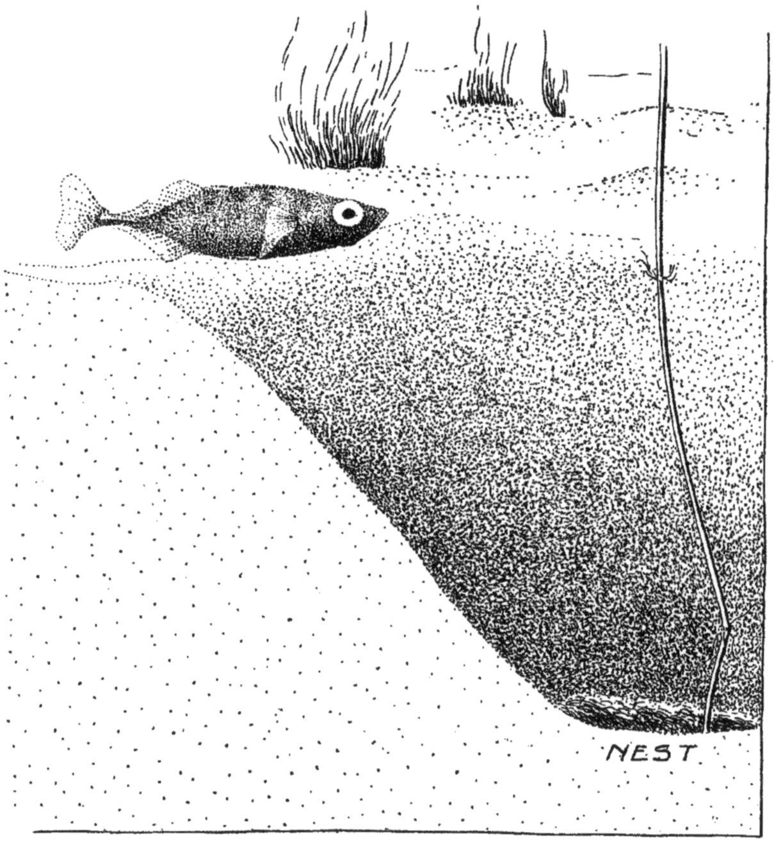

Fig. 2. Male with nest in deep depression made by displacement digging (male 1 from fig. 3).

× 50 cm) we could evoke complete displacement digging at will. This was due to the extremely small size of the individual territories, a condition that, just as in territorial birds, very strongly activates the fighting drive.

Such males spend their time in nearly continuous fighting and threatening. In the course of one or two days a male in such a situation would dig one or more (up till 8) deep holes (see figs 2 and 3). As long as the tank remained crowded with territorial males, boundary disputes with threatening and digging continued, resulting in enormous depressions at or near the nest. When these males were put apart each in a new tank where they were allowed to build a new nest alone, they naturally never showed threat behaviour and no such holes were formed.

These observations show, not only that the threatening movements must be interpreted as displacement digging instead of as displacement feeding, but also that KORTLANDT's and TINBERGEN's general conclusion that a

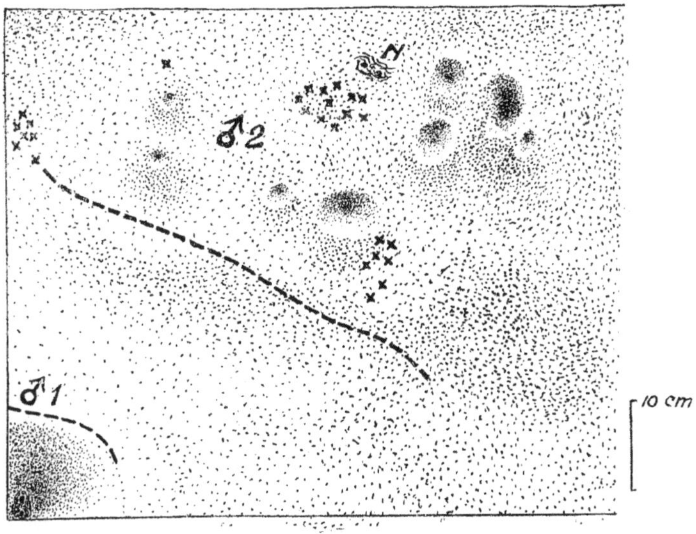

Fig. 3. Territories of two neighbouring males in "crowded" tank, with holes due to displacement digging. Broken lines: boundaries where the owners are threatening; the ground between them is no man's land. Crosses indicate *Elodea* and *Batrachium* vegetation. *N*: nest belonging to male 2.

displacement reaction is never as complete as its autochthonous example, does not hold good absolutely. When the obstructed urge is exceptionally strong, the resulting displacement reaction may become complete and practically identical with its example. The only difference is that the displacement reaction, in this case, is performed in a very "nervous" (hasty or jerky, intensive and emphasising) way.

In other cases too it has been observed that a displacement reaction may become as complete as the autochthonous movement. For instance, a lapwing disturbed at its nest will show displacement feeding. Sometimes this displacement feeding develops into actual catching and swallowing a worm (VAN SIJN in litt.).

Obstruction of the mating drive

When a male has finished its nest, it is ready to mate. It now responds to any pregnant female arriving in the territory by a series of sideward "leaps", each ending in a flash-like dart in the direction of the female and a subsequent sudden "dead stop". This whole ceremony, the "zig-zag dance", serves to stimulate the female to approach the male, thereby in her turn inducing him to lead her towards the nest. When the female does not follow at once, the male is unable to lead her, because his leading is absolutely dependent on her following behaviour, which acts as a signal. His mating drive is then discharged in three different displacement reactions, belonging to the nestbuilding and parental activities: he goes to the nest, pushes his snout into the roof ("pushing"), makes the peculiar ventilating movements, which normally aerate the eggs ("fanning") and in addition glues his kidney secretion to the nest ("gluing"), an activity normally serving to stick the algae and other plants together which are used in the construction of the nest. Fig. 4 shows the sudden change in activity provoked by the arrival of a female. Testing of nest material and digging stop abruptly and zig-zag dance, pushing, fanning and gluing are performed instead.

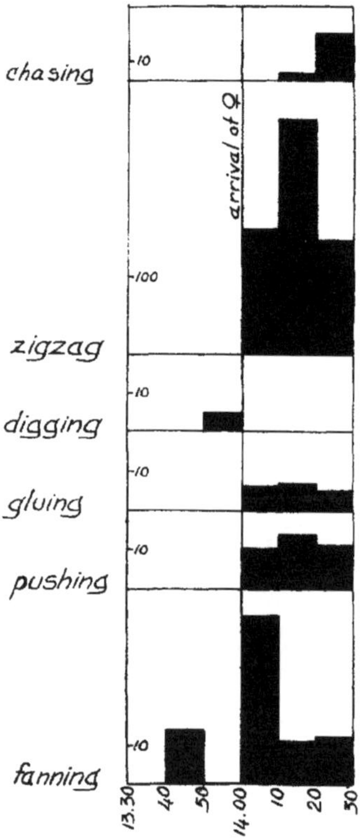

Fig. 4. Frequency of six activities of a male in sexual condition during 30 minutes before and 30 minutes after arrival of a female.

Thus there is a definite connection between digging and the fighting drive on the one hand, whereas pushing, fanning and gluing are outlets for the obstructed mating drive.

Comparison of displacement digging and displacement fanning

In several respects the displacement digging differs from its example, autochthonous digging. A threatening male always takes care to turn its broad lateral side towards the opponent, thus displaying the red parts. These, as experiments have shown, act as an intimidating signal. Further, the ventral spines, or more often one of them, *viz.*, that on the side turned towards the enemy, is erected. The ventral spines are actually used as weapons in one of the types of fighting. Lastly, the movement is "emphasised" by brisk shaking motions of the whole body as if the fish made an attempt to thrust its snout deep into the bottom, yet at the same time stops; something like "digging on the spot".

These three components have the effect of rendering the displacement digging more conspicuous, adding colour, movement and the display of an actual weapon to it. In other words, the activity is ritualised, it is adapted to its function as a signal. As mentioned above, its releasive function has actually been proved by experiment.

With displacement fanning the case is different. Only by careful observation one can observe a slight difference between authochthonous fanning and displacement fanning. We get the impression that in the latter case the frequency of the rhythmical movements of tail and pectorals is slightly higher than in autochthonous fanning. This seems to be dependent on the strength of the mating urge; the stronger the latter is, the faster the fanning movements seem to be. When fanning is performed as an autochthonous movement, the strength of the urge definitely increases the frequency, hence our cautious formulation of the difference between the two types of fanning. Displacement pushing and displacement gluing cannot be told from their autochthonous examples at all, as far as our experience goes.

Now, in spite of all attempts, we never found any evidence pointing to a signal function of these displacement reactions; it does not seem to make the slightest difference to a female whether the male shows displacement pushing, fanning or gluing, or omits them. The gluing does have a stimulating influence on the male himself, for a male that is not quite ready to lead a willing female nearly always leads her quite readily immediately after gluing; but we never had any indication of an influence on the female.

These observations, therefore, are striking proof of the correlation between ritualisation and signal function. Displacement reactions with signal function are ritualised, those without signal function are not.

We are of opinion that this correlation is not accidental and that there can be little doubt that ritualisation is the outcome of adaptation of the displacement reaction to its secondary function, *viz.* that of a signal or releaser.

LITERATURE

ARMSTRONG, E. A. (1942). Bird display. — Cambridge, Camb. Univ. Press, xvi + 381 p.

CARPENTER, C. R. (1934). A field study of the behavior and social relations of howling monkeys. — Comp. psychol. Monogr. X, 2, p. 1-168.

HUXLEY, J. S. (1914). The courtship habits of the great crested grebe (*Podiceps cristatus*) with an addition to the theory of sexual selection. — Proc. zool. Soc. Lond. XXXV, p. 253-291.

KIRKMAN, F. B. (1937). Bird behaviour. — London & Edinburg, T. Nelson, xv + 232 p.

KORTLANDT, A. (1940). Wechselwirkung zwischen Instinkten. — Arch. néerl. Zool. IV, p. 442-520.

MAKKINK, G. F. (1936). An attempt at an ethogram of the European avocet (*Recurvirostra avocetta* L.) with ethological and psychological remarks. — Ardea XXV, p. 1-60.

RAND, A. L. (1943). Some irrelevant behavior in birds. — The Auk XL, p. 168-171.

TINBERGEN, N. (1939). On the analysis of social organization among vertebrates, with special reference to birds. — Amer. Midl. Nat. XXI, p. 210-234.

——, (1940). Die Übersprungbewegung. — Z. Tierpsychol. IV, p. 1-40.

Retrospective: Beginnings

Desmond Morris

My scientific involvement with sticklebacks began suddenly one afternoon in 1950. I was a zoology student at Birmingham University where my professor, Peter Medawar, had noticed that I was particularly interested in the study of animal behaviour. He told me that he was going to attend a lecture on that subject by a Dutch ethologist called Niko Tinbergen, who had arrived in England only a few months ago to set up a research group at Oxford, and invited me to go with him. He drove me to the lecture himself and I was flattered to be given this special treatment. It occurred to me that not many busy professors would pinpoint the special interest of one of their pupils in this way, but then Medawar was an exceptional professor. He told me that Tinbergen was 'pathologically modest but very important' and thought that I might find his talk inspiring. And he was right. After listening to Tinbergen for one hour I was totally converted to ethology and wanted desperately to move to Oxford to join the new research group that was already forming there. I was particularly fascinated by the studies he had made in Holland on the three-spined stickleback. I still have the notes I took during that talk of his 59 years ago and see that I made a special comment saying: 'The female stickleback may adopt a head-up posture which is the opposite of the male's head-down threat posture. This submissive posture protects her from the male's aggression.' I found the idea of a functional interpretation of display postures particularly intriguing.

What had appealed to me so much about Tinbergen's talk was that he offered me the chance to become a professional zoologist without having to perform unpleasant experiments on animals. All through my childhood I had been obsessed with watching animals, especially fish, and I had been fortunate that my family owned a private lake that allowed me to do this uninterrupted. I had built myself a raft so that I could float gently around the lake, lying flat with my face pressed close to the surface of the water, studying the behaviour of the fish below. Young friends of mine loved fishing with

My favourite male *Pygosteus* in breeding colours that I sketched in the spring of 1952. Reprinted from my memoirs (Watching; Encounters with humans and other animals, Max Press, London, 2006).

a rod and line, but I knew my fish so intimately that the idea of disrupting their lives with a worm on a hook seemed barbaric. So from an early age I wanted to be a quiet observer of animal behaviour rather than a disruptive intruder into it.

As a zoology student, it seemed to me that observing animals in nature would prove difficult in what was clearly a laboratory-oriented, experimental science. Tinbergen, to my delight, offered me a solution to this dilemma. He took natural history and drug it into the scientific arena. In his one-hour talk he demonstrated how it was possible to make a respectable science out of naturalistic observations of animal behaviour by studying animals in their natural environment, quantifying their actions, and carrying out simple field tests on them. This came as a revelation to me and, with Medawar's help, I eventually managed to join Tinbergen's group in Oxford as one of his first batch of doctoral students.

In August 1951 I arrived in Oxford and Tinbergen gave me a sheet of paper with five possible research projects typed out on it. They were: The sexual behaviour of *Drosophila*; the reproductive behaviour of chaffinches; imprinting in ducks; the reactions of bumblebees to flowers; or the reproductive behaviour of the ten-spined stickleback, to compare this species with the already well-known three-spined species.

For me, the choice was easy. I had already been watching sticklebacks in the streams and rivers of my native Wiltshire, and had even kept the common three-spined ones in aquarium tanks at home. My only concern was that, to make really detailed observations, I would have to work with tanks in the laboratory, but Tinbergen said that, as the fish was so small, I could recreate a natural habitat for it if I used large enough tanks. I installed many such tanks and had one special aquarium made that was seven feet (213 cm) long in which I could plant natural vegetation and imitate a section of a small stream. Tinbergen insisted that I should use the first breeding season to compile an ethogram — a detailed list of every motor pattern, every action, every posture, every movement, every colour change, made by this species. Only when I had observed its complete behaviour repertoire would I have sufficient knowledge to design appropriate tests.

This intensely naturalistic approach to the study of animal behaviour, even in the laboratory, set the small ethology group against the Oxford animal psychologists. With the youthful arrogance of pioneers in a new discipline, we viewed the old, entrenched rat psychologists as the enemy, a bunch of misguided fools who did not realize that, unless a rat is allowed to establish a territorial burrow system, its behaviour in the laboratory is meaningless and their experiments inappropriate. Of course, we were focusing on rat behaviour and they were merely using rats as surrogates, but spurred on by Tinbergen's almost evangelical zeal, we simply regarded them as irrelevant to the true study of animal behaviour.

Tinbergen's research trinity was function, causation and evolution. We had to examine these three factors with every behaviour pattern we studied. To understand the evolution of behaviour patterns we had to study the behaviour of closely related species. The comparisons I was able to make between the ten-spined stickleback and its three-spined relative were significant in this respect. Other comparative studies were being done at the time with gulls and with ducks and our scientific discipline at the time was often referred to as Comparative Ethology.

Apart from Tinbergen, my other main influence was Konrad Lorenz. In May 1951, when Lorenz was thinking of moving his headquarters to Slimbridge, he gave a series of lectures at Bristol University that I was able to attend. To describe his performance as a tour de force is an understatement. Looking like a cross between God and Stalin, his presence was overpowering. "Contrary to your Shakespeare," he boomed, "there is madness in my method." And indeed there was. Almost all his discoveries were made by accident and his life consisted largely of a series of disasters with the menagerie of animals with which he surrounded himself. His understanding of animal communication and display patterns was revelatory. When he spoke about a fish his hands became fins, when he talked about wolves his eyes were those of a predator, and when he told tales about his geese his arms became wings tucked into his sides. He wasn't anthropomorphic, he was the opposite — theriomorphic — he became the animal he was describing.

Konrad Lorenz and Niko Tinbergen, the founding fathers of comparative ethology, were great friends but were very different personalities. Niko was the quiet, objective, studious one, forever concerned about the need to quantify his observations. Konrad was the flamboyant, charismatically theatrical one, always ready to dazzle you with yet another revealing anecdote. Konrad understood animals better than anyone I have ever met before or since, but quantification was not his strong point. He found it boring and left all that to the industrious Tinbergen. When the two of them worked in the field together, Niko told me, it was always he who had to do all the hard work, while Konrad lay in the long grass watching him approvingly. And Konrad Lorenz must surely be the only scientist who has won the Nobel Prize without ever publishing any quantified results.

So, although I learned an immense amount from Konrad concerning the way to observe animal behaviour, it was from Tinbergen that I learnt how to martial and measure these observations. And the stickleback gave me plenty to do. For a common little freshwater fish, it was full of complex behaviour sequences of the kind that provided endless material for analysis. Thanks to the stickleback I became fascinated by reproductive behaviour in general and animal display signals in particular. I soon branched out to study similar behaviour in birds and discovered a species that, like the stickleback, was ideally suited to study at close range — the Australian zebra finch. In 1954 I wrote the first scientific paper on this little bird and have been fascinated to see that, since then, it has become the subject of countless later studies. My goal, however, was to extend my behaviour studies to mammals. Tinbergen was uncomfortable with this, having limited all his own studies and those of his pupils to insects, fish and birds. To make the move to mammals I therefore had to leave Oxford and take up a post at the Zoological Society of London, but my five years with Niko Tinbergen had left its mark on me and when, in 1967 I eventually came to apply ethology to human behaviour in a book called *The Naked Ape*, the chapter headings were much the same as those of my stickleback thesis.

My later work with television and popular books owed a great debt to Lorenz and Tinbergen. If I had not written 48 scientific papers before I moved on to popular presentations I would have lacked the academic grounding that is a necessary background for anyone who is trying to interest a general audience in the subject of animal behaviour. I still wince whenever I hear a teleological argument or an anthropomorphic comment made by someone presenting animals on television. On the other side of the coin, I sigh when I open my latest copy of the journal *Animal Behaviour* and find there are no drawings or photographs to illustrate the animals discussed in the increasingly mathematical papers that are published there. The early days of stickleback research may have been statistically clumsy, but they were bursting with new ideas and I was immensely lucky to have been a part of that pioneering group in Oxford in the early 1950s.

„TYPICAL INTENSITY" AND ITS RELATION TO THE PROBLEM OF RITUALISATION

by

DESMOND MORRIS [1])
(Department of Zoology and Comparative Anatomy, University of Oxford)

(With 3 Figs)

(Rec. 15-VII-1956)

INTRODUCTION

It is a basic property of simple signals, when these are contrasted with other types of response, that they remain constant in form regardless of any changes in the circumstances which cause them. For example, a car moves faster, the harder the accelerator is pressed, but a telephone bell rings at the same speed, regardless of how urgent the call may be.

Superficially, it might appear detrimental that such a feature as signal rigidity should interfere with the simple direct relationship that usually exists between the strength of a stimulus and the strength of the response it provokes. However, the reduction in the amount of information provided by the signal about the exact state of the signaller, that is involved with this stereotyping, is apparently more than compensated for by the elimination of signal ambiguity. A signal that is constant in form cannot be mistaken.

Now, it is a fundamental characteristic of the communicatory behaviour of animals that the signal patterns used are derived from non-signal sources (see, in particular, DAANJE 1950, TINBERGEN 1952, and MORRIS 1956a). This signalisation may occur either in the phylogeny of an animal (when it is called ritualisation), or in its ontogeny (when it may be termed stylisation). In either case it results in the need for a fundamentally variable reaction becoming constant in form. These conflicting demands, of variability and stability, lead in many cases to a compromise: — T y p i c a l I n t e n s i t y.

1) I am greatly indebted to Dr N. TINBERGEN, Dr W. M. S. RUSSELL, Mr A. D. BLEST, Miss F. HALL, Mr P. GUITON, and various other colleagues, for reading the present paper in MS and/or for valuable discussions on the subject. I am particularly grateful to Dr W. M. S. RUSSELL, discussions with whom initially stimulated the train of thought which led to this paper. Thanks are also due to the Agricultural Research Council, from which the author received a grant during the above work.

THE RELATIONSHIP OF TYPICAL INTENSITY TO FREQUENCY

Frequency and intensity may be defined as follows: Frequency measurements refer to the number of times a specified response is shown in a given period; intensity measurements refer to the form which the response takes each time it occurs.

Fig. 1 shows three hypothetical graphs which illustrate I. Variable Intensity, II. Typical Intensity, and III. Fixed Intensity. In the first case, where there is an increase in stimulation the response increases both in frequency

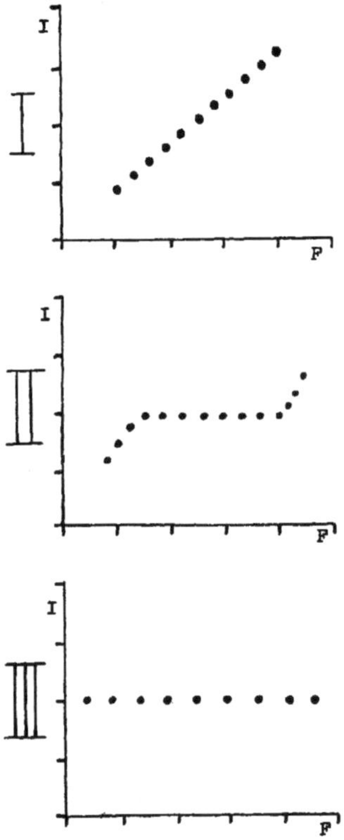

Fig. 1. The stereotyping of a response. I shows the response with a variable intensity. An increase in frequency (F) is accompanied by an increase in intensity (I). II shows the response with a typical intensity. At most frequencies, the intensity is the same; only at very high or very low frequencies do changes in intensity appear. III shows the response with a fixed intensity. Here, no matter how high or low the frequency, the intensity never varies.

Fig. 2. The low intensity ruffled courtship posture of the Cutthroat Finch male (left).

Fig. 3. The high intensity ruffled courtship posture of the Cutthroat Finch male (centre). This posture has developed a typical intensity, great variations in response frequency being accompanied by no more than these very small intensity differences. The small extent of these intensity differences can best be judged by comparing the two display postures with the posture of the non-courting male, seen on the left.

(F) and intensity (I). In the second case, where a flattening of the graph has occurred, the intensity of the response now only varies when the latter occurs with very low or very high frequencies. In the third case, the intensity has become constant at all frequencies; no matter how weak or how strong the stimuli now, the response, if it appears at all, will occur in a rigidly fixed form.

It should, however, be pointed out that a stereotyped response, of the latter kind, can still vary with varying stimulation, because its frequency of occurrence will still be varying. To return to the telephone analogy: if the call is urgent, the bell will ring no faster, but it will go on ringing for longer before the caller gives up. It is the existence of this other response variable, and its potential independence from the intensity variable, that makes it possible for a signal response to become fixed and unambiguous, without completely ceasing to reflect quantitative fluctuations in the mood of the signaller.

For example, an animal may reveal that it is very hungry by (a) eating fast and (b) eating for a long time; or that it is not very hungry by (a) eating slowly and (b) only eating for a short while. On the other hand, it may reveal that it is very highly motivated sexually by (a) taking up a characteristic courtship posture and (b) displaying thus to the female for a long time; or that it is only mildly sexual by (a) taking up an *identical, or nearly identical posture* and (b) displaying thus to the female for only a short period. The difference between the responses in the feeding and sexual contexts given here is obvious. In the sexual case, if the courtship display postures are identical at the two motivational levels, then a *fixed* intensity (fig. 1, III) has developed. More commonly, however, slight differences can be detected which reveal that instead only a *typical* intensity has developed. Usually, if the motivational state is extremely low or extremely high, minute changes in the intensity of even the most fixed response can be detected. The photographs in figs 2 and 3 illustrate this point. They show the courtship posture of the Cutthroat Finch (*Amadina fasciata*) (see MORRIS 1954c, 1955). In fig. 2 the low intensity form is seen and in fig. 3 the high intensity version is shown. Note how minute are the differences between them (especially when either is compared with the non-displaying male on the left in fig. 3). Once a male of this species has begun to court a female, he assumes this posture and no matter how long or short his period of courting may be, or how many times he may display in a given time period, he never shows much more variation in display intensity than is seen here. In other words, for enormous changes in response frequency, he will show only slight changes in response intensity. These latter changes are so slight that casual observation would lead to the conclusion that his display had a fixed intensity. However, as with most cases, a closer scrutiny reveals that it has, in fact, a typical intensity, albeit of an extreme kind.

Continuing with the Cutthroat F'nch as an example, it is instructive to look for a moment at the nature of its courtship dance. Each dance consists of a number of „inverted curtseys" (see MORRIS 1954c, 1956c). With each of these curtseys, the bird makes a stilted upward jump. Each jump has a highly characteristic form with a very typical intensity, so that if a jump is given at all it nearly always involves exactly the same amplitude, speed *etc.* This fact helps to give the dancing the distinctive rhythmic form so typical of most display movements. An important point emerges here, however, namely that although each inverted curtsey has a typical intensity, each dance, which is composed of a number of these curtseys, in this case does not. The bird may perform low intensity dances made up of only one curtsey, or high intensity dances made up of many. (It is nevertheless theoretically quite possible for the dance also to develop a typical intensity, so that it nearly always comprises the same number of curtsey movements. In many cases developments of this kind have taken place.) What is important here is the relationship, as regards frequency and intensity, between the courtship responses at the two levels of dancing and curtseying. For it will be seen that the frequency of curtseying 's also referred to here as the intensity of dancing.

I am grateful to Dr W. M. S. RUSSELL for bringing the significance of such relationships to my notice. He pointed out that the intensity of a response is also the frequency of the response at the level of integration immediately below it, and *vice versa.* (This point is briefly discussed in RUSSELL *et al.,* 1954, p. 197-8). As a further and more complex example of this, we may take the beak-cleaning pattern of birds:

A beak-wipe consists of (a) twist body (b) lower and rotate head (c) scrape (d) raise head and (e) twist body back.

A beak-wiping consists of a number of such wipes given in quick succession.

A beak-wiping-bout consists of a number of such wipings (in a series uninterrupted by other activities).

A full cleaning pattern comprises not only a number of (a) wiping-bouts, but also (b) shaking bouts, (c) preening bouts and (d) scratching bouts.

At each level of integration in a hierarchy of this kind, the frequency of the activity below is the intensity of a given response and the intensity of the activity above is the frequency. Thus, the intensity of a beak-wiping is given by the number of wipes it comprises, whilst the number of beak-wipings gives the intensity of a wiping bout. Now supposing, for the moment, that beak-wipings, as responses, developed a typical intensity; this would mean that beak-wipes would have developed a typical frequency. If four wipes nearly always occurred per wiping, giving the latter a typical intensity of four, then beak-wipes, which would now occur at this frequency regardless of fluctuations in their own intensities, would have a typical frequency, also of four.

It is, of course, a quite arbitrary choice as to whether one approaches each case as a typical intensity, or a typical frequency, since where there is one there must also be the other. For various reasons, however, I prefer the former. Frequency measures can sometimes be confusing owing to the fact that they may involve an artificial limiting time period, set by the experimenter. Intensity measures, on the other hand, provide a more reliable approach. An intensity measure is available every time an act occurs, but a frequency measure may be either a score of bout length, or the total number of actions per (arbitrary) time unit.

(See discussion on differences between Intensity and Tendency in Russell, Mead and Hayes, 1954. It should be noted that Act Tendency in their sense has four measures: frequency, latency, number, and duration. These are all being lumped together as frequency in the present paper. This does not, however, reflect any basic difference in approach, since, on page 166 of their paper the above authors say of their four categories that "Evidently all these measures can be reduced in principle to frequencies.")

TYPICAL INTENSITY IN AMBIVALENTLY MOTIVATED RESPONSES

Thus far, only those situatons in which one type of response is stimulated have been considered. Common in signal-producing situations, however, is the state of motivational conflict. With such a state one finds various important forms of compromise; for instance, alternating intention movements, ambivalent posturing, *etc.* In many of these cases typical intensities have developed.

The fighting behaviour of the Black-headed Gulls studied recently by Moynihan (1955) must be mentioned here. Moynihan found that, with increasing or decreasing frequencies of particular conflict states, he did not observe a whole range of as many intensities. Instead, only a few quite distinct conflict responses occurred, each having a typical intensity covering part of the total motivational range.

He also found that confl'ct responses provided rather special motivational problems, as follows. Each response was the result of a combination of so much 'attack' and so much 'escape'. It could vary motivationally, therefore, in two distinct ways. Firstly, the ratio of attack to escape could remain constant, whilst their absolute levels changed; this type of variation gives rise to no new problems. Secondly, the ratio of attack to escape could change; this does provide a new problem. In the first case, one can relate simple frequency changes to the change in the form of the response, and deal with the situation as if it were an instance of simple single motivation. All the previous comments concerning typical intensities apply here as before. In the second case, however, this is not possible, and a new concept of Typical Compromise has to be introduced. This complication arose in the case of MOYNIHAN's gulls, as it does in most cases, but a more suitable example for presentation here is offered by the lateral display common to so many species of fish during combat. The fighting males, which are both hostile towards and afraid of one another, either attack, flee, or show the lateral display. This means that for almost the entire time, when engaged in combat, the individuals are either facing directly towards or directly away from one another, or are precisely parallel. In other words, out of all the possible angles they could take up, only three basic ones are used. (There are a few species where this is not the case, but it is nevertheless the general rule.) This may mean that the combatants are always either (1) fearlessly aggressive, or (2) completely afraid, or (3) in an exact state of balance between the two. Alternatively, the lateral display could have developed as a Typical Compromise such that if the fish are both afraid and aggressive, they take up this position, even though they may be, at one time, slightly more aggressive than afraid and, at another, slightly more afraid than aggressive. In order to determine which of these two possibilities is relevant in a given case, it is necessary to check the variations in the ratios of the attack-developments and flight-developments from the lateral postures. If all fish showing lateral display are just as likely to attack as to flee, then, if there are no intermediate postures, the first type of modification must have occurred. If, on the other hand, laterally displaying fish may in some instances show a greater tendency to flee than attack and in other instances a greater tendency to attack than flee, then a typical compromise must have developed. Judging by the analyses carried out so far, the latter appears to be the more usual occurrence.

The important difference between typical intensities and typical compromises is that in the former case the intensities could be compared with frequency differences, whereas in the latter case the compromises must be compared with frequency-ratio differences. Functionally, however, typical compromises are identical in significance with typical intensities. They fix characteristic conflict signals and render them unambiguous.

With the development of typical intensity spreading its causal range (see MORRIS 1954d) in one way, and typical compromise spreading it in another, a conflict response can easily become so readily evoked that no fight can occur without it appearing. In some species actual contact fighting itself may even become extremely rare and disputes settled simply by prolonged bouts of performing the special conflict responses of the animals concerned. So far, only a single conflict response has been considered, but in most species there is a distinct set of these reactions, which form specific "threat codes" (see MORRIS 1954d, 1956b). The Mexican Swordtail (*Xiphophorus helleri*) provides a suitable example here. Whereas most species only show the three angles (towards; lateral; away) mentioned so far, when fighting, the Swordtail shows five: towards; positive-sigmoid; lateral; negative-sigmoid; away. In the positive sigmoid posture the fish holds its body parallel to the rival, as in the typical lateral display, but turns its head towards the rival and its tail away, as in the frontal posture. This gives its body a characteristic sigmoid shape, as does the negative-sigmoid display, in which the head points away from the rival, as if to flee from it. Fish which show the positive-sigmoid display are more likely to attack than flee, as might be expected, and the converse is true for the negative-sigmoid. In this species, then, there are three distinct conflict responses where, in others, there is only one — the simple lateral display. This means that the Swordtail responses have less chance of developing large causal ranges than the others. One way in which this difficulty is overcome in certain species (but not in the present case) is by the spreading of causal ranges to such an extent that the o v e r l a p with one another, giving rise to multiple conflict responses. This has been discussed fully elsewhere, however, and will not be dwelt on here (see MORRIS, 1954d, Chapter 4).

TYPICAL INTENSITY AND RITUALISATION

As will be clear by now, the development of typical intensity is a fundamental aspect of the process of ritualisation. During the latter a number of types of change can be observed. These have been discussed to some extent by DAANJE (1950) and will be reviewed in more detail shortly by BLEST (in prep.). They will therefore only be briefly mentioned here and will only be discussed with regard to their relationship to typical intensity.

Firstly, there are several ways in which patterns can be modified as signals, which do not touch on the problem of typical intensity. Included here are (1) Change of stimulus-valance, (2) Qualitative motivational changes ("neurophysiological emancipation" — see TINBERGEN, 1952), (3) Loss or modification of original orientation, and (4) Supplementing of response by addition of display structures, colour, *etc.*

Then there are the following changes which may or may not be connected with the development of typical intensities: (5) threshold lowering, (6) development of rhythmic repetition, (7) differential exaggeration of components, (8) omission of components, (9) "freezing" of movements, (10) change in sequence of components, (11) changes in component co-ordination, (12) increase or decrease in speed of performance, (13) change in vigour of movements.

Finally, of course, there is (14) development of typical intensity itself and, in special cases, (15) development of typical compromise.

No attempt has been made here to produce a comprehensive list of types of modification. Nor do the fifteen given above represent fifteen mutually exclusive categories. Rather, they present fifteen types of modification of the kind which is met with during the course of comparative ethological studies. The first group will not be mentioned further here, but some comments must be made on the second.

Firstly, lowering of response threshold, which so often goes hand in hand with other ritualising changes, can apparently be the result of one, or both, of two kinds of basic development. On the one hand, a particular form of a response may become more frequent when it develops as the typical intensity of that response. It will now occur as the reaction to a wider variety of motivational states and will therefore be seen more often. Its frequency is increased, so to speak, at the expense of the other forms of the response. On the other hand, it may appear more often because all forms of the response appear more. In the latter case the response need not have developed a typical intensity.

Which of these two possibilities is operating in a given case is easy to determine. If the particular form of the response concerned shows an increase in frequency independent of an increase in the frequency of the response as a whole, then typical intensity is responsible. If its increase is accompanied by an increase in all other forms of the response as well, then typical intensity is not involved.

Next, rhythmic repetition, a particularly common development, can be partially, if not wholly, explained by the development of typical intensity. The existence of the latter has two effects here. First, it makes each separate action in a rhythmic repetition the same strength and, second, it thereby increases the likelihood of the time intervals being similar between each act.

Differential exaggeration of components applies to multiform responses as opposed to uniform ones. In the former, the components are qualitatively different (A-B-C-D); in the latter each component is the same (A-A-A-A). The beak cleaning examples illustrate this: the beak-wipe response has multiform components (twist-lower-scrape-raise), whereas the beak-wiping response has uniform ones (wipe-wipe-wipe-wipe). In a ritualised multiform response, the development of a typical intensity at a certain stage through its sequence of components will give rise to differential exaggeration favouring that stage. For example, ritualised beak wiping itself in certain species consists of little more than the initial bowing components. These may now be given with low or high frequency, but hardly ever develop into a full wiping. (It should be noted, however, that here the differential exaggeration occurs as a negative process; if a component is positively exaggerated beyond anything ever seen in the original response, then this of course is a special ritualising development quite distinct from typical intensity.)

Leading on from this, the complete omission of components will take place as a further extension of the last process when differential exaggeration of the negative type becomes more extreme. An even further development of this is the "freezing" of movements. A pattern begins, gets to a given stage and then fixes there. It seems likely that typical intensity may be involved here as before. Also, the other modifications mentioned in the second major group (*i.e.* changes in component sequence, co-ordination, speed, or vigour) may be associated, to a greater or lesser extent, with the development of typical intensities, but they will not be further discussed here. (They will be more fully dealth with by BLEST, in his forthcoming review.)

TYPICAL INTENSITY IN NON-SIGNAL SITUATIONS

Although the development of typical intensity is so intimately connected with the process of signalisation, it is not altogether restricted to this context. Many non-signal responses have evolved typical intensities purely in relation to increased mechanical efficiency.

For example, in the fanning activity of the reproductive male stickleback (see VAN IERSEL, 1953), the amplitude and speed of the fin beats is fairly constant, regardless of motivational fluctuations, giving fanning a typical intensity. The important and striking variable in this activity comes at the next highest level, in connection with the intensity of fanning bouts, or, to express it another way, with fanning frequencies.

In contrast with this, the fanning response of the River Bullhead (*Cottus gobio*) (see MORRIS, 1954b) shows no typical intensity at any level. The fin-beats vary considerably in amplitude and in speed, giving no typical intensity to fanning here. In this species, in fact, both the fanning intensity and the fanning frequency are extremely variable.

A possible reason for this difference between the fanning of the two species is that the Bullhead fans while resting on the bottom, whereas the sticklebacks have to maintain themselves hovering in the water just in front of their nest entrances. In support of this view, there is the observation of VAN IERSEL (1953, p. 24) that, in the case of the Three-spined Stickleback (*Gasterosteus aculeatus*), the very slight decreases in fanning intensity which can sometimes be seen are accompanied by maladjustments in the orientation of the male to its nest. VAN IERSEL concludes "that the taxis component of the movement becomes incomplete when the intensity of the fixed motor pattern decreases." The taxis component of the Bullhead fanning would not suffer in this way, however, this species being supported by the substratum.

A second comparative example of this kind concerns the snail-eating behaviour of Song Thrushes (*Turdus ericetorum*) and Blackbirds (*T. merula*) (see MORRIS, 1954a). The former species has evolved a special shell hammering motor pattern, thus enabling it to extract and devour the contents. This movement appears to have been derived from the more basic actions of shake, peck, and wipe. All or some of these have been modified in such a way that a Song Thrush can now pick up a Snail in its beak and, with repeated rhythmic swings of its neck, batter the shell on a hard stone surface until it cracks open. The strength and speed of each swing have become stabilised so that hammering has developed a typical intensity.

This appears to have arisen because stronger or faster blows would make the shell difficult to hold and weaker ones would not crack it open so efficiently. The Blackbird, on the other hand, does not seem to have evolved this excellent technique. It tries hard sometimes to smash shells by using the old peck and shake actions in a vigorous and erratic fashion, but seldom with success. Although it is easy to see why the Song Thrush has evolved a typical intensity here, it is difficult to understand why the Blackbird has not.

One final example concerns a purely ontogenetic case. When parrots — which are very active birds — are kept in captivity, they are often chained to a single perch or placed in a ridiculously small cage. At first, the birds make wild and erratic movements in attempting to escape from their strict confinement. After a while, however, they develop stereotyped versions of their escape intention movements (see LORENZ, 1950, p. 55), which then have extremely typical intensities — almost fixed intensities. Regardless of the frequency of the performance, the actions are repeated rhythmically over and over again ad nauseum with almost exactly the same amplitudes and at almost identical speeds. It is of interest to note that the exact form of the stereotyped movements differs considerably between individuals. Of the four types which I know personally one consists of a figure-of-8-on-its-side movement of the head and neck. Another involves a stilted form of bowing. A third is primarily a lateral pivoting and the fourth a tipping of the head first to one side then the other. It is most interesting to find individual variations occurring in a strictly ontogenetic development, contrasting, as one would expect, with the phylogenetic cases referred to previously, where such variations are infinitesimal.

In conclusion, the functional aspect of the typical intensities in the parrot case presents somewhat of a problem. It would seem that typical intensity is not only valuable in connection with certain signal requirements, and as an aid to non-signal mechanical efficiency, but is also in any case a basically convenient mode of response performance for the nervous system.

SUMMARY

Where a response is acting as an agent of social communication, it is advantageous for it to possess a certain constancy of form. This eliminates signal ambiguity, and is achieved in many cases by the development of a frequency-intensity relationship in which very great changes in the frequency of a response are only accompanied by very small changes in its form, thus giving it a "Typical Intensity".

This process of response-stereotyping is a fundamental aspect of the phenomenon of ritualisation, but can also be shown to be operating in certain non-signal situations for different reasons. It is pointed out that a "Typical Intensity" may arise eiher phylogenetically, or ontogenetically, and examples of both are given.

LITERATURE

Blest, A. D. (in prep.). The concept of ritualisation and the evolution of behaviour patterns.

Daanje, A. (1950). On locomotory movements in birds and the intention movements derived from them. — Behaviour 3, p. 48-99.

Iersel, J. J. A. v. (1953). An analysis of the parental behaviour of the male Three-spined Stickleback (*Gasterosteus aculeatus* L.). — Behaviour, supplement 3.

Lorenz, K. Z. (1950).King Solomon's Ring. — London.

Morris, D. (1954a). The snail-eating behaviour of thrushes and blackbirds. — Brit. Birds 47, p. 33-49.

—— (1954b). The reproductive behaviour of the River Bullhead, *Cottus gobio* L, with special reference to the fanning activity. — Behaviour 7, p. 1-31.

—— (1954c). The courtship behaviour of the Cutthroat Finch (*Amadina fasciata*). — Avic. Mag. 60, p. 169—177.

—— (1954d). An analysis of the reproductive behaviour of the Ten-spined Stickleback (*Pygosteus pungitius* L). — Doctor Thesis, Oxford. (Also, Behaviour, supplement 6, in press.)

—— (1955). The markings of the Cutthroat Finch. — Birds Illustrated 1, p. 182-183.

—— (1955a). The feather postures of birds and the problem of the origin of social signals. — Behaviour 9, p. 75-114.

—— (1956b). The function and causation of courtship ceremonies. — Fondation Singer Polignac. Colloque Internat. sur L'Instinct, June 1954.

—— (1956c). The reproductive behaviour of the Bronze Mannikin (*Lonchura cucullata*). — In press.

MOYNIHAN, M. (1955). Some aspects of the reproductive behaviour of the Black-headed Gull. — Behaviour, supplement 4.

RUSSELL, W. M. S., MEAD, A. P. and HAYES, J. S. (1954). A basis for the quantitative study of the structure of behaviour. — Behaviour 6, p. 153-205.

TINBERGEN, N. (1952). Derived activities; their causation biological significance, origin, and emancipation during evoluton. — Quart. Rev. Biol. 27, p. 1-32.

RÉSUMÉ

Lorsque une réponse agit comme un agent de communication sociale, il est avantageux qu'elle possède une forme constante. Cette constance éloigne l'ambiguité des signaux et, dans beaucoup de cas, elle s'accompagne du développement d'un rapport entre la fréquence et l'intensité dans lequel de très grands changements dans la fréquence d'une réponse ne sont accompagnés que par un très petit changement de sa forme: ainsi elle acquiert une „intensité caractéristique".

Ce procédé de stéréotyper une réponse constitue un aspect fondamental de l'évolution des rites, ou a pu montrer qu'agir ainsi, dans certaines situations, est dû à des raisons différentes. C'est indiquer qu'une „intensité caractéristique" peut apparaître dans l'histoire phylogénique ou ontogénique d'une espèce et des exemples des deux cas sont présentés

THE REPRODUCTIVE BEHAVIOUR
OF THE
TEN-SPINED STICKLEBACK
(*PYGOSTEUS PUNGITIUS* L.)

BY

DESMOND MORRIS

DEPARTMENT OF ZOOLOGY AND COMPARATIVE ANATOMY,
OXFORD UNIVERSITY

WITH 50 FIGURES

LEIDEN
E. J. BRILL
1958

THE ORGANISATION OF THE CYCLE

The problem remains as to what can be concluded, on the basis of the above description, concerning the organisation of the cycle as a whole. Clearly, if all the details are to be taken into consideration, no very simple explanatory system will suffice.

It is necessary here to examine the hierarchical theory postulated by TINBERGEN (1942, 1950, 1951a), in connection with the reproductive behaviour of *Gasterosteus*. TINBERGEN postulates a top reproductive centre of neural organisation, to which are subordinated four sub-centres — fighting, building, mating, and care-of-offspring. To each of these sub-centres are subordinated a number of simple act centres. For example, chasing, biting and threatening, *etc.*, are subordinated to the fighting sub-centre, and so on. (See fig. 98 in TINBERGEN 1951a.)

As a general principle, this is undoubtedly of tremendous importance in understanding the way in which elaborate behaviour complexes are controlled and organised; but, in the case of the reproductive cycle of *Pygosteus*, it requires some adjustment and modification, when this behaviour complex is examined in detail. It is important to stress, however, that any such adjustments as are necessary, in no way invalidate the basis principle of hierarchical organisation as laid down by TINBERGEN.

Firstly, in what respects do the present findings agree with the hierarchy theory? *Pygosteus* can probably be said to have a top reproductive centre and subordinated to this a fighting and a mating sub-centre, in TINBERGEN's sense. The various fighting acts show perfect co-variance, as do the mating acts. I am less happy, however, about postulating a nesting, or a parental sub-centre, on the basis of the evidence given in the present report. (It is for this reason, incidentally, that I have concentrated on these latter activities in figs 47 and 48, and have omitted fighting and mating there.) The exact reasons for this must now be given.

In fig. 47 I have shown Collecting, Boring and Fanning fluctuations and it can be seen that they co-vary to some extent. All three show a marked increase during Phase C, when the nest is being built, and a marked decrease during the sexual phases D and E. After this, however, they show less co-variance. Fanning increases steadily during the parental cycle, whereas collecting and boring decrease. Boring does, however, show another peak at the time of the fanning peak, but collecting does not reach its last (minor) peak until the other two have subsided.

Of the other nest-actions (see fig. 48), Mending co-varies fairly accurately with parental fanning, being absent from the nest-building proper, and also not suffering from the suppressing effect which the parental cycle has on all other nest actions (except fanning, of course) during Phase H, which is the peak of fanning duration.

Insertion Glueing appears, rather unexpectedly, to co-vary with sex, rather than with any of the other nest-actions. It does not reach its peak until the nest is complete and the male is ready for a female, and it disappears completely (like sex) during the parental cycle. Superficial Glueing, however, co-varies with nest-building proper, only ever being seen in Phase C.

Retrieving the young co-varies with no other nest action, but takes over from parental fanning when the eggs hatch.

Summing up the above, it is clear that is it far from easy to state that this action is a nesting act and belongs to the "nesting sub-centre", or that that action is parental and belongs to the "parental sub-centre". This can be done in the case of courtship actions for example, or aggressive actions, but here, in the case of the nest-actions, the situation demands a different approach.

THE "CFB" UNIT

Collecting material, boring it into the nest, and then fanning, is a combination (CBF) which is frequently observed in *Pygosteus*. It may occur as one continuous complex action, or it may occur in different partial forms at different times. Now, although the three component actions have been observed to co-vary to some extent through the cycle, they do not do so perfectly, and it is valid to ask what form this imperfection takes.

It was noted that, as the cycle progressed, the relative frequencies of the different combinations of C, B and F varied in a rather special way. At the beginning of the cycle, C alone was seen; later CB became more frequent; later still CBF took over, then BF, and finally F alone. This can be diagrammatised as follows:

C		
C	B	
C	B	F
	B	F
		F

During the early nest-building phase, males often picked up and carried material, but then did not bore it into a host plant. Soon they did so, however, but did not fan after boring. As the nest progressed, fanning was seen to appear at the end of boring and when the nest was completed, collecting dropped out, but boring and fanning still occurred together. Not until the parental cycle had begun, did fanning appear without boring preceding it, to complete the shift from C, through B to F.

On the basis of this, it is tempting to postulate a CBF unit, or centre, which is activated as a whole during the reproductive cycle and which varies the exact nature of its output according to the external stimulus situation. A quantitative record of this variation is shown in fig. 49. This gives the percentage of CBF activities which are C, B and F respectively at the different stages of the reproductive cycle. It shows clearly the three peaks in C, B and F (marked with arrows), which would be expected from the above.

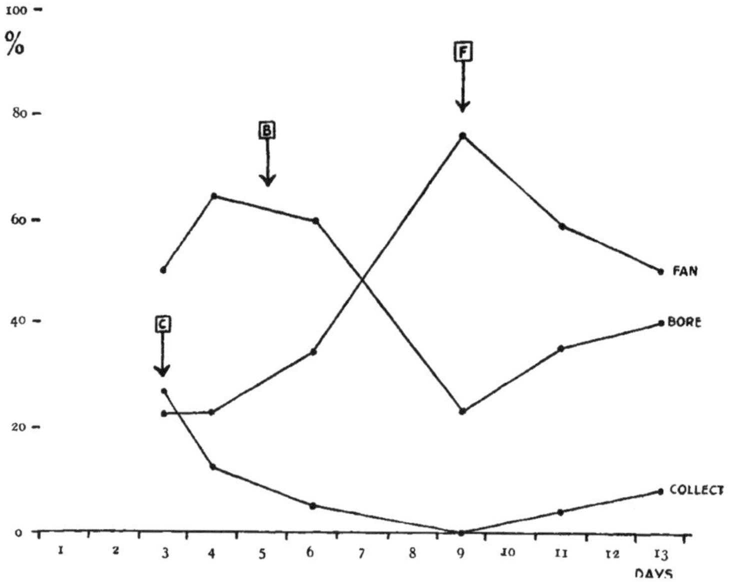

Fig. 49. Changes in proportions of the CBF complex. Explanation in text.

If the concept of a CBF unit is valid, it makes it impossible to consider pre-parental fanning in *Pygosteus* as displacement fanning. (This is certainly in line with all other indications in this species.)

I suggest that the evolutionary explanation of this is that sexual behaviour has appeared later and later in the reproductive cycle and has wedged itself in, so to speak between the earlier and later stages of behaviour associated with care of the offspring. It is plausible that eggs were once laid in dense clumps of weeds, where the male protected them from enemies. Later the male probably strengthened the host clump himself, eventually covering the eggs with a rough nest. Although this undoubtedly improved their protection, it must have increased their aeration requirements. Thus, building and ventilating (CBF) probably became intensified together in evolution. But selection may have favoured those males which began to build before they possessed eggs. (We have seen that it is more difficult to build a nest around eggs than it is to make a nest before acquiring eggs.) In this way, the care-of-offspring behaviour became partially separated into two phases — pre-parental and parental. This partial separation would explain many of the present difficulties.

Although I shall be comparing *Pygosteus* and *Gasterosteus* below, it may be said at this stage that this separation seems to have gone further in the latter species. (In birds, the separation of nesting from parental behaviour is much more distinctly marked.)

It must be borne in mind that the above evolutionary remarks and suggestions are extremely hypothetical, and must be regarded as highly tentative. Something of the kind is necessary, however, to emphasize the fact that nesting and parental behaviour patterns are far from distinct from one another in *Pygosteus*, and to stress that it is therefore dangerous to talk of any of their component actions as occurring as displacement activities, without very special confirmative evidence. When a bird regularly wipes its beak during courtship, the action of beak-wiping is so distinct from any sexual activity, that there is little danger in calling it a displacement activity, but when *Pygosteus* performs fanning during the pre-parental phase, it is very dangerous to call this displacement fanning, simply because no eggs are yet present.

Summarising, it may be said that, in the case of *Pygosteus*, it is not helpful to speak of a nesting centre or a nesting drive, or of a parental centre or a parental drive. Rather, it is better to consider separately each response which is functionally concerned with the care of offspring. This avoids a number of difficulties which arise if one attempts to fit all the results into a rigid hierarchy system.

COMPARISON WITH *GASTEROSTEUS*

Throughout the present report the reproductive behaviour of *Pygosteus* has been compared with that of *Gasterosteus*, with respect to various details of the cycle. It is fitting here to make a more general comparison of the two cycles as whole systems. This will be discussed only briefly, however, as a more elaborate comparative study of stickleback behaviour is at present being made by Dr F. HALL.

LEINER made the first serious attempt to compare the behaviour of the two species (1931a, 1934). He came to the conclusion that it was easier to derive the *Gasterosteus* reproductive pattern from the *Pygosteus* pattern than *vice versa*, and stressed that the behaviour of the former species is much more complicated and highly developed. He cites the anatomical evidence which suggests that the few-spined *Gasterosteus* evolved from a many-spined ancestor. (For details of the anatomical evidence, which concerns the formation of the dorsal bone plates, see BERTIN, 1925.) In fact, this anatomical data is probably the most important evidence in support of his conclusions, the behavioural data he presents being far less convincing. His arguments involve three forms of naivety: firstly, he assumes that the behaviour of contemporary *Pygosteus* is the same as that of the many-spined ancester of *Gasterosteus*. Secondly, he assumes that simplicity cannot evolve from complexity. Thirdly, he overlooks the possibility of convergence. Despite this, however, and despite the fact that I do not agree with some of the details of his arguments, I am inclined to agree with his general conclusions, for the following reasons.

In the appendix to MORRIS, 1954e, I have described experiments which show that the few large spines of *Gasterosteus* (see fig. 50) are more efficient defence weapons against predators than the many smaller spines of *Pygosteus*. This, coupled with the skeletal evidence that *Gasterosteus* has evolved from a many-spined ancestor, presents a strong case for what I consider to be the relevant fundamental evolutionary trend. My data suggests that, with increased defensive capacities, *Gasterosteus* was able to leave the shelter of the aquatic vegetation, and occupy the vast new ecological niche of sandy-muddy substrata, in the reproductive season. As I have pointed out in the appendix, *Pygosteus* compensates for its defensive shortcomings by its timidity, which keeps it within the protection of the dense vegetation.

I

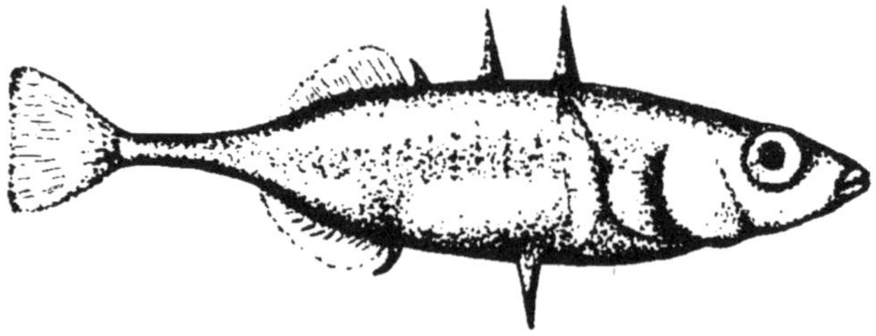

II

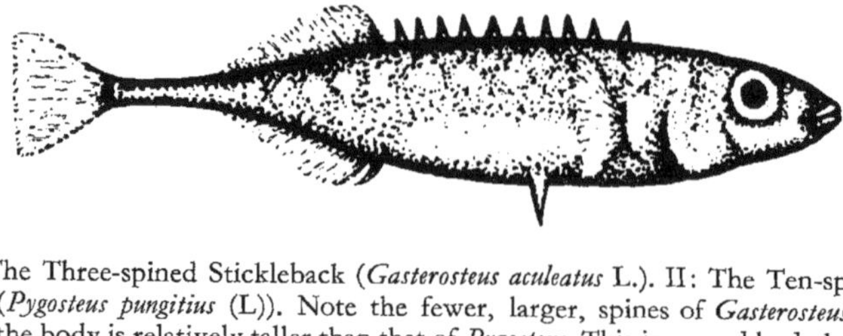

Fig. 50. I: The Three-spined Stickleback (*Gasterosteus aculeatus* L.). II: The Ten-spined Stickleback (*Pygosteus pungitius* (L)). Note the fewer, larger, spines of *Gasterosteus* and the fact that the body is relatively taller than that of *Pygosteus*. This increased body height separates the dorsal and ventral spines more than in *Pygosteus*, thus rendering them more formidable in the mouth or throat of a predator. In both species of stickleback, the anterior dorsal fin rays, the pelvic fins, and the first anal fin ray, have become modified to sharp spines.

I suggest that most of the behaviour differences between the two species can be thought of as the result of the fact that *Gasterosteus* was thus able to become more independent of the vegetation during the reproductive season. From this point of view, it is interesting to examine the different types of reproductive behaviour, one by one:

Fighting: There is little difference between the two species. Both show biting, carrying, roundabout fighting, head-down threat, ventral rolling, *etc.*; the minor differences that exist (in spine raising, for example) have yet to be studied. One interesting point is that a beaten male *Gasterosteus* is less inclined to show the tail-down submissive posture than a beaten male *Pygos-*

teus. This is correlated with the fact that the tail-down display of the ripe female *Gasterosteus* is far more exaggerated (and ritualised?) than the similar display in *Pygosteus*. It seems that the extra importance of this display as a sexual signal in *Gasterosteus* has been accompanied by a decrease in its importance in its original agonistic context.

The fact that intraspecific fighting is similar in the two species is not surprising, since there is little relevant difference in the two environments.

Nesting: The greatest difference between the two species concerns their modes of nest-building. Correlated with its emancipation from the weeds, *Gasterosteus* has modified its nest in the new environment. Most differences in nesting are purely quantitative. *Gasterosteus* and *Pygosteus* both possess sand-digging, boring and mending (pushing) movements, but *Gasterosteus* uses boring less and digging and pushing more, when building. Also *Gasterosteus* uses superficial glueing far more than *Pygosteus*, but possesses no insertion glueing. Differences in frequencies, sequences and orientations of the same motor patterns result in the formation of extremely different nests. Of the few qualitative differences, insertion glueing and looping, which are unique to *Pygosteus*, and sucking (see VAN IERSEL, 1953, p. 10), which is unique to *Gasterosteus*, can be mentioned. It is, nevertheless, variations in the same motor patterns which account for most differences, rather than the possession of different motor patterns. It is emerging from contemporary ethological studies that this last point is of extremely widespread application and of paramount importance in the understanding of the evolution of behaviour.

Courtship: Basically, the courtship sequences of the two species are very similar. In both species, the male dances, leads, shows-the-entrance, shivers and then spawns. The behaviour of the females is the same in both cases. The differences between the forms of the various male displays have already been discussed, but it is worth repeating here that the slower leading to the nest by *Pygosteus* and the fact that it does not repeatedly rush to the nest whilst courting, are correlated with the denser vegetation in which it performs.

Parental: The less favourable position for the eggs of *Gasterosteus* is correlated with more heavily protected eggs, and a more intense fanning cycle. The latter is longer and reaches a higher peak. A qualitative difference between the two species, in their parental behaviour, is the added ventilation afforded by the making of holes in its nest by *Gasterosteus* (see figs. 3 to 8 in LEINER, 1931b), towards the end of the fanning cycle.

The above specific differences, briefly presented as they are, nevertheless show clearly the way in which the two species are adapted to their different reproductive environments. It is one thing, however, to suggest that *Gasterosteus* has evolved "away from the weeds" and then to show how it is adapted to its new environment by comparing it with the weed-bound *Pygosteus*, as I have done here; but it is another thing altogether to attempt to derive the reproductive behaviour pattern of *Gasterosteus* from that of *Pygosteus*, as LEINER has done (1931a).

It is hoped that future research on other species of stickleback will make it possible to draw richer evolutionary conclusions. From this point of view, it is unfortunate that there are so few species of stickleback available, when they are such admirable subjects for ethological research. LEINER (1934) and SEVENSTER (1951) have already contributed to our comparative knowledge of one other species, namely *Spinachia*, the Sea Stickleback, but as yet we know hardly anything about the reproductive behaviour of either *Apeltes* or *Eucalia*. Both the latter species are now under observation by ethologists.

CONCLUSION

The study on which this report has been based has been carried out over a period of three years. The report has been shortened purposely, by the omission of a considerable amount of quantitative data which has been obtained, in an attempt to present it as a concise and, I hope, coherent story of a complex behaviour pattern. Much of the more specialised quantitative data (dealing with such problems as Typical Intensity, for example) will be given in subsequent publications, which will be concerned with special problems, rather than particular species. It was felt that it would be a mistake, when presenting a "species-study", as here, to include such material, for fear of obscuring the central theme, namely, that of the nature and course of the reproductive cycle of *Pygosteus*.

Whilst the work on *Pygosteus* was being carried out, the behaviour of a number of other species of vertebrates was also studied, in order to guard against unwarranted over-generalisation on the basis of uni-specific research. Fish and birds in particular, but also certain amphibia and mammals, were studied simultaneously, and much of the findings with these other species has already been published (MORRIS 1954a, 1954b, 1954c, 1954d, 1954f, 1955a, 1955b). The value of such widespread comparative studies cannot be over-stressed. It is ignored completely by the vast majority of animal psychologists, and, although it is part of the very essence of the ethological approach, a number of contemporary ethologists are far too narrow in the range of species which they study seriously. Only when one studies a number of very different species, can one begin to understand the way in which behaviour patterns in general have evolved. The FAM courtship hypothesis presented here, for example, would have been impossible without such wide comparative studies. (If shortcomings are found to exist in this hypothesis they will be found, I am sure, as a result of even wider comparative studies, rather than more detailed specific studies.)

Several of the studies on other species have concentrated on the reproductive behaviour patterns of the species concerned, and some of these findings are particularly relevant to the present study of *Pygosteus* (*e.g.* MORRIS 1954c).

A CAUSAL ANALYSIS OF A DISPLACEMENT ACTIVITY (FANNING IN *GASTEROSTEUS ACULEATUS* L.)

BY

P. SEVENSTER
(DEPARTMENT OF ZOOLOGY, LEIDEN UNIVERSITY)

(WITH 55 FIGURES AND 25 TABLES)

LEIDEN
E. J. BRILL
1961

CHAPTER I. INTRODUCTION

This study is the result of an investigation on the mechanism underlying the phenomenon of what are now generally called displacement activities. The phenomenon has been the subject of frequent discussion in Ethology. Although a number of observers (*e.g.* MAKKINK, 1936; PORTIELJE, 1936, 1938 [1]) had noticed such activities in the species they were concerned with, KORTLANDT and TINBERGEN were the first to describe them as a phenomenon of general importance and wide-spread occurrence. Moreover, they undertook an analysis of the circumstances in which displacement activities occur, and presented a hypothesis of their causation. For extensive information I refer to KORTLANDT (1940), TINBERGEN (1940 and 1952), and TINBERGEN and VAN IERSEL (1947), but as an introduction to the present study a brief survey of their line of thought seems indispensable.

They started from the observation that in a certain situation an animal may perform movements whose occurrence is at first sight unintelligible, whereas in other situations occurrence of the same movements is easily understood. To cite one of TINBERGEN's examples: snow bunting males, fighting and threatening at the boundary of their territories, may suddenly show pecking movements, as if feeding, even when no food is present. Such movements, when occurring in the manner described, are now called displacement activities and their basic characteristic is their "irrelevance". By this is meant that in the situation in which the activity (pecking, in our example) occurs, neither the internal state of the animal, as indicated by the rest of its behaviour (fighting and threatening), nor, usually, the external environment (absence of food) suggests an activation of the instinct to which the activity is related in the "normal", *i.e.* non-displacement, situation (feeding). The "normal" situation, ultimately, is taken to be the one in which the activity has an obvious biological function, to which it seems adapted in the first place, while in the displacement situation such a function seems absent or is at least less obvious. Consequently, the observer expects the activity in the normal situation, while he does not expect it in the displacement situation. This element of expectation, founded on the behavioural context, is the core of TINBERGEN's definition: "a displacement activity is an activity, belonging to the executive motor pattern of an instinct other than the instinct(s) activated" (TINBERGEN, 1952).

1) An extensive survey is given by TINBERGEN, 1940.

4 ANALYSIS OF A DISPLACEMENT ACTIVITY

According to the analysis of KORTLANDT and TINBERGEN a displacement activity takes place when the "expected" activities, *i.e.* those belonging to the executive motor patterns of the activated instinct(s), cannot occur. These activated instincts (or their "drives"), are then said to be "thwarted". Thwarting would arise either because there are two antagonistic instincts activated whose activities are incompatible, or because there is one instinct activated, while the external stimuli necessary to release the appropriate activities are absent. The essential element in both possibilities is expressed in TINBERGEN's formulation: "A displacement activity seems to appear, when an activated drive is denied discharge through its own consummatory act(s)" (TINBERGEN, 1952). In this view the displacement situation is characterised 1. by the occurrence of an activity, even though the instinct to which it belongs is not activated, and 2. by the activation of (an) instinct(s), even though the activities belonging to it (them) do not occur. The hypothesis of both TINBERGEN and KORTLANDT is, therefore, that the displacement activity results from the activation of the thwarted instinct(s) or, as TINBERGEN says: "that the displacement activities are outlets through which the thwarted drives can express themselves in motion" (TINBERGEN, 1952). In other words, while in the normal situation an activity depends on a particular set of internal and external factors, the same activity in the displacement situation depends on another set of such factors, *i.e.* those typical for the thwarted activities. In the terminology of KORTLANDT, finally, activities are activated "autochthonously" in the normal situation, *i.e.* by their own drive, and "allochthonously", *i.e.* by a strange drive, in the displacement situation (KORTLANDT, 1940).

It is further implied that an excessive activation of the thwarted instinct is necessary if a displacement activity is to occur. The thwarting process

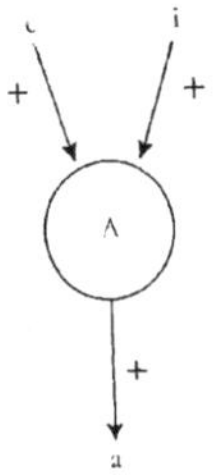

Fig. 1. The autochthonous occurrence of an activity *a*.

itself is held responsible for this excess of activation or, in the usual terms, this "surplus of drive" (TINBERGEN, 1952). Henceforward the above hypothesis about the causation of a displacement activity will be called the "surplus-hypothesis", as it presupposes this surplus of drive for which an outlet must be provided.

The essential features of the hypothesis can be represented in a simple diagram. In Fig. 1 the autochthonous occurrence of an activity *a* is schematised. This scheme does not pretend to convey anything else than the suppositions: 1) that the central nervous system must contain a coordinating mechanism *A*, enabling the executive organs involved to perform the activity *a*, and 2. that the state of activity of this mechanism is controlled by a number of external factors *e* and internal factors *i* [1]). Evidence for the latter supposition is, of course, to be found in the observation that although external stimuli are necessary to release the activity, the frequency and/or the intensity of its occurrence may vary considerably even when the external situation remains constant. In Fig. 2, the state of activity of *A* is controlled by that of another coördinating mechanism *B* and, therefore, by *e* and *i* of *B*. The activity *b* is supposed to be thwarted, and now *a* occurs allochthonously. The second diagram gives, I think, an adequate picture of the surplus-hypothesis for the present purpose, although I realise that in details a different representation would be possible [2]). It need not concern us here whether *a* and/or

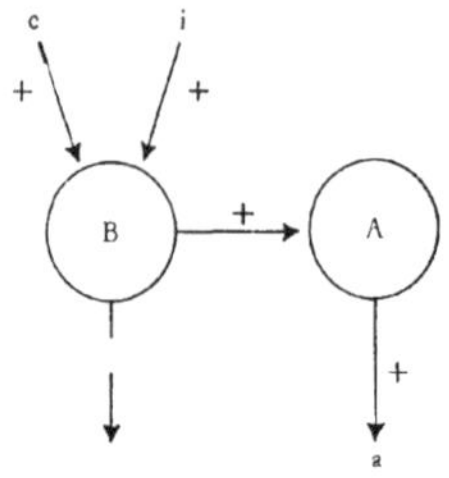

Fig. 2. The allochthonous occurrence of an activity *a*.

b are activities in a more restricted sense or rather groups of activities. This leaves room for the possibility that *A* and *B* are on different levels in the behavioural hierarchy (TINBERGEN, 1950).

The neurophysiological interpretation of the hypothesis leads to some difficulties. As we have seen, an excessive activation of the thwarted drive is required. If we consider the case in which only one drive would be activated in the absence of releasing stimuli, this would give rise to a vacuum activity,

1) Nothing is said about the nature of the interaction between external and internal factors except that they cooperate somehow in activating the coordinating mechanism.

2) Also: in a complete diagram external and internal inhibiting factors should be added which for the sake of simplicity have been omitted.

judging from general experience. Evidently we must assume that in the displacement situation vacuum activity is prevented by a special mechanism, the nature of which seems obscure. In the case of a conflict between two antagonistic drives, such a mechanism could be found in the antagonistic action of muscles or in an inhibitory relationship between the conflicting neural mechanisms. The former would lead to a tension in the executive organs and the latter would result in the mechanisms concerned being inhibited instead of excessively activated. How either state of affairs could give rise to a surplus of drive cannot be understood without special assumptions. The origin of the surplus of drive, therefore, is not clear.

In my opinion, however, this does not necessarily mean that we should discard the hypothesis. On the one hand, it seems quite possible that with some additional hypotheses the supposed mechanism could be accounted for in neurophysiological terms. On the other hand, it is conceivable that with ethological methods a mechanism has been discovered which has not yet been found by neurophysiologists. However, if we are to maintain the hypothesis the ethological evidence should be well established. A number of cases should be analysed precisely and, above all, quantitatively, to test the hypothesis as rigorously as possible.

The present study aims to contribute to a solution of the problem by an analysis of the fanning activity in the three-spined stickleback (*Gasterosteus aculeatus* L.) which has been cited as a typical example of a displacement activity (TINBERGEN and VAN IERSEL, 1947). The activity seemed particularly apt for such an analysis because the displacement situation can be produced easily and the autochthonous occurrence has been thoroughly investigated by VAN IERSEL (1953). Moreover, it can be accurately quantified.

TINBERGEN and VAN IERSEL (1947) attempted to apply the surplus hypothesis to displacement fanning. Their work is reviewed in Chapter IV, and their conclusions will be used as a starting point for my analysis.

Acknowledgements.

I am very much indebted to Dr J. J. A. VAN IERSEL for his help and advice at all stages of the work and for his permission to use his extensive data on the three-spined stickleback. Furthermore I am grateful to Mrs T. W. JASPARS-SCHRADER, Dr F. SIERKSMA, Mr R. BAMS, and Dr G. MEES for their assistance in the experimental work; to Dr F. SIERKSMA, Dr W. M. S. RUSSELL and Mrs N. ULMER for valuable discussions both of the subject itself and of its presentation; and to Mrs N. ULMER, Mrs I. SEEGER and Dr K. F. A. ROSS for their advice on the English text; and to Dr I. PRECHTL for her help with the German summary. Finally I want to thank Mr N. TEEGELAAR for the preparation of the figures.

ning in Chapter VIII and the frustration experiments in Chapter VI confirm the idea that the occurrence of sexual fanning, like that of courtship fanning, is quantitatively determined by the interaction of sex and aggression (unless strong activating factors bring about some degree of facilitation, as in the CO_2-experiments in Chapter XII).

5. SYNTHESIS

In concluding this chapter, a more general account of the disinhibition-hypothesis seems in place; especially, an attempt will be made to consider the hypothesis against the background of the male stickleback's reproductive cycle.

The evidence on which the disinhibition-hypothesis is based is threefold: 1. the parental centre is activated by parental factors in all occurrences of fanning (section 1); 2. the parental centre appears to be disinhibited during unfacilitated courtship fanning (section 2); 3. the occurrence of unfacilitated courtship fanning (and sexual fanning) is determined by the interaction of the sexual and aggressive drives (Chapters VIII and IX). The hypothesis concludes that the interaction of sex and aggression from time to time disinhibits the parental centre, thereby allowing the parental factors to reach the threshold for fanning. This disinhibiting effect is interpreted as a consequence of fluctuations in the sexual and aggressive drives (section 3). The hypothesis was found to apply also to sexual fanning (section 4). Consequently, it applies to all displacement fanning, and thus offers a solution to the problem which I set out to investigate in this study: the causation of displacement fanning in the three-spined stickleback.

The main advantage of the disinhibition-hypothesis, as opposed to the surplus-hypothesis, is that it presupposes no special neurophysiological mechanism (p. 5). The occurrence of displacement fanning is interpreted in terms of the activating and inhibiting factors which are known to affect the occurrence of parental fanning. As a consequence, no essential discrepancy arises between the causation of displacement fanning and that of parental fanning: occurrences of both can be accounted for on the scheme in Fig. 55. This scheme represents the inhibitory relationships between the three behaviour systems involved in the occurrence of fanning: sex, aggression, and parental care, each with its own activating factors (*cf.* Fig. 39).

The schematic representation in Fig. 55 allows a coherent interpretation of the occurrence of fanning throughout the reproductive cycle. Obviously both parental fanning and displacement fanning are thought to be determined by an interplay of the sexual, aggressive, and parental drives. Which type of fanning one is actually dealing with depends on the relative predominance among these drives. In the sexual phase the parental drive is still weak, and the sexual and aggressive drives, both comparatively strong, dominate the scene completely. Fanning can only occur by virtue of an occasional disinhibition and is, therefore, entirely determined by the sexual and aggressive drives, as is illustrated by, among other things, the absence of the "damming-up" phenomenon (p. 42). This situation may be symbolised by ASp [1]), to indicate the subordinate rôle of the parental drive. In the more advanced

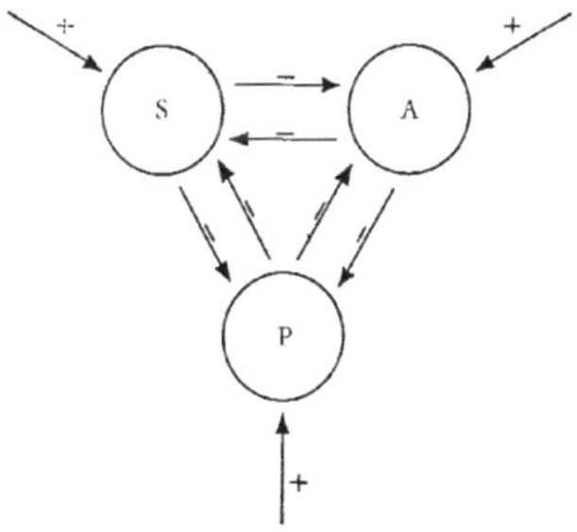

Fig. 55. The interaction of the sexual, aggressive, and parental systems.

stages of the parental phase, on the other hand, the sexual drive, and to some extent the aggressive drive, has decreased progressively, and now the parental drive, strongly increased, dominates. Fanning is no longer seriously limited by the sexual and aggressive drives; rather, it is determined by specific parental factors, as demonstrated by the phenomenon of "damming-up" (p. 37). The situation can be symbolised as asP. Thus, the two types of fanning reflect the changes in predominance in the three drives, which result from their long-range fluctuations in the course of the reproductive cycle, as illustrated in Fig. 4 (p. 19). From this illustration in Fig. 4, it will be clear that other situations may also occur: *e.g.* Asp in the beginning of the nest-building phase before the onset of fanning, or AsP towards the end of the parental phase when the sexual drive has disappeared irreversibly but the aggressive drive is increasing again after the hatching of the young. All these "situations" have to be taken as a general characteristic of certain stages of the reproductive cycle. Similar shifts in predominance occur, of course, on a smaller scale: *e.g.* in the "sexual fanning cycle", *etc.* The effects of such short-time shifts have been treated at length. At present we are only con-

1) The notation is borrowed from MORRIS (1957) who used it in comparing different types of courtship.

cerned with the two general situations mentioned: the one typical for displacement fanning: ASp, the other typical for parental fanning: asP.

The shifts in predominance are directed by certain external factors. Appearance of eggs induces the parental drive to develop and ultimately to establish its predominance (asP). The appearance of a female, by increasing the sexual and aggressive drives, reverses this shift (ASp), thus bringing the occurrence of fanning again under the control of sex and aggression (courtship fanning in sex tests in the parental phase).

Without further elaboration, it is clear that this approach leads to a unitary interpretation of all occurrences of fanning throughout the reproductive cycle. The consistency of this interpretation seems an advantage of the disinhibition-hypothesis.

The above interpretation seems to obliterate a sharp distinction between displacement fanning and parental fanning. The typical displacement occurrence of fanning (symbolised by ASp) would seem to merge gradually into the typically parental occurrence (asP) through all sorts of transitional situations. This, however, seems to be only partially true. Observation indicates that such transitional situations are comparatively rare. For it is usually not difficult to decide which type of fanning one actually observes at a given moment.

Even in the early stages of the parental phase, when such transitions would be especially expected, the behaviour shown is usually not of an intermediate type. Rather it consists of a sequence of alternating spells, one period resembling the sexual phase in that "fanning cycles" with creeping through and vacuum zigzagging are observed, another resembling the more constant occurrence of fanning characteristic of the later stages of the parental phase. Also, in the latter period, bouts are more frequent and of longer duration. Although I did not undertake a systematic analysis of this interesting process, the general impression is that in the course of time the periods of parental behaviour more and more supersede those reminiscent of the sexual phase (*cf.* also VAN IERSEL, 1953). Although I do not suggest that the distinction between the two periods remains clear-cut until the very end of this process, there seems to be a tendency for the interacting systems to operate one way or the other: at a certain moment the occurrence of fanning is either displacement (ASp) or parental (asP).

The facilitation of sexual fanning in my CO_2-water experiments (Chapter XII) points in the same direction. In spite of its initial increase, the fanning in these experiments remained under the control of sex and aggression, as demonstrated by its subsequent decrease and ultimate disappearance (p. 107 ff.).

Such a course of events was not found in the corresponding experiments on parental fanning (Chapter XI). The implication seems to be that the reaction to CO_2-water also is either of the displacement type or of the parental type.

Similarly, in courtship fanning, the frequency of bouts even in strongly facilitated sex tests is still determined by the interaction of sex and aggression (p. 148), and thereby reveals the displacement character of this fanning.

Thus, the difference between the two types of occurrence seems not to be a gradual one. To the contrary: the data suggest that the distinction may be rather sharp. An explanation of this seems to follow from the interplay of the systems involved (Fig. 55). Since it is known that the parental drive inhibits sex stronger than aggression (Chapter VII), an increase in the parental drive may be expected to favour the aggressive drive in its relation to the sexual drive. Furthermore, we have seen that CO_2-water not only increases the parental drive but also promotes the aggressive drive, whereas it depresses the sexual drive. Ignoring the question whether the two latter effects are an indirect consequence of the increase in the parental drive (p. 103), we may quite generally conclude that an increase in the parental drive goes together with a shift towards greater predominance of the aggressive drive in its relation to the sexual drive (I know of no data which contradict this statement). Such a shift, as we have seen, is unfavourable for the occurrence of displacement fanning. Now suppose we start from the typical displacement situation ASp. An increase in the parental drive, say by the exposure to CO_2-water, will at first result in (or be accompanied by) an increasing predominance of the aggressive drive relative to the sexual drive. As a consequence displacement fanning will decrease and finally disappear, which is actually demonstrated by the CO_2-water experiments in the sexual phase (p. 106 ff.). The situation Asp, characteristic of the nest-building phase, will arise: again the CO_2-water experiments in the sexual phase actually demonstrate an increased frequency of building activities after the decrease in fanning (p. 109). Only when a further increase in the parental drive succeeds in depressing the aggressive drive too (*cf.* the effect of a CO_2-bath in the parental phase: p. 101, small print), may the situation asP arise and fanning be expected to reappear, now of the parental type. It is very likely that one of the exceptional CO_2-water experiments in the sexual phase demonstrates precisely this sequence: an outburst of parental fanning was observed after the decrease in sexual fanning, the two waves being separated (p. 111, Fig. 49).

Aggression tests (and sometimes sex tests) in the parental phase often demonstrate a similar sequence. In the very first stages of the parental phase, an aggression test may still elicit a fair number of zigzags and, accordingly, a certain amount of courtship fanning, *i.e.* fanning of the displacement type, may be shown (see p. 82). With the advance of the phase, however, zigzagging soon disappears from the tests: these tests now only release biting, and fanning no longer occurs. Only at still later stages does the aggressiveness also decrease and fanning, presumably of the parental type, reappear in increasing amounts in the tests.

The interpretation of these sequences seems clear. We have repeatedly seen that the occurrence of displacement fanning depends on a certain balance of the sexual and aggressive drives; in terms of the disinhibition-hypothesis neither drive should be predominant if disinhibition is to occur. The present discussion suggests that the first step towards a dominance of the parental drive tends to disturb this balance by favouring the aggressive drive. Consequently, displacement fanning may disappear before it can be replaced by parental fanning. Meanwhile, it should be stressed that these sequences do not constitute a universal rule. Occasionally we see that strong facilitation may suddenly bring about a (temporary) predominance of the parental drive in a situation which is otherwise typical for displacement fanning (p. 144, small print). In those cases the transitional situation, in which no fanning occurs, is passed over. It is quite possible that similar outbreaks of the parental drive will also occur in the early stages of the parental phase, before a permanent dominance of the parental drive has been established (*cf.* the beginning of this paragraph). Even in these cases, however, the suddenness of such outbreaks illustrates once more that the difference between the two types of fanning is not a gradual one.

The theoretical implication of this discussion is important. Although all occurrences of fanning can be explained by the same causal factors, it nevertheless appears that in general these occurrences are of two distinct types.

The one type of occurrence is characterised by an almost complete dependence on the interaction of sex and aggression, the other by the almost complete absence of such a dependence. In both, the activation is "autochthonous". This term, therefore, no longer conveys a distinction between the two types of occurrence. Rather, the non-displacement occurrence could be called the "autonomous" occurrence in view of its independence of sex and agression (*cf.* p. 165).

Some aspects of the general formulations, presently discussed on the basis of Fig. 55, are suggestive of quite a different interpretation of the causation of fanning. Much of the evidence in this study, namely, could be understood on the assumption that fanning is an ambivalent movement, consisting of a sexual component and an aggressive component. In view of the form of the movement this idea seems not too unlikely. For in fanning the tail fin moves as in forward locomotion (though at a higher frequency), whereas the pectorals push the fish backward. Furthermore, the fanning fish is aiming at the nest entrance. Consequently, fanning can be regarded as the combination of a tendency to enter the nest and a tendency to avoid this, the two tendencies being exactly in equilibrium. Now, the first of these tendencies might well be connected with creeping through and hence with the act of fertilisation. Presumably, therefore, this tendency would have to be the sexual component, motivated by the sexual drive (even though in this study, it should be noted, the sexual drive in principle is measured by the tendency to lead). Obviously, on the above assumption the other tendency, *viz.* the one to "shrink back" from the nest, would have to be connected with the aggressive drive (as measured by the tendency to bite). Though this connection is not at all evident from the form of the movement, there is some evidence in favour of this view. On the one hand, I have observed that males will actually "shrink back" when the nest contains a particularly strong stimulus (*e.g.* an amount of $NaHCO_3$-powder); in such cases fanning may not occur, or, significantly, it may start at a greater distance from the entrance than is usual. On the other hand, we have seen that CO_2-water simultaneously increases the amount of fanning a n d the aggressive drive. Thus, it may not be altogether unlikely that there is some connection between the "shrinking-back"-tendency and the aggressive drive, both being promoted by "parental" factors such as CO_2-water. Summarising the present train of thought, we might say that fanning can be regarded as frustrated creeping through, the tail fin being controlled by the sexual tendency to creep through and the pectoral fins by the tendency to "shrink back", possibly connected with the aggressive drive.

This assumption, speculative though it may be, is treated at some length because it may be attractive from an evolutionary point of view. For it seems to provide a possible derivation of the fanning movement from more elementary tendencies. If, in the early stages of the evolution of the stickleback's rather exceptional reproductive pattern, the male already had the habit of creeping through its nest (or nest-like structure) regularly in connection with its tendency to fertilise, it might be imagined that the metabolical effects from the developing eggs (CO_2-production, for example) would induce or reinforce [1]) the antagonistic tendency to shrink back from the nest entrance. The movements of the pectoral fins involved in this tendency would inevitably contribute to the aeration and hence possibly to the survival of the eggs. Consequently, selection would favour a mutual adjustment of the two antagonistic tendencies so as to increase the likelihood of the exact equilibrium necessary for fanning.

Apart from this evolutionary background, the idea seems to be in agreement with many data on displacement fanning. It offers an explanation of the close connection between this fanning and the balance of sex and aggression, which we found repeatedly demonstrated. Further, the occurrences of creeping through in the "sexual fanning cycles" are a strong support: the increasing sexual drive produces an increasing amount of fanning until it culminates in creeping through (*cf.* especially the formulation on p. 62). The interpretation of these data would therefore be considerably simplified, because on our speculation a third system, *viz.* the parental system, need no longer be invoked. Serious objections, however, arise if the phenomena of facilitation and the occurrence of parental fanning are taken into consideration. The close connection between fanning and the balance of sex and aggression, which is postulated by the idea discussed, is not found in the occurrence of parental fanning. To the contrary, the amount of parental fanning was found to be highly independent of the sexual and aggressive drives (p. 42 ff.). It has already been pointed out that this high degree of independence supports the view that there is a separate parental system (*cf.* VAN IERSEL, 1953). Similarly, the

1) for it seems quite likely that even in the absence of eggs there will be a reluctance to creep through.

analysis of facilitation in courtship fanning led to the conclusion that with the same state as to sex and aggression, the amount of courtship fanning varied with the strength of the parental drive (p. 118). Furthermore, facilitation of courtship fanning also occurred with empty nests (p. 121), thus demonstrating the importance of internal parental factors. Also, the development in the parental phase is not intelligible on the assumption that fanning is an ambivalent movement in the manner discussed. For, on this assumption one would expect that the developing eggs continuously increase the "shrinking back" tendency (connected with the aggressive drive). It would seem that an ever-increasing tendency to creep through (connected with the sexual drive) is required to counterbalance this. Instead,we find that the sexual drive virtually disappears and that the amount of fanning nevertheless greatly increases (Fig. 4). This fanning, then, is apparently no longer connected with the tendency to creep through (as in the "fanning cycles" of the sexual phase). Finally, it would be difficult to account for the phenomenon of "damming up", which is a distinctive characteristic of parental fanning, on the assumption under discussion.

I conclude, therefore, that even though this assumption may throw some light on the evolutionary origin of fanning and some phenomena may still reflect this origin, we have to accept that the coordination of fanning has become emancipated from such an origin: it is now controlled by a separate system.

CHAPTER XVII. SOME GENERAL ASPECTS OF THE DISINHIBITION-HYPOTHESIS

It was the purpose of this study to contribute to our insight into the causation of displacement activities. In the particular case investigated, the disinhibition-hypothesis, unfolded in the previous chapter, appeared to cover the evidence collected. Even on a number of points which are still poorly understood, the hypothesis seems to offer at least a promising starting-point for further analysis. VAN IERSEL and BOL (1958) applied the disinhibition-hypothesis to displacement preening in terns and arrived at a satisfactory interpretation of their results. Since neither displacement preening in terns nor displacement fanning in the stickleback seem to differ in any fundamental aspect from other displacement activities, it may be expected that the disinhibition-hypothesis will appear to apply to yet other cases. It seems appropriate, therefore, to add some remarks on the applicability of the hypothesis to certain features generally held to be associated with displacement activities (*cf.* VAN IERSEL and BOL, 1958, who have already discussed some of the following points).

The occurrence of many (if not all) displacement activities is characterised by some sort of conflict between two antagonistic systems. No doubt the predictability of the displacement activity from the interaction of the conflicting systems has contributed strongly to the idea that the activity should be activated by one (or both) of these systems. The disinhibition-hypothesis disposes of such an "allochthonous" activation. It is assumed that the displacement activity is activated autochthonously, but this autochthonous activation is thought to reach its threshold only by virtue of "disinhibition". Since the occurrence of disinhibition is held to be determined by the interaction of the conflicting systems, the predictability of the displacement activity from this interaction is explained: although in the displacement situation three systems are activated, the behaviour shown is actually controlled by only two of them.

Under some (exceptional?) circumstances the occurrence of the displacement activity, though still largely controlled by the conflicting systems, may also be affected by the activation of its own system. This phenomenon of "facilitation" is easily understood on the disinhibition-hypothesis. Similarly, the possible influence of postural facilitation (TINBERGEN, 1952) and of certain external stimuli on the choice between alternative displacement activities (KORTLANDT, 1940; for a further discussion see VAN IERSEL and BOL, 1958) can be understood.

Displacement activities are often seen to be incomplete. This phenomenon follows quite evidently from the disinhibition-hypothesis. On this hypothesis the duration of the disinhibition obviously limits the duration of the displacement activity. Thus, the activity may be broken off quite arbitrarily at any point.

The intensity of a displacement activity depends, according to the disinhibition-hypothesis, on the degree of disinhibition and on the degree of (autochthonous) activation. As to the degree of disinhibition, the present investigation has shown that disinhibitions may be quite "drastic" (p. 144). Furthermore, the conflicting systems are usually so dominant in the displacement situation that it seems likely that inhibition from other systems can be practically neglected (p. 138). Also, the sudden disinhibition might produce a rebound effect by post-inhibitory facilitation. During a disinhibition, therefore, the threshold for the displacement activity might be lowered even more than is usual for a non-displacement occurrence. Thus, the disinhibition-hypothesis allows for the possibility that a displacement activity may be exceptionally intensive. Further, with the same degree of disinhibition and hence with the same interaction of the conflicting systems, the intensity of the displacement activity should be correlated with the degree of (authochthonous) activation. As, unfortunately, we did not succeed in measuring the intensity of fanning (p. 18), no confirmation is available on these points.

Displacement activities are often reported to be "frantic" in appearance. It seems to me that a combination of high intensity (owing to strong activation and/or drastic disinhibition) and incompleteness (owing to short conflicts) may account for some cases. For the "franticness" may well be a subjective impression based on the general experience of the observer that there is a connection between the intensity and the duration of a movement in the non-displacement situation. This connection is lacking in a displacement activity, the duration being determined by other factors than the intensity. Consequently, the displacement occurrences of an activity may look rather different from its normal occurrences. (This appears to be the case in displacement gluing in the three-spined stickleback: unpublished results.)

The degree of disinhibition and the strength of the internal activating factors would, of course, determine to what extent external stimulation is required for the occurrence of the displacement activity. A drastic disinhibition (especially when combined with strong internal activation) could be expected to give rise to a displacement activity "in vacuo". In this connection it seems again important that in the displacement situation inhibitory influences from other systems than the conflicting ones are probably negligible.

Finally, the disinhibition-hypothesis has some bearings on the terminology currently used in dealing with displacement activities. Obviously, the term "autochthonous" becomes superfluous since on the hypothesis the source of the activation is not a criterion for the distinction between the displacement and the non-displacement occurrences of an activity. In the case investigated, it has been suggested, the latter could be called the "autonomous" occurrence (p. 161), to indicate its independence from the interaction of the conflicting systems. Awaiting further evidence, however, the non-committal term "normal" occurrence is perhaps to be preferred.

Also, the term displacement itself would no longer be valid if it is taken to imply the displacement of activation from one behaviour system to another. It remains, however, quite possible to use the term in a purely descriptive sense, denoting the displacement of the activity from its normal context of occurrence. In my opinion, the term should definitely be retained even for cases in which the disinhibition-hypothesis has stood the most rigorous test. This is especially recommendable in view of the extensive literature in which the term has been used, and in view of its usefulness for comparative purposes. An attempt to redefine the phenomenon of displacement activities is also premature. The characteristics which are to be incorporated in such a definition should rest on a large number of cases. It has already been pointed out that the definition of TINBERGEN (1952), strictly speaking, would not be valid if the disinhibition-hypothesis is right in its assumption that the behaviour system (instinct) to which the displacement activity belongs is activated by its own factors (p. 125). In the case investigated, however, the definition would still hold, if the word "prominently" is inserted: "a displacement activity is an activity belonging to the executive motor pattern of an instinct other than the instincts prominently activated".

SUMMARY

The present study is concerned with the causation of displacement activities. The particular case analysed is displacement fanning in the three-spined stickleback (*Gasterosteus aculeatus* L.). A quantitative approach is followed whenever possible.

Fanning is characterised by a special coordination of the pectoral and caudal fins. The movement is performed at the nest and occurs in bouts of varying duration. Its occurrence can be quantified by the "amount of fanning", *i.e.* the total duration of all bouts in sec. per time-unit. Fanning occurs throughout the male stickleback's breeding cycle. The form of the movement is always the same. According to the behavioural context, however, three types of occurrence can be distinguished:

1. the occurrence in the parental phase, *i.e.* when eggs are present in the nest: "parental fanning".
2. the occurrence in the sexual phase (and in the nest-building phase), when no eggs are present: "sexual fanning".
3. the occurrence during courtship, with or without the presence of eggs: "courtship fanning".

The occurrence in the parental phase serves the function of ventilating the eggs, to which the movement is well adapted. Furthermore, this parental fanning is specifically activated by factors which are connected with the eggs' respiratory needs and hence with the need for ventilation. These factors are therefore "parental factors".

Sexual fanning and courtship fanning can be regarded as displacement activities. They serve no obvious function, and their occurrence is closely connected with sexual activities, while the factors which activate parental fanning seem to be absent. This has given rise to the idea that displacement fanning is activated allochthonously by the sexual drive (TINBERGEN and VAN IERSEL, 1947; VAN IERSEL, 1953). In the present paper this idea is referred to as the "surplus-hypothesis". Starting from this hypothesis, the present analysis concentrates on two points: 1. the relation of fanning to the sexual and aggressive drives (Part II), and 2. the effect of parental factors on fanning (Part III). These points are investigated separately for each of the contexts in which fanning occurs.

Retrospective: Habituation of conspecifics aggression in the three-spined stickleback (*Gasterosteus aculeatus* L.)

Article published in *Behaviour* volume 36 (1969), p. 232–245

Harman V.S. Peeke
(University of California San Francisco School of Medicine, USA)

Seldom have two fields of inquiry ostensibly directed at studying the same general phenomena been so dissimilar with regards to methods used, organisms studied and theoretical formulations guiding their research as were European ethology and American animal and comparative psychology in the middle of the last century. It was not until the early to mid-1960's that a rapprochement began to develop between these fields, which led to a more comprehensive approach to the study of animal behaviour. It was during this intellectually exciting period that my research (Peeke, 1969) was planned and carried out.

Until the mid- to late 1960's, American animal and comparative psychology was dominated by research into mechanisms of learning in laboratory animals (primarily rabbits, rats, mice and pigeons). The research was directed, to a great extent, at providing evidence for one or more competing theories of learning, the most prominent of which were those of Pavlov (1927), Hull (1951), Spence (1956), Guthrie (1959), Skinner (1959), and Tolman (1959). Whether the studies were directed at elucidating mechanisms involved in classical (Pavlovian) conditioning or instrumental (operant) learning or both, they were concerned with how two stimuli became associated. These studies addressed how a formerly 'neutral' (or ineffective) stimulus could come to exert influence on behaviour similar to that of the effective stimulus it had been paired with.

Changes in behaviour resulting from exposure to a repeated or constant stimulus were usually termed "habituation" (Humphrey, 1933; Thorpe,

1963). Some investigators doubted that habituation was a form of learning. They thought other phenomena, such as fatigue and sensory adaptation, might instead account for a response decrement to a constant stimulus. At quite a different level, even if habituation were clearly a form of learning, it lacked the excitement of being the fundamental process that promised to explain the behaviour of animals based on an increasing number of complicated interactions. I stepped away from the domesticated laboratory animals and the sterility of the laboratory methodologies used to study learning. I wanted to investigate the role that the simplest (but likely most pervasive) form of learning, habituation, plays in the modification of naturally occurring behaviours of a wild species.

I chose the three-spined stickleback (*Gasterosteus aculeatus*) because this small fish had already been the subject of more research, behavioural and otherwise, than any other. Its behaviour had been described in detail in many sources (e.g., ter Pelkwijk & Tinbergen, 1937; Tinbergen, 1951; van Iersel, 1953). It is plentiful in many regions of the northern hemisphere, including Southern California, where an artificially land-locked population was available for study. The rich tapestry of the reproductive behaviour, particularly of the male stickleback, can be captured in a ten gallon aquarium with a sandy substrate, string algae and aquatic plants. From these materials the male digs a short trench and constructs a nest. From nest completion until the first successful egg fertilization in the nest, the male's sexual and aggression 'drives' are high and stable (Sevenster, 1961).

These 'drives' are measured by stable and discrete responses, labelled by Lorenz (1950) as "fixed action patterns" or FAPs, and subsequently refined by Barlow (1968) as "modal action patterns," or MAPs. Each major 'drive' had one or more FAP that was a specific behavioural index of that drive. The FAP for 'sex,' or 'courtship,' elicited experimentally by a gravid female presented to the territorial male in a clear glass tube, is the 'zig-zag' dance. Substitute a nuptually coloured male for the female in the tube and you have an aggression test. The territorial male responds with 'bites' at the intruded male (Sevenster, 1961). It is common when the stickleback is responding at high rates for 'bites' to appear in a 'sex test' and similarly for a territorial male to 'zig-zag' to an intruded male. The presentations used for 'sex' and 'aggression' tests were brief (3–5 minutes) usually with a different stimulus fish on each occasion. Both precautions were presumably to avoid any confounding effects of variable stimulus familiarity or exposure time, which were precisely the variables I was interested in.

'Bites' at the conspecific stimulus fish (as well as to a less effective stimulus, a wooden model) clearly habituated over 10 days of 15 minute sessions with the same stimulus (Peeke, 1969). To identify habituation as a form of learning, the two alternative explanations for response decrements to repeated or constant stimuli, fatigue and/or sensory adaptation, had to be eliminated. Demonstrations of temporal persistence of response decrements between days provided convincing evidence against these two explanations.

I was especially interested in the increased number of 'zig-zags' directed at the live male stimulus during the initial days of habituation of aggressive behaviour, and the increased number of 'zig-zags' directed at the gravid female during daily brief 'sex tests.' A popular theory of 'drive interaction' at that time was that each drive, fuelled by both internal (e.g., hormonal) and external (e.g., temperature, conspecifics) sources, maintained mutually inhibitory influences on other drives in direct proportion to the strength of the drive. Assuming, as most do, that habituation is an inhibitory process, it might be predicted that as the stickleback inhibited his 'bite' response to the intruded male stimulus, the same stimulus might begin to elicit courtship responses (zig-zags) as the inhibition from the 'aggression drive' on the 'sexual drive' weakened due to habituation. The courtship response to the male stimulus quickly habituates as it is a weak courtship stimulus. The continuing increased 'zig-zag' responses during short daily sex tests provided evidence that the sexual drive continues to increase as aggression is habituated (Peeke, 1969).

The drive centre theory, particularly as it was related to 'displacement behaviours,' was championed by the European school of ethology, particularly by scientists trained in the Netherlands. It passed from the scene more from changes in emphasis in the study of behaviour than from any fatal structural flaw. In a sense, the drive centre theory was successful in that it led to its own demise as it generated new lines of research.

Acknowledgements

Virtually simultaneous with the publication of my paper (Peeke, 1969), J. van den Assem &, J. van der Molen (1969) also published a paper in Behaviour reporting similar findings in the same species using quite different but equally valid methods. I wish to posthumously thank Bill Rowland of Indiana University for his decades of collegial conversations and friendship. Much of our research bears striking similarities, despite the fact that his degree was in biology from New York and mine in psychology from California. It must have been because we had the same patient mentor, Everett J. Wyers. I would also like to thank my career-long collaborator and friend Michael Figler of Towson State University, a comparative psychologist whom I spent 25 years trying to make into a "muddy-boots biologist." I failed.

References

van den Assem, J. & van der Molen, J.N. (1969). Waning of the aggressive response in the three-spined stickleback upon constant exposure to a conspecific. I, A preliminary analysis of the phenomenon. — Behaviour 34: 286-323.

Barlow, G.W. (1968). Ethological units of behaviour. — In: The central nervous system and fish behaviour (Ingle, D., ed.). University of Chicago Press, Chicago, IL, p. 217-237.

Guthrie, E.R. (1959). Association by contiguity. — In: Psychology: a study of a science. Vol. 2, General systematic formulations, learning and special processes (Koch, S., ed.). McGraw-Hill, New York, p. 158-195.

Hull, C.L. (1951). Essentials of behavior. — Yale University Press, New Haven, CT.

Humphrey, G. (1933). The nature of learning and its relation to the living system. — Harcourt, New York.

van Iersel, J.A.A. (1953). An analysis of the parental behaviour of the male three-spined stickleback. — Behaviour, Supplement III: 1-159.

Lorenz, K. (1950). The comparative method in studying innate behaviour patterns. — Sym. Soc. Exp. Biol. 4: 221-268.

Pavlov, I.P. (1927). Conditioned reflexes. G.V. Aurep (Trans.). — Oxford University Press, Oxford, UK.

Peeke, H.V.S. (1969). Habituation of conspecifics aggression in the three-spined stickleback (*Gasterosteus aculeatus* L.). — Behaviour 35: 137-156.

ter Pelkwijk, J.J. & Tinbergen, N. (1937). Eine rizbiologiscle analyse einiger verhaltensweisen von *Gasterosteus aculeatus* L. — Z. Tierpsychol. 1: 193-204.

Sevenster, P. (1961). A causal analysis of a displacement activity (fanning in *Gasterosteus aculeatus*). — Behaviour, Supplement IX: 1-170.

Skinner, B.F. (1959). A case history in scientific method. — In: Psychology: a study of a science. Vol. 2, General systematic formulations, learning and special processes (Koch, S., ed.). McGraw-Hill, New York, p. 350-379.

Spence, K.W. (1956). Behavior theory and conditioning. — Yale University Press, New Haven, CT.

Thorpe, W.H. (1963). Learning and instinct in animals, 2nd edn. — Methuen, London, UK.

Tinbergen, N. (1951). The study of instinct. — Oxford University Press, Oxford.

Tolman, E.R. (1959). Principals of purposive behavior. — In: Psychology: a study of a science. Vol. 2, General systematic formulations, learning and special processes (Koch, S., ed.). McGraw-Hill, New York, p. 92-157.

HABITUATION OF CONSPECIFIC AGGRESSION IN THE THREE-SPINED STICKLEBACK (*GASTEROSTEUS ACULEATUS* L.)

by

HARMAN V. S. PEEKE [1] [2]
(University of Southern California, Los Angeles, Calif., U.S.A.)

(With 5 Figures)
(Rec. 15-VI-1969)

INTRODUCTION

Recent research with birds (HINDE, 1954 a, b; 1961; MELZACK, 1961) and with fish (CLAYTON & HINDE, 1968; PEEKE *et al.*, in press) has demonstrated waning of a repeatedly elicited fixed action pattern in vertebrates. The present study investigates the course of response decrement (habituation) associated with repeated elicitation of aggressive responses directed toward a conspecific male stimulus in the three-spined stickleback *(Gasterosteus aculeatus* L.).

A stimulus that reliably elicits a response in an organism upon frequent successive repetitions will usually result in a decrement in frequency and/or magnitude of that response. The reason for this decrement may be sensory adaptation, fatigue or habituation. It has been customary to call the process responsible for the response decrement habituation only in cases in which both sensory adaptation and fatigue may be ruled out.

Habituation, Fatigue and Sensory Adaptation.

Fatigue is usually considered an effector-motor process. Hence if fatigue were responsible for the decrement in responsiveness observed in an experiment no stimulus would be able to elicit the response. The decrement would

---

1) This research is part of a Thesis submitted to the Department of Psychology, University of Southern California in partial fulfillment for the degree of Doctor of Philosophy. Thanks are due to the members of my dissertation committee, Dr HENRY SLUCKI (Psychology) and Professor JAY SAVAGE (Biology), and particularly to my Chairman and preceptor, Professor EVERETT WYERS.

This research was supported in part by Grants # G-18891 and # GB-4363 from the National Science Foundation to Professor WYERS.

2) Now at the Department of Psychiatry, University of California Medical Center, San Francisco, and the Laboratory of Psychobiology, Langley Porter Institute, San Francisco.

not be "stimulus specific", but rather "response specific". If successive repetition of the stimulus results in receptor change, such as diminution in sensitivity, this diminution would effect all stimuli impinging upon that receptor organ, not just the specific stimuli (and those very similar to them) that brought about the adaptation. The decrement would be "modality specific" but not "stimulus specific". Thus stimulus specificity may be used as a criterion of habituation as opposed to sensory adaptation and fatigue.

The effects of both sensory adaptation and fatigue are usually considered transient, *i.e.*, the decremental effects of these processes should dissipate rapidly, whereas habituation, like many forms of learning, persists. Thus temporal persistence emerges as the major criterion of habituation. Studies of habituatory response decrement have only rarely adhered to procedures that would clearly allow unambiguous interpretation as habituation differentiated from fatigue and adaptation.

Long Term Temporal Persistence.

Temporal persistence has been demonstrated in a number of ways in previous studies of habituation. FLUERE & WALTON (1907) demonstrated habituation of a tentacle response in the sea anemone to nonneutrative paper that persisted from day to day. HINDE (1954b) has shown persistence of the decrement in the mobbing response of the Chaffinch over periods of 24 hours, and PEEKE *et al.* (in press) have shown day to day decrements of biting responses directed at models of male fish by territorial three-spined sticklebacks. The importance of such demonstrations is to show that the effects of experience with the stimulus in the production of response decrement are not simply a matter of fatigue or sensory adaptation, both of whose effects should be transitory as pointed out above. Previous studies of habituation concerned with this problem typically interpolate a rest period either in the form of between-stimulus-blocks rest intervals or between-days recovery periods. Adaptation independent of fatigue and sensory deficit effects is then demonstrated in one of two ways: either through failure to find full recovery of response magnitude (or response probability) from day to day, or in instances in which initial recovery may return the response to the previous level, subsequent habituation is more rapid during each successive stimulus presentation period after the recovery period.

Components of a Behavior Class.

In a number of studies of habituation of complex responses indexing a class of behavior (such as "aggression"), it was found that components of the behaviors in question did not habituate at the same rate; that is, some components disappeared before others. This effect was reported in birds by HINDE (1961) and MELZACK (1961) and in spiders by DREES (1952), PRECHT (1958) and SZLEP (1964). In the present study the major responses that have previously been identified as being components of aggression in the stickleback were recorded in order to test the generality of this finding.

In addition to recording and reporting component responses of aggressive behavior, other responses which are considered component responses of other behavior classes (such as parental and sex) were also recorded in order to investigate the possibility that they might covary with the expected changes in aggression brought about by the operations in this experiment.

One requirement for a study of the course and the modification of the course of aggressive behavior in the stickleback is the selection of a period in the reproductive cycle during which aggression is both stable and at a high level. Such a period is found during the "sexual stage" of the cycle.

The Reproductive Behavior of the Three-Spined Stickleback.

Several complete and authoritative descriptions of the complete reproductive behavior of the three-spined stickleback are readily available in the literature — TER PELKWIJK & TINBERGEN (1937a; b) and VAN IERSEL (1953) — hence the briefest sketch of the pertinent material is presented here.

The reproductive cycle is characterized by several distinct stages through which the male passes. The first, Nest-building, is self-explanatory, characterized by behaviors involved in excavation and construction of the nest on the substrate. At the completion of the nest, the male stickleback passes into the sexual stage.

The sexual stage is characterized by the readiness of the male to court a gravid female. The female stickleback displays a characteristic head up tail down posture and a large, swollen abdomen when she is gravid. The courting response of the male to such a stimulus is the "zigzag dance". The male moves toward the female in a series of rapid forward, then sideways leaps. A single "zigzag" consists of one forward and one sideways leap; the dance consists of a series of these movements. In this study the sexual response measured was the single "zigzag". During the sexual (and the nest building) phase the male stickleback defends his nest and surrounding territory against conspecific intruders, particularly other males. This is done by a number of aggressive responses such as "threatening", "charging" at the other male, "spine fighting" and "biting". Rarely is damage done and the territory is almost universally defended. In the present study, two aggressive responses are recorded: "charging" and "biting".

by an analysis of variance and yielded a significant value ($p < .05$). Subsequent analysis between individual groups indicated that group RM gave more "zig-zags" toward the female during the daily sex tests than did the group that received only the sex test during the ten days of the experiment (group C) ($p < .01$). Similarly, group MM gave more responses toward the female than group C ($p < .03$). Thus, in terms of the overall differences between groups comparing the total number of "zig-zags" summed over all ten days, both groups RM and MM were more responsive than group C.

An analysis of each group, comparing the frequency of "zig-zags" on day one with the number on day ten, demonstrated that only group RM showed a reliable increase from day one to day ten ($p < .01$). Group RM shows a relatively steady increase over the course of the experiment in the level of sexuality displayed toward the female stimulus. Group MM shows a less marked increase up to day five, but then returns to only a very slightly higher level than it manifested on day one. Group C shows no real change or trend over the duration of the experiment.

The sex test results demonstrated that daily fifteen minute presentations of a real male stimulus over a period of ten days resulted in an increase in the level of "zig-zag" responses directed at female stimulus. Neither administration of the model male nor the no-male treatment resulted in any sustained change over the ten days in the level of "zig-zags" during the sex tests.

DISCUSSION

The results of this study indicate that both groups presented with aggression-eliciting stimuli (real male or model) for fifteen minutes over ten days show a decrement in aggression over those ten days. This decrement may be termed "habituation" insofar as in both cases the decrement persists over the interval between daily presentations, thus displaying the temporal persistence criterion of habituation. Such temporal persistence would not be seen if the decrement were due to either fatigue or sensory adaptation. Furthermore, the group presented with the model shows stimulus specificity of habituation as shown by the day eleven, post-test presentation of the real male which elicited a high level of aggression.

It was found that the two components of aggression, charges and bites, habituated at differential rates; that is, charges habituated earlier than did bites.

These results support the findings of HINDE (1954a, b; 1961), MELZACK (1961) and SZLEP (1964), where FAP's [1]) were found to habituate with

1) Fixed action pattern (FAP) is used descriptively here in the manner as defined by HESS (1962).

repeated elicitation. The stimulus specificity of habituation in group MM and the temporal persistence found with both groups reflects the basic similarity between habituation of FAP's and simpler reflexive responses. Further, this study extends the demonstration of habituation to phyletic level where, until recently very little work has been done with this phenomenon (THORPE, 1963).

The locomotor "tail-flip" reflex of the goldfish has been habituated in response to a repeated pressure wave set up by a tap on the side of an aquarium, and also in response to a shadow stimulus passing overhead (RODGERS *et al.*, 1963). A similar type of response has also been habituated to a shadow stimulus in the Guppy (*Lebistes reticulatus*) (RUSSELL, 1967). Habituation of aggressive responses of a predatory nature was obtained by TRIPLETT (1901) in the Perch, *Morone americanus*. This study separated the Perch from its prey, the minnow, by a glass partition. The attempts of the Perch to attack the minnows waned and when the partition was removed the minnows were spared. A study of BAENNINGER (1966) suggests habituation of conspecific aggression in *Betta splendens*, but the method used by that author is not sufficient to permit the conclusion that the demonstrated decrement in approach to a conspecific stimulus was habituation rather than simple avoidance (fright) of the stimulus. In addition, there was no demonstration of temporal persistence or stimulus specificity. Recently CLAYTON & HINDE (1968) demonstrated waning of the aggressive display in *Betta splendens*. They found that the fish's aggressive display did wane over days in response to its own mirror image presented constantly over a ten day period. They found that the response did not habituate completely and that recovery never reached prehabituation levels. This is impressive evidence for habituation of an aggressive response in *Bettas*, but it does use a method, that of the mirror image stimulus, that represents a peculiar situation where the response of the animal is also a visual stimulus to the same animal. PEEKE *et al.* (in press) found that the aggressive response (biting) of the stickleback (*Gasterosteus aculeatus*) waned over days when the stimulus was a crude wooden model of another male stickleback. These papers, in conjunction with the present report, provide strong evidence that, in fish, both reflexes and FAP's habituate with repeated elicitation.

MELZACK *et al.* (1958) and MELZACK (1961) found that the fright responses of chicks waned with repeated elicitation from a silhouette of a "hawk-like" shape flown overhead. However, he found that the chicks did not stop orienting toward the stimulus, but while they no longer responded with overt fear, they continued to remain vigilant. In the present study an attempt was made to measure orienting responses in the hope of providing a measure comparable to vigilance in this situation. It was found that there was a decrease in the number of orientations throughout the experiment. However, this decrement may have reflected only a change in the aggressive responses which were almost always preceded by an "orientation" insofar as the after-effects of the response (*e.g.*, "biting" or "charging") usually resulted in either the stimulus fish inside the tube moving, even if slightly, or, in the case of the model, the mechanical propulsion of the model required pre-response reorientation by the subject fish. The aggressive responses were reduced to very low levels by repeated elicitation through stimulus presentations whereas the orientations never reached a very low level. It would be speculation to suggest that they never would have waned completely, but it does seem that they are much more resistant to habituation than overt aggressive responses. The persistence of vigilance responses, *i.e.*, orientations toward a conspecific aggressor, appears similar to the different-species predator in MELZACK's study. SZLEP (1964), studying prey catching responses in the spider to vibratory stimulation of the web, found that orientations ("perceptions of the stimulus") would wane completely. Orientations were the final response to wane in the complex of responses directed at the vibratory stimulus designed to imitate prey caught in the web. In SZLEP's study, in contrast to HINDE (1954a, b; 1961), MELZACK *et al.* (1958), MELZACK (1961) and the present experiment, the "habituated" stimulus was the prey and the organism studied, the predator.

One marked difference between the results of this study of habituation of a FAP and the usual findings with reflexive response habituation is that the stimulus eliciting the greatest aggression habituated at a faster rate than the less adequate stimulus. The typical finding with reflexive responses is that the more intense stimuli habituate less rapidly. Aggressive responses to the real male declined at a much steeper gradient and finally reached almost the same terminal level as the responses to the model. This would suggest that there might be at least one major difference in the habituatory process between FAP and reflex habituation.

A second difference between habituation of FAPs and simple reflexes is that FAPs appear to take much longer to habituate. This may be misleading however, and may not really reflect a basic difference between reflex and FAP habituation. When the habituation of a FAP is discussed, as in the present context, these components of the behavior habituate in an ordered fashion; that is, one component of aggression (*e.g.*, charges) will wane before the next (*e.g.*, bites). This kind of ordered relationship was a prominent finding in the detailed work with the spider reported by SZLEP (1964). If it is assumed that each component in this ordered relationship habituates in sequence and if the speed and course of habituation of each component were studied separately it might well be that the component FAPs of a behavior and reflexes would show similar courses of response decrement. Such a result would suggest a common mechanism.

In addition to the finding that aggressive behavior waned as a function of stimulation from an aggression eliciting stimulus, sexual responsivity increased as a negative function of the decrement in aggression. This was shown as an increase in sexual responses both toward the normal aggression eliciting stimulus and toward a female stimulus presented each day. This occurred only in the group stimulated with the real male, the most adequate releaser for aggression.

Such a reciprocal relationship between sex and aggression is postulated in the theoretical scheme advanced by SEVENSTER (1961). (This theory is primarily directed at explaining the occurrence of displacement activities.) This theory quite clearly postulates an inhibitory interrelationship between drives. "It has been shown that sex and aggression form a system of mutual inhibition: with a given absolute level of the sexual drive, its relative level is determined by the absolute level of the aggressive drive, and *vice versa*". (SEVENSTER, 1961, p. 146). The "relative level" of a drive is what is seen in the animal's behavior. It seems reasonable that any method that might reduce the level of one of the drives would distort the normal balance of the motivational system. Such a reduction in one of the systems might be obtained by habituation of aggressive behavior, assuming that such a decrement reflects an inhibition ("internal inhibition" in Pavlovian parlance) of the motivation system (reducing the "absolute level" of aggression). The results of the present study fit within such a framework very nicely when we consider the group stimulated with a real male. However, it does not hold for the group habituated to the model. This is not too surprising when we consider that the initial level of aggression elicited by the real male was approximately four times as great as that elicited by the model and if we assume that the amount of "internal inhibition" generated during habituation is some positive function of the magnitude of the change in the primary response (*i.e.*, aggression). From this point of view the "internal inhibition" generated during habituation to the model was really very minimal compared to that generated by the real male and hence had no real effect on the postulated inhibitory interrelationship between drives. The results with the real male group can be considered evidence only for the absolute level of aggression having an effect on the relative level of sex. The reverse relationship could not be tested in this study. Such a relationship, that of an increase in aggressive responses as a function of waning sexual behaviors, requires an experiment in which the male is stimulated with a gravid female each day for several days, and is tested with a male stimulus for a short period each day to chart the course of aggressive behaviors.

SUMMARY

1) Male three-spined sticklebacks which were clearly in the sexual phase of their reproductive cycle, were presented with aggression eliciting stimuli for a period of fifteen minutes each day for ten days. One group was stimulated with a live, nuptially colored male stickleback in a clear plastic tube each day. Another group was presented with a wooden model of a male stickleback each day. The third group received no aggression eliciting stimulation. Both groups which received stimuli showed a decrement in responding to the stimuli over the ten days; the more dramatic decrement was found with the group presented with a live stimulus since this stimulus elicited the greatest aggression initially.

2) Aggressive responses did not all wane at the same rate, but rather one, charges at the stimulus, dropped out first, followed by bites. Orientations toward the stimulus decreased over the ten days but never reached a low level.

3) All three groups were presented with a gravid female stimulus for five minutes each day. The group presented with a model of a male stickleback and the group presented with no male stimulus both showed no change in the level of sexual response directed toward the female stimulus. The group presented with the live male stimulus demonstrated an increasing number of sexual responses directed at the gravid female.

BIBLIOGRAPHY

Baenninger, R. (1966). Waning of aggressive motivation in *Betta splendens*. — Psychol. Sci. 4, p. 241-242.

Clayton, F. L. & Hinde, R. A. (1968). Habituation and recovery of aggressive display in *Betta splendens*. — Behaviour 30, p. 96-106.

Drees, O. (1952). Untersuchungen über die angeborenen Verhaltensweisen bei Springspinnen. — Z. Tierpsychol. 9, p. 169-207.

Hess, E. H. (1962). An approach toward the complete analysis of behavior. — In: New directions in psychology, Holt, Rinehart and Winston, pp. 157-266.

Hinde, R. A. (1954a). Factors governing the changes in strength of a partially inborn response, as shown by the mobbing behavior of the Chaffinch (*Fringilla coelebs*) I. The nature of the response, and an examination of its course. — Proc. Roy. Soc. B, p. 306-331.

— (1954b). Factors *etc.* II. The waning of the response. — Proc. Roy. Soc. B, p. 331-358.

— (1961). Factors *etc.* III. The interaction of short term and long term incremental and decremental effects. — Proc. Roy. Soc. B, p. 398-420.

Iersel, J. A. A. van (1953). An analysis of the parental behaviour of the male three-spined stickleback. — Behaviour Suppl. 3.

Melzack, R., Penick, E. & Beckett, A. (1959). The problem of "innate fear" of the hawk shape: an experimental study with mallard ducks. — J. Comp. Physiol. Psychol. 52, p. 694-698.

— (1961). On the survival of mallard ducks after "habituation" to the hawk-shaped figure. — Behaviour 17, p. 19-26.

Peeke, H. V. S. (1966). Habituation of aggression in the three-spined stickleback (*Gasterosteus aculeatus* L.) — Doctoral Dissertation, University of Southern California.

—, Wyers, E. J. & Herz, M. J. (in press). Waning of the aggressive response to male models in the three-spined stickleback. — Anim. Behav.

Retrospective: Measures of the aggression of parental male three-spined sticklebacks

Article published in *Behaviour* volume 40 (1971), p. 228–262

R.J. Wootton
(IBERS, Edward Llwyd Building, Aberystwyth University, Aberystwyth, SY23 3DA, Wales, UK)

In 1965, after graduating in the U.K. with a degree in Zoology, I took up a Commonwealth Scholarship in the Department of Zoology, University of British Columbia (UBC) in its PhD programme. My choice of university had been dictated by two considerations. One of my undergraduate tutors, R.M. Holmes, had stimulated my interest in the biology of fishes and UBC promised to be a good university at which to pursue that interest. Equally important, I was attracted by the combination of sea and mountains provided by British Columbia, having spent three years in the fens of eastern England as an undergraduate.

My initial aim had been to study an aspect of fish migration, but after some vacillation and discussions with potential supervisors and faculty on my doctoral committee, I developed a research proposal focused on aggression. The origin of this proposal probably lay in lectures on animal behaviour given by R.A. Hinde that I attended as an undergraduate.

Hinde was interested in the problem of quantifying behaviour, particularly in the context of the intervening variables or "drives" that were used by ethologists to provide causal explanations for species typical behaviour. One example of such a use of intervening variables was Leyhausen's (1956) interpretation of the display of the domestic cat in terms of "fear" and "aggressiveness." For my research, a more relevant example was Sevenster's (1961) analysis of displacement fanning in the male three-spined stickleback, *Gasterosteus aculeatus*, in terms of the mutually inhibitory relationships between parental, courtship and aggressive drives.

Hinde (1961, 1966) argued that for the concept of an intervening variable to be useful in a causal analysis of behaviour, the correlations between various possible measures of the intervening variable should be high and consistent. A key reference in this argument is Miller (1957), who reviewed a series of experiments on the correlations between different methods of measuring the intervening variables of hunger and thirst in rats.

Given the example of Sevenster (1961) and the interest in *G. aculeatus* at UBC at that time (see Hagen, 1967), I decided to investigate the unitary nature of the intervening variable of aggression in breeding male *G. aculeatus*. A study by Segaar (1961) had suggested that the aggression directed by a male *G. aculeatus* at another male changed over the course of the parental cycle. On the basis of the studies of Sevenster (1961) and Segaar (1961) and preliminary observations, I identified several possible measures of aggression, including both frequency and duration measures, and three different contexts in which aggression directed towards another male could be assessed over the parental cycle (Wootton, 1971). These three contexts were firstly, a mature male was presented in a tube 200 mm from the male's nest for five minutes once a day; secondly, a mature male was visible through a transparent partition over the entire parental cycle; thirdly, wax dummies of fish painted to resemble a male stickleback were presented to the male for two minutes per day. The main aim of the study was to evaluate the concept of aggression as an intervening variable of use in the analysis of *G. aculeatus* behaviour, rather than describing, quantitatively, the behavioural changes of the male over the parental cycle.

The study identified several variables that were highly correlated and showed similar trends over the parental cycle. This suggested that aggression could be used as an intervening variable in the causal analysis of stickleback behaviour. The research was a ground-clearing rather than a ground-breaking study. It provided evidence for the validity of measures of aggression that were being used in the quantitative analyses of male *G. aculeatus* behaviour.

Perhaps the most important consequence of the research resulted because Dr. J.M. Cullen of Oxford University was the external examiner of my thesis. The research became one of the several strands that led a PhD student of Cullen, F.A. Huntingford, to utilize multivariate analyses of aggression and anti-predator behaviour in male *G. aculeatus*. These analyses identified the dimensions of "boldness" and "timidity" (Huntingford, 1976a, b). Huntingford's studies led to the concept of behavioural syndromes, which are still a topic of active research.

FitzGerald & Kedney (1987) extended the approach of using different operations to evaluate the aggression of male *G. aculeatus*. They compared measures of aggression with the ability of a male to dominate a rival male in a dyadic interaction or to establish a territory in a group of 20 males. Their results failed to identify simple relationships between these operations to evaluate the ability of a mature male to obtain and defend a territory, illustrating the problem of quantifying the concept of aggression. Sadly, the early death of G.J. FitzGerald meant that he did not follow up his study.

The successful completion of my thesis owed much to the quietly effective support of my supervisor, Prof. N.R. Liley, and the advice of my doctoral committee. In the 40 years since my thesis was completed, the emphasis in behavioural studies has shifted massively to the function of behaviour. A study of intervening variables has now lost much of its relevance. However, the strategy of adopting a multivariate approach to quantifying biological phenomena of interest, including behavioural sequences, is still a relevant and developing area of science.

References

FitzGerald, G.J. & Kedney, G.I. (1987). Aggression, fighting, and territoriality in sticklebacks: three different phenomena? — Biol. Behav. 12: 186-195.

Hagen, D.W. (1967). Isolating mechanisms in a threespine stickleback (*Gasterosteus*). — J. Fish. Res. Bd. Can. 24: 1637-1692.

Hinde, R.A. (1961). Unitary drives. — Anim. Behav. 7: 130-141.

Hinde, R.A. (1966). Animal behaviour. — McGraw Hill, New York.

Huntingford, F.A. (1976a). The relationship between anti-predator behaviour and aggression among conspecifics in the three-spined stickleback, *Gasterosteus aculeatus*. — Anim. Behav. 24: 245-260.

Huntingford, F.A. (1976b). An investigation of the territorial behaviour of the three-spined stickleback (*Gasterosteus aculeatus*) using principal components analysis. — Anim. Behav. 24: 822-834.

Leyhausen, P. (1956). Verhaltensstudien bei Katzen. — Zeits. f. Tierpsychol., Beiheft 2.

Miller, N.E. (1957). Experiments on motivation. — Science 126: 1271-1278.

Segaar, J. (1961). Telencephalon and behaviour in *Gasterosteus aculeatus* males. — Behaviour 18: 256-287.

Sevenster, P. (1961). A causal analysis of a displacement activity (fanning in *Gasterosteus aculeatus* L.). — Behaviour, Supplement IX: 1-170.

Wootton, R.J. (1971). Measures of the aggression of parental male three-spined sticklebacks. — Behaviour 40: 228-262.

MEASURES OF THE AGGRESSION OF PARENTAL MALE THREE-SPINED STICKLEBACKS

by

R. J. WOOTTON [1] [2]
(Department of Zoology, University of British Columbia, Vancouver, Canada)

(With 9 Figures)
(Rec. 28-XII-1970)

INTRODUCTION

In the causal study of behaviour standard methods and stimuli are often needed to obtain reproducible, quantitative measures of a particular type of behaviour. Frequently workers studying the same species have used different methods to measure the same behavioural phenomenon. Rarely considered is whether the different methods yield the same results for a given experiment.

Thus, in the study of the reproductive behaviour of the three-spined stickleback, a number of different methods have been used to obtain a measure of aggression. In one method a tube containing a fish is placed within the territory of a solitary male stickleback, usually for three or five minutes. The frequency of bites directed at the fish in the tube by the territorial male is recorded. VAN IERSEL (1958), SEVENSTER (1961), SEGAAR (1961) and SEVENSTER-BOL (1962) have all used this method to measure the aggressive drive of the male. SYMONS (1965) also used this method to investigate the relationship between aggression and spine-raising. BAGGERMAN (1966) used a different method in her study of the effect of hormones on the reproductive behaviour of the male stickleback. The experimental male was separated from another male by a glass partition, and could see this other male throughout the experiment and not just at the time of testing. A record was made of the behaviour of the experimental male directed towards the other fish. MORRIS (1958) used a set of coloured models to measure changes in the aggression

1) Present address: Department of Zoology, University College of Wales, Aberystwyth, U.K.

2) I wish to thank Dr N. R. LILEY for his helpful supervision of this work, Dr J. M. CULLEN for his comments on an early version of this MS., and R. BLACK and D. KRAMER for many useful discussions. M. WOOTTON also gave much technical help.
This study was carried out while the author held a Commonwealth Scholarship at the University of British Columbia.

of the male ten-spined stickleback (*Pygosteus pungilius*). Again the frequency of bites was regarded as a measure of aggression. The test period was one minute long. Models have also been used by TINBERGEN (1951) and MUCKENSTURM (1967, 1968) in the study of the stimuli which the male responds to by fighting and courtship.

This study compares these three methods of measuring aggression. Each method is used to follow the changes in aggression which occur during the parental period of male sticklebacks.

In many studies only one measure has been taken as representative of the aggressive drive. This has been critisised by MILLER (1957). He writes, "... the type of analysis required ... to make intervening variables, such as drives, meaningful is one in which it is possible to compare the effects of a number of different experimental operations on a number of different measures of the variable". He suggests that if discrepant results are found, that is some measures are not highly correlated, the discrepancy can be explained in two ways. Either the drive is not a single unitary variable but a cluster of interrelated variables, or the measures are impure and are affected by factors other than the drive.

HINDE (1959) has reviewed the concept of unitary drives. He suggests that the concept hinders a causal analysis of behaviour because a unitary drive may be regarded as an unanalysable entity in the causal network. Therefore, he suggests that in the causal analysis of a complex sequence of activities, unitary concepts such as drives be abandoned.

Both MILLER and HINDE imply that more than one measure of a behaviour sequence should be taken to test the assumption that the behaviour is a unitary phenomenon. This has not often been done in studies on the three-spined stickleback. The purpose of this study then is to compare selected measures of behaviour obtained by different test methods during the parental phase of the male stickleback. It also relates measures not used in previous studies to measures such as frequency of biting which have regularly been used.

Life history of the three-spined stickleback.

HAGEN (1967) has shown that in south-western British Columbia the non-migratory *leiurus* form of the three-spined stickleback is to a large extent reproductively and ecologically isolated from the migratory *trachurus* form. Since this isolation may have led to some divergence in the behaviour of the two forms, care was taken to use only the *leiurus* form from one population.

The *leiurus* form remains in freshwater for the whole of its life. Little is known about the social behaviour of a population in the winter. In the spring the mature males are territorial. The male builds a nest within its territory. When the nest is complete the male courts gravid females and induces them to spawn in the nest. During the courtship encounter the male's back, normally dark, blanches to a pale creamy brown. This value of about 12% of the total test time on Day 8. Total time at the nest excluding fanning decreased from a mean of 6% of the test period on Day 1 to 0% on Day 9.

Oriented time in the partition tests accounted for a third or less of the test period. Fanning, especially in the middle of the parental phase, was an important component and reached a maximum of about 30% of the test period on Day 7. Nest-directed activities followed a comparable curve to that found in the model tests but the percentages were higher.

H. PARENTAL FANNING

The continual presence of another male in the partition arrangement has a important effect on the amount of fanning shown by the parental male. Evidence for this comes from measurements of fanning for two groups of parental males. One group, those used in Experiment 1 and 3, were isolated except for the short test periods. Measurements of total fanning duration and the number of bouts of fanning were made daily on this group of fish at least an hour after the test period. Measurements of fanning duration were also made for fish used in Experiment 2 (the partition experiment).

For the first nine days after fertilization the isolated fish fanned for a longer duration than did the partition males (Fig. 9). In both cases the maximum duration was reached on or about Day 6. Over the final three days there was no difference between the two groups. Since the number of bouts of fanning was approximately the same for both groups, it follows that the isolated fish tended to fan in longer bouts than the partition males, and to show a shorter inter-bout period.

DISCUSSION AND CONCLUSIONS

A. CHANGES IN RECORDED BEHAVIOUR

The two distinctive features of the 12 day parental period were the U-shaped (quadratic) trend shown by a number of measures of the response of the experiment fish to the test stimulus, and the inverted U-shaped trend shown by the measures of fanning. Each test method gave a quadratic trend for Bi, Ch and Bi/O.T. But only the methods in which fanning became a significant component of the test periods gave quadratic trends for T.O.T., B.O.T. and Bi/B.O.T. Fanning and behaviour oriented towards the test stimulus are incompatible. Any time spent fanning in a test period is not available for behaviour oriented towards the test stimulus.

This incompatability provides one explanation for the inverse relationship between the trend for measures of fanning and that for Bi or Ch. In the tube tests, however, fanning is a negligible component of the test periods, the percentage of time spent oriented towards the test stimulus remains relatively constant over the twelve day period and yet Bi and Ch both show the typical U-shaped trend. Perhaps at the neural level there is an inhibitory relationship between fanning and aggression, or the hormonal state of the parental male may be incompatible with high levels of aggression. A further possibility is that the high levels of fanning seen in the isolated males may cause muscular fatigue and all other activities show a reduced intensity.

The results of this study confirm and extend the results of previous workers. SEGAAR (1961) had found a clear U-shaped trend in biting at a test male in a tube 23 cm from the nest site. He recorded for the first nine days after fertilization. SYMONS (1965) presented results for the first six days of the parental period. The test male was placed at 10 cm, 100 cm and 250 cm from the nest site. Only at the two larger distances was there a clear decline in the frequency over the six day period. The inverse relationship between fanning and aggression was studied by SEVENSTER (1961). He found a weak negative correlation between fanning and biting. VAN DEN ASSEM (1967) showed that in the partition situation the presence of the second male affected the amount of fanning of the parental male. A finding confirmed by this study.

In populations of sticklebacks aggression is a component of territorial behaviour. Are there changes in territorial size that can be correlated with the changes in behaviour measured in this study? MORRIS (1958) recorded an increase in the size of the territory held by a male three-spined stickleback after its eggs had hatched, that is in the period that the frequency of biting and charging were found to be increasing in this study. BLACK (1969) has shown that the size of the territory held by a parental male decreases over the first half of the parental period. Again this change parallels a change, a decrease, in some of the measures used in this study. Considerable decreases in territorial size have been seen in wild populations of sticklebacks, the decreases in territorial size followed increases in fanning (unpublished personal observations). All this suggests that decreases in aggression measured under laboratory conditions are comparable to changes occurring in the wild.

VAN DEN ASSEM & VAN DER MOLEN (1969), PEEKE (1969) and PEEKE, WYERS & HERZ (1969) have all shown that the aggressive response of male sticklebacks to a test stimulus wanes with time. That is the stickleback becomes habituated to the test stimulus at least with regard to the aggressive response. This waning may be a factor in the decrease in the aggressive response over the first days of the parental period. But in all three experiments described in this study, the testing began several days before the start of the parental period. Any major waning of the responses because of habituation, had probably occurred before the onset of the parental period. In the partition tests, those males which did not have eggs in their nests did not show the U-shaped trend in measures of biting and charging. This suggests that this trend is typical of parental males and occurs even in situations where habituation of the aggressive response would be expected.

B. SPINE-RAISING AND THREATENING

SYMONS (1965), in a detailed study of the relationship between spine-raising and other behaviour of the male, showed that the raising of the dorsal spines was positively associated with zig-zagging, and negatively associated with biting if the approaches that ended with a bite also included zig-zagging. Raising the dorsal and ventral spines together showed a positive relationship with both biting and zig-zagging. Because zig-zagging occurred so rarely in the partition and tube tests, it was not possible to confirm these correlations. However, both the percentage of approaches with various combinations of spines raised and the trends these percentages followed in the parental period (Fig. 7) agreed closely with SYMONS' results, except for the tests with models.

In most approaches to models both dorsal and ventral spines were kept erect. Is the spine-raising so evident because the experimental fish has had little previous experience of such artificial objects as models? The "novelty" of the test situation does seem to affect the amount of spine-raising. Thus, SYMONS found that at the start of a tube test, a higher proportion of charges were made with spines raised than towards the end of the test period. Early tests in a series of tube tests tended to produce a higher incidence of spine raising (unpublished personal observation). In contrast, in the partition tests the amount of spine-raising remains at low, relatively constant levels. Testing in the partition situation does not involve adding or removing anything from the normal environment of the experimental male.

A related aspect of the behaviour of the experimental *leiurus* males was the scarcity of head-down threats. In observations on wild populations of both *trachurus* and *leiurus* fish living in clear water it was found that the head-down threat was frequently seen in encounters between territorial males. Especially in encounters between the *trachurus* males the erect red, ventral spines were very conspicuous. Reports by VAN DEN ASSEM (1967) and BAGGERMAN (1966) have both indicated that threatening is commonly seen between *trachurus* males separated by a glass partition. But BLACK (personal communication) also found that *leiurus* males from the Little Campbell River showed little threatening even when there was free contact between them. The water of the Little Campbell River is tea-coloured and the vegetation is very thick. So perhaps visual threats under these conditions are ineffective and are loosing their importance in aggressive encounters. LEA (1968) found that *leiurus* males from the Little Campbell River had relatively short spines and a majority of the males lacked red pigment on the ventral spines. A comparative study of the habitats and behaviour of the various races of sticklebacks might well prove valuable as a study in the evolution of behaviour.

C. RELATIONSHIPS BETWEEN SELECTED MEASURES

The correlations between measures fell into two groups: those that were similar regardless of the test method and those that varied between test methods. In the first group were the correlations between biting (Bi) and charging (Ch) and biting and the rate of biting per minute of oriented time (Bi/O.T.). In the second group were included the correlations between biting and total oriented time (T.O.T.) and biting and the proportion of charges ending in a bite (Bi/Ch). In all three methods Bi is a function of the rate of biting within the oriented time. Since this latter measure is independent of the length of the test period it is an excellent measure to use when making comparisons between individual fish or between test methods.

T.O.T., itself, was very dependent on the test method because of the incompatability between orientation towards the test stimulus and fanning. Bi/Ch was also a measure strongly influenced by the test method. In the tube and partition tests it was highly correlated with Bi. SYMONS (1965) had already shown that the correlation between biting and the percentage of charges ending in a bite was high ($r = 0.716$). But in model tests, Bi/Ch remained high irrespective of the frequencies of charging or biting. This may be because the males could come into physical contact with the object attacked, or because the rate and the amount of movement of the models differed from that of the live test males.

In both the partition and tube methods, Bi was correlated with the mean number of bites per bout of oriented time (Bi/B.O.T.). That is fish with high levels of Bi/B.O.T. tended to have high values for Bi. Inspite of this, in the tube method, Bi/B.O.T. did not show the U-shaped trend shown by Bi. This suggests that whatever inhibits biting does not do it by altering the number of bites delivered once a bout of orientation and biting has been initiated.

There appeared to be a difference in the relationships between T.O.T., B.O.T., and M.B.O.T. for the tube and partition tests (c.f. Fig. 2 and Fig. 4). A simple explanation of this difference would be that it is a consequence of the different ranges of T.O.T. found in the tube and partition tests, and that if the two graphs are considered together they provide a picture of the relationships between the variables over the whole range of T.O.T. This interpretation fits the changes in B.O.T. and M.B.O.T. but does not fit the changes in the mean bout length of non-oriented time. A second explanation for the difference can be obtained using CANE's (1961) analysis of the possible relationships between three interpendent variables. On this basis Fig. 2 would be interpreted to show that the probability of orientating towards the test stimulus in the tube situation changed, but the average length of orientation and non-orientation remained constant. Fig. 4, on this interpretation, would show that in the partition tests, the average length of non-orientation changed but the average length and probability of orientation to the stimulus remained constant. Since non-orientation includes time spent fanning and other time at the nest, this interpretation would indicate that these nest directed activities alter the relationships beween T.O.T., B.O.T., and M.B.O.T.

Relationships between other activities seen during male-male interactions will not be discussed except to note that neither threatening nor any particular pattern of spine-raising followed the trends found for Bi, Ch and Bi/O.T.

D. COMPARISON OF TEST METHODS

In spite of differences between the three test methods in the test stimulus, the length of the test period and the amount of exposure to the test stimulus, all three methods gave comparable changes in biting and related measures during the parental period. This was true even though there were differences in the levels of some measures between the three methods. For example Bi took far higher mean values in the tube method than in either the partition or the model tests. But in all cases a quadratic trend in Bi was found.

Two factors did tend to produce quantitative differences in the results obtained by the three methods. The first factor was the importance of fanning in many of the records from the partition tests. In some test periods there was little time left available for interactions with the male on the other side of the partition. The other factor was that some fish were not consistent in their response to the models. On one day a fish might attack the model whilst on the following day the fish would 'freeze', hanging motionless in the water with tail down and slightly curled to one side, while the model was in the tank. This observation and the spine-raising seen in model tests suggest that models are not equivalent stimuli to live fish. But the results obtained using Model S/R do suggest that some models can be used to measure changes in the aggression of the male stickleback and can be used as adequate substitutes for live test fish when this is desirable.

This study has shown that the three methods give comparable results when used to study changes in aggression. For some purposes there may be practical reasons for choosing one method in preference to another. Thus the tube method gave high values for Bi, Ch and T.O.T., but fewer bouts of nest-oriented behaviour. In the partition tests a broader range of behaviour regularly occurred during the test periods but levels of Bi, Ch and T.O.T. were relatively low. Ideally the behaviour of the test fish on the other side of the partition or in the tube should be recorded simultaneously with that of the experimental male so that records for the experimental fish could be corrected for changes in the behaviour of the test fish.

E. CONCLUSIONS

At the start of this study it was thought that by using a number of different methods to quantify the aggressive behaviour of the male stickleback, the concept of a unitary aggressive drive would be found to be inadequate. Contrary to this assumption it was found that the most obvious measures of aggression behaved in very similar ways and that the results from the methods used were comparable. This suggests that there is a common causal factor for the changes in aggression seen during the parental phase. It now remains to see whether indirect measures of aggression will also show similar changes during the parental phase. There is evidence that territorial size changes in a way that parallels the changes in the more direct measures of aggression. Other indirect measures might include an operant response *e.g.* SEVENSTER (1968), or the strength of an aversive stimulus the male will overcome in order to attack another male.

SUMMARY

This study compared three methods used to measure aggression in the male three-spined stickleback by using them to measure changes in behaviour over the twelve days following the fertilization of eggs. These methods have been used in various studies without the certainty that they measure the same behavioural phenomena. In addition a number of different, possible measures of aggression were taken in the hope that this would show whether a unitary drive concept of aggression was realistic.

The methods consisted of recording the behaviour of experimental males in one of three situations: (i) a tube containing another male was visible for five minutes, (ii) another male was visible through a glass partition at all times, (iii) a fish-shaped wax model was visible for two minutes.

Each method showed that in the twelve days after the fertilization of eggs, there was a U-shaped trend in the frequency of biting, frequency of charging and in the rate of biting per minute of time spent oriented towards the test stimulus. In Method (i), the total oriented time took up about 70% of a test period, was not correlated with frequency of biting and did not show a U-shaped trend. In Method (ii) and (iii) the total oriented time formed an average less than 50% of a test period, was correlated with frequency

Retro- and perspective: Determinants of dominance

Theo C.M. Bakker[1)]
(Institute for Evolutionary Biology and Ecology, University of Bonn, An der Immenburg 1, D-53121 Bonn, Germany)

In the seventies of the last century, research on the function and causation of aggression was promoted in the Netherlands by a priority programme of the Netherlands Organization for the Advancement of Pure Research (ZWO, nowadays called NWO; see Wiepkema & van Hooff, 1977). Geert van Oortmerssen, a pupil of Gerard Baerends of the University of Groningen, was involved in this programme. In the mid-seventies, I was doing a Masters project in his research group concerning the genetics of aggressiveness in wild house mice, *Mus musculus domesticus*, by performing artificial selection experiments. These selections led to the so-called SAL (short attack latency) and LAL (long attack latency) mice (van Oortmerssen & Bakker, 1981), which would become a classical example of behavioural syndromes ("animal personalities").

Geert van Oortmerssen evoked my interest in behaviour genetics and aggression. After my MSc examination in 1978, I inquired after a PhD position at the University of Leiden. I was encouraged by Geert van Oortmerssen, who informed me that a ZWO grant application of Maarten 't Hart, nowadays a famous Dutch literary writer, and Piet Sevenster on aggressiveness in sticklebacks had recently been awarded. I was very eager for a PhD study, and this project would be perfectly tailored to my interests. Moreover, in addition to the universities of Groningen and Utrecht, the University of Leiden was a Mecca of ethology in the Netherlands. I obtained the position and moved with my wife Petra, also a biologist, and my three-year-old son, Baldwin, to Leiden. The ethology research group at the University of Leiden was at that time led by Jan van Iersel and after his retirement in 1984 by

1) e-mail: t.bakker@uni-bonn.de

Piet Sevenster, who had been Niko Tinbergen's assistant during his first year in Oxford. Piet Sevenster would be my supervisor. This made me a proud, second generation pupil of Tinbergen. It was fitting that I received the Niko Tinbergen prize of the "Ethologische Gesellschaft" (the society of ethologists of German speaking countries) in 1990 for my PhD dissertation!

I learned a great deal from Piet, not only about sticklebacks but also about observing animals, data interpretation, and ethology. He was also a great naturalist and art lover. The PhD project concerned the genetic basis of aggressiveness in three-spined sticklebacks, *Gasterosteus aculeatus* L. (Bakker, 1986). Mine was a very elaborate project involving six artificially selected lines and a control line. It was subsidized for three years but continued for some more years. Piet Sevenster had organized grant money for an extension of the project by another three years, but this was conditionally bound to the appointment of another PhD student than me, who would then take over the project. Piet Sevenster refused to do so because he was convinced that only I could handle the project. Thus we did not get the money for the extension and some very tough years of unemployment followed. My family consisted in the meanwhile of four persons: our son Oliver was born one year after we moved to Leiden.

It is a great pity that we did not conserve the selection lines; molecular genetics was not common practice at that time. During the first three years of the project, I had the luxury of a full-time personal assistant, Els Dekker, who was a biologist, and helped with breeding, maintenance of the fish, and data collection. Without her help this ambitious research programme would have been impossible. Her tasks would later kindly and very competently be taken over by Enja Feuth-de Bruijn, a biologist with a permanent position as a technical assistant in the research group of Piet Sevenster.

The ethology research group of the University of Leiden had a long tradition in stickleback research and produced many seminal works (e.g., ter Pelkwijk & Tinbergen, 1937; van Iersel, 1953; Sevenster, 1961; van den Assem, 1967). It had great aquarium facilities, which were partly reorganized to house the many small 10 litre tanks that I used for the selection experiments. My start in stickleback research was also the start of a friendship with Tijs Goldschmidt, who did field studies on sticklebacks in close connection with Piet Sevenster. Later, Tijs completed a PhD study on haplochromine cichlids in Lake Victoria and wrote about his experiences in a successful book *Darwin's Dreampond* (Goldschmidt, 1996, translation of the original 1994

Dutch book *Darwins Hofvijver*). This book marked the beginning of his career as a literary writer. Tijs helped me to catch the base population of stickleback for my PhD research. He knew all the places where stickleback could be found near Vaassen, in the middle of the Netherlands. He had many good connections with local people, so we easily obtained permission to sample fish on private areas. We sampled sticklebacks from many different places in an area of about 35 km^2 in order to collect a heterogeneous sample to begin the artificial selection. Tijs surprised me not only by his great knowledge of sticklebacks and behavioural ecology but also by not hesitating whatsoever to step into brooks with his normal daily clothes and shoes as if water was part of his daily life.

The 1983 *Behaviour* paper (Bakker & Sevenster, 1983) on the determinants of dominance was my first stickleback paper and my second paper in a refereed journal. Although I was the first author, the correspondence to the journal was handled by Piet Sevenster. There was no discussion about that and hardly about the choice of the journal. *Behaviour* was *the preferred* journal for Dutch ethologists. The managing editor at that time was Gerard Baerends from the University of Groningen. I remember that he wrote to Piet Sevenster that the writing was not the high quality he was used to from Piet. Piet did not like that at all and he further improved my not perfect English. I cannot remember other comments as I did not get the correspondence. After one round of revision, the manuscript was readily accepted. The publication was well received by the scientific community. It is well cited: between one and five times almost every year since the year of its publication. In the year 2000 it was cited ten times, probably reflecting the growing interest in intra-sexual selection after two decades of a focus on inter-sexual selection by female mate-choice.

What was new in our paper and why was it cited so well? The strength of our paper was, I think, the exploration of and evidence for the multifarious influences on dominance in male sticklebacks: genetics, experience, aggressiveness, and red nuptial colouration. The analyses were facilitated by the discovery that isolated, reproductively active males can be ranked into a linear order of dominance based on the outcomes of dyadic dominance tests among them. Dominance tests implied the introduction of two reproductively active males in a neutral tank that is only large enough for the establishment of one territory. The rank in the dominance "hierarchy" was called dominance ability.

Genetics

Genetic influences were only briefly described in the 1983 paper by referring to my forthcoming PhD dissertation (Bakker, 1986; see section Behavioural Genetics, Phylogenetics & Speciation). Based on three generations of artificial selection for high and low dominance ability, the realized heritability for dominance ability was estimated at 0.34 (Bakker, 1986). This estimation was of similar magnitude as the heritability of various forms of aggressiveness (juvenile, male territorial, female) in sticklebacks of this population (Bakker, 1986, 1994a, b), and of the mean heritability of behavioural traits in animals in general (e.g., Roff, 1997). The data pointed to an asymmetry of the response to selection as only the selection for lowered dominance ability was successful in comparison to an unselected control line.

After my PhD study, I extended the behaviour-genetics of aggressiveness in sticklebacks by a comparison of lab-bred progeny from two Dutch stickleback populations with a different life history and their reciprocal crosses (Bakker et al., 1988; Bakker, 1994a). I was assisted by Enja Feuth-de Bruijn. This study provided further evidence for genetic variation in dominance ability. Males from the same freshwater population as we had used for our selections won about three quarters of dyadic combats against lab-bred males from an anadromous population (Bakker, 1994a). This outcome again confirmed that fish from the freshwater population had been selected for high dominance abilities under natural conditions. I am not aware of other genetic studies on stickleback dominance.

Experience

In addition to genetic influences, experience also has a decisive effect on the probability of winning a dominance fight. We experimentally showed in our 1983 paper that previous dominance or inferiority experience is also a determinant of dominance. Thereafter we extended the study of experiential effects on dominance by prolonging the interval between experience and the following dominance test to three or six hours (Bakker et al., 1989). Now an asymmetrical effect of prior winning and losing on dominance became evident: the effect of prior losing was stronger and longer lasting than that of prior winning. This is not unique for sticklebacks, but quite a general phenomenon and may be explained by different physiological mechanisms involved (e.g., Hsu et al., 2006; Oliveira et al., 2009).

Experiential effects on dominance are furthermore revealed by the almost linear ranking of a group of isolated males when tested in all pair-wise combinations for dominance ability (Bakker & Sevenster, 1983; Bakker, 1986). There are several recent theoretical papers that predict a linear dominance order by only winner and loser effects (reviewed in Hsu et al., 2006; Dugatkin & Dugatkin, 2007; Hock & Huber, 2009). Interestingly, the stability of the dominance hierarchy in a group of non-reproductive sticklebacks was affected by environmental disturbances such as increased turbulence, lowered water levels (simulated drought), or hypoxia (Sneddon & Yerbury, 2004; Sneddon et al., 2006).

Aggression

To our surprise, aggressiveness was not consistently associated with dominance ability (Bakker & Sevenster, 1983; see also FitzGerald & Kedney, 1987; Rowland, 1989). This was affirmed by the fact that the successful selection for dominance ability was not accompanied by significant changes in territorial aggressiveness (Bakker, 1986, 1994b). In agreement with this is the absence of a significant correlation between aggressiveness as determined in standard aggression tests (Bakker, 1986) and dominance ability in the population comparison (Bakker, 1994a; Bakker & Feuth-de Bruijn, unpubl. data: freshwater population: $r_s = 0.25$, $n = 10$, ns; anadromous population $r_s = 0.19$, $n = 10$, ns).

Colouration

The fourth determinant of dominance that we considered in our 1983 paper was the degree of red nuptial colouration. Because the lines selected for high and low dominance ability diverged significantly for redness (Bakker, 1986, 1994a), we became aware that the red breeding colouration of stickleback males might be an important determinant of dominance by its intimidating effect on rivals. Also within generations of the high and low dominance lines, dominance correlated significantly with redness (Bakker & Sevenster, 1983). We also found parallel changes in redness and dominance of lines selected for high and low territorial aggressiveness (Bakker, 1986, 1994a, b). Remem ber that parallel changes between dominance and territorial aggressivenes were absent in the lines selected for dominance (Bakker, 1986, 1994a, b) sug gesting two genetically distinct physiological pathways for colour change in the dominance lines and the territorial aggression lines. There were no significant correlations between redness and territorial aggressiveness in the successive generations of the dominance lines (Bakker & Sevenster, 1983 or in the successive generations of the other selection lines (Bakker, unpubl data). This contrasts to some other studies (Rowland, 1984; McLennan & McPhail, 1989, 1990). Regardless, the association between redness and dom inance that we assessed was of a correlative nature.

We conducted some preliminary experimental tests inspired by method used by Semler (1971), an early experimental study on sexual selection by female mate-choice. We coloured males artificially with nail-polish (Bakke & Sevenster, 1983). Although the effects of the manipulation on dominance pointed in the expected direction, we were not convinced by the data because "the colour was rather too flashing, the paint came off during the tests and the fish seemed to be ill affected" (Bakker & Sevenster, 1983). Only in 199 was a convincing experimental test published that showed that red belly colouration of male sticklebacks functions as a threat signal (Baube, 1997).

Charles (Charlie) L. Baube was a PhD student of Bill Rowland. I me Charlie while I was a guest in Bill's lab in 1992, where he introduced me to and helped me with reflection spectrophotometry of stickleback throats He applied and extended a method employing coloured light that Milin ski and I had used in 1990 in a mate-choice context. Charlie performed similar dominance tests as we did in our 1983 paper but now under dif ferent illumination. Illumination was manipulated by theatrical gels such that it optimized the stimulation of desired photopigments (L: sensitive to long wavelengths; M: sensitive to medium wavelengths; S: sensitive to shor wavelengths) while minimizing stimulation of the remaining photopigments Ultraviolet-sensitive photoreceptors in sticklebacks were only described in 2004 (Rowe et al., 2004), so Charlie concentrated on the three photorecep tors that were sensitive in the wave range visible to humans. Dominance tests were conducted under illumination that maintained male red coloura tion (W: white light of the same intensity as the other treatments by using neutral density filters; SML: provided S, M and L wavelength light) or elim inated male red colouration (SM: provided S and M wavelength light; L: pro vided L wavelength light). Only under W and SML illumination was male redness significantly positively correlated with the probability to initiate and win a dominance fight. This is a solid experimental demonstration that redder males have greater dominance abilities because of their redness.

The recognition that red colouration is important in male-male competition dates back to Tinbergen (e.g., ter Pelkwijk & Tinbergen, 1937; Tinbergen, 1948, 1951), who strongly promoted that red serves as a sign stimulus eliciting territorial aggression. Early experiments with dummies of male stickleback presented inside the territory of a male showed the aggression-releasing effect of a red belly (ter Pelkwijk & Tinbergen, 1937; Tinbergen, 1948, 1951). Crude red-bellied dummies elicited more attacks from the territory owner than did more realistic dummies that lacked a red belly. The results were so clear-cut (Tinbergen claimed that even red postal vans visible from the tank were attacked) that no need was felt for a quantitative analysis. Nevertheless, in 1949 such an analysis was made by Collias while he was a guest in Tinbergen's laboratory in Leiden, but the results were published much later (in Baerends, 1985, reproduced in this section, and Collias, 1990).

Collias's experiments affirm the statements of Tinbergen, although not as clearly as Collias (1990) suggested because his analysis suffers from pseudoreplication (Milinski, 1997). A red-bellied or grey-bellied dummy was presented for five minutes very close (7–13 cm) to a male's nest. After a pause of ten minutes the alternative dummy was presented. This was repeated several times per male using ten different males. However, two males had eggs in their nests. The repeated presentation of dummies may have caused sequence effects (Milinski, 1997). But the most serious problem is that all bites against the red or the grey dummy were summed up, causing pseudoreplication. If we do the most conservative analysis (only first tests with the red and grey dummy, only males without eggs), then males are not addressing significantly more bites to the red dummy than to the grey one (Wilcoxon matched-pairs signed-ranks test, $n = 8$, $P = 0.61$)! If we take the mean number of bites per male (but exclude males with eggs), then the red-bellied dummy received more bites than the grey one, indeed (Wilcoxon matched-pairs signed-ranks test, $n = 8$, $P = 0.012$). Yet this was not that self-evident, because several other studies failed to show an aggression-releasing effect of a red belly (for a review see Rowland, 1994). For example, Rowland (1982; Rowland & Sevenster, 1985) found that when he simultaneously presented a dummy with a red belly and another one without, the latter was significantly more often attacked by the territory owner.

Figure 1. Dummies of male three-spined sticklebacks used in behavioural research by W.J. Rowland in the 1980's–1990's and by N. Tinbergen and co-workers in the 1930's–1940's.

One explanation for this discrepancy concerns differences in the red colouration used in the various studies to test its aggression-releasing effect. Tinbergen and co-workers made their dummies out of wax and the red belly was applied with a mix of red powder and shellac. I was very lucky and surprised to find the remainder (something more than half a fish measuring about 4.4 cm; Fig. 1) of such a dummy of a stickleback male during the preparations for a renovation of the aquarium rooms in Leiden. I was in charge of the renovations, and in about 1985 I found the jewel somewhere in a dusty corner of the big (70 m^2) aquarium room on the fifth floor of the tower ("Torengebouw") of the Zoology building (van der Klaauw Laboratory).

Piet Sevenster identified it as a wax dummy of a male used by Niko Tinbergen and his students. Piet must have foreseen this find when he wrote in our 1983 paper: "Some of Tinbergen's early paraffin dummies are still available in our laboratory." Traces of the red colour that was applied with shellac were still visible on the ventral side. Piet assured me that the intensity and hue of the red colour were still very well comparable to the colouration at the time of usage. I carefully dropped the dummy in an envelope, wrote "Tinbergen dummy" on it and put it in a cupboard in my room on the fourth floor. When I moved to Bern, Switzerland in 1988 to join the behavioural ecology research-group of Manfred Milinski, I left the dummy in the cupboard in Leiden. That was a big mistake. The dummy got lost, Enja Feuth-de Bruijn told me, likely after the retirement of Piet in 1989. Stickleback research disappeared in Leiden (and from the Netherlands until recently: Niels Dingemanse introduced it again at the University of Groningen, e.g., Dingemanse et al., 2009) and was replaced by research on birds. The aquarium rooms were made suitable for bird keeping. Unfortunately, during the transformation of research topics and researchers the dummy was lost.

There is a striking difference between the red of Tinbergen's wax dummy and the red of Rowland's plastic dummy (Fig. 1). "Rowland's red colour appears stronger, mainly due to a difference in contrast with the body colour, in hue (Rowland's red is less orange than Tinbergen's colour) and to some extent in colour intensity" (Bakker & Milinski, 1993). The strong red of Rowland's dummy had an intimidating effect on the territory owner rather than an aggression-releasing effect. Rowland admits that "It is thus possible that the colouration of these dummies was supernormal" (Rowland et al., 1995). The intimidating effect of red was also noted by Tinbergen (1948) when he replaced a territory owner by a dummy. A dummy with a red belly was more effective in inhibiting rival males from intruding into the territory than a non-red dummy was. Male red colouration thus has a dual effect: provocation and intimidation by activating both aggression and fear.

Whether aggression or fear dominates depends on many factors, like the relative redness of the rival (e.g., Sevenster, 1949; Fig. 1), how close the rival is to the nest as aggression increases closer to the nest (van Iersel, 1958; Rowland, 1994), and whether the male is inside his territory or on neutral ground (Rowland et al., 1995; Bolyard & Rowland, 1996). The context-dependent response to red was nicely demonstrated by Rowland and co-workers. They quantified male aggression against video playbacks that were presented on monitors manipulated for the intensity of the red channel (Rowland et al., 1995; Bolyard & Rowland, 1996).

It is obvious that the aggression-releasing effect of a red belly cannot account for the evolution of the red belly by intra-sexual selection and must be a secondary effect; it may be rather disadvantageous to its bearer to provoke aggression of rivals. Yet male-male competition might have played a role in the evolution of the male's red breeding colouration through its intimidating effect on rivals. The advantage of a red belly in intra-sexual selection is evident from the increased probability of winning of males with high red scores in male-male competition when new territories are settled (Bakker & Sevenster, 1983; Bakker, 1994a; McKinnon, 1996; Baube, 1997; but see FitzGerald & Kedney, 1987; Rowland, 1989; and Bakker, 1994a). Red has an intimidating effect on rivals, so that redder males attack faster (Baube, 1997). The male that attacks first has a greater probability of winning a dominance fight (FitzGerald & Kedney, 1987; Bakker et al., 1989; Baube, 1997; Guderley & Couture, 2005). Thus being red has clear fitness advantages.

The advantages are further augmented by female mating preferences for redder males (Milinski & Bakker, 1990; for a recent review see McLennan, 2007). Interestingly, male-male competition amplifies the differences in redness between males (Candolin, 1999a) and increases the honesty of the red signal with respect to parental ability (Candolin, 2000a, b). Redness is associated with various male qualities (see McLennan, 2007 for a review), though not universally so. Especially its relationship with physical condition (e.g., Milinski & Bakker, 1990; Bakker & Mundwiler, 1994) and aggressiveness (e.g., Rowland, 1984; McLennan & McPhail, 1989; McKinnon, 1996) makes redness an honest intimidating factor signalling resource holding power. McLennan et al. (1988) suggested, however, that the evolution of colour patterns in the Gasterosteidae is more strongly correlated with inter- than with intra-sexual selection.

One issue regarding redness is still unclear and should be tackled in future research. Most studies measured one aspect of redness. It is therefore equivocal which aspect is important in intra- and inter-sexual selection. Some researchers measured red intensity, others red area (relative to lateral body area). Of course, the assessment of colour variables has become more accurate since our 1983 study. In the 1980's human-based colour scales were practised, followed in the 1990's by digital photography and measurement of the colour variables using a densitometer or image analysis software. In the past decade, reflection spectrophotometry has been more often applied. The latter method is the most accurate because it is independent of human vision. Colour vision in sticklebacks, however, is roughly comparable to human vision (Rowe et al., 2006), so that, at least for the judgment of red colouration, different methods may give a comparable ranking of males (Bakker, 1993; Rowe et al., 2006). One should realize though that less accurate methods may be too conservative to find significant effects. In our 1983 paper and in Bakker (1986, 1994a) I used a four-point colour scale that predominantly ranked red area but also included red intensity. Rowland (1984, 1989) probably used a similar ranking system. Baube's (1997) nine-point colour scale was solely based on red area. In some studies, females preferred males with the most intense red colouration (e.g., Milinski & Bakker, 1990; Bakker, 1993; Bakker & Mundwiler, 1994), but males with the relatively largest red area in others (e.g., Candolin, 1999a). Candolin (e.g., 1999a, b) measured various aspect of redness but got the most consistent effects for red area. In inter-sexual selection, the exact hue of the orange-red belly seemed less important in one study (Baube et al., 1995) but explained choice behaviour in another (Bakker et al., 1999). Thus there is an obvious need for studies that compare the importance of various aspects of redness, when possible combined with ultraviolet signals, in intra- and inter-sexual selection. Most ideally these studies should cover inter-population variation as well.

Other determinants of dominance

There are probably more determinants of dominance in sticklebacks that we did not consider in our 1983 paper. The most obvious one is body size, which was excluded by matching the contestants as much as possible for body size in our and almost all other stickleback studies on dominance. Surprisingly only one experimental study considered body size difference (reviewed in Rowland, 1994), which is probably the most universal determinant of dominance in the animal kingdom. Body size was an important determinant of dominance when differences in body mass between stickleback males exceeded 15% (Rowland, 1989).

Another potential determinant of dominance may be ultraviolet (UV) reflections of male sticklebacks. UV signals have been excluded thus far as dominance tests were performed under artificial illumination that hardly contains UV wavelengths. Recent studies found that sticklebacks possess UV photoreceptors (Rowe et al., 2004), and that male sticklebacks reflect in the UV (Rick et al., 2004). Territorial aggression tests revealed that males directed more aggression to opponents under full-spectrum light including UV than to males seen under illumination excluding UV (Rick & Bakker, 2008a). So UV signals may be another determinant of dominance that should be investigated, especially because UV signals are at least as important as red colouration in female mate-choice of sticklebacks (Rick & Bakker, 2008b).

There is no evidence that in sticklebacks other sensory systems are involved in establishing dominance relationships (see excellent review of McLennan, 2007). However, a potential candidate may be olfactory cues. It was generally believed for a long time that olfaction does not play a prominent role in stickleback life (reviewed in Rowland, 1994) but this view is changing (e.g., Reusch et al., 2001; see review by McLennan, 2007). For instance, in a mate-choice context kin recognition by females is triggered by olfactory cues (Mehlis et al., 2008). Whether olfactory cues also play a role in male-male competition was tested by quantifying aggression of territorial males against two identical computer animations (the modern equivalent of dummies: Künzler & Bakker, 1998) that differed in the scent in front of the animations. In front of one animation water conditioned by a familiar brother was introduced, and in front of the other animation water of an unfamiliar unrelated male was introduced (Mehlis et al., in press). Males were, however, equally aggressive against both animations. Yet familiarity may play a role in a competitive context (Utne-Palm & Hart, 2000).

In conclusion, proximate studies on dominance in sticklebacks are accumulating, so our 1983 paper marked rather the beginning of a revived interest in proximate factors. I expect in the future a further intensification of studies on dominance in sticklebacks, especially of genetic studies that will explore the molecular basis of complex traits using genome sequence data (Kingsley & Peichel, 2007). These will then form the basis to study genetic population differences in order to understand adaptive evolution of dominance in sticklebacks.

Acknowledgements

I want to dedicate this contribution to two persons who decisively influenced my scientific interests. First, I would like to dedicate it to the memory of William (Bill) J. Rowland, who suddenly died in 2004. He very much stimulated my interest in sexual selection in sticklebacks. Second, I would like to dedicate it to the 85th birthday of Piet Sevenster, who taught me everything about sticklebacks and science.

References

van den Assem, J. (1967). Territory in the three-spined stickleback, *Gasterosteus aculeatus* L.: an experimental study in intra-specific competition. — Behaviour, Supplement XVI: 1-164.

Baerends, G.P. (1985). Do dummy experiments with sticklebacks support the IRM-concept? — Behaviour 93: 258-377.

Bakker, T.C.M. (1986). Aggressiveness in sticklebacks (*Gasterosteus aculeatus* L.): a behaviour-genetic study. — Behaviour 98: 1-144.

Bakker, T.C.M. (1993). Positive genetic correlation between female preference and preferred male ornament in sticklebacks. — Nature 363: 255-257.

Bakker, T.C.M. (1994a). Evolution of aggressive behaviour in the threespine stickleback. — In: The evolutionary biology of the threespine stickleback (Bell, M.A. & Foster, S.A., eds). Oxford University Press, Oxford, p. 345-380.

Bakker, T.C.M. (1994b). Genetic correlations and the control of behavior, exemplified by aggressiveness in sticklebacks. — Adv. Study Behav. 23: 135-171.

Bakker, T.C.M., Feuth-de Bruijn, E. & Sevenster, P. (1988). Population differences in aggressiveness of sticklebacks (*Gasterosteus aculeatus* L.). — Behav. Genet. 18: 706 (abstract).

Bakker, T.C.M., Feuth-de Bruijn, E. & Sevenster, P. (1989). Asymmetrical effects of prior winning and losing on dominance in sticklebacks. — Ethology 82: 224-229.

Bakker, T.C.M., Künzler, R. & Mazzi, D. (1999). Condition-related mate choice in sticklebacks. — Nature 401: 234.

Bakker, T.C.M. & Milinski, M. (1993). The advantages of being red: sexual selection in the stickleback. — Mar. Behav. Physiol. 23: 287-300.

Bakker, T.C.M. & Mundwiler, B. (1994). Female mate choice and male red coloration in a natural stickleback population. — Behav. Ecol. 5: 74-80.

Bakker, T.C.M. & Sevenster, P. (1983). Determinants of dominance in male sticklebacks (*Gasterosteus aculeatus* L.). — Behaviour 86: 55-71.

Baube, C.L. (1997). Manipulations of signalling environment affect male competitive success in three-spined sticklebacks. — Anim. Behav. 53: 819-833.

Baube, C.L., Rowland, W.J. & Fowler, J.B. (1995). The mechanisms of colour-based mate choice in female threespine sticklebacks: hue, contrast and configurational cues. — Behaviour 132: 979-996.

Bolyard, K.J. & Rowland, W.J. (1996). Context-dependent response to red coloration in stickleback. — Anim. Behav. 52: 923-927.

Candolin, U. (1999a). Male-male competition facilitates female choice in sticklebacks. — Proc. R. Soc. Lond. B 266: 785-789.

Candolin, U. (1999b). The relationship between signal quality and physical condition: is sexual signalling honest in the three-spined stickleback? — Anim. Behav. 58: 1261-1267.

Candolin, U. (2000a). Male-male competition ensures honest signaling of male parental ability in the three-spined stickleback (*Gasterosteus aculeatus*). — Behav. Ecol. Sociobiol. 49: 57-61.

Candolin, U. (2000b). Changes in expression and honesty of sexual signalling over the reproductive lifetime of sticklebacks. — Proc. R. Soc. Lond. B 267: 2425-2430.

Collias, N.E. (1990). Statistical evidence for aggressive response to red by male three-spined sticklebacks. — Anim. Behav. 39: 401-403.
Dingemanse, N.J., Oosterhof, C., van der Plas, F. & Barber, I. (2009). Variation in stickleback head morphology associated with parasite infection. — Biol. J. Linn. Soc. 96: 759-768.
Dugatkin, L.A. & Dugatkin, A.D. (2007). Extrinsic effects, estimating opponents' RHP, and the structure of dominance hierarchies. — Biol. Lett. 3: 614-616.
FitzGerald, G.J. & Kedney, G.I. (1987). Aggression, fighting and territoriality in sticklebacks: three different phenomena? — Biol. Behav. 12: 186-195.
Goldschmidt, T. (1994). Darwins hofvijver: een drama in het Victoriameer. — Prometheus, Amsterdam, The Netherlands.
Goldschmidt, T. (1996). Darwin's dreampond: drama in Lake Victoria. — The MIT Press, Cambridge, MA, USA.
Guderley, H. & Couture, P. (2005). Stickleback fights: why do winners win? Influence of metabolic and morphometric parameters. — Physiol. Biochem. Zool. 78: 173-181.
Hock, K. & Huber, R. (2009). Models of winner and loser effects: a cost-benefit analysis. — Behaviour 146: 69-87.
Hsu, Y., Earley, R.L. & Wolf, L.L. (2006). Modulation of aggressive behaviour by fighting experience: mechanisms and contest outcomes. — Biol. Rev. 81: 33-74.
van Iersel, J.J.A. (1953). An analysis of the parental behaviour of the three-spined stickleback (*Gasterosteus aculeatus* L.). — Behaviour, Supplement III: 1-159.
van Iersel, J.J.A. (1958). Some aspects of territorial behaviour of the male three-spined stickleback. — Archs néerl. Zool. 13, Supplement 1: 383-400.
Kingsley, D.M. & Peichel, C.L. (2007). The molecular genetics of evolutionary change in sticklebacks. — In: Biology of the three-spined stickleback (Östlund-Nilsson, S., Mayer, I. & Huntingford, F.A., eds). CRC Press, Boca Raton, FL, USA, p. 41-81.
Künzler, R. & Bakker, T.C.M. (1998). Computer animations as a tool in the study of mating preferences. — Behaviour 135: 1137-1159.
McKinnon, J.S. (1996). Red coloration and male parental behaviour in the threespine stickleback. — J. Fish Biol. 49: 1030-1033.
McLennan, D.A. (2007). The umwelt of the three-spined stickleback. — In: Biology of the three-spined stickleback (Östlund-Nilsson, S., Mayer, I. & Huntingford, F.A., eds). CRC Press, Boca Raton, FL, USA, p. 179-224.
McLennan, D.A., Brooks, D.R. & McPhail, J.D. (1988). The benefits of communication between comparative ethology and phylogenetic systematics: a case study using gasterosteid fishes. — Can. J. Zool. 66: 2177-2190.
McLennan, D.A. & McPhail, J.D. (1989). Experimental investigations of the evolutionary significance of sexually dimorphic nuptial coloration in *Gasterosteus aculeatus* L.: temporal changes in the structure of the male mosaic signal. — Can. J. Zool. 67: 1767-1777.
McLennan, D.A. & McPhail, J.D. (1990). Experimental investigations of the evolutionary significance of sexually dimorphic nuptial colouration in *Gasterosteus aculeatus* (L.): the relationship between male colour and female behaviour. — Can. J. Zool. 68: 482-492.
Mehlis, M., Bakker, T.C.M. & Frommen, J.G. (2008). Smells like sib spirit: kin recognition in three-spined sticklebacks (*Gasterosteus aculeatus*) is mediated by olfactory cues. — Anim. Cogn. 11: 643-650.
Mehlis, M., Bakker, T.C.M., Langen, K. & Frommen, J.G. (in press). Cain and Abel reloaded? Kin recognition and male-male aggression in three-spined sticklebacks (*Gasterosteus aculeatus* L.). — J. Fish Biol.

Milinski, M. (1997). How to avoid seven deadly sins in the study of behavior. — Adv. Study Behav. 26: 159-180.
Milinski, M. & Bakker, T.C.M. (1990). Female sticklebacks use male coloration in mate choice and hence avoid parasitized males. — Nature 344: 330-333.
Oliveira, R.F., Silva, A. & Canário, A.V.M. (2009). Why do winners keep winning? Androgen mediation of winner but not loser effects in cichlid fish. — Proc. R. Soc. Lond. B 276: 2249-2256.
van Oortmerssen, G.A. & Bakker, T.C.M. (1981). Artificial selection for short and long attack latencies in wild *Mus musculus domesticus*. — Behav. Genet. 11: 115-126.
ter Pelkwijk, J.J. & Tinbergen, N. (1937). Eine reizbiologische Analyse einiger Verhaltensweisen von *Gasterosteus aculeatus*. — Z. Tierpsychol. 1: 193-200.
Reusch, T.B.H., Häberli, M.A., Aeschlimann, P.B. & Milinski, M. (2001). Female sticklebacks count alleles in a strategy of sexual selection explaining MHC polymorphism. — Nature 414: 300-302.
Rick, I.P. & Bakker, T.C.M. (2008a). Males do not see only red: UV wavelengths and male territorial aggression in the three-spined stickleback (*Gasterosteus aculeatus*). — Naturwissenschaften 95: 631-638.
Rick, I.P. & Bakker, T.C.M. (2008b). Color signalling in conspicuous red sticklebacks: do ultraviolet signals surpass others? — BMC Evol. Biol. 2008, 8: 189.
Rick, I.P., Modarressie, R. & Bakker, T.C.M. (2004). Male three-spined sticklebacks reflect in ultraviolet light. — Behaviour 141: 1531-1541.
Roff, D.A. (1997). Evolutionary quantitative genetics. — Chapman & Hall, New York, USA.
Rowe, M.P., Baube, C.L., Loew, E.R. & Phillips, J.B. (2004). Optimal mechanisms for finding and selecting mates: how threespine stickleback (*Gasterosteus aculeatus*) should encode male throat colors. — J. Comp. Physiol. A 190: 241-256.
Rowe, M.P., Baube, C.L. & Phillips, J.B. (2006). Trying to see red through stickleback photoreceptors: functional substitution of receptor sensitivities. — Ethology 112: 218-229.
Rowland, W.J. (1982). The effects of male nuptial coloration on stickleback aggression — a re-examination. — Behaviour 80: 118-126.
Rowland, W.J. (1984). The relationships among nuptial coloration, aggression, and courtship of male three-spined sticklebacks, *Gasterosteus aculeatus*. — Can. J. Zool. 62: 999-1004.
Rowland, W.J. (1989). The effects of body size, aggression and nuptial colouration on competition for territories in male threespine sticklebacks *Gasterosteus aculeatus*. — Anim. Behav. 37: 282-289.
Rowland, W.J. (1994). Proximate determinants of stickleback behaviour: an evolutionary perspective. — In: The evolutionary biology of the threespine stickleback (Bell, M.A. & Foster, S.A., eds). Oxford University Press, Oxford, p. 297-344.
Rowland, W.J., Bolyard, K.J. & Halpern, A.D. (1995). The dual effect of stickleback nuptial coloration on rivals: manipulation of a graded signal using video playback. — Anim. Behav. 50: 267-272.
Rowland, W.J. & Sevenster, P. (1985). Sign stimuli in the threespine stickleback (*Gasterosteus aculeatus*): a re-examination and extension of some classic experiments. — Behaviour 93: 241-257.
Semler, D.E. (1971). Some aspects of adaptation in a polymorphism for breeding colors in the threespine stickleback (*Gasterosteus aculeatus*). — J. Zool. (Lond.) 165: 291-301.

Sevenster, P. (1949). Modderbaarsjes. — Lev. Nat. 52: 184-189.
Sevenster, P. (1961). A causal analysis of a displacement activity (fanning in *Gasterosteus aculeatus* L.). — Behaviour, Supplement IX: 1-170.
Sneddon, L.U., Hawkesworth, S., Braithwaite, V.A. & Yerbury, J. (2006). Impact of environmental disturbance on the stability and benefits of individual status within dominance hierarchies. — Ethology 112: 437-447.
Sneddon, L.U. & Yerbury, J. (2004). Differences in response to hypoxia in the three-spined stickleback from lotic and lentic localities: dominance and an anaerobic metabolite. — J. Fish Biol. 64: 799-804.
Tinbergen, N. (1948). Social releasers and the experimental method required for their study. — Wilson Bull. 60: 6-51.
Tinbergen, N. (1951). The study of instinct. — Oxford University Press, Oxford.
Utne-Palm, A.C. & Hart, P.J.B. (2000). The effects of familiarity on competitive interactions between threespined sticklebacks. — Oikos 91: 225-232.
Wiepkema, P.R. & van Hooff, J.A.R.A.M. (eds) (1977). Agressief Gedrag: Oorzaken en Functies. — Bohn, Scheltema & Holkema, Utrecht, the Netherlands.

DETERMINANTS OF DOMINANCE IN MALE STICKLEBACKS (GASTEROSTEUS ACULEATUS L.)

by

TH. C. M. BAKKER and P. SEVENSTER[1])

(Dept. of Ethology, Zoological Laboratory, University of Leiden, The Netherlands)

(Acc. 15-III-1983)

Introduction

Encounters between adult male sticklebacks in reproductive condition usually lead to aggresive interactions. When a territorial male, i.e. one defending an area around its nest, meets a rival inside his territory, the owner of the territory will inexorably attack the intruder and practically always the intruder will promptly flee and leave the territory. If, however, the intruder is prevented from fleeing *e.g.* because he is enclosed in a glass tube, the territory-owner keeps attacking him with some persistence and intensity. This situation has been used as an "aggression test" to measure the aggressiveness of territorial males (VAN IERSEL, 1958; SEVENSTER, 1961; *etc.*). In such tests a male of bright colouration and standard size is put in a standard glass tube (which inevitably causes some darkening and fading of the colouration!) and presented for a few minutes to the male to be tested, at a standard distance from the nest and (by various means) at a standard height above the bottom. Variations may be imposed by the nature of the research in question. In any case the number of bites delivered at the tube towards the intruder or "test male" is counted and taken as a measure of the aggressiveness of the male concerned (at the site and time of the test). Other behaviour elements can be incorporated in the outcome of a test. In spite of some disadvantages (SEVENSTER, 1961) the method has yielded useful data in many studies of stickleback behaviour (SEVENSTER-BOL, 1962; MUCKENSTURM, 1968,

[1]) The authors are very grateful to Miss E. J. DEKKER for her able and careful assistance and her keen observations. Of all others who in one way or another contributed, we should like to mention W. TWISK, J. VAN DER LINDEN, TH. van Es and H. de SWART for carrying out a number of important tests.

Part of the work was supported by the Foundation for Fundamental Biological Research (BION), which is subsidised by the Netherlands Organisation for the Advancement of Pure Research (Z.W.O.).

1979; PEEKE *et al.*, 1969; HUNTINGFORD, 1976; ROWLAND, 1982; WOOTTON, 1971, *etc.*). In the experiments here presented, a modification of the standard method has been used.

The aggressiveness, as measured by such tests, declines with increasing distance from the nest (VAN IERSEL, 1958). Consequently males introduced into an area already occupied by a territorial male, will tend to build a nest, if they succeed at all, in settling at the greatest possible distance from the nest of the first occupant (VAN DEN ASSEM, 1967). Once two neighboring territories have been established, overt fighting between the two males gradually wanes and is replaced by threatening displays and other ambivalent behaviours at the border (VAN DEN ASSEM & VAN DER MOLEN, 1969), but these will not be considered in this paper.

There is one interesting form of fighting, however, with which we will be occupied in some more detail. This is the "spine-fighting", as described by VAN IERSEL, 1953, in which two males circle around each other rapidly, often with open mouth and with spines erected. A description is also found in WOOTTON, 1976, who calls it "roundabout fighting" to avoid the suggestion that the spines are actually used to hit the opponent. This suggestion is the more plausible since usually only the ventral spine directed towards the opponent is erected. Actually, however, it has never been seen that the opponent was even touched by the spine. In this roundabout fighting then, the two fishes seem to chase one another, each going for the other's tail. The fast circling which ensues may continue for many minutes without interruption, there being obviously an extremely stable and delicate mutual adjustment of the fishes' movements. Intense and persistent roundabout fighting is seen especially when two rather aggressive males meet in a situation where neither has priority over the other. In the wild (personal observations by one of us, P.S.) it is commonly observed between adjacent territories when a sharp boundary has not been established yet and a strip of no man's land is still being disputed. Further, in the general turmoil after a predator or a human observer has chased off the entire population for a while so that the returning males have to reestablish their territories. Similar intensive disputes with roundabout fighting occur in the commotion during or after an outbreak of nest raiding in which a number of neighbouring males took part. Finally, roundabout fighting is occasionally seen in the wild when for some reason an obstacle separating the territories (row of plants, stone, *etc.*) has disappeared. In a similar way it is easily provoked in the laboratory by suddenly removing an opaque partition which separated two nest-owning males. This almost invariably releases fights with frequent and prolonged bouts of roundabout fighting. In such experiments still another, extremely rare, form of fighting was observed on some occasions and in one of them repeatedly. As far as we know, this has not been described so far. The two males suddenly seize one another by the mouth and remain more or less motionless for a while with the jaws interlocked like in the "mouth-fighting" of Cichlids. This "mouth-fighting" in the three-spined sticklebacks is also shown in a film on animal behaviour by SIELMANN (in the French version: "Comportements innés et acquis")[1].

As might be predicted from the above, roundabout fighting is also frequently seen, when two nest-owning males are introduced simultaneously into a strange area which is not large enough to accommodate two territories. One male will become "dominant", and eventually build a nest, the other will give up fighting and become the "subordinate" or "inferior", hiding somewhere or keeping quiet so as not to elicit attacks from the dominant.

What determines the outcome of this type of contest? What decides who is going to be dominant? These are the central questions with which we are concerned in this article.

Material and methods

For the research reported, sticklebacks were caught from brooks in the neighbourhood of Vaassen (the Netherlands), where a population of the non-migratory forma *leiura* (BERTIN, 1925) is found. From these wild-caught specimens several generations were bred in the laboratory for use in an extensive program of genetical selection on aggressive behaviour. The results of the selection experiments will be published elsewhere and can be disregarded in the context of this article. The fish were kept under highly standardised conditions: three weeks after fertilisation the young were isolated in small plastic aquaria of 34 × 17 × 20 cm with sand, a few plants and some nest material. Temperature was kept at 18-20°C and the light schedule at 16 hours light against 8 hours dark. Food was provided twice a day and consisted of live *Tubifex* worms, live *Artemia*, live *Chironomus* larvae and frozen *Mysis* or *Artemia*.

Aggression test.

After nestbuilding (at an age of 3½ to 6 months) the aggressiveness of the males was assessed by means of an aggression-test. In these tests a rival male was enclosed in a plexiglass cell of 20 × 6 × 6 cm, which was hung against the front of the aquarium, *i.e.* at a distance of 10 to 30 cm from the nest of the male to be tested. If the nest concerned was close to the frontpane the test cell was presented at the greatest possible distance from the nest site. The cell was left there for at least 5 minutes. If the experimental male did not approach the test male within this period, the cell was removed, but if he did the subsequent behaviour was recorded for 5 minutes from then on. The outcome of a test can be expressed as the percentage of time spent biting or bumping at the test male. In this article the mean score of four weekly tests in four successive weeks is used as an assessment of the territorial aggressiveness of an individual experimental male.

---

[1]) Ministerie van Nationale Opvoeding en Kultuur, Brussels, nr 6313.

TABLE 8. Incidence of roundaboutfighting (+ or -R fights) and mean duration of decision (in sec.)

Test	Mean duration of decision	
	D + F	E + G
+ R fights (first tests)	1000.0	501.4
+ R fights	315.0	212.4
- R fights	41.0	120.4

curs, the duration of the test is much prolonged: Table 8. Especially in first encounters roundabout fighting can be long-continued. This phenomenon is most obvious among the males selected for high dominance (DH).

Discussion

First of all in this discussion, some remarks on our use of the words "dominance" and "rank" are in place. The concept of the linear dominance hierarchy was developed by SCHJELDERUP-EBBE (1922) for chickens. It applies and should be restricted to a group of individuals living (constantly or part of the time) in a common area, and recognising one another individually. When groups are too large, stable hierarchies are not established, because individual recognition fails and disputes continue (SCHJELDERUP-EBBE, 1922).

Such stable hierarchies, in the stricter sense, are not found in the three-spined sticklebacks. Some sort of hierarchy was described by VAN DEN ASSEM (1967) in a stable situation with a number of rivalling territorial males. He found a general, positive correlation between the size of the territory and the number of attacks initiated by the owner of that territory. Those with the largest territory were seldom attacked by their neighbours, those with the smallest territory did not initiate attacks and were continuously on the defensive.

For the moment, these hierarchies should be well distinguished from the linear order of dominance as we have used, though they do probably correspond to some extent. Between tests, we kept the males in isolation and, in the tests, each individual male was confronted with another particular individual only once. So individual recognition was ruled out in our concept of dominance. Confrontations were unique and the rank of a male in our linear order of dominance was purely determined by the number of confrontations in which he won and in which he lost. It does not seem unlikely, however, that this order of dominance would be reflected in the hierarchy of a rival situation like those described by VAN DEN ASSEM (1967).

It was shown that the colouration of a male three-spined stickleback is an important determinant of his rank in the order of dominance. For the moment we leave aside the unlikely possibility that an unknown factor predictably correlated with colour is responsible for the male's rank. SEMLER (1971) however studied the effects of artificially coloured males on females in courtship. Encouraged by his article we did a few pilot experiments with males artificially coloured according to his recipe. In 14 dominance tests one of the males randomly chosen was painted red with with nail-polish (Yves Rocher, Papaye CO854202), the other was treated with colourless polish (Carmen): in 8 of the tests the red male won, in 2 the colourless, and 4 tests were undecided. The tests were still unsatisfactory: the colour was rather too flashing, the paint came off during the tests and the fish seemed to be ill affected. Yet, the evidence seems to support our conclusion that colour as such is the effective factor. This reminds us of the classical examples of colour as an indicator of social status. MARLER (1955) proved experimentally that the difference in colour between the sexes in the chaffinch (*Fringilla coelebs* L.) is responsible for the difference in dominance in mixed flocks. Similarly in the Harris sparrow (*Zonotrichia querula*) variations in the amount of black on throat and crown are important determinants of social status (ROHWER, 1975), but there is an interaction with the accompanying behaviour. If subordinates were dyed and given larger black bibs, their social status did not increase. Instead they were attacked and persecuted. If testosterone was administered to subordinates, their status did not increase either. Only if subordinates were both dyed and injected with the hormone, their social status dramatically increased (ROHWER & ROHWER, 1978). The congruence between the behaviour and the colouration appeared to be the crucial factor. The colouration in itself seems to elicit aggression in the conspecific opponents, but if supported by the appropriate behaviour its intimidating effect has the upper hand.

A similar situation seems to obtain in the three-spined stickleback. In several studies the "releasing" effect of the red belly has been emphasized strongly (TER PELKWIJK & TINBERGEN, 1937; TINBERGEN, 1948, a.o.): dummies with a red underside were attacked more frequently and more intensively by a territorial male than dummies without any red. But there are also many reports of similar investigatisons where an eliciting effect of the red colour could not be demonstrated (PEEKE *et al.*, 1969; MUCKENSTURM, 1968, 1979). Recently ROWLAND (1982) repeated such dummy presentations once more and convincingly demonstrated an intimidating effect of red dummies as compared to grey dummies. Consequently the latter were attacked more. This is in agreement with an explanation given by SEVENSTER (1949) for similar phenomena in the ten-spined stickleback, *Pungitius pungitius* (L.). In this species males in reproductive condition become dark, or even almost black. Dark dummies released most bites in territorial males, as compared to other colours, but absolutely pitch-black dummies did not release as much biting in some males. It was not that these males did not pay much attention to such dummies, but they were obviously inhibited to approach them, so that it was conluded that an intimidating effect of black in these tests outweighed the releasing effect. This principle might well account also for the disagreement in the reports on the three-spined stickleback (see ROWLAND, 1982). The intensity of the colours used on the dummies cannot be read from the publications concerned. Some of TINBERGEN's early paraffin dummies are still available in our laboratory. They were painted with shellack and powder of various colours. This produced a rather transparent, soft colour without striking contrasts. From some illustrations in the literature we get the impression that other authors may have used dummies with more saturated colours and more marked contrasts, and therefore may have increased their dummies' intimidating effects. Somehow future workers in this field should make an effort to specify their dummies in this respect as accurately as possible. The experience of the males may be another source of variation in dummy experiments as our data in this article would seem to predict. Whether the experimental male has been exposed to conspecifics or has been raised in isolation, whether he has been provoked to fight a dummy before or alternatively has been beaten up with a dummy, may have consequences for his future reactions to dummies. This has not been investigated in a standardised way, but seems to follow from many observations. The most extreme form of the influence of experience on later reactions to dummies is the conditioned reaction to the green tip of a rod (courting or fighting as the case may be) in sticklebacks (SEVENSTER, 1968, 1973). Finally there are quite a few males which fail to react to dummies for reasons not even vaguely understood.

The intimidating effect of the colouration may well account for the consistent correlation of rank with colour in our dominance tests (Table 4). When in a first dominance test two unexperienced males are simultaneously placed in the strange tank, which is new to both, the brightest male is generally the first to attack the other, duller one, and is then usually going to win. Since there is no experience (leaving aside the confrontation with the test male in an aggression test) and since the correlation of dominance with aggressiveness is but erratic, the most likely explanation seems to be that the duller male is intimidated by the sight of his bright opponent, whereas the latter is not (as much) intimidated by the sight of the duller one. Consequently one would expect the brighter male to initiate the attack. From then on experience becomes an additional factor in the rest of the encounter and in later encounters. The detailed course of the behaviour in the dominance tests has yet to be studied (with more easily recognizable individuals!) before this explanation can be confirmed. However, the overall effect of colour and experience as the result of an interaction of two separate factors is well established by the data in Table 5.

The possible role of aggressiveness, as assessed in our aggression tests, is more difficult to trace in the available data. For one thing, the analysis is handicaped by the uncertain relationship of colouration and aggressiveness: Table 9. In one line (DL) the two factors seem to be correlated (though not significant) and the brighter males tend to be more aggressive. In the other line (DH) no such correlation can be demonstrated. Since the groups, even within DL, are not at all homogeneous the evidence remains inconclusive, but in any case the correlation, if any, is not very important. Therefore we have tried to play off colour against aggressiveness like we did with colour and experience.

In Table 10 first dominance tests are arranged according to the difference in colour of the two males and/or to their difference in aggressiveness. When colours are different, it appears that with one exception the brightest always wins the contest, even when he has the lowest score for aggressiveness (4 out of 10, in 2 cases aggression scores are equal). Even when colours are equal, it seems that aggressiveness does not decide the outcome of a contest: in 1 out of 3 cases the most aggressive wins, but in the other 2 the aggression scores are equal, and in 2 more cases the contest remains undecided, even though in 1 case there is a pronounced difference in aggressiveness.

Though further evidence is certainly needed, our final conclusion is that aggressiveness as defined and measured by us, plays an almost negligible role in determining the outcome of a dominance test, which is easily masked by colour and experience. This conclusion is in agreement with the fact that our genetical selection for dominance has not brought

TABLE 9. Spearman rank correlation coefficient between territorial aggression score and colouration score

Group	Correlation aggression-colouration	
D	+ 0.35	+ 0.00 (DH-line)
F	- 0.51	
E	- 0.26	- 0.36 (DL-line)
G	- 0.69	

All coefficients are not significant.

TABLE 10. Possible influence of territorial aggression score on the outcome of first dominance tests (Groups D, E, F and G)

Colouration		Aggression	
dominant	inferior	dominant	inferior
1	2	31.8	60.3
2	4	38.2	62.0
3	4	24.7	59.5
2	3	27.9	41.6
2	4	32.9	34.6
2	1	52.6	45.4
1	3	70.6	44.7
2	4	72.1	6.1
2	4	30.0	10.2
2	4	44.6	43.3
2	4	57.6	26.7
2	3	47.2	27.2
1	2	50.8	7.4
2	2	62.2	48.7
2	2	57.7	59.4
2	2	45.4	47.7
Undecided			
2	2	43.3	37.7 a)
2	2	35.6	18.7 b)

a: males still roundaboutfighting after 45 min; b: no sign of aggression after 20 min.

about a clear shift in aggressiveness, but has caused a conspicuous change in colouration.

In the introduction "mouth-fighting" was described. It was a surprise to find that this rare type of fighting was observed on several occasions in the dominance tests of DH and DL, but always in connection with roundabout fighting. Remarkably, both types of fighting showed little, if any, influence of colouration, but were strongly dependent on the experience of the eventual loser. It has already been mentioned that roundabout fighting is more frequently observed in the DH line. For these regularities no explanation can be offered.

Summary

When two males of the three-spined stickleback (*Gasterosteus aculeatus* L.) in reproductive condition are placed in a tank of small size, one becomes dominant over the other in a very short time. In this way a group of isolated reproductive males were tested pairwise on their capacity to become dominant. It appeared that the males could be arranged in a linear order of dominance. Their rank was to some extent correlated with their territorial aggressiveness as assessed in other tests. Their experience in previous dominance tests played a more important role. The brightness of their colouration, however, was probably the most consistent and decisive determinant of dominance. The interaction of colour and experience could be traced in the outcome of the dominance tests. The occurrence of roundabout fighting is discussed.

References

Assem, J. van den (1967). Territory in the three-spined stickleback, *Gasterosteus aculeatus* L. An experimental study in intra-specific competition. — Behaviour Suppl. 16, p. 1-164.

—— & Molen, J. N. van der (1969). Waning of the aggressive response in the three-spined stickleback upon constant exposure to a conspecific. I. A preliminary analysis of the phenomenon. — Behaviour, 36, p. 286-324.

Bertin, L. (1925). Recherches bionomiques, biométriques et systématiques sur les épinoches (Gasterosteidés). — Ann. Inst. Océanogr. Monaco 2, p. 1-204.

Gandolfi, G., Gandolfi, R. & Le Moli, F. (1973). Short-term individual recognition in *Padogobius martensi* (Teleostei, Gobiidae). — Atti Accad. Naz. Lincei, 54, p. 281-285.

Huntingford, F. A. (1976). The relationship between anti-predator behaviour and aggression among conspecifics in the three-spined stickleback, *Gasterosteus aculeatus*. — Anim. Behav. 24, p. 245-260.

Iersel, J. J. A. van (1953). An analysis of the parental behaviour of the male three-spined stickleback (*Gasterosteus aculeatus* L.). — Behaviour Suppl. 3, p. 1-159.

—— (1958). Some aspects of territorial behaviour of the male three-spined stickleback. — Archs néerl. Zool. 13, Suppl. 1, p. 381-400.

Marler, P. (1955). Studies of fighting in chaffinches: (2) The effect on dominance relations of disguising females as males. — Brit. J. Anim. Behav. 3, p. 137-146.

Muckensturm, B. (1968). La réaction des epinoches aux leurres ne provient pas d'une confusion avec un congénère. — C.R. Acad. Sc. Paris 266, (Série D), p. 2114-2116.

—— (1979). Les réactions agressives des épinoches en lumières colorées. — C.R. Acad. Sc. Paris 289, (Série D), p. 1065-1067.

Peeke, H. V. S., Wyers, E. J. & Herz, M. J. (1969). Waning of the aggressive response to male models in the three-spined stickleback (*Gasterosteus aculeatus* L.). — Anim. Behav. 17, p. 224-228.

Pelkwijk, J. J. ter & Tinbergen, N. (1937). Eine reizbiologische Analyse einiger Verhaltenswiesen von *Gasterosteus aculeatus* L. — Z. Tierpsychol. 1, p. 193-200.

Rohwer, S. (1975). The social significance of avian winter plumage variability. — Evolution 29, p. 593-610.

DO THE DUMMY EXPERIMENTS WITH STICKLEBACKS SUPPORT THE IRM-CONCEPT?

by

GERARD P. BAERENDS[1])

(Zoological Laboratory, University of Groningen, The Netherlands)

(With 4 Figures)

(Acc. 7-IX-1984)

Introduction

When in the beginning of this century the wish arose to study animal behaviour with objective scientific methods a need was soon felt for obtaining insight into the sensory capacities of various animals and in the use they actually make of the information present in the environment. Von Frisch in particular was a pioneer in developing methods to let animals make known whether they are sensitive to specific physical and chemical factors and to which degree. Moreover, in his dispute with von Hess (1913) about the colour sensitivity of bees, von Frisch (1914) was among the first to teach us that an animal may use a specific sensory modality in one functional context while not in another. But it was von Uexküll (1909) who in particular opened our eyes to the fact that between animal species important differences exist in their responsiveness to the various elements of the same surrounding world. Part of these differences are due to the capacities of the sensory organs, but many of them have also to be attributed to stimulus selection by more centrally located mechanisms processing the information received.

After it had occurred to Lorenz that the structures and activities used by many animals to communicate with each other ('Auslöser' or social releasers) had during evolution developed under the selection pressure of the perceptual processes mentioned above, he coined the name 'auslösender Mechanismus' (releasing mechanism) to denote the machinery underlying these processes in the reactor (Lorenz, 1932,

[1]) I feel greatly indebted to Dr Nicholas E. Collias for his efforts to recover his 1949 data, for his permission to publish and use them in the present paper and for his critical comments on a draft for this manuscript. I further wish to thank Dr R. H. Drent for critically reading the manuscript, and Mrs H. Lochorn Hulsebos for typing it.

1935, 1937). He assumed this machinery to develop independently of experience with the adequate object and hence he attached the adjective innate to it (innate releasing mechanism, IRM). The concept was soon extended beyond the context of communication, to the machinery enabling an animal to respond to any stimulus situation in its environment vital for its survival, such as food items or nesting material.

The formulation of the IRM-concept initiated in the thirties many attempts at defining its properties in specific cases. For such experiments it was necessary to control and vary the features of the situation or object to be perceived and evaluated by the animal. An obvious way of doing this is replacing it by dummies of which the features can easily be manipulated and changed. Already von Uexküll's pupil Lissmann (1933) used this technique in studying stimulus control in fights of the fish *Betta splendens*.

In 1933 Niko Tinbergen was charged with the task of developing the new discipline of ethology at the University of Leyden. He chose his subjects of study from species readily available in the environment and with which he already had considerable experience. The three-spined stickleback and the herring gull were on the top of his list and the causal analysis of releasing situations was one of the main issues for investigation.

When in 1934 the experiments with sticklebacks started, a young student and good friend of Niko and his younger brother Lukas, Jan Joost ter Pelkwijk, very enthusiastically assisted in the proceedings. His artistic gifts paved the way to the fabrication of wax dummies of all sorts and in various degrees of perfection. It was, however, a surprise to find that for evoking social responses a correct representation of the features of a potential rival or mate was not a prerequisite. Even very rough dummies, hardly resembling a real fish, could elicit social responses in male and female sticklebacks, provided they exhibited a few morphological features characteristic to the species and were moved about in a specific way (ter Pelkwijk & Tinbergen, 1936, 1937). Which features were required can nowadays be found in almost any textbook on ethology. To elicit in a ripe female 'following towards the nest' a dummy had to 'zigzag' and to show red. Red was also found to induce attack in territorial males, especially when the dummy was moved up and down in a vertical position. Dummies with red were most effective when this colour was present on the ventral side. It was suggested that a blue dorsal side and an iridescent eye patch would add importantly to the value of a dummy, but to my knowledge this has never been properly checked. To be courted by a male a dummy should lack red and rather be entirely silvery coloured. A swollen abdomen considerably increased the effectiveness of such a dummy.

I remember that these results came out very convincingly when the test fish were first selected according to their readiness to perform in dummy experiments, and when these puppets were played in the correct manner and with a good deal of variation. It was probably because of the unambiguity of the results obtained in this way that no need was felt to report in detail on these experiments. Although already at that time TINBERGEN used to stress the importance of standardization and statistics in the design of such experimental work, quantitative data in support of the conclusions have never been published; neither are they in file in the Leyden laboratory.

Later, several investigators have, largely independently, tried to reproduce and extend these experiments in a rigidly designed way. The IRM-concept indeed deserved and needed much further testing. Would the IRM really select only a very limited number of features of the environment the animal's sensory organs were able to detect? Was it sensitive to 'Gestalt'? What were the rules along which the different features for which the IRM was sensitive were combined into the final output? The stickleback seemed an excellent subject for obtaining answers to these questions. I have myself tried such experiments with different co-workers, but we never obtained unambiguous results. I know of colleagues who had similar experiences.

TINBERGEN's assertion that dummies showing red are—irrespective of their shape—more vigorously attacked than dummies lacking this colour (TINBERGEN, 1939, 1942, 1948a, b, 1951), was quantitatively tested by Dr Nicholas E. COLLIAS when in 1949 he spent some time as a visiting scientist in TINBERGEN's laboratory at Leyden. There he worked in close contact with Jan VAN IERSEL and, remembering the assistance he received from Jan during that period, Dr COLLIAS very kindly looked up the data he collected at that time and put them at my disposal for use in this symposium. The results of four other studies dealing with the effect of the red belly in releasing aggression have been published (CHAUVIN-MUCKENSTURM, PEEKE *et al.*, ROWLAND, and WOOTTON; see below). In a number of aspects these results seem to disagree with TINBERGEN's conclusions. In the following I shall first discuss the data obtained in these studies, considering their value as arguments pro or contra the IRM-concept. Next I shall try to interpret the apparent controversies by means of a comparison of the findings in sticklebacks with those I obtained in a study of egg-

TABLE 1. Dr COLLIAS' data on the influence of a red belly on the release of attack in three-spined stickleback males

Individuals	a	b	c		d	e	f	g		h	i		j	
Red	—	—	59	—	—	—	83	—	1	6	0	—	7	—
No red	0	0	52	49	57	2	63	0	3	3	1	5	2	0
Red	0	0	171	118	37	3	85	59	19	1	4	7	13	2
No red	0	0	144	100	17	2	9	0	3	4	0	2	0	0
Red	0	7	194	113	45	3	20	30	22	2	—	7	—	0
No red	—	—	101	82	—	—	6	0	0	1	—	—	—	—
Indiv. totals red:no red	0:0	7:0	655:446*		82:74	6:4	188:78	131:7		9:8	18:8		22:2	

Each column gives for each of the 10 males (a-j) the number of bites directed at the red-bellied and the control dummy in successive tests. (* = Last test omitted to keep number of tests for each dummy equal).

recognition in the herring gull—another analysis of a releasing situation initiated by Niko TINBERGEN and carried out in a species selected by him. The results of that study led to a revised version of the classic IRM-concept, which in my opinion fits all stickleback data in discussion and accounts for the discrepancies with some of TINBERGEN's statements reported by other authors.

Stickleback data

Dr COLLIAS' data are summarized in Table 1; on the one hand they contain evidence confirming TINBERGEN's conclusion, but on the other hand I think that they illustrate well the reasons for scepticism about the general validity of the IRM-concept. Two types of dummies were used, both roughly fish-shaped, one with and one without red on the ventral side. These dummies were, suspended on a thin wire but not moved, presented to 10 males in 66 tests, 7-13 cm from the nest. The tests lasted 5 min and were carried out at 10 min intervals. The two dummies were alternately presented so that, to minimize the influence of fluctuations in responsiveness, the rate of biting against a red-bellied dummy could be compared with the average biting rate in the preceding and the following presentation of the neutrally coloured one.

Of the 18 comparisons possible 15 are consistent with respect to the aggression releasing effect of red. Also, in the 32 presentations of the red-bellied dummy it got bitten 118 times, whereas the control dummy was bitten 708 times in 34 presentations. Of the 10 males used 9 attacked the red-bellied dummy more often than the neutral one, but in 4 of these males (d, e, h, i) this difference cannot be considered significant. The 10th male (a) did not bite either dummy, but he did lunge repeatedly, showing incipient attack, only at the red-bellied dummy.

Dr COLLIAS felt that his data confirmed TINBERGEN's statement about the releasing value of red for attack. However, the very considerable quantitative differences in responsiveness between individual males, the apparent indifference towards red in some males (d, e, h), and the high number of biting responses received by the neutral dummy (39%), indicate that the response to a rival must be controlled by more factors than merely the input from an IRM sensitive to red. It is clear that the validity of the IRM-concept will become more convincing when the causes for the variability in the test results could be understood.

Almost 20 years later MUCKENSTURM (1967a, b, 1968a, b, 1976) again tested in various ways the effectiveness of red for inducing attack. She obviously found it difficult to accept the simple IRM-model of the 'objectivists school', a critical attitude which was at least partly inspired by her experiences with the high capacity for visual discrimination of the stickleback (such as responses to the very thin thread to which a dummy was suspended or to inconspicuous markings attached to a fish to facilitate individual recognition by the observer) and to its ability to learn about and cope with changes in the external situation.

In an early series of experiments (MUCKENSTURM, 1967b) she presented in successive tests spool-shaped, either evenly red or evenly grey dummies, and fish-shaped dummies painted either in naturalistic fashion (*i.e.* ventrally red and dorsally bluish and with a brightly blue eye) or with an unnatural colour combination (violet and yellow). She found the naturalistically coloured dummy to receive the smallest number of attacks, only about 50% of the number directed at each of the others. During presentation of the red-bellied dummy the incidence of 'interruptive' or 'conflict' activities was very much higher than with the other ones. Later (MUCKENSTURM, 1968a, b, CHAUVIN-MUCKENSTURM, 1976) she compared fish-shaped models painted ventrally either red, violet, yellow or grey. When the data for all fish were taken together, the violet dummy turned out to have received the highest number of attacks. The yellow dummy came second and the red one, with a much lower score, third. This time grey obtained the least responses. When approach was taken as criterion, yellow scored above violet. Individuals varied in preference, but tended to stick to their initial choices in later series of tests. The author came to the conclusion that the results of the dummy tests gave no information about how rival conspecifics are recognized, but only on the relative effectiveness of the different stimuli for attracting attention in exploratory behaviour.

Rowland (1982) compared, in simultaneous and in successive presentations, the incidence of bites directed to uniformly grey (Munsell 6/0) dummies (castings) and similar shaped dummies in which throat and forebelly were red (Munsell 5R4/14). In each of these series an overall analysis of the data showed a significantly greater number of bites at the plain grey dummies. The apparent contradiction of his results with those of Chauvin-Muckensturm may be due to the absence of a coloured alternative for red in Rowland's tests. The responses varied considerably between individuals. In about half of his fish Rowland did not find a significant preference; only one of them exhibited (in accordance with theory) a significant preference for red.

Considerable individual differences were found by Chauvin-Muckensturm (1976) in the tendencies to exhibit aggressive, escape, exploratory or interruptive behaviour during the tests. To what extent such differences could explain the individual variations in response, can unfortunately not be deduced from her data. However, the high incidence of interruptive behaviour in tests in which a red bellied dummy was presented, indicate that red induces a conflict between opposing tendencies. This suggests that differences in motivational state were responsible for the individual differences both authors established.

Data on the effect of the degree of extension of the red colour on a dummy were obtained by Peeke, Wyers & Herz (1969) by means of measuring the waning of biting at 5 differently coloured spindle shaped dummies (no red, red chin, red belly, three quarter red, totally red) across 12 days after the test male had completed a nest. The experiments were based on the idea that habituation is stimulus specific. No significant differences were found between the responsiveness to those different models, either with respect to initial aggression on the first day, or in the rate of habituation on the following days. However, to me the habituation curves suggest that the rate of waning for the dummy without red was somewhat slower than that for the others. These authors also found a considerable number of non-responders, which all exhibited escape behaviour.

Tinbergen's experiments had indicated that a red belly would elicit more attack responses than a red back. Wootton (1971) tried to apply this finding when, for the purpose of comparing different measures for aggressiveness, he carried out series of tests in which different kinds of rivals were presented to a territorial male on each of 12 days after fertilization. In one of these tests the rival was either a fully red, a fully silver, a red-backed or a red-bellied dummy. The outcome with the red-bellied rival was closer to that with a live male than the results obtained with the other dummies. Again considerable differences in the responsiveness to the dummies between the individual subjects were established. CHAUVIN-MUCKENSTURM (1976) also compared responsiveness to a red back and a red belly. She did so in a complicated series of experiments in which in addition two other variables (experience with conspecifics, and position of the longitudinal axis of the model) were also manipulated, and in which information on agonistic, sexual, exploratory and interruptive behaviour was collected. The overall results of these experiments do not reveal a difference in the attack evoking effect of ventral *versus* dorsal red. However, red-bellied dummies more often induced exploratory behaviour than red-backed ones. We shall return to this finding in the end of this paper.

CHAUVIN-MUCKENSTURM (1976) confirmed the statement of TER PELKWIJK & TINBERGEN (1936, 1937) that features of shape of a dummy are of little importance in determining attack responses, although she found an anaesthetized fish more effective than any dummy. She also confirmed that the vertical posture of a threatening male is strongly attack inducing. She found this position far more effective in releasing attack than a neutral horizontal posture. Consequently, Mrs CHAUVIN's critique of the IRM-concept mainly refers to the role of red in releasing aggression in stickleback males. The apparent inconsistencies in the results of experiments on this feature need to be understood and their consequences for the IRM-concept considered. In my opinion our work on 'egg-recognition' in the herring gull can help us to an understanding of the inconsistencies. This work has led to the proposal of some modifications of the IRM-concept by which the difficulties can be met.

Herring gull data

Our investigation as to the features of the herring gull egg which are effective in evoking incubation behaviour had the technical advantage that movement was not involved. Moreover, after pilot experiments with dummies made of different kinds of material and painted with various sorts of dyes had convinced us that chemical stimuli and specific weight were very unlikely to play a rôle, we could restrict the analysis to visual stimuli. Two-dimensional objects were never treated as eggs and three-dimensional transparent ones only rarely. A hard surface resisting means of the IRM only. The more apprehensive it is, the more it will rely on detailed knowledge stored in its memory. Only in rapid, often preliminary, responses it may exclusively use the IRM and thus respond to roughly made dummies. At such times it will also be sensitive to supernormal stimulation. This makes it understandable that animals with highly developed sensory capacities and memories, including humans, can be susceptible to supernormal stimulus situations. The disadvantage that in supernormality responses the chance of a 'false start' is greater than when a comparison with the engram is made first, may often outweigh the risk of the more time consuming information processing procedure involved in the latter. However, we do not know how great this time difference is; it should be emphasized that it is certainly of a much smaller order than the latencies we measured, and which, therefore, may only be used as a parameter for motivational state, not for the complexity of processing.

If this theory is true we can only study the IRM in most vertebrates after we have cut out the influence of memory. We have done this in the herring gulls by selecting for our experiments gulls that were relatively little scared, by counteracting conditioning to any dummy during the experiments, and by presenting to each gull a bewildering range of differently looking dummies.

Application of the herring gull model to the stickleback case

In the little standardized, early stickleback experiments the fishes that were found suitable for the tests are likely to have been similarly 'brainwashed'. Those who became not sufficiently adapted to the tests and retained a certain degree of shyness went on using their memory, while the input from IRMs serving social functions was throttled down. I suspect that these were the fish that made it difficult for some of the authors mentioned above to believe in the existence of IRMs.

So far the discrepancies with TINBERGEN's statements only concerned the effect of red on the responses of males. In discussing this responsiveness one seems to have insufficiently realized that, if the red belly has evolved for the function of repelling other males, it should on the whole reduce the number of attacks released by and directed at a red coloured male and only increase it when the rival operates from a relatively strong, if not superior, position. It is not surprising that many males in the tests avoided the red dummy; they are likely to have had a relatively high tendency to escape. By selecting those males that are not intimidated by the manipulation of the dummies, one works with test fish in which the input from the selector is interfered with little or not at all. SEVENSTER (1949) recognized this phenomenon when he compared the incidence of biting at an optimal (black body with white spines) and a suboptimal (black without spines, or grey) dummy by different males of the ten-spined stickleback (*Pygosteus pungitius*), and interpreted the individual variations he found in essentially the same way.

We should now ask why such controversial results as obtained for biting at red were not found for attacking a dummy presented in a vertical position, thus in what is generally believed to be a threat posture. This might be due to this display not only giving information about sex and condition of the performing fish, but also about its intention to stay at and defend a particular area or place, thus to a readiness to actively challenging competitors.

With my thesis that, as a rule, no experienced vertebrate is likely to rely on IRMs only when reacting to situations in its environment, I do not mean to suggest that the IRM would be an artifact. In my opinion, it does exist, but to study it, it first needs—in a sense—to be unpacked. So far we have done this unintentionally; it will, however, be better to do it with carefully designed methods. At the least one should no longer lump the results obtained in all similar tests and with all individuals undiscriminately, and one should monitor the motivational state of the animal during the tests. Special attention should be given to the development of methods for obtaining reliable results and conclusions from individuals.

Further study of the information processing in the machinery implied by the term 'IRM' is still of great importance. One should no longer see it as a mechanism functioning beside and independently of memory. If the hypothesis given here is in principle correct, the IRM has in many animals become a means by which information about the environment of a species, encoded in the genome through the process of selection and mutation (LORENZ, 1965), can come to expression. The analysis of CHAUVIN-MUCKENSTURM (1976) contains several indications that fit our hypothesis. She found, for instance, that naive stickleback males show relatively more exploratory behaviour towards a dummy which they have earlier encountered in the vertical threat position than in the neutral horizontal one. Moreover, if this dummy had a red belly, the amount of exploratory behaviour was greater than when the back had been red. Thus, features TINBERGEN attributed to the IRM she indeed found to have a guiding or selective influence on learning about the environment.

The three-spined stickleback has proved to be a suitable subject for dummy experiments, certainly because of the dominant influence vision seems to have in controlling its behaviour, and probably because its tendency to flee is on the average relatively low. It is regrettable that the IRM-experiments of TINBERGEN and TER PELKWIJK have never been really extended and deepened; I hope that the ideas laid down in the model proposed above will help to take them up again, with more sophisticated methods.

Summary

Several investigators have found it difficult to produce in systematic series of tests satisfactory quantitative data in support of the initial findings of TER PELKWIJK & TINBERGEN (1936, 1937) on the properties of IRMs supposed to be involved in evoking social responses in the three-spined stickleback. In particular with respect to the effectiveness of red in releasing aggressive responses, the results were often inconsistent; this paper gives a survey of the data obtained.

On the basis of a comparison of these stickleback data with data obtained in a study of egg-recognition in gulls (BAERENDS & DRENT, 1982), it is suggested that the inconsistencies are due to variation in the motivational state of the test fish, especially in the degree to which a tendency to escape is activated. It is argued that as a rule an experienced animal will in a particular situation not act on the basis of input from an IRM only, but in addition on relevant information encoded in the memory. It is conceived to be the principal function of the IRM to enable a naive animal to roughly identify at first encounter particular stimulus situations in its environment, which are essential to its survival. The animal would further extend its knowledge about these situations through learning.

In the gulls, anxiety was found to reduce the influence of the IRM and increase the relative importance of memorized knowledge. If this conclusion may be generalized to the stickleback case, the failure of individual males in responding aggressively to red is likely to have been due to a dominant influence of memorized knowledge in anxious animals. This implies that to study the content of an IRM it has often to be 'unpacked' first, by reducing the escape factor as much as possible.

References

BAERENDS, G. P. (1962). La reconnaissance de l'oeuf par le Goéland argenté. — Bull. Soc. Sc. Bretagne 37, p. 193-208.

—— (1970). A model of the functional organization of incubation behaviour in the herring gull. — In: BAERENDS, G. P. & DRENT, R. H. (eds) (1970), p. 265-310.

—— (1982). A model of the information processing mechanism. — In: BAERENDS, G. P. & DRENT, R. H. (eds) (1982), p. 339-346.

——, BLOKZIJL, G. J. & KRUIJT, J. P. (1982). The processing of heterogeneous information on egg features inducing the retrieval response. — In: BAERENDS, G. P. & DRENT, R. H. (eds) (1982), p. 212-224.

——, BRIL, K. A. & BULT, P. (1965). Versuche zur Analyse einer erlernten Reizsituation bei einem Schweinsaffen. — Z.f. Tierpsychol. 22, p. 394-411.

—— & DRENT, R. H. (eds) (1970). The herring gull and its egg. Part I. — Behaviour suppl. 17, 312 pp.

—— & —— (eds) (1982). The herring gull and its egg. Part II. — Behaviour 82, 416 pp.

—— & KRUIJT, J. P. (1973). Stimulus selection. — In: HINDE, R. A. & STEVENSON-HINDE, J. (eds) (1973), p. 23-50.

INTERTIDAL BREEDING AND AERIAL DEVELOPMENT OF EMBRYOS OF A STICKLEBACK FISH (GASTEROSTEUS)

by

J.F. MACDONALD, J. BEKKERS, S.M. MACISAAC[1]) and D.M. BLOUW[2,3,4])

(Department of Biology, St. Francis Xavier University, Antigonish, Nova Scotia, Canada B2G 2W5)

(With 7 Figures)
(Acc. 25-V-1995)

Summary

Males of the newly discovered 'white' stickleback disperse their embryos over the filamentous algae in which they nest and, unlike any other stickleback, they provide no subsequent parental care. Previously known populations of white sticklebacks nest only in shallow subtidal waters where filamentous algae are abundant. Our purpose in this paper is to describe highly divergent populations at Spry Bay, Nova Scotia, Canada, in which males build nests on bare rock in the subtidal and intertidal zones, and disperse their embryos over bare rock. Intertidal embryos are exposed to air and are vulnerable to desiccation at low tide. Field studies reveal that nests are built from locally available materials directly on the rock substrate. They are constructed very quickly and tend to be less substantial than subtidal nests. Internest distances are shorter in the intertidal zone than in adjacent subtidal areas. Dispersed intertidal embryos tend to settle into crevices between stones where the microenvironment remains moist and temperate between tides due to the presence of organic detritus and shade. Such embryos survive, develop, and hatch. Similar behaviours occur at other sites in Nova Scotia, and we know that it has persisted for at least four years at Spry Bay. We conclude that breeding and dispersing embryos over bare rock substrate, and intertidal breeding, are novel but stable breeding repertoires. We hypothesize that they have evolved secondarily to the evolution of emancipation from parental care in the white stickleback.

Introduction

Male parental care was thought to be universal and invariant in stickleback fishes. However, the discovery of a new form with brilliant, irides-

[1]) New address: Department of Biology, University of Ottawa, Ottawa, Ontario, Canada K1N 6N5.

[2]) To whom correspondence should be addressed.

[3]) New address: Biology Programme, University of Northern, British Columbia, 3333 University Way, Prince George, British Columbia, Canada V2N 4Z9.

[4]) We are very grateful to T. Bakker and P. Sevenster for their helpful comments on the manuscript and for organizing a successful and enjoyable conference. V. Grant and A. L. MacDonald provided excellent assistance in the field and in the lab respectively. Financial support was from the St. Francis Xavier University Council for Research and NSERC (A2324 to DMB and an Undergraduate Summer Research Award to SMM).

cent white breeding colours, the 'white' stickleback, shows that this is not the case (BLOUW & HAGEN, 1990; BLOUW, in press). White stickleback males typically compete for territories, build nests above the substrate in filamentous algae, court females, and then fertilize the eggs. However instead of guarding and aerating the embryos as do all other sticklebacks, white stickleback males actively remove the embryos from the nest and disperse them over the surrounding filamentous algae. The embryos are abandoned and develop without parental care (BLOUW, in press). This emancipation behaviour enables white stickleback males to remain sexual for long periods during the breeding season. In contrast, the threespine stickleback, *Gasterosteus aculeatus*, endures forced asexual parental periods that may extend for several weeks between short sexual bouts (JAMIESON *et al.*, 1992a).

Mate choice experiments and field observations show that the white stickleback is reproductively isolated from its presumed immediate ancestor, *G. aculeatus*. There is no single morphological (BLOUW & HAGEN, 1990) or electrophoretic (HAGLUND *et al.*, 1990) trait which enables unambiguous discrimination between them. The behavioral variants reported here contribute further evidence of the extensive behavioral and ecological differences that have evolved between the white stickleback and *G. aculeatus*. The white stickleback has not been formally recognized as a distinct taxon in keeping with the practice of including new variants, even those known to be reproductively isolated, within the '*Gasterosteus aculeatus* complex' (MCPHAIL & LINDSEY, 1970; MCPHAIL, 1994).

BLOUW & HAGEN (1990) found the white stickleback to be widely distributed in brackish and salt water throughout northeastern Nova Scotia, and they found that white males always nest above the substrate in filamentous algae. Experiments have shown that the algal environment is permissive of the evolution of emancipation from parental care because filamentous algae provide conditions in which survivorship of embryos may be high (BLOUW, in press). Accordingly it was thought that the presence of filamentous algae is essential for successful reproduction.

However, in 1991 we identified an unusual population of white sticklebacks at Spry Bay, Nova Scotia, Canada, where large numbers of males establish territories and breed on bare rock substrate. This population immediately captured our attention because, unlike all other localities where we had studied the breeding behaviour of the white stickleback, the nest sites at Spry Bay are largely devoid of filamentous algae or any other vegetation. Males nest directly on rock and disperse their embryos over bare rock. Most surprisingly, breeding also occurs in the intertidal zone at Spry Bay; intertidal males abandon their territories and re-establish them at each tide cycle. The embryos of intertidal males are dispersed and stranded intertidally, above the low water line. Thus embryos abandoned in the upper half of the intertidal zone are exposed to the aerial environment longer than to an aquatic environment during their development.

These unexpected observations prompted us to undertake a field study to characterize the breeding repertoires of sticklebacks of the genus *Gasterosteus* at Spry Bay. We have now identified a total of five different repertoires:

i) *Gasterosteus wheatlandi* has broadly similar subtidal breeding behaviours to *G. aculeatus* (McInerney, 1969; Wootton, 1976). However, because *G. wheatlandi* is readily distinguishable from the other sticklebacks and because its specific breeding behaviours are clearly distinct from those reported here it will not be considered further.

ii) Classic breeding by the threespine stickleback, *Gasterosteus aculeatus* (van Iersel, 1953; Wootton, 1976), in the subtidal zone, nesting directly on substrate, with highly developed parental care.

iii) Subtidal nesting on filamentous algae with no parental care by the white stickleback, hereafter called the '*algal-breeding white stickleback*' (as described by Blouw & Hagen, 1990; Blouw, in press).

iv) A previously undescribed repertoire which includes subtidal nesting on bare rock substrate, without parental care, by sticklebacks with vivid white dorsal nuptial colours ('*rocky subtidal white stickleback*').

v) A second undescribed repertoire with intertidal nesting on bare rock substrate, also without parental care, by sticklebacks with vivid white dorsal nuptial colours ('*rocky intertidal white stickleback*').

The purpose of this paper is to describe the highly unusual breeding behaviours of the sticklebacks at Spry Bay and to characterize the environments in which these behaviours occur. We focus principally on the most divergent form, the rocky intertidal white stickleback.

Methods

Study site.

The study area is a tidal basin with two freshwater streams entering from the north. It is separated from the ocean by the raised bed of a highway (Fig. 1). Tidal water circulates freely between the ocean and the basin through a channel formed by a timber crib-work populations, but they are useful here to contrast the breeding repertoires of the sticklebacks at Spry Bay in broad terms.

Breeding *G. aculeatus* males turn bluish-green dorsally and red ventrally as they begin to defend territories and build nests (day 0; Fig. 2). The nest is built on the substrate in an area devoid of algae (Jamieson *et al.*, 1992b) and is assembled from bits of vegetation carried to the nest site. After nest construction is complete (day 2), the male performs territory and nest maintenance and courts gravid females (day 3-7 or 8) with a series of zigzag swimming patterns. After the first spawning the male resumes courtship of prospective mates for three or four days. Thereafter, he suspends courtship and devotes himself exclusively to guarding and aerating the embryos (day 8 or 9-13) and, after hatching, to guarding the fry (day 14-23) (McLennan & McPhail, 1989; Wootton, 1976).

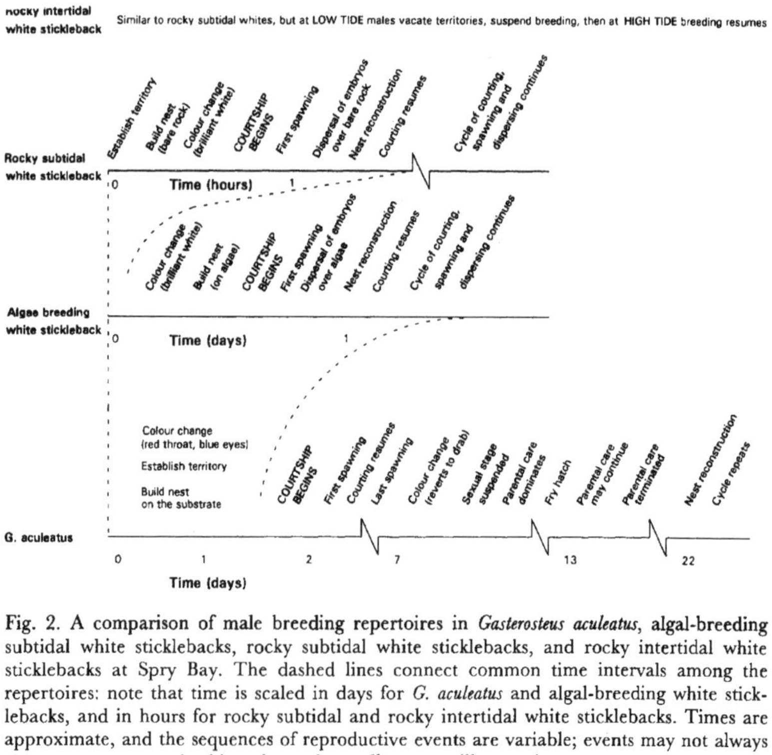

Fig. 2. A comparison of male breeding repertoires in *Gasterosteus aculeatus*, algal-breeding subtidal white sticklebacks, rocky subtidal white sticklebacks, and rocky intertidal white sticklebacks at Spry Bay. The dashed lines connect common time intervals among the repertoires: note that time is scaled in days for *G. aculeatus* and algal-breeding white sticklebacks, and in hours for rocky subtidal and rocky intertidal white sticklebacks. Times are approximate, and the sequences of reproductive events are variable; events may not always occur in this order and not all events will occur in every case.

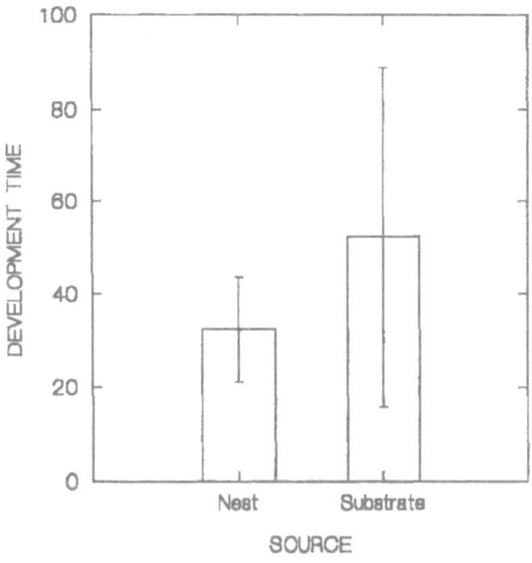

Fig. 7. Mean (± range) development time of embryos (N = 211) recovered from nests and from substrate (N = 69) in the intertidal zone.

opment of the notochord, were observed); 29% had presumably hatched (only empty chorions remained); while only 13% had desiccated and 10% died of unknown causes. Thus embryos can survive and develop to hatching in the intertidal zone despite the periodic absence of water.

In a separate investigation at site 1 samples of substrate including rocks, stones, detritus, and other materials were scooped from the intertidal zone in an area where males were seen to be breeding. In the laboratory the samples were examined for the presence of embryos. A total of 69 embryos, all single and in living (before preservation) condition, were found. Development time was estimated to be a maximum of 92 hours, a minimum of 20 hours and a mean of 52.3 hours. The embryos in the substrate samples showed greater development than those found within nests, but there was also more variation in stages of development (Fig. 7). We cannot exclude the possibility that both are due, in whole or in part, to differences in temperature regime between the two sampling periods.

Discussion

The behavioral diversity we observe in *Gasterosteus* at Spry Bay is greater than any reported previously (reviewed by Foster, 1994). At least three reproductively isolated forms occur within the tidal basin (*G. wheatlandi*, *G. aculeatus*, white stickleback). Among the white sticklebacks at Spry Bay some breed in, and disperse embryos over, filamentous algae. Others breed on bare substrate in the subtidal zone and disperse their embryos over rocky substrate. Still others breed and disperse embryos in the intertidal zone where they develop at least part of the time in an aerial environment. Although it is unknown whether either the rocky subtidal white stickleback or the rocky intertidal white stickleback mate assortatively, experimental evidence (MacDonald *et al.*, this volume) shows that breeding habitat selection, which may be either a prelude to or an integral component of assortative mating (Diehl & Bush, 1989), does occur.

The polygynous, sex-role reversed, male parental care breeding system of the sticklebacks has been evolutionarily conserved for a long period of phylogenetic history (Brooks & McLennan, 1991). Therefore it is of considerable interest to discover the reasons for the renewed variance and diversification in the reproductive behavioral traits that we observe in the white stickleback. Blouw (in press) provides evidence that the filamentous algae in which the algal-breeding white sticklebacks nest is permissive of the evolution of emancipation from parental behaviour. This is because embryos that are dispersed and abandoned in filamentous algae seem to survive well and develop normally. But which evolutionary factors facilitated the evolution of dispersal of embryos over bare substrate in the absence of the nurturing filamentous algae? And which factors facilitated the evolution of intertidal breeding and aerial development of embryos?

We offer the hypothesis that the evolution of these latter behaviours was conditioned by the evolution of emancipation from parental care of the white stickleback, and that it occurred in a stepwise manner. First, the white stickleback differentiated from *G. aculeatus* in environments with abundant filamentous algae (Blouw & Hagen, 1990) and, in these environments, the evolution of male emancipation occurred because of the permissive qualities of the filamentous algae (Blouw, in press). Emancipation behaviours may have evolved either contemporaneously with divergence of the white stickleback from *G. aculeatus*, or after the evolution of reproductive isolation and preference for breeding above the substrate in filamentous algae. Second, the white, algal-breeding stickleback with male emancipation evolved to nest on bare substrate and to disperse embryos over bare substrate. Third, the rocky subtidal white stickleback evolved to invade the intertidal zone. We suggest, then, that the rocky intertidal white stickleback is the most recently diverged form of *Gasterosteus* at Spry Bay.

What is the evidence for this evolutionary scenario? For the first step, BLOUW & HAGEN (1990) and BLOUW (in press) provide evidence that is consistent with the hypothesis that the white stickleback evolved in environments dominated by filamentous algae, and that the evolution of emancipation was made possible by the qualities of filamentous algae which suit it admirably as a surrogate parent to stickleback embryos.

The second evolutionary step, breeding on bare substrate by emancipated males, may have occurred because of the limited availability of habitats with an abundance of filamentous algae. The density of white algal-breeding sticklebacks is very high during the breeding season (BLOUW & HAGEN, 1990) and, although we have not made quantitative estimates in any geographic area, habitats with bare substrate appear to be more common than ones dominated by filamentous algae in coastal Nova Scotia. Therefore white algal-breeding males may experience severe competition for a limited number of nest sites on filamentous algae. If rocky subtidal white sticklebacks evolved from algal breeding white sticklebacks by competitive exclusion then we predict that, in controlled laboratory experiments, the rocky subtidal males should be competitively inferior. MACDONALD *et al.* (this volume) show that this is so with fish from Spry Bay. Algal breeding males from the seaward side of the highway behaviourally dominate rocky intertidal and subtidal males so completely that the latter are most often unable to establish a territory or nest when housed in the same aquarium. In no case did a rocky subtidal or rocky intertidal male secure a spawning when in competition with an algal-breeding white male.

An alternative hypothesis for abandoning algae in favour of spawning on bare substrate relates to the fact that filamentous algae, as living organisms, are not seasonally constant in abundance. If there is a fitness advantage to reproducing early in the season, perhaps because offspring produced early suffer less severe competition than those produced later, the filamentous algae may not yet be abundant and selection may favour males that breed where there is little or no algae. JAMIESON *et al.* (1992b) provide evidence that this might occur in some populations of white sticklebacks.

The next evolutionary step, abandoning embryos in the highly variable and seemingly inhospitable intertidal zone, at first appears to be highly prejudicial to fitness. However our evidence suggests that spawning in the intertidal zone may not be maladaptive, largely because the crevice environment into which the embryos fall is more temperate than the surface environment and because it remains moist due to the presence of abundant organic detritus. We suggest that there may be several potential advantages to spawning in the intertidal zone.

First, if breeding space is limiting as argued similarly for the algal-breeding white stickleback (above), then the intertidal zone may offer reproductive opportunity for males that would otherwise not be able to attain a territory to reproduce. Possession of a territory does not guarantee reproductive success, however it is a requisite. Therefore a territory in the intertidal zone may be better than no territory at all if progeny produced in the intertidal zone survive. Our results show that intertidal embryos do survive and hatch.

Second, access to females may increase in the intertidal zone. As the tide floods females move with it, possibly to take advantage of an expanding feeding area. If, as seems likely in the absence of parental care, the number of females available is a limiting factor in male reproductive success then, by following females into the intertidal zone and courting them there, intertidal males may secure more spawnings than subtidal males. In other words at high tide the intertidal zone may offer a habitat with a more favourable operational sex ratio for male fitness than the adjacent subtidal zone.

Third, intertidal spawning may be advantageous by accelerating embryo development. Development rate is related to temperature, but extreme high temperatures are fatal (VAN IERSEL, 1953). Our results suggest that the temperature in intertidal crevices is higher than that in the adjacent aquatic environment, but it is not extreme like that on the adjacent rock surface. Assuming that rapid progression through the vulnerable embryonic stage is advantageous, the intertidal crevice environment into which embryos settle when dispersed may therefore enhance fitness. Further comparative experiments on the fate and performance of dispersed embryos in different environments are clearly required to evaluate relative fitnesses. It is also possible that, when exposed to the aerial environment, oxygen may be more readily transported into the developing embryo. If so, developmental processes may also be accelerated. This assumes that oxygen diffusion through the aqueous boundary layer at the chorion is greater when exposed to air than water. Experiments in which embryos are exposed to varying periods of aerial development at the same temperature would be informative.

Behavioral precedents for intertidal nesting occur in the ancestral *G. aculeatus*. JAMIESON *et al.* (1992b) noted that at Little Narrows, Nova Scotia, *G. aculeatus* may nest in very shallow water and that, at low tide, the backs of males attending nests, and even some nests, were out of water. P. SEVENSTER (pers. comm.) has observed similar occurrences in Porvoo, Finland, and similar stranding of nests has been reported for *Spinachia spinachia*, the fifteenspine stickleback which is found only in European marine waters (BREDER & ROSEN, 1966; WOOTTON, 1976). BREDER & ROSEN (1966, p. 291) indicate that *S. spinachia* nests may be placed "between the tide marks, the male retreating only as far as necessary and returning when the tidal stage permits". However the embryos are retained within the nest and parental care is well developed.

Given a long phylogenetic stasis in stickleback parental behaviour, embryo dispersal may be viewed as an evolutionary novelty, defined by MAYR (1963) as "any newly acquired structure or property that permits the performance of a new function, which in turn will open a new adaptive zone". We hypothesize (above) that the evolution of embryo dispersal has created evolutionary opportunity to expand from habitats with filamentous algae to those without, and from the subtidal environment to the intertidal. These highly unusual breeding behaviours have been sustained for at least four years at Spry Bay, and the numbers of sticklebacks undertaking them each year is very large. We have observed evidence of intertidal breeding at other sites in Nova Scotia, both on the eastern Atlantic (Jeddore Oyster Ponds) and the northern Gulf of St. Lawrence (French River estuary) coasts. The temporal stability and large numbers of fish undertaking these repertoires at Spry Bay, and their occurrence elsewhere, suggest that the rocky subtidal and rocky intertidal breeding behaviours are stable and adaptive in the environments in which they occur.

Organisms that occupy the intertidal zone must be adapted both physiologically and behaviourally to highly fluctuating environmental conditions (GIBSON, 1982). Others must evacuate as the tide recedes and are classed as 'tidal visitors' (GIBSON, 1982). Intertidal white males fall into this class; they vacate their breeding territories at each low tide. However the embryos remain behind, stranded in a potentially inhospitable environment. It would be interesting to investigate whether specific adaptations have evolved in the embryos of rocky intertidal white sticklebacks to resist desiccation.

It is expected that in highly variable environments parental care should be particularly well developed (CLUTTON-BROCK, 1991). Accordingly, we might expect parental care to be even more highly developed in the variable intertidal zone than in adjacent subtidal areas. But of course it is impossible for the white sticklebacks that breed in large numbers at high tide to provide parental care at low tide – the adults must retreat from the intertidal zone with the water. Nevertheless it is interesting that invasion of the intertidal zone has not been achieved by *G. aculeatus* with its highly developed parental care, but by the white stickleback which disperses embryos. This suggests that embryos aggregated in a nest will have lower fitness in the intertidal zone than those that are individually dispersed. This might be because embryos aggregated in a nest are a more rewarding food resource for intertidal predators than the same embryos when they are spatially separated, or they may be more susceptible to desiccation or suffocation when aggregated in the nest. It may also be because spatially separated embryos suffer a lower rate of fungal infection than eggs that occur in clumps (BLOUW, MACISAAC, GRANT and BEKKERS, unpublished data).

The role of nest quality in female choice of mating partners would be rewarding to investigate at Spry Bay. Since males construct nests from locally available materials (even spruce needles and bits of plastic were noted in some nests) this suggests that females are tolerant of a wide variety of nesting materials when choosing mates. Intertidal nests are often constructed very quickly, and are frequently unsubstantial and flimsy in comparison with those of algal breeding white sticklebacks and *G. aculeatus*. These observations suggest that male characteristics and/or the location of the nest may be more important than nest size or nest composition in female choice. Given their brilliant, iridescent breeding colours, their emancipation from parental care, and their propensity to build nests in highly exposed, visible locations, it seems likely that sexual selection has been an important force in shaping the breeding behaviours of the white stickleback. Accordingly an experimental analysis of the male phenotypes that are preferred by females among the behavioural variants at Spry Bay, and the heritable basis of such preferences, could lead to new insights into evolution through sexual selection.

White stickleback males at Spry Bay eat embryos, and they eat more embryos at mid-season than at other times. Presumably it is at mid-season that males are in greatest need of energy, and that the largest supply of embryos is available. Males ate significantly more dead and undeveloped embryos than well-developed ones at all times during the breeding season. This could be because well-developed embryos are more pigmented and probably more cryptic against the dark rock or detritus substrate. Another possibility is that, if most embryos are eaten shortly after fertilization, perhaps while being dispersed, then most of those in the stomachs will be undeveloped. Finally, it may be that males actively avoid live and developed embryos because their destruction represents a greater fitness cost. Note however that this cost occurs only if males eat their own embryos; if males can discriminate and selectively prey on the embryos of rivals there may be a fitness gain. We have not observed nest raiding by white stickleback males at Spry Bay, so it seems reasonable to conclude that the embryos in the digestive tract are obtained from ingestion of a male's own embryos, or from foraging on dispersed embryos. Given the extended sexual period of the white stickleback and the virtual absence of parental investment in the embryos it is perhaps not surprising that embryo cannibalism is common.

References

Blouw, D.M. (in press). Evolution of offspring desertion in a stickleback fish. — Ecoscience.

— & Boyd, G.J. (1992). Inheritance of reduction, loss, and asymmetry of the pelvis in *Pungitius pungitius* (ninespine stickleback). — Heredity 68, p. 33-42.

— & Hagen, D.W. (1990). Breeding ecology and evidence of reproductive isolation of a widespread stickleback fish (Gasterosteidae) in Nova Scotia, Canada. — Biol. J. Linn. Soc. 39, p. 195-217.

Breder, C.M. Jr. & Rosen, D.E. (1966). Modes of reproduction in fishes. — T.F.H. Press, Neptune City, N.J.

Brooks, D.R. & McLennan, D.A. (1991). Phylogeny, ecology and behavior. — University of Chicago Press, Chicago.

EXPERIMENTS ON EMBRYO SURVIVORSHIP, HABITAT SELECTION, AND COMPETITIVE ABILITY OF A STICKLEBACK FISH (GASTEROSTEUS) WHICH NESTS IN THE ROCKY INTERTIDAL ZONE

by

J.F. MACDONALD, S.M. MACISAAC[1]), J. BEKKERS and D.M. BLOUW[2,3,4])

(Department of Biology, St. Francis Xavier University, Antigonish, Nova Scotia, Canada B2G 2W5)

(With 2 Figures)
(Acc. 26-V-1995)

Summary

Like other 'white' sticklebacks, those nesting intertidally at Spry Bay, Nova Scotia, Canada, remove fertilized eggs from the nest, disperse them over the surrounding area, and provide no subsequent parental care. However the breeding substrate at Spry Bay is devoid of the filamentous algae which are the normal breeding substrate of the white stickleback and which elsewhere facilitate survival of abandoned embryos. Experiments show that embryos dispersed in the intertidal zone at Spry Bay survive surprisingly well despite the complete absence of parental care and the periodic absence of water. Intertidal embryos generally develop faster than embryos in the adjacent subtidal zone, but there are exceptions to this pattern. Intertidal males show site fidelity to their nests despite interrupted access to territory, and they show consistent preference for nesting in the intertidal zone in the field. In the laboratory, and in the absence of tidal flux and competition, males taken from the rocky intertidal zone generally prefer to breed on bare rock substrate even when filamentous algae are freely available. When white stickleback males from an adjacent population which breeds subtidally in filamentous algae are introduced into the experimental aquaria as competitors, they exclude the males that breed on bare rock. The surprisingly high survivorship of embryos in the rocky intertidal zone, the habitat selection of males for bare substrate, and their exclusion by males that breed subtidally on algae, all suggest that the highly unusual breeding behaviours of intertidal white males form an effective alternative breeding repertoire. Because breeding habitat selection has implications for mate selection, the possibility exists that sticklebacks with different breeding repertoires at Spry Bay may mate assortatively.

[1]) New address: Department of Biology, University of Ottawa, Ottawa, Ontario, Canada, K1N 6N5.

[2]) To whom correspondence should be addressed.

[3]) New address: Biology Programme, University of Northern British Columbia, 3333 University Way, Prince George, British Columbia, Canada V2N 4Z9.

[4]) We thank Theo Bakker and Piet Sevenster for helpful comments on the manuscript; the comments of Theo Bakker on analysis and presentation of some of the data were particularly helpful. A. L. MacDonald and V. Grant provided excellent assistance. Financial support was from the St. Francis Xavier University Council for Research and NSERC (A2324 to DMB, and an Undergraduate Research Award to SMM).

 Behaviour 132 (15-16)

Introduction

An unprecedented number of *Gasterosteus* breeding repertoires occur at Spry Bay, Nova Scotia (MacDonald *et al.*, this volume). These include classic *G. aculeatus* and *G. wheatlandi* breeding (as described by Wootton, 1976); subtidal algal breeding of the white stickleback (as described by Blouw & Hagen, 1990; Blouw, in press); subtidal breeding of the white stickleback on bare rock substrate ('rocky subtidal males' of MacDonald *et al.*, this volume); and intertidal breeding of the white stickleback on bare rock substrate ('rocky intertidal males' of MacDonald *et al.*, this volume).

White sticklebacks differ from all other sticklebacks in that they disperse their embryos and do not provide parental care to developing offspring (Blouw, in press). In most places in Nova Scotia white sticklebacks disperse their embryos over the filamentous algae in which they breed, but the rocky subtidal and rocky intertidal white sticklebacks of Spry Bay are highly unusual in that they disperse their embryos over bare rock substrate where there is virtually no vegetation. In the case of intertidal males, this means that the embryos are left stranded at each low tide (MacDonald *et al.*, this volume). Observational evidence shows that despite the seemingly inhospitable conditions of the intertidal zone, embryos may be capable of surviving and developing in detritus-filled crevices that remain moist and temperate at low tide (MacDonald *et al.*, this volume).

The purpose of this paper is to supplement our descriptive and observational studies at Spry Bay with experimental evidence. First we examine the fate of embryos dispersed and abandoned in the intertidal zone by monitoring their survivorship and development when reared in experimental enclosures in the field. Second, we examine whether rocky intertidal males are distinct from rocky subtidal males in terms of habitat selection and site fidelity. Third, we test whether male choice of nesting substrate (filamentous algae or bare rock) in the lab is consistent with choices made in the field.

Methods

Intertidal embryo survival and development.

To experimentally evaluate the effect of intertidal conditions on embryo survival and development, we populated plastic frames surrounded with fine mesh (soft plastic mosquito other was an elaborate nest with much algae removed from the clump and transported to the nest site. The four rock-breeding males that managed to construct nests in competition with algal-breeding males did so only after the algal-breeding males had first established their nest sites. The nests of the rock-breeding males were small, not very elaborate, and built on substrate at the end of the tank farthest from the nest of the dominant algal-breeding males.

Despite displaying courtship behaviour in some trials, no rock-breeding male was able to secure a spawning; females always chose to breed with the dominant algal-breeding male. Note that in most trials there may have been little true choice because only the dominant algal-breeding male provided a nest.

Discussion

The results of the experiments reported here bolster the conclusion that spawning on, and dispersing of embryos over, bare rock substrate by the white stickleback in both the subtidal and intertidal zones at Spry Bay are successful breeding repertoires (MacDonald *et al.*, this volume). The evidence in support of this conclusion includes:

(i) that the behaviour has persisted for at least the four years over which we have made observations at Spry Bay,

(ii) embryos stranded in the intertidal zone develop quickly and they survive well,

(iii) the same behaviours occur at other sites in Nova Scotia (MacDonald *et al.*, this volume),

(iv) males who have made a choice of breeding substrate in the field persist in their choice after disturbance (marking) and removal to some distance from their nest site,

(v) males in the field show site fidelity to their nest site, even the intertidal males who must vacate their territories at each low tide and,

(vi) fish brought into the lab show the same nesting habitat preference as they displayed in the field more often than could be due to chance alone.

MacDonald *et al.* (this volume) offer the hypothesis that breeding on bare rock substrate by the white stickleback, and breeding in the intertidal zone, evolved secondarily to the evolution of emancipation from parental care.

The results indicate that once a male has made a choice between the intertidal and subtidal zones, his choice persists. We found that rocky intertidal and subtidal white sticklebacks not only return to the same habitat but can also relocate their former territory and nests, thereby displaying homing ability. Evidence for homing ability and site fidelity has been found for many intertidal fish species, and it has been argued that homing ability in intertidal fishes has survival value; not all areas of the intertidal zone are equal in quality, thus the ability to recognize and home to a particular location could be essential for survival (Gibson, 1967, 1969, 1982; Mayr & Berger, 1992; Yoshiyama *et al.*, 1992; Setran & Behrens, 1993). In the case of white sticklebacks it is essential to reproductive fitness that males hold a territory and construct a nest. Presumably homing and reoccupation of previously held territories is more efficient than *de novo* scramble competition for territories and nest sites at each tide cycle. It would be informative to test experimentally whether neighbouring males, if they have shared prior territorial and competitive experience, have on average higher reproductive success during a tide cycle than males who have no prior competitve and territorial experience with each other.

At Spry Bay the two highly contrasting breeding substrates (bare rock, filamentous algae) from which we took males for laboratory experiments are separated by a very short distance (< 200 m). In laboratory tests with males from each of the two sites, the choice of breeding substrate was generally consistent with choices made in the field when there was no competition. Thus even when given the opportunity, rock-breeding males do not show a preference for breeding in algae, and they secured spawnings and dispersed embryos on bare rock substrate. In competion however, rock-breeding males failed to establish territories and were consistently displaced, both from nesting and spawning, by algal-breeding males. Thus preference for breeding on bare rock substrate may have evolved because, despite being displaced from nesting sites with filamentous algae, rock breeding males reproduce successfully in what was previously marginal habitat.

This shift in breeding niche, and concomitant habitat selection, may be evolutionarily coupled with other adaptations to breeding on bare rock and breeding in the intertidal zone. It would be informative to assess whether correlated adaptations have evolved in tandem with the habitat shift. For example, it might be predicted that the embryos of intertidal males are more resistant to dessication (MARLIAVE, 1981) and/or temperature fluctuation (VAN IERSEL, 1953) than those of subtidal males or algae-breeding males. Other experimentally testable predictions can be made for other traits. Tests of such predictions should enable assessment of the evolutionary sequence of behavioural diversification at Spry Bay that is hypothesized by MACDONALD *et al.* (this volume).

Traits involved in the diversification of a species to occupy different habitats are likely to be the same traits that can result in reproductive isolation. BUSH (1994) reviews the arguments surrounding the likelihood that sympatric speciation occurs in animals and concludes that, when mate choice is associated with habitat choice, then persistent positive assortative mating is likely to arise among those individuals that make similar habitat choices. If sufficient genetic and behavioural differences accumulate between those making different breeding habitat choices, then the potential exists for sympatric speciation.

It is clear that white stickleback males make different choices among contrasting breeding substrates that are separated by very short distances and with no barriers to stickleback movement between them at Spry Bay. We have shown that homing is well developed and that once a breeding habitat choice is made, it tends to persist. Clearly then, the potential exists that assortative mating has developed, or is developing, among the sticklebacks with different breeding repertoires at Spry Bay. Perhaps the most rigorous test of assortative mating would be to take nesting males and gravid females from the contrasting breeding habitats in the field (bare rock, filamentous algae) and artificially cross and rear their offspring in the laboratory. If assortative mating has evolved in tandem with breeding substrate preference then female progeny from parents captured in bare rock habitats should spawn disproportionately often with male progeny that nest on bare substrate, and females from algal breeding parents should prefer algal breeding males. By artificially crossing parents of known habitat preference, and rearing their progeny uniformly in the laboratory, the confounding effects of learning and prior breeding experience on substrate preference and mate preference behaviours can be diminished.

References

BLOUW, D.M. (in press) Male emancipation in a stickleback; the evolution of offspring desertion. — Ecoscience.

NUPTIAL COLOUR LOSS AND SIGNAL MASKING IN GASTEROSTEUS: AN ANALYSIS USING VIDEO IMAGING

by

C.G. MCDONALD[1]), T.E. REIMCHEN and C.W. HAWRYSHYN[2])

(University of Victoria, Department of Biology, Victoria B.C., P.O. Box 1700, V8W 2Y2 Canada)

(With 5 Figures)
(Acc. 29-III-1995)

Summary

Threespine stickleback vary greatly in male nuptial display, from full red expression between the throat and pelvis to total melanism without any traces of red. The dominant expression for any population is correlated with the underwater spectral irradiance and many populations with melanistic nuptial colour occur in bog habitats where the aquatic spectrum is shifted to long wavelengths. Field observations indicate that in these bog habitats, the typical red nuptial signal is 'spectrally masked' and this might reduce the effectiveness of display during male-male and male-female interactions. We used video imaging techniques to test whether the spectral composition of the visual background influences female preference for nuptial colouration in male threespine sticklebacks. Throat colour preference tests were carried out in which females chose between video images of red and black-throated males superimposed on either a blue or a red background. Despite belonging to a red-throated population, females were found to respond preferentially to the black-throated male image viewed against a red background, indicating that preference was not determined by throat patch hue. This suggests that preference for an epigamic optical signal is likely a function of its efficacy in generating visual contrast rather than its intrinsic spectral characteristics.

Introduction

On the western coast of North America, populations of threespine stickleback (*Gasterosteus aculeatus*) vary greatly in ventral nuptial display, from

---

[1]) Corresponding author; e-mail: mcdonald@uvvm.uvic.ca
[2]) We thank Victor Rush, Charles Baube and Jeff McKinnon for useful discussion and Siobhan Cowley for comments on the manuscript. Charles Baube kindly made stickleback spectral reflectance data available to us. Roberto Racca provided helpful comments on experimental design. Dorothy Paul and Zen Faulkes kindly provided essential video equipment. Field operations were supported by a NSERC operating grant awarded to TER. Video processing hardware and software development (Racca Scientific Consulting) were supported by a NSERC operating grant awarded to CWH. The underwater spectroradiometer (LiCor 1800) used in this study was purchased with a NSERC equipment grant awarded to CWH.

 Behaviour 132 (13-14)

full red expression on the throat and abdomen to total melanism without traces of red. A number of hypotheses have been proposed to explain this variation, including differential predation pressure (McPhail, 1969; Semler, 1971; Moodie, 1972) and selection for convergent threat display with the black Olympic mudminnow (*Novumbra hubbsi*) (Hagen & Moodie, 1979; Hagen *et al.*, 1980). Stickleback populations from the Queen Charlotte Islands, an archipelago off the coast of British Columbia, often lack red nuptial display in stained (dystrophic) habitats independent of the presence or absence of vertebrate predators (Reimchen *et al.*, 1985). Instead, reproductive males tend to be melanistic. Melanistic nuptial colour may be a highly conspicuous optical signal suited to stained habitats, while red nuptial display may be 'masked' by the ambient spectra (Reimchen, 1989). This hypothesis, signal masking, is consistent with the sensory drive hypothesis, which suggests that signal design is largely determined by the sensory perception of the signal receiver (Endler, 1992). Signal masking can be defined as a reduction in signal contrast resulting from the signal and the visual background both possessing similar spectral characteristics.

Several studies on a number of species indicate that sensory bias of the signal receiver plays a role in determining optical signal design (Endler, 1983, 1991; Fleishman, 1988). However, few studies have attempted to manipulate preference for epigamic signals by imposing a sensory constraint on the signal receiver (but see Sigmund, 1983; Long & Houde, 1989; Milinski & Bakker, 1990). In the present study, a video imaging technique was used to constrain the ability of female sticklebacks to discriminate between the red throat colour of video recorded males and the visual background. By experimentally altering the visual background to produce a sensory bias, it was possible to investigate whether signal masking (*i.e.* reduced colour contrast) would result in a loss of female preference for red nuptial colour in a population that is monomorphic for this trait. The experimental design allowed for females to choose between spectrally modified mirror images of a single male, thus making it possible to isolate signal contrast as the sole variable related to nuptial colour preference. In this work, we also present data on the spectral characteristics of clear and stained aquatic habitats and discuss this data in relation to optical signal perception.

TABLE 2. Female responses to red and black-throated male video images as a function of background colouration

Variable	Treatment Red background	Blue background
Red throat	0	8
Black throat	5	3

Fisher's exact p = 0.0128, 1-tailed.

ment, the female preferentially tracked the red-throated male image before (76.9%) and after (100%) throat colours were switched. These findings provide evidence that side bias was not a confounding factor. As females chose between mirror images of the same male, it is highly probable that preference was associated solely with throat colour contrast.

In an attempt to determine whether females which did not display tracking behaviour responded preferentially to a particular throat colour, a time per side analysis was carried out. Only females which exhibited swimming behaviour directed towards the video screen (but did not track the images) were used for this analysis. The amount of time each female spent on each side of the test chamber swimming at the video screen was recorded (for both blue and red background treatments). Under both the red ($N = 16$, $T^+ = 55.5$, $p > 0.50$, 1-tailed) and blue ($N = 14$, $T^+ = 47$, $p > 0.50$, 1-tailed) background treatments, time spent per side did not differ significantly.

Discussion

These results demonstrate that female preference for throat colour of video recorded male threespine sticklebacks (*Gasterosteus aculeatus*) can be altered by manipulating throat colour contrast against the visual background. Despite belonging to a red-throated population, females were found to respond preferentially to the black-throated male image viewed against a red background, indicating that preference was not determined by throat patch hue. Throat colour providing luminance contrast (*i.e.* the black throat patch viewed against the red background) and possibly colour contrast (*i.e.* the red throat patch viewed against the blue background) was effective in eliciting female responses. We can not be sure, however, that under the blue background treatment, red throat colour provided strong colour contrast as we did not compensate for the subjective brightness of the emissions of the different phosphors. These findings suggest that preference for an epigamic optical signal is likely a function of the signal's efficacy in generating visual contrast rather than its intrinsic spectral characteristics.

Our experimental conditions did not permit complete assurance that the observed behaviours can be interpreted as sexual responses. It is possible, for example, that the tracking response we observed was investigatory behaviour rather than courtship. Nevertheless, our findings strongly suggest that a male expressing a nuptial signal which confers greater detectability relative to other males should be more successful in attracting and maintaining the attention of females. Put differently, males which are more readily perceived can be expected elicit a preferential response. Thus, upon first encountering a courting male, female preference may be determined more by male conspicuousness than by a sexual response to a particular trait.

A number of studies lend support to our findings. For example, Blouw & Hagen (1990) found that during the initial stages of courtship, female threespine sticklebacks responded preferentially to males of the newly described white stickleback species (*Gasterosteus*). The white stickleback is a close relative of the threespine stickleback and reproductive males of this species are highly conspicuous by virtue of their iridescent, white dorsal colouration, which threespine stickleback males do not possess. Blouw & Hagen (1990) found that there were no backgrounds that consistently matched the iridescent white colour at the nesting sites they examined, suggesting that white stickleback males are conspicuous in their natural habitat. Female threespine sticklebacks stopped responding to white stickleback males only when they were introduced to the nest which, unlike that of the threespine stickleback, is situated above the substrate and constructed of fine filamentous algae. It appears then that differences in the position and composition of the nest, rather than nuptial colouration, were factors related to the ultimate rejection of white stickleback males as potential mates. In another study by Baube *et al.* (this volume), it was found that female threespine sticklebacks preferred red-throated dummies to green-throated dummies. Using a colour vision model, Baube *et al.* calculated that red throat colour should provide greater colour contrast against both the dorsal colouration of the dummies and the visual background used in their experiment. These findings (Blouw & Hagen, 1990; Baube *et al.*, this volume; present study) all indicate that the primary determinants of nuptial colour hue are a function of the necessity to increase the detectability of courting males.

A number of studies on species other than the threespine stickleback also report findings consistent with those presented here. For example, in a study by Sigmund (1983) on visual display in an *Anolis* lizard (*Anolis carolinensis*), it was found that contrast of a male's dewlap with background colouration can have a significant influence on female mate choice. Females were shown to exhibit no preference between males with red or green dewlaps, provided that dewlap colour contrasted with the background. Since only red dewlaps occur in the population examined by Sigmund, one may conclude that female choice was based upon the conspicuousness of the nuptial signal and not a result of a fixed preference for a particular hue. As females rarely chose a male with a green dewlap if it did not contrast with the background, it is unlikely that green dewlaps were chosen simply because of their novelty. Sigmund suggested that these results provide evidence that red dewlap colouration evolved to enhance contrast against the green foliage in front of which *Anolis* lizards typically display.

Similarly, Endler (1983) demonstrated that female guppies (*Poecilia reticulata*) tend to favour males which contrast with background colour over those which resemble it. He was also able to show, using a multigenerational study, that selection operates to increase the frequency of males which contrast with the background. Thus, it seems likely that nuptial colouration in the guppy has evolved to enhance the detectability of males, and furthermore that a given photic regime can influence optical signal design.

Although there is evidence to suggest that there is genetic variation for the strength of female preference for red nuptial colour in threespine sticklebacks (Bakker, 1993), it is apparent that the efficacy of signal transmission and detection must also play an important role in determining colour preference. Indeed, the findings presented here indicate that female preference for red nuptial colour can be significantly reduced by its superimposition on a low contrast visual background. However, it remains to be determined whether red nuptial colour is spectrally masked in natural habitats.

It is important to note that although red nuptial colour would provide minimal hue contrast in red-shifted habitats (Reimchen, 1989), it would conversely provide high luminance contrast if expressed laterally (rather than ventrally). There are two reasons for this. First, reflective surfaces with a spectral distribution similar to that of the ambient spectra reflect relatively more light than surfaces that are spectrally offset (Lythgoe, 1979). Second, laterally situated nuptial colour would receive substantial illumination from the downwelling irradiance. Since sidewelling irradiance is substantially lower than downwelling irradiance (Fig. 3), laterally expressed red would appear considerably brighter than the background (at least at proximal distances). However, Reimchen (1989) found that in red environment populations where red nuptial colour was present, expression was limited to the mid-ventral region of the throat. Upwelling irradiance in dystrophic systems is generally extremely low (Fig. 3), and thus would probably not render red ventral colouration conspicuous.

It has been hypothesized that the melanistic appearance of reproductive males from red environment systems may provide a high contrast optical signal (Reimchen, 1989). Specifically, Reimchen suggests that melanistic nuptial colour would result in reversed countershading (Hailman, 1977). However, it should again be noted that ambient light levels in dystrophic systems are extremely low (Fig. 3 and 4). Hence, melanistic nuptial expression may provide poor contrast against the low backwelling irradiance. There are, however, some scenarios where the efficacy of melanistic nuptial expression would likely be quite high. Firstly, the nesting sites of some of the Queen Charlotte lakes are in littoral regions where the substrate is highly reflective (pers. obs.). Melanistic males nesting in microhabitats where ambient light levels are elevated might very well appear quite conspicuous. Secondly, differences in the spatial stereometry of courtship in dystrophic systems may influence the conspicuousness of melanistic nuptial colour. As far as the authors are aware, this possibility has not been addressed. Here it is significant to note that male ninespine sticklebacks (*Pungitius pungitius*) possess melanistic nuptial colour and that females view courtship behaviour largely from a position below the male (Morris, 1958). A melanistic male silhouetted against the downwelling light would certainly provide a highly conspicuous visual stimulus.

In conclusion, it is evident that the behavioural results presented here indicate that signal masking can influence colour preference and that signal design is likely largely determined by parameters related to the generation of visual contrast. These findings are consistent with the sensory drive hypothesis (ENDLER, 1992). However, it is also evident that the conspicuousness of red nuptial colour in dystrophic habitats must be quantified to determine if signal masking does indeed occur in nature.

References

BAKKER, T.C.M. (1993). Positive genetic correlation between female preference and preferred male ornament in sticklebacks. — Nature 363, p. 255-257.

BAUBE, C.L., ROWLAND, W.J. & FOWLER, J.B. (1995). The mechanisms of colour-based mate choice in female threespine sticklebacks: hue, contrast and configurational cues. — Behaviour 132, p. 979-996.

BLOUW, D.M. & HAGEN, D.W. (1990). Breeding ecology and evidence of reproductive isolation of a widespread stickleback fish (Gasterosteidae) in Nova Scotia, Canada. — Biol. J. Linn. Soc. 39, p. 195-217.

ENDLER, J.A. (1983). Natural and sexual selection on color patterns in *Poecilia reticulata*. — Env. Biol. Fishes 9, p. 173-190.

—— (1991). Variation in the appearance of guppy color patterns to guppies and their predators under different visual conditions. — Vision Res. 31, p. 587-608.

—— (1992). Signals, signal conditions and the direction of evolution. — Am. Nat. 139, p. S125-S153.

FLEISHMAN, L.J. (1988). Sensory influences on physical design of a visual display. — Anim. Behav. 36, p. 1420-1424.

HAGEN, D.W. & MOODIE, G.E.E. (1979). Polymorphism for breeding colors in *Gasterosteus aculeatus*. I. Their genetic and geographic distribution. — Evolution 33, p. 641-648.

——, —— & Moodie, P.F. (1980). Polymorphism for breeding colors in *Gasterosteus aculeatus*. II. Reproductive success as a result of convergence for threat display. — Evolution 34, p. 1050-1059.

HAILMAN, J.P. (1977). Optical signals. — Indiana University Press, Bloomington.

LONG, K.D. & HOUDE, A.E. (1989). Orange spots as a visual cue for female mate choice in the guppy (*Poecilia reticulata*). — Ethology 82, p. 316-324.

LYTHGOE, J.N. (1979). The ecology of vision. — Oxford University Press, Oxford.

MCDONALD, C.G. (1994). The influence of the spectral quality of underwater light on visual sensitivity and the perception of optical signals in the threespine stickleback (*Gasterosteus aculeatus*). MSc thesis. — University of Victoria, Victoria.

—— & HAWRYSHYN, C.W. (1995). Intraspecific variation of spectral sensitivity in threespine stickleback (*Gasterosteus aculeatus*) from different photic regimes. — J. Comp. Physiol. A. 176, p. 255-260.

MCPHAIL, J.D. (1969). Predation and the evolution of a stickleback (*Gasterosteus*). — J. Fish. Res. Board Can. 26, p. 3183-3208.

MILINSKI, M. & BAKKER, T.C.M. (1990). Female sticklebacks use male coloration in mate choice and hence avoid parasitized males. — Nature 344, p. 330-333.

MOODIE, G.E.E. (1972). Predation, natural selection and adaptation in an unusual stickleback. — Heredity 28, p. 155-167.

STICKLEBACK MALES, ESPECIALLY LARGE AND RED ONES, ARE MORE LIKELY TO NEST CONCEALED IN MACROPHYTES

by

SARAH B.M. KRAAK[1], THEO C.M. BAKKER[2] and SANDRA HOČEVAR[3]

(Abteilung Verhaltensoekologie, University of Bern, Wohlenstrasse 50A, CH-3032 Hinterkappelen, Switzerland)

(Acc. 25-II-2000)

Summary

Previous studies have shown (1) that stickleback males enjoy increased mating success when their nest is concealed, (2) that males reduce their courtship in the face of predation risk to a lesser extent when their nest is concealed, and (3) that eggs in concealed nests have higher hatching chances. Here we test the prediction that male sticklebacks prefer to establish a territory at a site with a macrophyte under which they can conceal their nest. We planted macrophytes at half of the potential nest sites at two depths, in a section of a channel in which sticklebacks naturally occur. Subsequently, we found significantly more nests with eggs at the sites concealed by macrophytes than at the control sites, suggesting that wild sticklebacks preferred to build their nests at sites that offer concealment. At the shallow depth, males occupying a site with a macrophyte were larger and redder than males at control sites, but not at the deeper level. This suggests that males of higher competitive ability and greater conspicuousness were more likely to settle at shallow sites where predation risk by the grey heron is high.

Keywords: three-spined stickleback, *Gasterosteus aculeatus*, nest site choice, nest concealment, competition, predation risk, large size, red coloration.

[1] Current address: Department of Biology, University of Leicester, University Road, Leicester LE1 7RH, UK.

[2] Current address: Institut für Evolutionsbiologie und Ökologie, University of Bonn, An der Immenburg 1, D-53121 Bonn, Germany.

[3] We thank the Swiss National Science Foundation for financial support, C.D.K. Cook for the determination of the macrophyte, and Willem Steenge for all the practical help in and outside the field. The manuscript benefited from comments made by Ulrika Candolin, Mats Lindén, N.R. Liley, Eric Taylor, and an anonymous reviewer.

 Behaviour 137, 907-919

Introduction

Over the past few decades, female mate choice has been demonstrated to be widespread in diverse species across many taxa. Traditionally, female mate choice has been invoked to explain the evolution of extravagant male traits by sexual selection (Darwin, 1871; Andersson, 1994). Relatively few studies consider the possibility that females choose characteristics of the territory that a potential mate defends, at least in fish (Sargent, 1982; Thompson, 1986; Hastings, 1988a, b; Warner, 1988; Kraak *et al.*, 1999a). Certain aspects of a territory may be associated with direct fitness benefits to a female. For example, if the female or her young have to feed within the territory, it is important for a female to assess the quality of the resources held within the territory. Similarly, if territories vary in the degree to which they offer protection, from predators or any other detrimental agent, to the female or her young, a female should assess this aspect of the territory. If certain territories enhance offspring survival more than others, females and males should have the same territory preferences. If the availability of these good quality territories is limiting, males will compete for them, resulting in a correlation between male competitive ability and territory quality. Females may then either base their choice on male competitive quality, or directly on territory quality, depending on what is the most reliable correlate of offspring survival.

Fishes with male parental care for eggs in a nest provide a suitable system for investigating these issues. An example of such fishes is the three-spined stickleback *Gasterosteus aculeatus*, where the male defends a territory in which he builds a nest. The female oviposits her eggs in the nest and leaves them to the care of the male. Nest sites may vary in how well the female is protected from disturbances and predation during oviposition. And, most importantly, nest sites may differ in how well the eggs are protected, resulting in variation in the chances of survival to independence of the offspring. Stickleback males have to oxygenate their eggs by fanning, and defend them against egg predators, which are often cannibalistic conspecifics (FitzGerald, 1991). The male himself may also be safer in some types of territory than in others, and if a caring male is preyed upon, his offspring will certainly fall victim to egg predators.

Kraak *et al.* (1999a) found in a field study that males and their nests were less likely to disappear when they were concealed by a macrophyte. A study in wading pools (Sargent & Gebler, 1980) showed that male sticklebacks at concealed nests suffered fewer intrusions and devoted more time to fanning, resulting in higher hatching rates. Male sticklebacks with nests concealed by macrophytes also received more eggs (Kraak *et al.*, 1999a), implying more mates, than males with exposed nests. Apparently females found hidden nests attractive oviposition sites, probably because the survival chances of their offspring were higher in those nests.

Since males are more likely to breed successfully, i.e. have higher mating success and higher hatching success, when they have a concealed nest, one might expect that males prefer to settle and build their nest at sites where they can conceal their nest, *e.g.* under a macrophyte. If macrophytes are rare, one might also expect that males compete for these sites, and that males with traits that are associated with high competitive ability are more likely to defend such a site, *e.g.* bigger males. Furthermore, conspicuous males might be more prone to predation by visual predators, or to nest destruction by conspecifics that are attracted to the nest by visual cues. In that case males with brighter breeding coloration have more to gain from choosing sites where they can hide themselves and their nest.

In the present study we address two questions. Do male sticklebacks prefer to settle and build a nest at a site with a macrophyte? And, are males with certain traits more likely to have their nest at a site concealed by a macrophyte?

Methods

We performed our study in June and July 1997, at the same site as our previous studies (Kraak *et al.*, 1997, 1999a, b), at Roche (near Montreux, Switzerland, 46°26′N, 6°55′E) in the drainage system of the Rhone near Lake Geneva. Our study site was restricted to a stretch (of about 90 m length) of a small channel. The channel is boarded with blocks of concrete, 35 cm by 60 cm, along the sloping walls, and it has a wooden boarded floor. Sticklebacks nest in cutouts that exist in one corner of each concrete plate, or under the boards that line the concrete slopes at the bottom. Macrophytes, mainly small-leafed *Berula erecta*, occur in the channel sparsely, and even more so at shallow depths. These plants are not used by sticklebacks as nesting material. The most common predator of fish at Roche is the grey heron *Ardea cinerea*. Male sticklebacks nesting at or near the bottom are out of reach of this predator; shallower nest sites are at risk. During parental duties on a nest which is concealed under a macrophyte, stickleback males are visually hidden from this predator, and from conspecifics.

For the experiment we chose two stretches of channel without any vegetation, and at which there were no nesting males at the start of the experiment. These stretches had lengths of 8.2 m

Discussion

First of all we need to justify why we chose to score male nest site preference as presence of nest with eggs. The objection to this criterion is, that it can be influenced as well by female preferences (for nest site and/or male) and such risks as nest raiding. We chose to use presence of nest with eggs, because only of those nests we could be sure that they actually were completed nests, and not an abandoned or unfinished nest or a clump of nesting material. Moreover, if we observed a nest without eggs over *e.g.* three days in succession at the same site, we could not know whether this represented one nest or three different successive nests. Nests with eggs did not pose this problem because if a nest contained *e.g.* three day old embryos the nest must have been in place already for three days. Finally, of nests without eggs we could not reliably determine the owner, because in our population males did not guard their nests very well yet when empty. Sometimes we found more than one different male associated with a particular empty nest(-like) structure in quick succesion; this was not the case when the nest contained eggs. However, if we consider all observations of nest(-like) structures (with and without eggs), which form a dataset consisting of unreliable and non-independent data points, we find that out of 314 observations 245 were at a site with a macrophyte and 69 at a site without. This ratio is similar to the one we found for nests with eggs only, and deviates so much from an expectation of 1:1 ($\chi^2 = 98.6$, $p < 10^{-22}$) that we believe our results regarding male nest site preference are robust. Since we could not catch owners of nests without eggs, we can not check for robustness in a similar way regarding the male traits. We have to take this into account when interpreting the results.

Our field experiment revealed a clear preference of male three-spined sticklebacks for concealed nest sites, which tends to be more pronounced at shallow depths. In sticklebacks, this is the first time that such a preference was experimentally assessed in the field. Our results agree with experimental laboratory studies (Hagen, 1967; Kynard, 1979; Sargent & Gebler, 1980; Cleveland, 1994), and with correlational field data (Hagen, 1967; Black & Wootton, 1970; Moodie, 1972; Kynard, 1978; FitzGerald, 1983; Mori, 1993). It is likely that this preference exists due to benefits of concealed nest sites either for the males themselves, for their embryos in their nest, or for their offspring after hatching. The evidence so far points to an increased reproductive success of males with concealed nests due to a greater mating success (Sargent & Gebler, 1980; Sargent, 1982; Candolin & Voigt, 1998; Guderley & Guevara, 1998; Kraak *et al.*, 1999a, b) and/or greater paternal success as measured by the smaller likelihood of nest abandonment before completion of the parental cycle (Moodie, 1972; Kynard, 1978; Mori, 1993; FitzGerald, 1993; Kraak *et al.*, 1999a), or higher hatching rates (Sargent & Gebler, 1980). A further advantage of concealed nests is better defense against sneakers (Sargent & Gebler, 1980; Sargent, 1982) but too much concealment could also promote the opposite (Mori, 1995).

One of the factors that may influence nest-site choice is predation risk. Recently, Candolin & Voigt (1998) experimentally showed that the visual presence of a predator made more males build their nest close to vegetation. Such a predator-induced nest-site choice is also known for other fishes, for instance *Gobiomorphus breviceps* (Hamilton, 1998). In our study, males that nested closer to the water surface tended to choose relatively more often nest sites that were concealed than males that nested at deeper levels. Males at the shallow sites were probably more liable to predation by the grey heron, the main stickleback predator in the channel. In another study of the same population, more exposed nests disappeared at shallow levels than at deeper levels before the completion of the parental cycle (Kraak *et al.*, 1999a).

Another factor that influences nest-site choice in sticklebacks is male-male competition. When the availability of nest sites is limited or when nest sites differ in quality, one expects males with high competitive abilities to be at an advantage. In several fish species, larger males are better competitors for nest sites (*e.g.* Bisazza & Marconato, 1988; further examples in Turner, 1993). In sticklebacks, redder and larger males are better competitors for territories (Bakker & Sevenster, 1983; Rowland, 1989; Bakker, 1994; Baube, 1997). However, there is a scarcity of data on traits associated with males that outcompeted other males for concealed nest sites. In Kynard's (1979) experiments, males with different numbers of lateral plates differed in their competitive abilities when competing for a limited number of concealed nest sites. In Sargent & Gebler's (1980) study, better competitors were not larger but other traits were not scored. In a previous field study of the same population, males with concealed nests did not differ in color traits or body size from males with more exposed nests (Kraak *et al.*, 1999a). Field studies of FitzGerald (1993), Mori (1995), and Guderley & Guevara (1998) suggested that redder males, or males with more skin carotenoids, nest at more concealed sites. In the present study, we found that males at concealed nest sites were redder and larger suggesting that these males might have outcompeted other males. Differences between males nesting at concealed sites and those nesting at more exposed sites were most pronounced at the shallow level. The shallow level may be attractive for nest building to males, provided that they are protected against predators: water velocity in the channel is high (up to 45 cm/s) and water temperature is low (maximum 14°C), but are less extreme at the more shallow level (Bakker *et al.*, unpubl. study). Lower current velocity and higher water temperature may reduce energetic costs of the male's activities and may be beneficial to his hatched offspring, which concentrate near the water surface. Furthermore, fewer sticklebacks pass the nest sites per unit of time at the shallow level (Kraak *et al.*, 1999a), thus reducing disturbances during courting, mating and caring at the shallow level. Mating opportunities did not seem to be affected: males nesting at the shallow level would go down to court passing females as often as males at the deep level would (S.B.M. Kraak, unpubl. data).

The differences in traits between males that nest at shallow sites that differ in concealment may thus be explained by differences in competitive abilities of males with a common nest site preference. This explanation requires that attractive nest sites were limiting. The difference with our previous study, in which no associations between male traits and nest concealment were found, is that in the present study we manipulated nest site concealment. Concealed nest sites are normally very rare at shallow levels in the channel, but in the present study half of the available nest sites were made concealed by placing macrophytes at potential sites that were randomly assigned. Not all concealed sites were used, suggesting that they were not limiting in the present study. This argues against different competitive abilities as an explanation for the observed results. However, the unused concealed sites may have been unattractive for reasons that we are unaware of. In that case, males may have competed for a limited number of attractive concealed sites.

An alternative explanation for the observed results is that males with different traits have different nest site preferences. Such variation in nest site preferences in sticklebacks with respect to concealment is suggested by variation in nest site choice in the absence of competitors (van Iersel, 1958; van den Assem, 1967; Hagen, 1967; Jenni *et al.*, 1969; Jenni, 1972; Candolin & Voigt, 1998). The benefits of concealed nests are likely to be different for males that differ in the intensity of red breeding coloration; redder males are probably more conspicuous to visual predators, and also to nest raids by conspecifics if these use visual cues. Greater vulnerability of redder males to predation was suggested in a previous study of the same population because males at exposed nest sites tended to be redder at the deeper level where the predation risk is lower than those at the shallow level were (Bakker *et al.*, unpubl. study). Red males have a potentially high mating success, because females find red males attractive (Bakker & Mundwiler, 1994), but if they have to reduce their courtship because of predation risk (Candolin & Voigt, 1998) they might not be able to actually exploit this advantage. Redder males, with their greater vulnerability to predation, may therefore gain more from having a concealed nest, and this may have resulted in them being more likely to occupy these sites.

An interesting alternative explanation is that males adjusted their expression of red coloration after nest site choice, according to the riskiness of their chosen site. We consider this unlikely, because we have no indications that males adjust the expression of their red coloration, and because male red coloration before breeding generally correlates well with male red coloration after having built a nest in an aquarium (T.C.M. Bakker, unpubl. data), implying that (relative) male red coloration is already largely determined before nest site choice. Furthermore, Guderley & Guevara (1998) found that males at concealed nests had more skin carotenoids; it is unlikely that males can adjust the level of skin carotenoids so quickly following the choice of a nest site.

Finally we must consider the possibility that our finding that males at concealed nest sites were redder and larger might be restricted to nests with eggs, as was mentioned at the start of the discussion. Maybe all males were equally likely to settle at sites with macrophytes, but female choice differed between concealed and unconcealed sites, for example perhaps females preferred duller or smaller males among those whose nest was not concealed. Or perhaps non-concealed nests were less likely to be raided when the owner was dull or small.

The results of the present study shed some light upon the discrepancy between field and laboratory findings with respect to the impact of intensity of red breeding coloration in sexual selection in this population (Kraak *et al.*, 1999a). The absence of a greater reproductive success of redder males in the field (Kraak *et al.*, 1999a) may be due to the lack of concealed nest sites in the channel close to the water surface. Greater reproductive success of redder males may be found in other parts of the water system where shallow concealed nest sites are not limited. The ecological context in which a population occurs, such as the local abundance of macrophytes and predators, may thus determine the direction and intensity of sexual selection.

References

Andersson, M. (1994). Sexual selection. — Princeton University Press, Princeton, New Jersey.

van den Assem, J. (1967). Territory in the three-spined stickleback, *Gasterosteus aculeatus* L. An experimental study in intra-specific competition. — Behaviour, Suppl. 16, p. 1-164.

Bakker, T.C.M. (1994). Evolution of aggressive behaviour in the threespine stickleback. — In: The evolutionary biology of the threespine stickleback (M.A. Bell & S.A. Foster, eds). Oxford University Press, Oxford, p. 345-380.

— — & Mundwiler, B. (1994). Female mate choice and male red coloration in a natural stickleback population. — Behav. Ecol. 5, p. 74-80.

— — & Sevenster, P. (1983). Determinants of dominance in male sticklebacks (*Gasterosteus aculeatus* L.). — Behaviour 86, p. 55-71.

Baube, C.L. (1997). Manipulations of signalling environment affect male competitive success in three-spined sticklebacks. — Anim. Behav. 53, p. 819-833.

Bisazza, A. & Marconato, A. (1988). Female mate choice, male-male competition and parental care in the river bullhead, *Cottus gobio* L. (Pisces, Cottidae). — Anim. Behav. 36, p. 1352-1360.

Black, R. & Wootton, R.J. (1970). Dispersion in a natural population of three-spined sticklebacks. — Can. J. Zool. 48, p. 1133-1135.

Candolin, U. & Voigt, H.-R. (1998). Predator-induced nest site preference: safe nests allow courtship in sticklebacks. — Anim. Behav. 56, p. 1205-1211.

Cleveland, A. (1994). Nest site habitat preference and competition in *Gasterosteus aculeatus* and *G. wheatlandi*. — Copeia 1994, p. 698-704.

Darwin, C. (1871). The descent of Man, and selection in relation to sex. — John Murray, London.

FitzGerald, G.J. (1983). The reproductive ecology and behaviour of three sympatric sticklebacks (Gasterosteidae) in a saltmarsh. — Biol. Behav. 8, p. 67-79.

— — (1991). The role of cannibalism in the reproductive ecology of the threespine stickleback. — Ethology 89, p. 177-194.

— — (1993). The reproductive behavior of the stickleback. — Sci. Amer. 268, p. 80-85.

Frischknecht, M. (1993). The breeding colouration of male three-spined sticklebacks (*Gasterosteus aculeatus*) as an indicator of energy investment in vigour. — Evol. Ecol. 7, p. 439-450.

Guderley, H. & Guevara, R.C. (1998). Physiological correlates of paternal care in male threespine stickleback. — Ecoscience 5, p. 438-447.

REFLECTANCE SPECTRA FROM FREE-SWIMMING STICKLEBACKS (GASTEROSTEUS): SOCIAL CONTEXT AND EYE-JAW CONTRAST

by

VICTOR N. RUSH[1,2], **JEFFREY S. MCKINNON**[3,4], **MICHAEL A. ABNEY**[1,5] **and R. CRAIG SARGENT**[1,6]

([1] Biology Department and Center for Ecology, Evolution and Behavior, University of Kentucky, Lexington, KY 40506-0225, USA; [3] Department Biol. Sci., University of Wisconsin-Whitewater, Whitewater, WI 53190, USA)

(Acc. 7-VII-2003)

Summary

The color patterns of many organisms change rapidly with social context but such dynamic signals have been little studied with current methods. In this study we applied objective spectrophotometry techniques to the color displays of unrestrained male threespine sticklebacks, to assess the influence of social context on coloration. Analyzing our data with a color space model based on stickleback visual physiology, we found that unrestrained males enhanced saturation of both their blue eye and red jaw color in response to the presence of a mature male or female conspecific. Divergence between the eye and the jaw lead to enhanced contrast, likely increasing conspicuousness. We found little relationship between measures of color saturation and condition, but the color of males in better condition varied more with social context. This study is the first to evaluate contrast between stickleback color pattern elements quantitatively and the first in which reflectance spectra were collected from free-swimming fish. The methods presented here could be used in future studies of sticklebacks and could potentially be adapted to other animals.

2) Present address: Tucker-Davis Technologies, 4637 NW 6th Street, Gainesville, FL 32605, USA.

4) Corresponding author's e-mail address: mckinnoj@uww.edu

5) Present address: US Fish & Wildlife Service, 1339 20th Street, Vero Beach, FL 32960-3909, USA.

6) This research was supported by NSF 9728892 to Sargent, an NIH grant to Sargent and Rush and a University of Kentucky major equipment grant to Sargent. McKinnon was funded by an NSF Research Opportunity Award and by NSF 9981638. Charlie Baube and two anonymous reviewers provided helpful comments on the ms.

Behaviour 140, 1003-1019
Also available online -

Introduction

In fishes and other taxa, social stimuli may act through neural and hormonal mechanisms to trigger changes to pigment-bearing cells, and thus color patterns, within minutes or even seconds (Fuji & Novales, 1969; Bagnara & Hadley, 1973). Color patterns are clearly important in behavioral interactions and to fitness (Andersson, 1994), yet socially mediated color change has been surprisingly under-emphasized in recent behavioral studies. This is in part because it is difficult to analyze color patterns during social interactions with the techniques now recommended for visual signals (Endler, 1990; Bennett *et al.*, 1994). In the work that has been done, most ethologists have either assessed the color of unrestrained animals subjectively or immobilized experimental subjects with restraints or anaesthetics. The limitations of the former method are well documented (Endler, 1990; see below also) but the immobilization necessary for collecting recommended spectral data may also be problematic, since it can cause potentially misleading changes to pigmentation patterns (McLennan & McPhail, 1989a). In this study, we used a novel technique to collect spectral data from unrestrained, social sticklebacks. This made it possible for us to conduct analyses incorporating models based on stickleback visual physiology. This is the first evaluation of social influences on color expression, in a species with dynamic coloration, that uses current spectrophotometry methods and physiologically-based visual analyses.

The red-dominated nuptial coloration of the male threespine stickleback is both an example of a socially influenced display trait and one of the best known and best studied color signals in nature (reviewed by Rowland, 1994). Studies of stickleback coloration have built on classic ethological investigations (Tinbergen, 1951) to yield many important results in sexual selection (*e.g.* Milinski & Bakker, 1990; Candolin, 1999a) and speciation (*e.g.* McPhail, 1969; Boughman, 2001). Most such studies have treated social context only incidentally, but work by McLennan & McPhail (1989a, b, 1990) and Candolin (1999a, 2000) has shown that both red jaw intensity and blue eye coloration are adjusted in response to conspecifics, and that such adjustments may be adaptive. These important findings are limited, however, by either their reliance on human-based assessment of male coloration (directly with human eyes or using photographic systems that mimic human vision) or restraint of the subject males, or both.

It is important to analyze stickleback coloration in terms of the stickleback visual system because color vision in this species differs from that of humans in potentially significant ways. The key differences are the position of the long wavelength cone, which is shifted to a longer wavelength in sticklebacks; the increased separation between the middle and long wavelength cones (the close positioning of which, in humans, enhances detection of subtle shades of yellow and orange); and the addition of a fourth cone with peak sensitivity in the UV (Baube & Loew, unpubl.). Clearly there are also substantial commonalities between human and stickleback color vision (reviewed by Frischknecht, 1993; McKinnon, 1995), so some methods involving technologies designed to mimic the human eye or evaluations by humans provide useful information (see also Bakker & Mundwiler, 1994). Methods more closely allied to the stickleback visual system are preferable, however, and may lead to novel insights (Endler, 1990; Baube *et al.*, 1995).

We used a color space model (Neumeyer, 1992) to examine how changes in the reflectance of the jaw and eye might be detected as differences in the proportional response of stickleback visual pigments. This simple model also allows us to test the possibility that contrast between elements of the stickleback color pattern is heightened during social interactions. Contrast between color pattern elements may be critical to color pattern conspicuousness (Endler, 1990; Brooks & Endler, 2001), but previously this aspect of coloration had not been quantitatively evaluated in sticklebacks. Our approach also allows us to begin assessing the role that UV photoreceptors might play during social interactions (Cuthill *et al.*, 1999; Losey *et al.*, 1999; Smith *et al.*, 2002).

Social interactions may influence not only coloration but also the relationships between coloration and condition or coloration and other traits of interest to conspecifics (Candolin, 2000). Consequently, we also utilize spectrophotometric data to re-evaluate widely cited but contentious earlier results (Milinski & Bakker, 1990; Rowland, 1994) on the relationship between condition and color and to explore the influence of social context on the dynamics of condition-dependent signaling in sticklebacks (Candolin, 1999b).

Methods

Experimental subjects

Anadromous threespine sticklebacks were collected early in the spring breeding season from the Little Campbell River, Surrey, British Columbia (Hagen, 1967). They were transported by air to the Sargent laboratory in Lexington, Kentucky, after a brief stay in the McKinnon

Discussion

Our study provides the first documentation, not reliant upon human-based indices, of social effects on stickleback coloration and the first such data from unrestrained males. We find that social stimuli affect both eye and jaw coloration. In the case of the eye, the reflectance pattern shifts to a shorter peak wavelength and relatively stronger stimulation of the S ('blue') cone when a conspecific is present. This result also demonstrates that stickleback eye coloration is highly labile and may be difficult to measure reliably from restrained fish. Our analyses of jaw reflectance confirm previous reports of an increased intensity of male red coloration in the presence of conspecifics (*e.g.* McLennan & McPhail, 1990). Our observation that males elevate their eye and jaw color saturation in response to both male and female conspecifics is consistent with other evidence that stickleback coloration functions in both intersexual and intrasexual interactions (McLennan & McPhail, 1990; Rowland, 1994; Baube, 1997).

Most studies of sticklebacks have emphasized red coloration or have considered jaw and eye coloration in isolation, but our analyses suggest that interactions between red ventro-lateral areas and blue eye color may be important in the perception of each and in maximizing overall conspicuousness. We find that as the eye becomes more blue and the throat more red with social stimulation, the contrast between the two is also elevated. Because our measure of contrast is based on the visual physiology of the fish, the result is probably real in terms of stickleback perception. Further, our results suggest that the most intensely red males also possess the most saturated blue eyes (though the result was significant only for the female stimulus), consistent with the hypothesis that contrast between the two is selected for when conspicuousness is favored as well as with some earlier data (McLennan & McPhail, 1989a, b). Other color pattern elements may also play a substantial role in enhancing the apparent intensity of red coloration and overall conspicuousness. In some red-throated stream populations, non-red lateral areas may develop an intense blue sheen and in some cases become black (Bell, 1982; McKinnon, 1995; Von Hippel, 1999). Thus, a more comprehensive evaluation of additional elements of stickleback color patterns may be important in future work, as with guppies (Endler, 1991; Brooks & Endler, 2001).

Ganglion cells, one of the color coding units of the fish visual system, may be a mechanism by which enhanced contrast between color pattern elements stimulates the stickleback eye and brain. Some ganglion cells in goldfish respond preferentially to red coloration in the center of a circle and green coloration in the surrounding area. Other ganglion cells respond strongly to blue and green in the center of the field and red in the surround (Daw, 1968). The shift in coloration that we see would maximally stimulate a ganglion cell that has its center around the eye and its broad surround near the throat; or vice versa. While our results are only qualitative with respect to ganglion cell function, they suggest a degree of sensory exploitation of stickleback visual processing by the male signal. Detailed analysis of the stickleback visual system would clearly go a long way to fully testing this idea.

Our results do not suggest a special or unique role for UV reflectance from the stickleback eye and jaw, unlike some other fishes (Kodric-Brown & Johnson, 2002; Macías Garcia & de Perera, 2002) and birds (Bennet *et al.*, 1996; Sheldon *et al.*, 1999). Certainly UV reflectance is present, especially for the eye, and it could influence potential mates. However, UV-reflectance is at a lower level than for human-visible wavelengths, and there are no obvious changes in reflectance in the UV, simply continuation of patterns present in the visible spectrum. Also, the relative stimulation of the UV cone did not change with social context in our analyses, in contrast to results for those cones that are most sensitive at human-visible wavelengths. It is also noteworthy that L, M and S cones exhibit substantial sensitivity in the UV, but relative sensitivities of these cones are more similar in the UV than for visible wavelengths (Baube & Loew, unpubl.). Sensitivity in the UV accounts in part for the high relative stimulation obtained for the L cone even with light reflected from the blue eye of the stickleback. UV reflectance from other areas of the stickleback body, and at still shorter wavelengths, remains to be investigated.

The relationships between condition, health and red coloration in sticklebacks have been much studied and discussed, but remain contentious (*e.g.* Milinski & Bakker, 1990; Rowland, 1994). In this, the first study of these relationships using reflectance data and a color vision model based on stickleback visual physiology, we did not find a positive relationship between condition and red intensity. Despite our small sample size, however, we found support for a more dynamic signal in high condition males, much as Candolin (1999b) documented in response to predators. Thus, high condition males reduce their red intensity more in the absence of a female conspecific.

1016 RUSH, MCKINNON, ABNEY & SARGENT

It remains to be seen whether this result is obtained because high condition males are of intrinsically superior quality and therefore investing differently in present and future reproduction — or whether the low condition males have simply been through more breeding cycles and color both intensifies and loses flexibility throughout the breeding season for possibly non-adaptive reasons (Candolin, 1999b). It is also interesting to speculate whether females can detect the ability of males to alter their coloration, and whether such flexibility is attractive.

Our results suggest that meaningful reflectance data can be collected from unrestrained threespine sticklebacks. Thus, it is possible to obtain measures of coloration that allow evaluation of the influence of social context and other factors associated with transient color changes. Moreover, reflectance data can be analyzed in terms of their perception by conspecifics or other species such as predators, even for light not visible to humans. In future studies, male coloration might be analyzed during more complete interactions with females and perhaps other males, in which case it may not be necessary to use parental males as experimental subjects. They were used here in order to insure that males returned to the nest for color-measurement when standardized stimuli were presented. In addition, it may be possible for similar methods to be applied to different fish species or other vertebrates. Protocols could be developed that allow individuals to be coaxed into particular positions in which standardized color measurements can be taken; or some training of experimental subjects could be conducted.

These analyses omit several details, such as ocular absorbance and wavelength-specific absorption of light by water, which might influence relative cone stimulation (Endler, 1990, 1991; Baube *et al.*, 1995). Addition of such details to future studies would be valuable. We would be surprised, however, if there were substantial changes to our most central finding, that social stimuli affect stickleback coloration so as to cause greater differential cone stimulation and higher 'chroma' for each color patch as well as higher contrast between the jaw and eye. On the other hand, a more thorough understanding of the details of stickleback color vision, and in particular how color perception varies among populations and environments (McDonald & Hawryshyn, 1995; Boughman, 2001), would greatly enhance the interpretation of our reflectance results. It would also be worthwhile to investigate the effect of angles of illumination and observation on reflectance from the eye, and whether aquarium glass could have more subtle effects on reflectance data than simply absorbing light at some wavelengths.

Homosexuality, Cannibalism & Sexual Strategies

Frank A. von Hippel

> Certain fishes, belonging to several families, make nests; and some of these fishes take care of their young when hatched... The males of certain fishes do all the work, and afterwards take exclusive charge of the young. This is the case... with the sticklebacks (*Gasterosteus*), in which the males become brilliantly coloured during the spawning season. The male of the smooth-tailed stickleback (*G. leiurus*) performs during a long time the duties of a nurse with exemplary care and vigilance, and is continually employed in gently leading back the young to the nest when they stray too far. He courageously drives away all enemies, including the females of his own species. It would indeed be no small relief to the male if the female, after depositing her eggs, were immediately devoured by some enemy, for he is forced incessantly to drive her from the nest. — (Darwin, 1871)

Cannibalism of embryos in a male's nest may be performed by the father (filial cannibalism) or by competing males or females out for a bite to eat of unrelated young (heterocannibalism). Cannibalistic females may raid a nest in a bold assault, or they may adopt a sneak attack by courting a male in order to access the embryos he already has in his nest. Clearly the male stickleback lives in a world of stressors, and not the least of his problems is the intention of females entering his territory. Males kept with females lose mass quicker, court less, fan less, complete fewer reproductive cycles, and die sooner — the femme fatale effect (de Fraipont et al., 1992). Kynard (1978) observed males grasping females on the caudal peduncle and carrying them out of their territory to evict them. It is no wonder that the early ethologists focused so much attention on the competing drives to mate, to fight or to flee. In his memoirs, Desmond Morris (2006) described the essence of stickleback behaviour in this way: "Someone had once written that 'Fish are such incredibly boring creatures that I feel sure they must go to heaven when they die.' In that case, I thought, my sticklebacks will certainly go to hell."

Morris (1952) made one of the first remarkable observations regarding individuals at a nest who did not belong there when he noted that nesting males often lacked discrimination between the sexes and would dance to both females and males. "Sexually frustrated males," who did not have a nest, were cryptically coloured and sometimes adopted female sexual behaviour. "The first indication of [a homosexual response] was when one of the cryptic males was seen creeping stealthily through the weeds towards a nest at which the owner was showing the entrance to a female." Furthermore, Morris noted that "if the female managed to get into the nest and lay its eggs before the pseudofemale advanced to the entrance, it appeared as if the latter was waiting for the eggs to be laid, and was then passing over them and fertilizing them before the male owner could interfere. If the pseudofemale pushed past the female into the nest, it appeared as if it was competing with the female for her role. If the pseudofemale pushed through the nest when both the male and the female had left the site, then the activity appeared to be something akin to masturbation." Some of the pseudofemales are cannibalistic "egg thieves," but Morris considered this a consequence of their female-mimicking behaviour rather than a cause of it. Similarly, Morris noted that "if they do ejaculate sperm when they are 'pretending to lay eggs', then although the eggs may get fertilized in this way, it is practically certain to be an accidental outcome of the homosexual pattern." Therefore, Morris considered pseudofemale behaviour by sexually frustrated males to be "an outlet that serves to relieve frustration," and homosexual behaviour could be viewed as a "displacement pattern."

Morris's (1952) paper on homosexuality in male stickleback both opens this section of the book and serves as the launching point for a fascinating series of papers exploring nest raiding, cannibalism and alternative mating tactics. It is followed by Morris's (1955) follow-up paper in which he described homosexual behaviour in female stickleback, and related patterns of homosexuality in animals to human homosexuality. "It would appear then," wrote Morris (1952), "that complex homosexual pseudofemale behaviour has only been observed in Sticklebacks, Cichlids, Anatids, Rats, Monkies, and Man. The fact that the species in which the phenomenon has been observed are just those species which the Comparative Ethologists and the Comparative Psychologists have studied particularly elaborately, is perhaps an indication that such behaviour is of a much wider occurrence in the animal kingdom than was previously believed, and that it is only revealed after a detailed study of the animal concerned has been carried out."

Morris's papers are followed by van den Assem's (1967) important study on territoriality in the threespine stickleback. van den Assem provided a remarkable treatment of territorial behaviour, covering in detail and at length nearly every conceivable topic associated with the male's territory. Of the 169 pages of text, the exploration of sneaking behaviour is the main element republished here. van den Assem found that sneaking behaviour by rival males is common, even by territorial males, and that in about half of the cases the sneaker male was able to enter the nest following a spawning female and fertilizer her eggs before the nest owner could do so. van den Assem's study demonstrated a clear competitive advantage to "pseudofemale" behaviour on the part of rival males; this behaviour allows the sneaker male to position himself near a spawning female and thereby steal a fertilization — something that Morris had noticed but believed to be a consequence of homosexual behaviour rather than a sexual strategy. However, it was not until about the time of van den Assem's study that scientists began to think of animal behaviour in terms of competing strategies. van den Assem's results led him to conclude that, "sneaking is not restricted to inferior males, and has in fact nothing to do with inferiority, nor with long term sexual frustration."

Like Morris (1952), van den Assem found that egg-stealing males without a nest consumed the stolen eggs, while those with a nest cared for them in the usual manner. Morris considered two functions of sneaking, first as a means of obtaining eggs to eat when food is limiting, which he viewed as a means of regulating population density, and second as a "displacement outlet relieving sexual frustration." van den Assem demonstrated that sneaking behaviour leads to stolen fertilizations, which has a direct fitness benefit, but he asked the question why not leave the eggs in the nest of the cuckolded male? That male would provide parental care, leaving the sneaking male in the ideal situation of stealing fertilizations and then leaving someone else to care for the resulting offspring. And yet van den Assem found that sneaker males stole eggs from the nest and returned them to their own nest. van den Assem considered several explanations for egg stealing, and concluded that the most likely function was to synchronize parental cycles in a population. van den Assem also considered the stickleback results to support the hypothesis that territorial behaviour is an agent for limiting the number of reproductive individuals in a population. However, van den Assem did not agree with

Wynne-Edwards's (1962) well-known, but subsequently ridiculed, idea that inter-group selection is stronger than selection at the level of the individual. "Regulation of numbers is a by-product of selection for something else of advantage at the individual level — contrary to Wynne-Edwards who states that this regulatory mechanism is selected *because* it regulates numbers."

The early ethological literature is replete with such group selection ideas. For example, Tinbergen (1951) wrote in *The Study of Instinct*, "We can safely conclude that [social releasers] are adaptations serving to promote co-operation between the individuals of a conspecific community for the benefit of the group." Later in the book he wrote, "While many behaviour elements are directly advantageous to the individual, there are other elements that are of no direct benefit for the individual but are of advantage to a group of individuals. In social species each instinct may contain such elements. In sexually reproducing species the reproductive instinct contains few elements that have survival value for the individual. Therefore one has to distinguish in each instinct between 'individual elements' and 'social elements'. The social elements [are] activities serving to maintain the species." Group selection arguments such as these were not thoroughly debunked until George Williams (1966) published his classic book *Adaptation and Natural Selection*. Although van den Assem's (1967) treatise on territoriality came out subsequent to Williams's (1966) book, Williams's refutation of group selection took a few years to solidify into the destructive blow that it was. Therefore, van den Assem still felt the need to argue Wynne-Edwards's idea at some length.

Next in this section are excerpts from Robert Wootton's (1972) and Boyd Kynard's (1978) studies of stickleback reproduction in a natural environment, each accompanied by a retrospective. Until the 1970s, nearly all stickleback studies had been conducted in laboratories. Wootton (1972) and Kynard (1978) provided some of the first "ground truthing" that well-publicized behavioural patterns of the stickleback reproductive cycle were not laboratory artefacts, including sneaking behaviour and nest raiding on the part of males. Later, other researchers, such as Seiichi Mori (e.g., Fig. 1) and Susan Foster (see below) demonstrated geographic variability in stickleback behavioural patterns in the wild.

Kynard found that sneaking behaviour by neighbouring males in Wapato Lake was employed to cannibalize eggs, rather than steal fertilizations. Kynard also documented the first observations of "raider packs" of females and

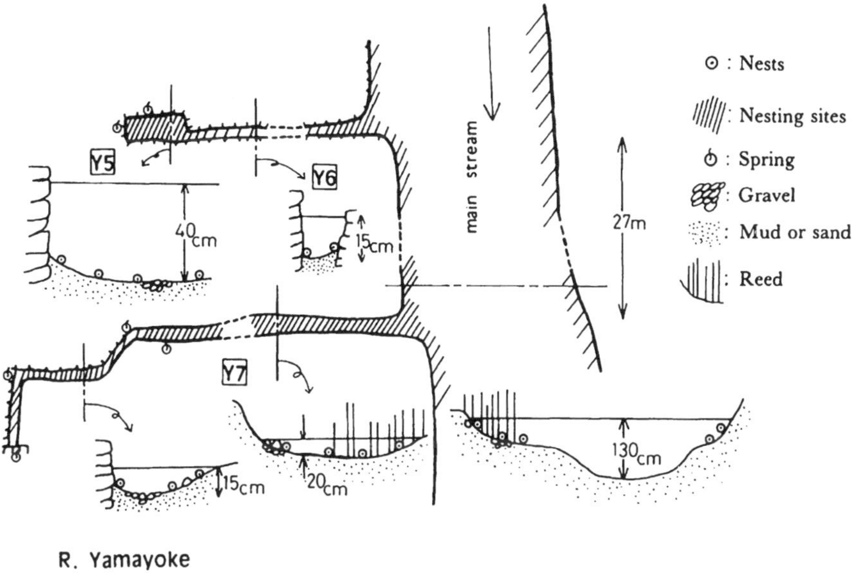

Figure 1. "Examples of study-stations at the main stream and the spring areas on the river Yamayoke. The oblique dashes indicate the nesting areas; they indicate the density of nests by the distances between dashes. Situation in late March-early April." From Mori (1985).

non-territorial males raiding nests. All of this nest raiding led to a behavioural defence: Wapato Lake males concealed their nests with debris for the first two days of the parental cycle, when the risk of egg losses to cannibals was highest, and they avoided fanning their nest during this time, which of course would reveal its location. Wootton (1972) found a similar nest-concealment defence: "Initially a completed nest was almost entirely covered with silt, spat over the nest by the male." Kynard also documented a population decline due to an introduced species; this would prove to foreshadow the loss of many stickleback populations around the world, including many of the Dutch populations used in the classic ethology experiments (reviewed by Foster et al., 2003; von Hippel, 2008). Simultaneous to Kynard's study, Li and Owings (1978) cited evidence for intrasexual selection driving nest raiding, with raiding males being relatively more attractive to females.

The key unsolved issue from this series of studies was why do males steal eggs (which they may have fertilized themselves) and return them to their own nest? None of the explanations put forward were convincing until Mark Ridley and Catherine Rechten (1981) published the finding, in the next paper of this section, that female stickleback prefer to spawn with males whose nests already contain eggs. "If females used the presence of eggs as a cue to choose males," wrote Ridley and Rechten, "males with empty nests should kidnap eggs, which they do." Mori (1995) studied nest-raiding males in a stream environment and found that they were typically territorial males without eggs in their nest, and yet they were more likely to produce young than males who did not raid.

So the stickleback male is faced with the daunting task of attracting females, who may only feign love in order to eat the male's existing embryos, or who may want to mate, but perhaps only if he already has eggs in the nest. If he is without eggs, he may be compelled to sneak to a neighbouring male's nest in order to steal some (which he might as well fertilize) in order to make his own nest more attractive. He may have just finished a reproductive cycle spent parenting day and night with little chance to forage, and so hunger may drive him to eat rather than to save the stolen eggs. Meanwhile, if he is gone from his nest for more than a minute or two, chances are good that the same fate is meeting any eggs he already has in his nest. In 1879, the Reverend W. Houghton wrote about the male's dilemma in this way:

> The male fish alone protects the nest; if it were not for his fatherly care, I suspect that the race of Sticklebacks would in time become extinct, the ova and young fry falling easy prey to the other members of the family. Some years ago I had an opportunity of observing how necessary for the protection of a young Stickleback family is the presence of the male parent. I noticed a brilliant fish hovering over his nest, and fanning the water incessantly with his fins. Having captured him, I placed him with one or two others into my collecting bottle. Here I kept him for about half an hour, whilst I amused myself by watching the manners and customs of the fish in their natural haunts. My eye was soon arrested by the spectacle of a large crowd of hungry marauders in the shape of other Sticklebacks of all ages and both sexes that had gathered around the nest of the very parent whom I had a prisoner in my bottle. They rushed at the nest like terriers at a badger, and began to pull it in pieces, knowing there was something good inside. Conscience-stricken that I was the author of this terrible catastrophe — for as long as the father-fish was present to protect his property the alien cannibals dared not approach — I restored my prisoner to the water, and gently put him in over the spot where about thirty

remorseless strangers were devastating the nest and devouring the contents. For the space of about half a minute my liberated captive hardly seemed to know where he was, or what he had been doing. Soon, however, he collects his scattered senses, and discovering the appalling nature of the fact, rushes to the rescue; first one and then another invader is attacked, and compelled to beat a hasty retreat, and wonderful to relate, in the space of about ten minutes not a foe was to be seen, and the brave defender was left in undisturbed possession of the field. What was to happen next? The conqueror surveys the ruined state of affairs, "Haec loca vi quondam et vastâ convulse ruinâ." and hastens to repair the fearful breaches which the besiegers had made. This he does by bringing mouthfuls of weed and bits of rotten twigs and other things, which he places upon the nest, using his nose to hammer the materials together.

Given that territorial males are also nest raiders and cannibals, how is it that they avoid eating their own offspring, especially their fry, which they have to retrieve with their mouth to secure them in a defendable cluster? In the next paper of this section, Enja Feuth-de Bruijn and Piet Sevenster (1983) conducted an experiment to test this question, and found that as the embryos and fry develop, the male experiences a competition between eating and retrieving young. The outcome of this competition depends on the hunger of the male and whether the male has previous experience eating stickleback young. Feuth-de Bruijn and Sevenster presented young ten-spined stickleback (also known as ninespine stickleback) to parental male threespine stickleback and observed the males to engage in persistent "mumbling," which did not occur when they were presented with young of their own species, or when they were presented with young perch (which they ate). Feuth-de Bruijn and Sevenster concluded that this mumbling occurs because the male threespine stickleback has difficulty distinguishing between the young of his own species and that of another stickleback species, and therefore has difficulty deciding whether to eat or to retrieve. "The long bouts of 'mumbling'," they wrote, "strongly suggest that the stimuli which release retrieving are of a chemical nature and that by 'mumbling' the males are actually sampling the young gustatorily."

Salfert and Moodie (1985), in the next paper of this section, expressed doubt that egg stealing is a courtship strategy, and rather saw cannibalism as a much more common motivation for nest raiding. But filial cannibalism, they argued, may serve as a reproductive strategy by eliminating a small brood when the male has the opportunity to obtain a much larger brood of eggs. This reproductive function is not mutually exclusive with a nutritive function.

The courtship patterns of sticklebacks and other species were viewed by early ethologists as "invariant components of the behavioural repertoires of all of the members of a given species"; in the final paper of this section, Susan Foster (1995) wrote of the courtship patterns in the threespine stickleback, "Ironically, the species with which Tinbergen had played as a child, and studied so effectively in later years, is proving to be among the most variable species ever studied." Foster compared cannibalistic behaviour across populations and found that populations which feed on large benthic (bottom-dwelling) invertebrates are highly cannibalistic, while planktivorous-feeding (limnetic) populations do not show cannibalism. This behavioural difference between the two ecotypes has far-reaching consequences. In cannibalistic populations, courtship is less conspicuous; for example, the zigzag dance is often absent or reduced, which lowers the probability of the nest being found by cannibals, while dorsal pricking is more prominent, which likely allows the male to inspect his surroundings for cannibalistic groups before leading a female to his nest. Foster also found that males in cannibalistic populations engage in diversionary displays to distract cannibals from their nest, while males in non-cannibalistic, limnetic populations do not. Foster concluded that deep, steep-sided lakes lead to the evolution of a plankton-feeding stickleback; planktivory leads to the loss of cannibalistic behaviour, which in turn leads to the loss of diversionary displays and the emancipation of lively courtship behaviour.

Courtship behaviour may also be emancipated in the laboratory as an artefact of housing males in isolation away from the threat of nest raiding and cannibalism. But the very fact that these more interesting behaviours occur in the lab, when fish are housed together in competitive arenas, is one of the features that make the stickleback such a great model for behavioural studies. Without this tendency of the stickleback to behave in an aquarium much as it does in the wild, it would not have been selected by Tinbergen as a key research animal, and the studies of homosexuality, nest raiding, cannibalism and sexual strategies would not have followed the historical course that they did — leading to an understanding of stickleback behaviour couched in ecological context and social competition.

References

van den Assem, J. (1967). Territory in the three-spined stickleback, *Gasterosteus aculeatus* L. An experimental study in intra-specific competition. — Behaviour, Supplement XVI: 1-164.

Darwin, C. (1871). The descent of man, and selection in relation to sex. — Murray, London.

Feuth-de Bruijn, E. & Sevenster, P. (1983). Parental reactions to young in sticklebacks (*Gasterosteus aculeatus* L.). — Behaviour 83: 186-203.

Foster, S.A. (1995). Understanding the evolution of behavior in threespine stickleback: the value of geographic variation. — Behaviour 132: 1107-1129.

Foster, S.A., Baker, J.A. & Bell, M.A. (2003). The case for conserving threespine stickleback populations: protecting an adaptive radiation. — Fisheries 28: 10-18.

de Fraipont, M., FitzGerald, G.J. & Guderley, H. (1992). Femme fatale — the case of the threespine stickleback. — Ethology 91: 147-152.

von Hippel, F.A. (2008). Conservation of threespine and ninespine stickleback radiations in the Cook Inlet Basin, Alaska. — Behaviour 145: 693-724.

Houghton, W. (1879). British fresh-water fishes. — W. Mackenzie, London.

Kynard, B.E. (1978). Breeding behavior of a lacustrine population of threespine sticklebacks (*Gasterosteus aculeatus* L.). — Behaviour 67: 178-206.

Li, S.K. & Owings, D.H. (1978). Sexual selection in the three-spined stickleback: II. Nest raiding during the courtship phase. — Behaviour 64: 298-304.

Mori, S. (1985). Reproductive behaviour of the landlocked three-spined stickleback, *Gasterosteus aculeatus microcephalus*, in Japan. — Behaviour 93: 21-35.

Mori, S. (1995). Factors associated with and fitness effects of nest-raiding in the three-spined stickleback, *Gasterosteus aculeatus*, in a natural situation. — Behaviour 132: 1011-1023.

Morris, D. (1952). Homosexuality in the ten-spined stickleback (*Pygosteus pungitius* L.). — Behaviour 4: 233-261.

Morris, D. (1955). The causation of pseudofemale and pseudomale behaviour: a further comment. — Behaviour 8: 46-56.

Morris, D. (2006). Watching; encounters with humans and other animals. — Max Press, London.

Ridley, M. & Rechten, C. (1981). Female sticklebacks prefer to spawn with males whose nests contain eggs. — Behaviour 76: 152-161.

Salfert, I.G. & Moodie, G.E.E. (1985). Filial egg-cannibalism in the brook stickleback, *Culaea inconstans* (Kirtland). — Behaviour 93: 82-100.

Tinbergen, N. (1951). The study of instinct. — Oxford University Press, Oxford.

Williams, G.C. (1966). Adaptation and natural selection. — Princeton University Press, Princeton, NJ.

Wootton, R.J. (1972). The behaviour of the male three-spined stickleback in a natural situation: a quantitative description. — Behaviour 41: 232-241.

Wynne-Edwards, V.C. (1962). Animal dispersion in relation to social behaviour. — Oliver & Boyd, Edinburgh.

HOMOSEXUALITY IN THE TEN-SPINED STICKLEBACK (*PYGOSTEUS PUNGITIUS* L.) [1]

by

DESMOND MORRIS

(Department of Zoology and Comparative Anatomy, University of Oxford)

(With 15 figures)

(Rec. 6-VIII-1952)

INTRODUCTION

Comparative Ethologists have spent a number of years studying the behaviour of the Three-Spined Stickleback, *Gasterosteus aculeatus* L., and its reproductive behaviour patterns are known in some detail (LEINER, 1929, 1930, 1931a, 1934; WUNDER, 1928, 1930; TER PELKWIJK and TINBERGEN, 1937; TINBERGEN and VAN IERSEL, 1947; VAN IERSEL, unpublished; and also in a number of papers by TINBERGEN, especially 1940, 1942, 1950, 1951). Other species of stickleback, of which there are several, have been little studied until very recently. Comparative studies are now being carried out, and this paper is part of a general analysis of the behaviour of the Ten-Spined species. It is being published separately from the main body of the work because it deals with a piece of abnormal behaviour that has been observed, whilst the major investigation is concerned with the normal patterns of behaviour. However, as there is no modern description of the normal reproductive behaviour of this species available in English (there are two earlier papers in German by LEINER, 1931b and 1934, and a good short account in Dutch by SEVENSTER, 1949), it has been thought necessary to briefly outline it here. It will be reported in detail in later papers.

The Ten-Spined Stickleback inhabits small rivers and streams in many parts of the British Isles. It differs from the more common Three-Spined Stickleback in, amongst other things, the location of the nest site. Whereas the Three-Spined species constructs a covered depression in the sand or mud of the river bed, the Ten-Spined Stickleback selects water-weeds in

1) My thanks are due to Dr N. TINBERGEN for his helpful advice during the above work, and for preparing the German summary, and to the Nature Conservancy for sponsoring this research.

which to build its nest. This nest is roughly spherical and is pierced by a bent tunnel, the tunnel exit being higher than the entrance (see figs 1 and 9).

The area around the nest site is patrolled and defended by the male owner, other fish, particularly males of the same species, being driven away. Normally there are a number of such territories together in a stretch of weed in the stream. At the beginning of the season the boundaries are not clearly defined, and males are constantly trespassing onto one anothers' nesting areas and a considerable amount of fighting ensues. Gradually the boundaries become more clearly defined, and correlated with this there is less actual fighting and more threatening.

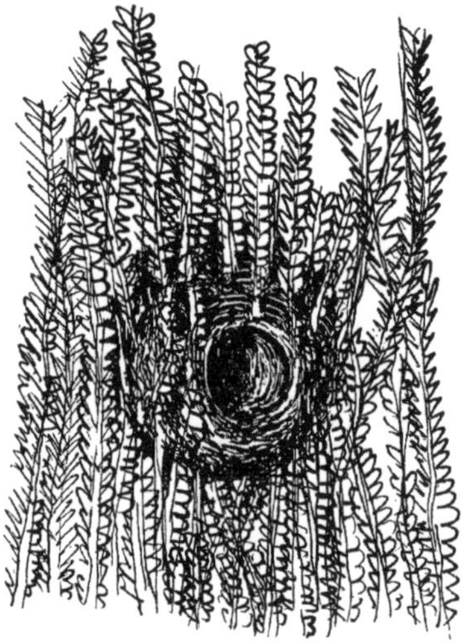

Fig. 1. The Nest of the Ten-Spined Stickleback.

When the females are ripe they swim through the territories and are courted by those males which are ready to fertilise. If the females do not respond to the courting males in a certain sexual manner, they, like other males, are attacked also. So the area around the nest of a male is forbidden ground to all fish save a female which is ready to lay eggs.

As it is extremely difficult to make observations on the reproductive behaviour of these fish in their natural state, the following observations were carried out in the laboratory using glass aquarium tanks. A male was introduced into a tank that had been well planted with weeds, and when it had constructed its nest a female was then introduced. Once she had laid her eggs she was removed as she had then completed her sexual role, and the male was left to rear the young. If fighting and territory demarcation were to be observed, two males were introduced together into a tank. They were introduced together for, if one was already present it might have assumed the whole tank as its territory, in which case the second male would be so badly attacked that it would rarely settle and build anywhere in the tank. The size of the tanks used was normally in the region of 2 × 1 × 1 feet. Practically all observations were made under such conditions. In one tank, however, the population density was increased to study aggressive behaviour which, it was thought, would be more frequently displayed if the males were crowded. A tank of the dimensions 42 × 18 × 18 inches was employed for the purpose and five males were introduced together (fig. 2).

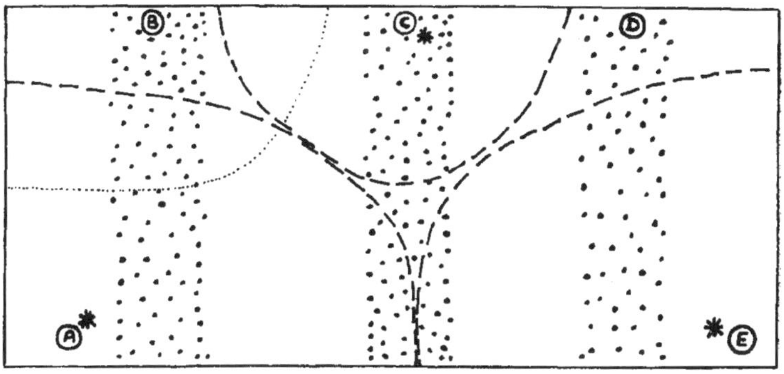

Fig. 2. Tank Plan showing territory areas at a particular stage of the experiment. The dotted areas are the three hedgerows of Willowmoss. The 5 points A B C D & E, are the positions of the five clumps of Myriophyllum. The broken lines represent the territory boundaries, and the asterisks show the positions of the three nests present at the time. There were five male fish in the tank, and one of the nestless males spent much of its time at B. It was unable to build, as it was repeatedly molested by the male at A, but when the latter's nest was damaged and abandoned, the fish at B quickly built and formed the territory shown by the fine dotted line. Later a similar change-over occurred between fish at D & E.

As was expected, a great deal of fighting and threatening was seen as a result of this crowding, but another particularly interesting behaviour pattern was also accidentally brought to light, apparently as a result of this particular population density. This was the phemomenon of homosexuality on the part of certain males. This homosexual behaviour took the form both of lack of discrimination where a sexual partner was concerned (on the part of dominant males), and of the inversion of the complete sexual pattern (on the part of dominated males). Before this is described in detail, the normal courtship pattern of the species will be reported.

It should perhaps be mentioned here that the term "Homosexual" is being used throughout this paper in its widest sense, to mean any sexual performance involving members of the same sex. For a discussion of the terminology, see last section of the paper.

NORMAL COURTSHIP BEHAVIOUR

When a female is ripe it has a swollen belly which is silvery in colour. Its dorsal surface and sides are cryptically marked with a broken pattern. If it

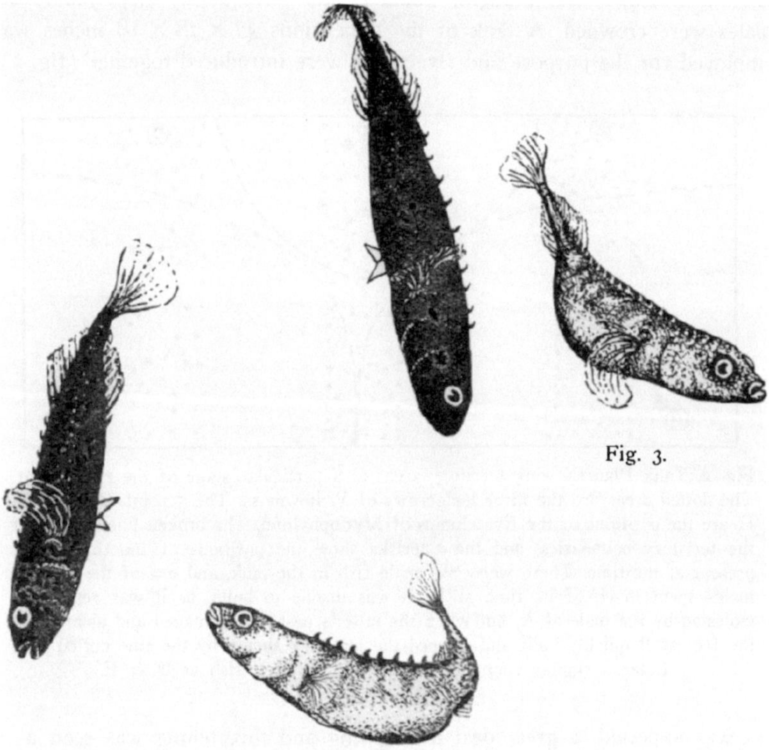

Fig. 3.

Fig. 4.

Fig. 3. Courtship A. Female-oriented dancing on the part of the black male.
Fig. 4. Courtship B. The ripe female responds to the dance of the male by turning towards it.

comes on the territory of a male which is ready to fertilise, it is courted by the male, which reacts to the swollen belly. There are four stages in the courtship of the male. They are as follows:

1. Dancing oriented to the female (Fig. 3).

2. Dancing oriented to the nest (Fig. 6).
3. Showing the nest entrance (Fig. 7).
4. Shivering along the tail of the female (Fig. 8).

The dancing of the male consists of a number of short jumps with the body held head downwards, the body axis being at an angle of approximately 60 degrees from the horizontal. (This angle is very variable and is dependent on several factors that need not be discussed here.) If the female flees at this stage or any other, it is chased and bitten. If it remains still in the water, the male dances around it until it makes a movement to follow (Fig. 4). The male reacts to such a movement by dancing off in the direction of the nest. The female then follows close behind with its body tilted slightly upwards at the anterior end, so that its nose is near to the two white ventral spines of the male which show up clearly against the jet black ventral surface (Fig. 5). These spines are erected throughout the courtship. Once the male has reached its nest it stands still, at the same angle, with its nose at the entrance hole. It then fans [1]) in this position and the female responds to this by lowering its nose from the region of the ventral spines down to the nest entrance. The female then reacts to the entrance hole by pushing into it a number of times. The courtship usually breaks down at this stage and the female swims off. The male usually chases after it and bites it on the tail. If the female is highly motivated sexually, it will respond to this bite by turning back towards the male, whereupon the male dances back to the nest again. The sequence of events, up to the stage of the female pushing into the nest entrance, may be repeated in this way a number of times. After a while the female develops its pushing movement into entry of the nest, in which it lies with its head protruding from the exit, and its tail from the entrance. The male now shivers its nose along the side of the protruding

1) F a n n i n g: This has already been described in detail, in the literature, for the Three-Spined Stickleback, and basically it is performed in the same way, and is functionally similar, in this species. It is a combination of fin movements that results in a current of water being passed through the nest, and is normally performed by a nest-owner during the parental phase. The pectoral fins throw a current of water forwards, and the tail beats hard at the same time, sending a current of water backwards, thus enabling the fish to hover at the nest. Fanning has the function of ventilating the nest and thereby aerating the developing eggs. It is also seen to occur during the sexual phase before eggs are present in the nest. At this stage it has been shown (van Iersel, unpublished) to act as a displacement activity, but I have recently observed that, for the Three-Spined species at least, it also has the function during this phase, of preventing silting up of the nest. In courtship, when the male is showing the nest entrance to the female, fanning is certainly functioning as a displacement activity, but here it is combined with the tilted posture characteristic of the male dance, instead of being performed more or less horizontally, as it is at other times.

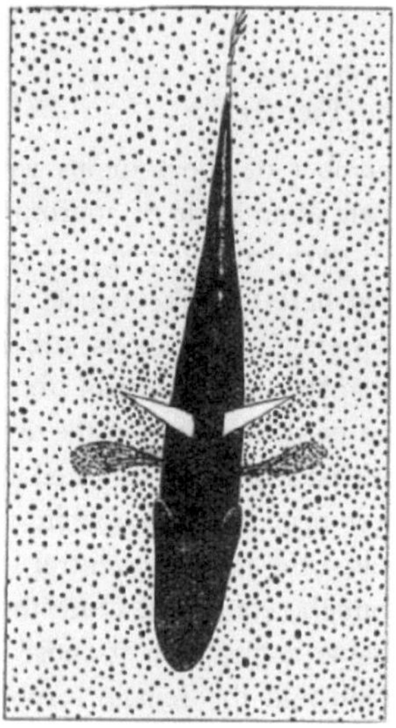

Fig. 5.

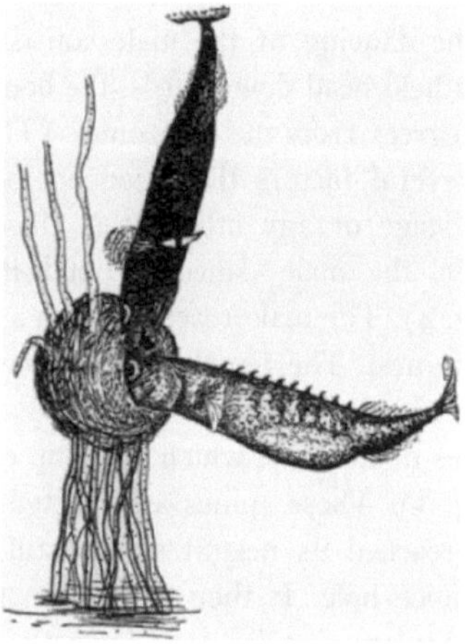

Fig. 7.

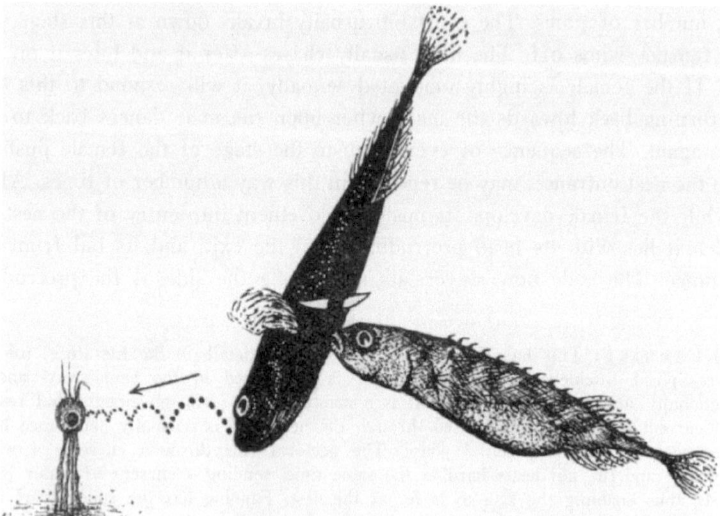

Fig. 6.

Fig. 5. The ventral surface of the dancing male as seen by the female, showing the conspicuous white spines against the black body.

Fig. 6. Courtship C. Nest-oriented dancing. The male swims to the nest in short inclined jumps, followed closely by the female.

Fig. 7. Courtship D. The male shows the nest-entrance to the female by fanning in the tilted position. The female pushes into the entrance with her nose.

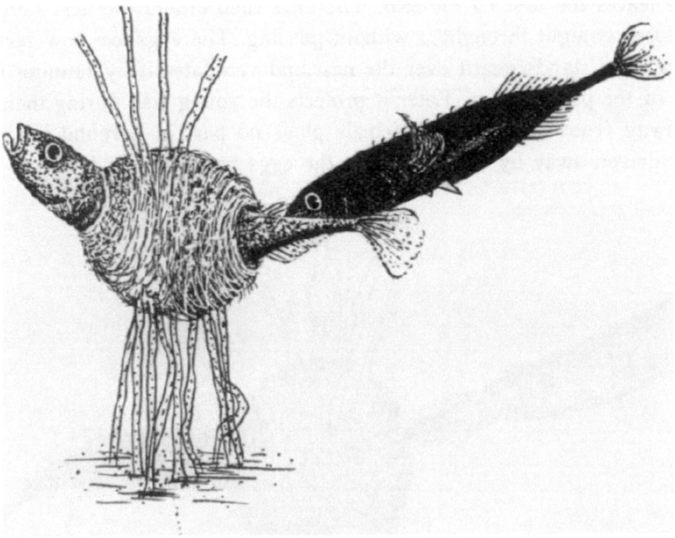

Fig. 8. Courtship E. Shivering. The female having entered the nest the male vibrates its nose along her tail.

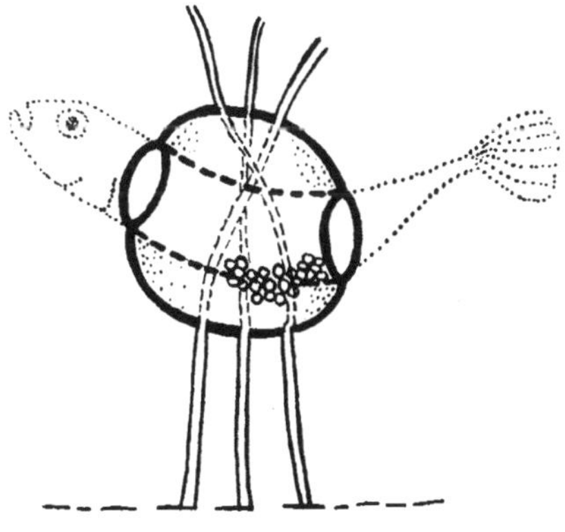

Fig. 9. A section through the nest at the time of spawning, showing the eggs.

tail of the female. After a few seconds the eggs are laid (Fig. 9) and the female leaves the nest by the exit. The male then enters the nest (Fig. 10) and passes straight throught it without pausing. The eggs are now fertilised and the male stands guard over the nest and ventilates it by fanning movements of the pectoral fins. Later it protects the young fish during their first days away from the nest. The female plays no part in parental behaviour, and is driven away by the male after the eggs have been laid.

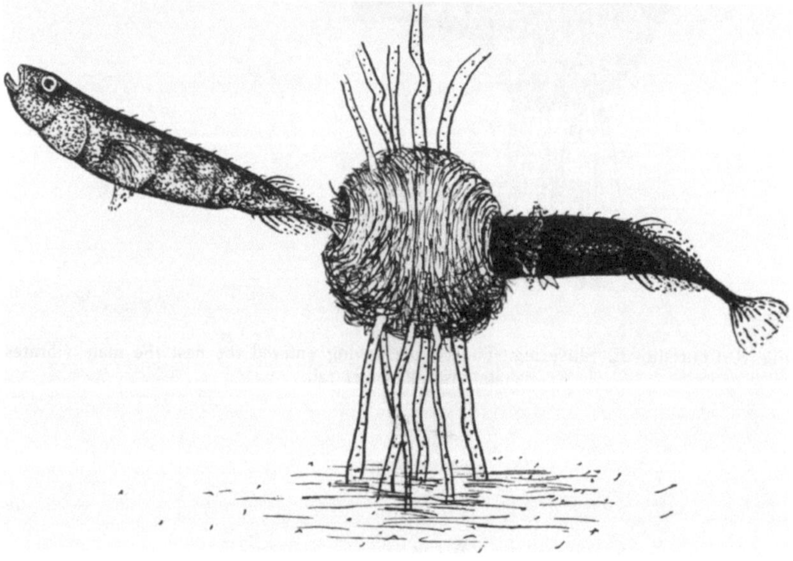

Fig. 10. Fertilisation. The female has spawned and is leaving the nest by its exit. The male is about to enter the nest and pass over the eggs, fertilising them as it does so.

There are many minor variations in this courtship pattern, but these will be discussed more fully in subsequent papers. Sufficient detail of the normal behaviour has been given above to make possible a comparison here between it, and the abnormal homosexual behaviour that has been observed.

HOMOSEXUAL BEHAVIOUR

CONDITIONS PRODUCING HOMOSEXUAL BEHAVIOUR

In the experimental tank in question two nests were built during the first few days, the owners of these nests taking on the full reproductive colour, a jet black which covers almost the entire body surface. The other three males were also seen to be aggressive, but to a lesser degree than the nest owners. These fish without nests were usually partially black but assumed cryptic coloration when repeatedly attacked and beaten. They did not, however, lose the white colour of the ventral spines and this could be used as a method of sex identification. The cryptic coloration is identical for both sexes, a fact which will be seen later to be of importance in the homosexual behaviour. It could be seen that these fish without nests were restricted in respect of the areas of the tank which they occupied, but their partially formed territories were only vaguely defined. During the following days two things became obvious concerning the nestless males. Firstly, they had little chance to settle and build nests because of the repeated attacks of the nest owners, and secondly, they were highly motivated sexually. Several observations led to this second conclusion. On occasions when females were introduced for a short while, these nestless males were several times seen to court, and lead the females to imaginary nests. Also glueing [1]), an activity concerned exclusively with nest-building, was observed, although there was no nest to be glued. Boring [2]) was seen, and this activity also is normally only performed into the nest entrance, but was in this case performed into a small lump of weed that looked like a nest, but which had, in fact, only just been put into the tank as nest material for the fish. Also, as mentioned, the fish were occasionally aggressive and were attempting to form territories.

It was concluded from these observations, and will be confirmed below, that the tank contained fish, all of which were sexually motivated to a high degree, but some of which were sexually frustrated, being prevented from fully expressing themselves sexually by the presence of the dominant males.

Throughout the weeks during which the tank was under observation the number of nests, and the owners of the nests, varied from time to time, but at no stage were all five males nest owners (Fig. 2). The volume of water in the tank was not great enough to permit the presence of five full terri-

---

1) G l u e i n g: "Glue" is a kidney secretion produced during the breeding season, which is used in nest building. The nature of the glueing movement differs from species to species. The Ten-Spined species glues in two different ways. It can either deposit the glue directly onto the nest surface as it moves its body over the nest, or it can exude a blob of glue some distance from the nest, taking this blob up in its mouth with a circling movement of its body, and then swim with it to the nest. It was this second type of glueing that was observed in the instances mentioned above. When produced by these nestless males, and taken up into the mouth, it was then swallowed, there being no nest into which it could be pushed.

2) B o r i n g: This is a strenuous pushing movement of the snout into the nest entrance, and is nearly always performed in combination with fanning, which follows it. It has the function of keeping the tunnel clear.

tories. The result was that the tank always contained some sexually frustrated males. It was the interaction between these frustrated males and the nest owners that provided the demonstration of homosexual behaviour. Had the tank contained fewer males there would have been no frustrated males, and the homosexual behaviour would not have occurred. Had there been a large number of males in the tank, it is probable that no territories would have been set up at all, as no single male would under those circumstances be able to keep part of the tank clear of fish long enough to settle and build. The reason why this homosexual behaviour has not been described before is probably because its occurrence depends on the presence of a particular population density, which is halfway between the two densities normally employed in the laboratory. For workers in this field normally use a very high density if they wish to prevent sexual behaviour, and a very low density if they wish to encourage it.

Concerning the conditions which provoke homosexual behaviour, it remains only to add two points. Firstly, the amount of water-weed present in the tank is a factor that must be taken into consideration, as this will control the amount of building space and cover. WUNDER (1930) found that, with the Three-spined species, artificial shelters put into a tank, in the form of screens, increased the number of territories. The tank employed here was planted with Willowmoss (*Fontinalis antipyretica*) in three hedgerows across the tank, covering about one half of its floor space. Willowmoss provides dense vegetation to a height of about six inches. Amongst these hedgerows were planted a number of taller strands of *Myriophyllum spicatum,* which rose the eighteen inches to the surface of the water. The fish build in both types of vegetation, but appear to prefer the willowmoss. Secondly, the nest owners were probably frustrated themselves, as females were only presented for short observation periods of an hour or so, during several weeks.

INDISCRIMINATE COURTING

Homosexuality on the part of the nest owners took the form of lack of discrimination between males and females. These nest owning males danced not only to females when they were introduced, but also to other males. This occurred most frequently when they had been dancing to females which were then removed. If at that point a cryptically coloured male came into the territory, it was also danced to. This courtship of cryptic males by black nest owners was also seen occasionally at times when no female had been introduced, although more often in such cases the intruding male was attacked in the normal way and driven off. Only black nest owning males courted homosexually in this way, and they nearly always confined their attentions to other males that were not black, although on rare occasions one black male was seen to court another black male.

If these males were presented with male and female models there were times when they courted one as frequently as the other.

In these cases the cryptically coloured males never responded by following, so that the courting nest-owners were never able to reach the nest-entrance-showing stage of courtship. When the cryptic males did not follow they were invariably bitten, whereupon they fled.

FRUSTRATED MALES BEHAVING AS FEMALES

So far there was no indication that there was any reciprocation of homosexual behaviour on the part of the courted males. It will become clear that

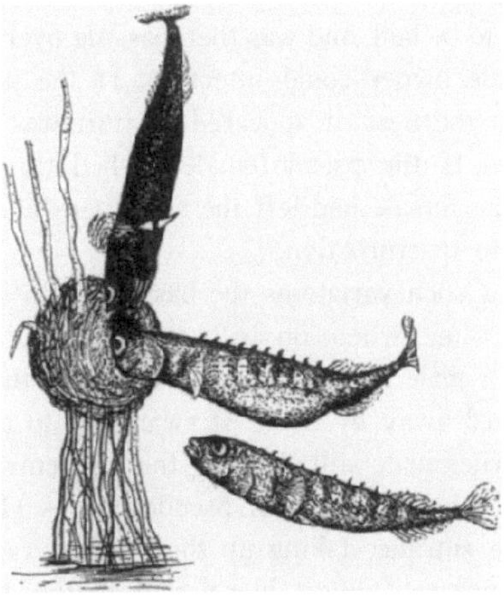

Fig. 11. A pseudofemale fish (lowest one in figure) watches as a ripe female pushes into the nest entrance in response to a male's courtship movements.

this was because in cases of indiscriminacy the courting males did not reach stages three or four (showing or shivering) of the courtship. Only these later stages of courtship initiated homosexual responses on the part of the frustrated males. The first indication of this was when one of the cryptic males was seen creeping stealthily through the weeds towards a nest at which the owner was showing the entrance to a female (Fig. 11). The cryptic male lay still in the weeds for some seconds, quite near to the nest, but was then seen to approach the entrance quickly. It then pushed the female out of the way, entered the nest, and passed through it.

It was at first thought that the intruding male was attempting to 'steal' a fertilisation, and that it was trying to take the place of the male owner, but further observations and tests revealed that it was, in fact, taking on the role of a female. I shall call such a male showing the female pattern a p s e u d o f e m a l e.

This homosexual pattern inversion was repeated a number of times on different occasions and gradually a more detailed picture of its nature was constructed. This involved many observations, for the abnormality of the activity rendered it extremely variable, there being a minimum of three individuals involved—the male owner, the female, and one or more pseudofemales—and a slight change in the time sequences of the performances of their various actions produced an apparently different phenomenon. For example, if the female managed to get into the nest and lay its eggs before the pseudofemale advanced to the entrance, it appeared as if the latter was waiting for the eggs to be laid, and was then passing over them and fertilising them before the male owner could interfere. If the pseudofemale pushed past the female into the nest, it appeared as if it was competing with the female for her role. If the pseudofemale pushed through the nest when both the male and the female had left the site, then the activity appeared to be something akin to masturbation.

From a number of such variations the basic pattern was deciphered. The first important clue came, when, on an occasion where three pseudofemales were advancing on a male and a female courting at the nest entrance, the female was frightened away by them. It swam off to the other end of the tank and left the owner male still showing the nest entrance, but showing it now, not to a female, but to a trio of pseudofemales (Fig. 12). These fish clustered around the entrance taking up the position vacated by the female. From above they appeared almost identical with the departed female, and the owner male treated them as if they were females. The movements of the pseudofemales were still stealthy, which is not surprising as they were in the most defended area of a territory belonging to an all-black male. They were seen to respond to the showing movements of the owner by pushing into the nest entrance with their noses, in exactly the same way as the female. On several occasions when they moved away from the nest a short distance, the owner ceased showing, and danced to them, whereupon they followed in the female manner with their noses between the white ventral spines. (It will be remembered that earlier it was stated that dancing alone was insufficient stimulus to initiate homosexual pattern-inversion, and yet here, following in response to dancing is reported. This is not, however, the contradiction it appears to be, for in this case the homosexual behaviour had already been initiated by the entrance-showing, and once initiated, it will then occur in response even to the dancing movements.) Also, when the pseudofemales reached the entrance again, they lowered their noses from the ventral spine region to the region of the nest entrance, — again a female reaction. Furthermore, when one of them entered the nest, it lay in it while the owner shivered on its tail. This pausing in the nest is characteristic of the

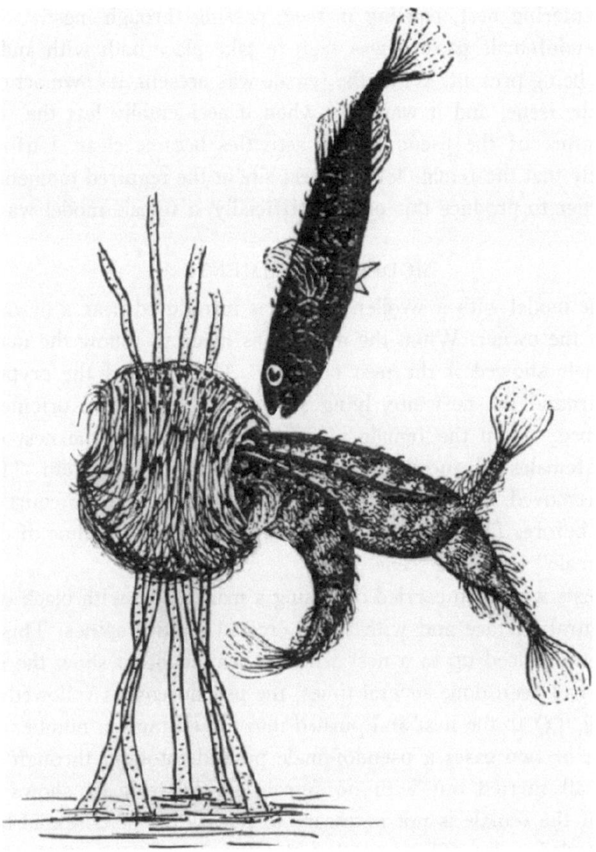

Fig. 12. Three pseudofemale individuals attempt to enter the nest, having driven away the ripe female as a result of competing with her for her position.

female passage through the nest when it lays its eggs, the fertilisation movement of the male being performed without a pause inside the nest. Pauses of up to 40 seconds have been recorded for these pseudofemales, which means that the length of these pauses is similar to the length of the female egg-laying pauses.

This instance and subsequent similar instances revealed that the pseudofemales would perform the complete female pattern under these circumstances, with the single obvious exception of egg-laying. Furthermore they were not taking on a passive role, but an active performance of six female movements, namely: following, nose-lowering to entrance, pushing into entrance, entering nest, pausing in nest, passing through nest.

This pseudofemale pattern was seen to take place both with and without the female being present. When the female was present, its own actions often confused the issue, and it was only when it accidentally left the scene that the real nature of the pseudofemale activities became clear. Unfortunately it was rarely that the female left the nest site at the required moment. Therefore, in order to produce this effect artificially, a female model was used.

MODEL EXPERIMENTS

A female model with a swollen belly was introduced near a nest, and was courted by the owner. When the model was made to follow the male to the nest the male showed it the nest entrance. At this point the cryptic males gathered around the nest site, lying still on the floor, and oriented to the nest entrance. When the female model was pushed into the nest entrance, the pseudofemales advanced and competed with it for position. The model was then removed, and the pseudofemales took its place, the courtship proceeding as before. This technique enabled me to control the time of departure of the "female" from the scene.

Model tests were also carried out using a male model with black coloration on the ventral surface and with white erected ventral spines. This was introduced and danced up to a nest where it was made to show the entrance. After this had been done several times, the pseudofemales followed the male model (Fig. 13) to the nest and pushed into the entrance a number of times, and in one or two cases a pseudofemale pushed into and through the nest. This was all carried out with no female in the tank. It shows that the presence of the female is not necessary to release the homosexual behaviour of the pseudofemales. The fact that the female is present when this behaviour occurs is purely incidental, as it is only then that the male owner reaches the later stages of courtship, at the nest, which produce the initial homosexual reponses. Without these model tests it would have been impossible to say whether the presence of the female was essential or not.

Her presence, although proved to be inessential, does nevertheless appear to help produce the response, for if a female model is introduced and brought up to, and pushed into, an artificial nest, where there is no male owner, then its presence does attract the pseudofemales. An artificial nest has to be used in this case, in order to ensure the absence of a male owner. Artificial nests used in these and other experiments were constructed from thin wire painted green, and green cotton. At the end of a length of wire a wire cage is made and green cotton is densely wound around this leaving a central tunnel.

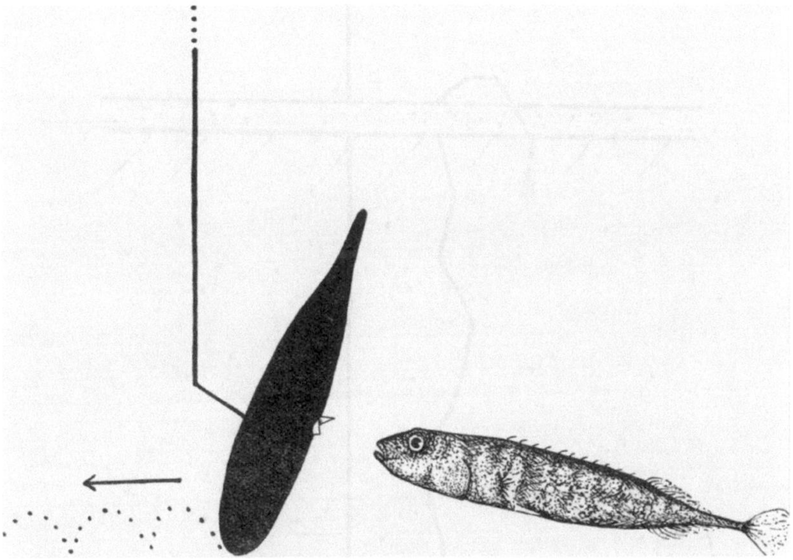

Fig. 13. A pseudofemale fish following a model of a male that is made to dance.

In this case, where there is no male showing the entrance to the female model, which is being pushed into the nest entrance, it is possible to observe whether the presence of the showing male in the other cases was inhibiting the cryptic males from taking up the male position. But even here they compete with the female model for her position at the nest, despite the fact that the space just above the model, which a male would normally occupy, is vacant (Fig. 14). This helps to confirm the interpretation that the behaviour pattern of the cryptic males is essentially female and not male. It also appears to complicate the issue concerning what mechanism is involved (see later section on mechanism).

Fig. 14. A model of a ripe female is pushed into the entrance of an artificial nest. Pseudofemales competed with this model for position, despite the fact that the normal male position—as shown by the fish figured with dotted lines— was vacant in this case.

FRUSTRATED FEMALES BEHAVING AS MALES

Ripe females kept in a tank without males do not show any homosexual behaviour, they simply shed their eggs after a while and then immediately eat them. Only one male activity has been seen to be performed by a female, and that is fanning. (This has been observed several times by the author. It was first seen by TINBERGEN who has a cinematic record of it.) Fanning is normally exclusively performed by the male, where, when authochthonously expressed, it is part of the parental pattern, and, as mentioned before, is concerned with egg ventilation.

The female is only seen to fan on rare occasions and under specifically frustrating conditions. When males are without nests they may court females if they are very highly motivated sexually. On these occasions they may even show the female the entrance to an imaginary nest, which they have yet to build, in the weeds. The female can be seen to respond to the imaginary-entrance-showing by pushing into the weeds. It is then that the female may be seen to fan sometimes. Such fanning may also occur if there is a nest present which is blocked in some way so that the female cannot enter. The female fanning movement is, in any case, a response to being unable to enter the nest at the appropriate moment.

THE FUNCTION OF HOMOSEXUAL PATTERN-INVERSION

It seems almost certain that the function of this behaviour complex is that of an outlet which relieves sexual frustration, but two other possibilities must be considered:

1. Egg Stealing. Very often, as a result of the pushing through the nest by the pseudofemales, the eggs in the nest, if there were any present, fell into view. Often they fell completely out of the nest onto the floor of the tank. They were then eaten by the pseudofemales that remained at the nest site, despite the attacks of the owner. It might be assumed from this that the cryptic males behaved as females when a real courtship was in progress, because this was the only way in which they could approach close to the nest without being molested by its owner, and that they only came close to the nest at this point in search of eggs. To put it in another way it might be assumed that they are egg-thieves that find themselves in a situation were submissive behaviour is called for, and this takes the form of female behaviour. Even if this is so, it does not alter the basic fact that they can and do behave as females.

It appears more likely, however, that the egg stealing is brought about accidentally by the female behaviour. It is certainly true that whenever the female managed to lay her eggs before the pseudofemales invaded the nest, the eggs were always knocked from the nest and eaten. But it was more frequently the case that the pseudofemales competed with the female for her position below the male at the entrance, and pushed her out of the way and entered the nest themselves. If the function of pseudofemale behaviour was to steal eggs, the pseudofemales would certainly have waited until the female had been through the nest before pushing into it. Also, if they behaved as females until they reached the nest which contained eggs, or was to contain eggs, it would then be a simple matter to pull the eggs, when laid, from the nest without passing through it. Further, if egg-stealing is the function, there is no point in a pseudofemale reacting to a courting male owner in the absence of a female.

It should also be mentioned at this point that these fish were fed regularly with *Tubifex, Daphnia,* and chopped Earthworm throughout the whole period during which the observations reported in this paper were made.

2. Female 'stealing'. This possibility is virtually ruled out by the evidence now assembled, although, as above, it may occur as the accidental outcome of the behaviour pattern. It is unfortunate that it is practically impossible to tell if such pseudofemale passages through the nest involve the ejaculation of sperm. If they do, and the owner manages to save the eggs from being eaten, which it may possibly do sometimes, then the pseudofemale will, in fact, have stolen a fertilisation. But it is unlikely that this is the function of the pattern because of the decidedly female nature of the behaviour of the cryptic males, even, as model tests showed, in the absence of an owner male. Furthermore, if the same argument is applied here as before, that the males can only reach the nest where they can steal the fertilisation by behaving as females, this possibility is still unlikely because of the pause in the nest made by the pseudofemale. It is true that males may pause in their own nests when passing through them during nest-building, but they never pause in the nest when they are actually fertilising eggs. So, if they do ejaculate sperm when they are "pretending to lay eggs", then although the eggs may get fertilised in this way, it is practically certain to be an accidental outcome of the homosexual pattern.

It would appear to be the case then, that pseudofemale behaviour on the part of sexually frustrated males is an outlet that serves to relieve frustration. In all essentials it is similar, as an outlet, to a displacement activity, except that it is more complex than most other recorded displacement activities. It is more than a displacement activity, it is a displacement pattern; but then it is not an act, but a whole pattern that is being thwarted. It should also be noticed that whereas other displacement activites "spark-over" from Drive A-male to Drive B-male (or from Drive A-female to Drive B-female), in this case there is sparking-over from Drive A-male to Drive A-female.

THE POSSIBLE MECHANISM INVOLVED

It might be expected that males frustrated in the manner described would attempt to express themselves sexually by stealing a nest and territory from a dominant male and driving it away, rather than employ a dormant female pattern. But the importance of relative dominance appears to be so great to such a fish that it is easier for it to call upon a contra-sexual pattern that it would never normally perform. It is well known for a number of species that an injection of female sex hormones into a male will result in the performance of the female sexual pattern, so that the nervous equipment for female sexual behaviour is present in the male, but is not used because the hormone balance favours the male pattern. In the case of the pseudofemale sticklebacks, the power of the dominant male sex hormone is over-ridden by the external inhibition of the activities normally resulting from the action of the male hormone.

The speed of the change-over from male to female behaviour on the part of the thwarted males is so great that it seems unlikely that a hormone change has taken place. For example, a cryptic, thwarted, nestless male may court a female for a few seconds sometimes, before it is interrupted by a black nest-owner. The latter, dancing more intensely, leads the female off to its nest, and straight away, in a matter of a few seconds, the previously courting cryptic male will follow the now courting couple and behave as a pseudofemale. This very quick change-over in a few seconds is almost certainly under nervous control, rather than as the result of some change in sex hormone balance.

It has been suggested to me by Tinbergen however, that if this is so, it is strange that, in the case of the model experiment utilising a female model and an artificial nest, the pseudofemales do not — in the absence of a male owner — at once switch over to male behaviour. At other times, as stated above, they will, if unmolested for a moment in the presence of a ripe female, snatch that moment and express themselves as males, even though they have no nest. Why then, when a female model is being pushed into an artificial nest, with no male owner in sight, should they compete with the model, instead of snatching this moment also for male expression? The answer, I feel, is not that the latter case reveals some hormonal change, but rather that it reveals that the "nest which is not one's own" is sufficiently intimidating, even without the presence of its owner. And, in a way, the fact that a ripe female is pushing into the entrance indicates that there is an owner, even though no owner is to be seen.

FORD and BEACH (1952) reporting instances of pseudofemale behaviour in rats (see section in this paper on other species) state that "The physiological basis ... is not completely understood, but it does not appear to involve hormonal abnormalities". One pseudofemale rat was castrated and the result was that both male and pseudofemale responses disappeared in a few days. Male hormone was then injected daily and the outcome of this was that b o t h male and pseudofemale behaviour was shown again. One can conclude from this that pseudofemale behaviour c a n be under nervous rather than hormonal control. Further evidence that this is so in the case of the sticklebacks is that the individuals used were successful at mating as males and rearing young, both before and after their spell in the crowded tank where they behaved as pseudofemales. The rare cases of individuals of various species, where there is some hormonal abnormality that is causing pseudofemale (or pseudomale) behaviour, are usually incapable of functioning as their real sex, either before, or after, or both.

ECOLOGICAL SIGNIFICANCE

It may appear at first that the crowded conditions produced artificially in the laboratory would never exist in natural surroundings, the argument being that sticklebacks finding themselves this crowded would tend to spread out. But it must be remembered that not only do sticklebacks often become isolated in small stretches of water, but also that the areas of a river in which the Ten-Spined species can breed are limited. The conditions required are profuse fresh vegetation of a type in which nests can be built, water which is not very fast moving (as this would cause too much movement of the weeds and endanger the nest), water which is well oxygenated and of a particular pH, and water which does not dry up in the possible high temperatures of late Spring and early Summer.

The rivers from which the fish used were collected, for example the River Kennet in Wiltshire, are in most places too swift-moving for breeding purposes, and the fish are mostly found in slow-moving backwaters. It is quite possible that the population density in these backwaters does rise some years to the level employed in the laboratory in this case. Certainly, judging from the large numbers caught in one small backwater this Spring, the population density, it appears, is quite likely to rise to that level. It is true that the fish in such a backwater could migrate to another suitable area, but it will be seen that if the whole river population reaches a certain level, there will be a season or two when the homosexual pattern-inversion behaviour does take place in the natural state, and with it the accompanying phenomena of egg stealing and eating. In one or two seasons this might bring the population density down considerably, to a level where homosexuality and egg stealing would no longer occur. Although this species is difficult to watch in its natural state, it is hoped that the ecological significance of the phenomenon will be tested by field observations in the future.

Although there is little observational information concerning the behaviour of this species in the wild, there is a certain amount known about its ecology, and that of the Three-Spined Stickleback. As will be mentioned in the next section, the latter species has been observed to exhibit pseudofemale behaviour also, but it has so far been seen in less detail. All indications, however, point to the phenomenon being much the same in both species.

It has been pointed out by HEUTS (personal communication) that with the Three-Spined species there is often crowding on the breeding grounds, in different parts of Europe. He reports having obtained, in certain seasons, several hundred sticklebacks with one haul of a net on the breeding grounds, although he admits that these were exceptional years. But the very fact that this happens, and what is more, does not happen each year, adds weight to the suggestion that pseudofemale behaviour could occur significantly in the wild state. HEUTS also reports that examinations of stomach contents that he has carried out, often reveal the presence of eggs. Furthermore he has been able to identify these as being Stickleback eggs. So that it can be said that both crowding and egg eating do occur in the wild state.

A certain amount of quantitative evidence is available concerning this. BLEGVAD (1917) examined the stomach contents of 427 specimens of *Gasterosteus aculeatus,* and 112 specimens of *Pygosteus pungitius,* taken from Danish waters. In the case of *G. aculeatus,* 87 % of the food eaten was Arthropodal, 8 % was Stickleback eggs and larvae, and 5 % was various other food. For *P. pungitius* the figures were 72 % for Arthropods, and 28 % for Stickleback eggs and larvae. So it appears that not only do these fish eat their own eggs in nature, but that these eggs can make up an important part of their diet. More recently HYNES (1950) has investigated the problem, and his report provides two relevant points that are not available from BLEGVAD. Namely, that although both eggs and larvae are eaten, the latter are relatively rarely found in stomach contents. This means that BLEGVAD's figure of 28 % for eggs and larvae for *P. pungitius* probably consisted mostly of eggs. Also HYNES found that the majority of fish that had eaten eggs, were males. It is important to have this information, because, as already mentioned females may shed their eggs (if no mate is available at the required moment) and eat them, and it is necessary to be certain that all egg-eating in nature is not of this type. Another criticism might be that the eggs eaten in the wild by male fish could be devoured as a result of the habit of such fish of removing and swallowing an egg, from their own nest, that is not developing satisfactorally. But the evidence HYNES presents does not point to this, for many eggs were often found in one stomach, and this leads HYNES to say that it seems as if non-breeding males raid nests. HYNES also states that in the streams with which he was working, there was a very rich prey fauna, and apparently no scarcity of food.

This ecological evidence may be summarised as follows:

1. Crowding does occur in nature.
2. Egg eating does occur in nature.
3. The egg eating is probably egg stealing.
4. There is plenty of food available other than eggs.

This, coupled with the laboratory observations described in this paper, points to the fact that pseudofemale behaviour probably occurs in nature. What the ecological evidence does not do is solve the problem of whether the pseudofemale behaviour is an elaborate egg-stealing device, or whether it is a displacement-like outlet relieving sexual frustration, with egg stealing as an accidental outcome. It will be seen that it is a question of whether the level of crowding necessary to produce pseudofemale behaviour and egg stealing is high enough to reduce the normally abundant food supply to a low enough level relatively, to make it an advantage to have a mechanism to obtain eggs as food. It is true that the laboratory observations have lead me to the opinion that the phenomenon is a displacement-like outlet, but it would nevertheless be valuable to have some field observations on this point.

HOMOSEXUALITY IN OTHER SPECIES

Similar homosexuality (both indiscriminacy and pattern inversion types) has been observed for the Three-Spined Stickleback, in similar circumstances. However, as mentioned above, in this species it has been observed in less detail. It was in this species that I first observed stickleback pseudofemale behaviour, but I did not at the time understand what I was seeing. I thought then that it was fertilisation-stealing as a result of frustration. It is perhaps worth recording here what pseudofemale behaviour has been seen for the Three-Spined species. The following is a condensed version of notes made at the time:

Two males placed in a tank of dimensions 24 × 12 × 15. Both take up territories straight away. A is smaller than B, which is constantly attacking it. B's territory is much larger than A's. (See Fig. 15 for tank plan.). B builds a nest in a corner of the tank. A builds no nest. A ripe female is introduced and is courted by A despite the fact that it has no nest. B takes over the female and courts her repeatedly. A never has another chance of courting the female, but watches the courtship of the other two. When the courtship between B and the female is at the nest entrance stage, A is seen

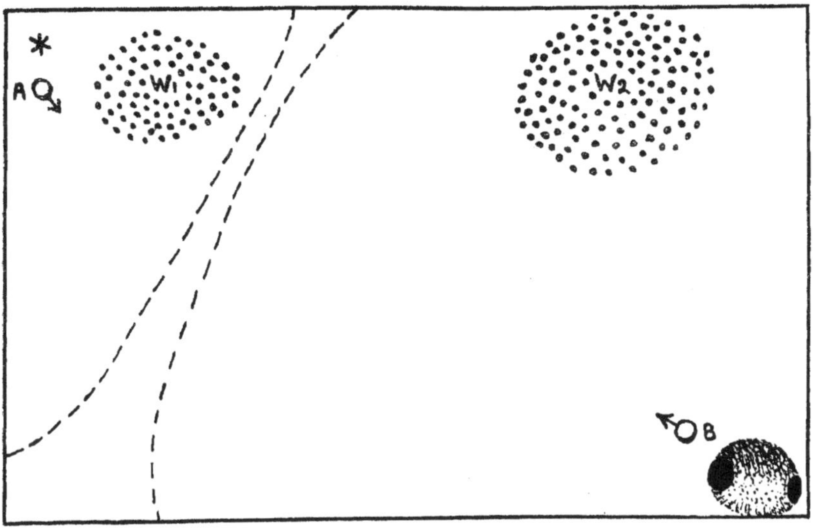

Fig. 15. Tank plan showing territories of two males A & B. The boundaries are shown by broken lines. W1 and W2 are two clumps of water-weed. Male B was dominant, and built a nest as shown. Male A attempted unsuccessfully to build at the point marked with an asterisk.

to creep stealthily along the floor of the tank in the direction of the nest, often lying quite still for some seconds. On such occasions B always notices A, and chases it away.

After a number of courtships the female enters the nest and lays her eggs. She is followed through the nest by the owner, B, but while B is passing through the nest A rushes across the floor of the tank and pushes into the nest. B hovers outside the nest and watches A, who is now lying in it like a female. After some seconds, A leaves the nest and is chased and savagely bitten by B. The eggs are still safe in the nest. (They subsequently developed and were reared.). A, although nestless, returned to its corner of the tank and immediately performs the following movements: a) sand-digging, "nest"-mending, c) glueing. B, at the same time, was at its nest performing: a) fanning, b) boring, c) "nest"-mending, d) glueing.

It seems fairly clear from this that pseudofemale behaviour will prove to be much the same for both species. Two interesting new points arise from the above notes. Firstly, the eggs were not destroyed in this case, whereas the Ten-Spined pseudofemales always destroyed them. There are three factors that probably contribute to this difference. There was only one Three-Spined pseudofemale, whilst there were several Ten-Spined pseudofemales. Three-Spined male B was very aggressive towards male A, probably because in this species the pseudofemale coloration is less like the real female coloration, than it is in the Ten-Spined species. The nest of the Three-Spined species is a covered depression in the sand, and it is therefore less likely that eggs will be knocked out.

The second interesting point from these notes is the observation that immediately after behaving as a pseudofemale, male A performed nest-building movements. It had not built a nest, and had never been seen to perform any nestbuilding movements before, but as soon as it had behaved as a pseudofemale, there was an outburst of nestbuilding activity. This was never seen in the Ten-Spined species.

Homosexual behaviour has been observed for a number of other species, but it is fair to say that in most cases it takes the form of a sexually active male courting another male instead of a female. It would appear that where one sex performs the pattern belonging to the other sex, it is more frequently the female that performs the male pattern, rather than *vice versa*. Observations of males performing female sexual actions are not common, and reports of males performing the whole, or almost the whole, sexual pattern of the female are rare. (It must be remembered that the sexual behaviour of the male is nearly always more marked and easier to detect than the female sexual behaviour, so that pseudofemale activities may, in fact, occur as frequently as pseudomale activities, but be less often noticed as such.).

There is little information available concerning pseudofemale behaviour in other species of fish, although a few points can be extracted from the Cichlid work of Baerends and Baerends (1950). It appears that similar territorial conditions can exist for a tank of Cichlids, as have been described here for Sticklebacks, with several dominant territory owners and several territoryless males in one tank. Under these conditions homosexual pattern-inversion was seen on one occasion by Baerends and Baerends. On this occasion an intruding male behaved very much like a female, performing three active female courtship movements (following to pit, circling, and snapping up: see Baerends.). Concerning the two male fish involved in this case, they mention that, "The resident male, as well as the intruding one, had been kept isolated during a rather long time." There are two other points from this paper which, although not actually dealing with homosexual behaviour, are nevertheless relevant. Firstly, on one occasion when eggs had just been laid after a courtship, a second male rushed across the tank and took part in the fertilising of them, and although attacked by the owner, could not be driven away. Secondly, it was observed that two frontally fighting male Cichlids were often assisted by "territoryless congeners". When the fight was over, the territory owners were seen to drive their "helpers" away. Concerning this the authors state that "It looks as though the helpers take the opportunity to give expression to their fighting urge which is otherwise not possible to them, having no territories." It is not intended to discuss these observations here except to say that it is quite obvious from the work of BAERENDS and BAERENDS that it would be rewarding to study this aspect of behaviour in more detail, not only in the Sticklebacks, but also in the Cichlids.

In certain species of bird the reversal of pairing position has been seen to occur during courtship. This has been reported for the Great Crested Grebe, *Colymbus cristatus* (SELOUS 1901; HUXLEY 1914), the water hen, *Galinula chloropus* (SELOUS 1901), and the Pigeon (WHITMAN 1919). Homosexual "pairs" are well known amongst the Anatids.

STONE (1924) and BEACH (1938) have independently reported rare cases of male Rats exhibiting the female pattern, and in these reports there appears to be a situation akin to the one related here. In the first place, male rats with a high level of sexual motivation may attempt copulation with other males. In a few isolated cases the males that are being courted respond with the female pattern, although in most instances they fight off the other male. These pseudofemale rats behave as males when placed with a receptive female (as do the Sticklebacks), and it has been observed that the switch-over from one pattern to the other may occur in a few seconds. This has lead BEACH to call the phenomenon "reversible inversion" of the usual sexual role.

Pseudofemale behaviour has been reported in some detail for certain Monkies and Man. Here the effects of experience complicate the issue. WILLIAMS (1944), in a study of homosexuality in Man, employs the two designations FMH and MMH for Feminine Male Homosexual and Masculine Male Homosexual respectively. (FMH = pseudofemale). It has been shown that for both types, homosexual behaviour may be p r e f e r r e d to heterosexual behaviour, even though the latter is available to the individuals concerned. In these cases the inversion of the usual sexual role has become irreversible as a result of conditioning. (It should be remembered that such irreversibility may also be caused, in certain individuals, by some hormonal abnormality.)

It would appear then that complex homosexual pseudofemale behaviour has only been observed in Sticklebacks, Cichlids, Anatids, Rats, Monkies, and Man. The fact that the species in which the phenomenon has been observed are just those species which the Comparative Ethologists and the Comparative Psychologists have studied particularly elaborately, is perhaps an indication that such behaviour is of a much wider occurrence in the animal kingdom than was previously believed, and that it is only revealed after a detailed study of the animal concerned has been carried out.

DISCUSSION OF TERMINOLOGY

Throughout this paper the term Homosexual has been used in its widest sense to refer to any sexual relationship between two individuals of the same sex, regardless of the details of the case. As will have been seen by now, there are several distinct ways in which homosexual behaviour can occur:

I. Only sensory field involved.

A. Lack of discrimination: The male may fail to identify the sex of the indended partner.

B. Sufficiency of sub-optimal stimulus: The male may identify the sex of its intended partner as being male, but still react sexually to it.

C. Preference for own sex. The male may prefer a sexual partner of its own sex, as a pseudofemale partner.

II. Reversal of motor pattern involved.

D. Inversion of sexual pattern. The male may exhibit the female sexual behaviour pattern in response to the advances of males of types A, B, or C.

E. Preference for inversion. The male may prefer to behave as a pseudofemale (*i.e.* type D) despite the availability of heterosexual outlets.

(The above list applies equally well, of course, to the female sex.)

Type A homosexuality may be the result of the behaviour or appearance of the intended partner being, accidentally or otherwise, similar to that of the opposite sex. (For example, an over-fed male fish with a swollen belly may be mistaken for a female with her belly swollen with eggs.) Both type A and B homosexuality may be the result of an individual being very highly motivated sexually, so that the combination of this strong internal stimulus, with the small external stimulus provided by another member of the same sex, will be sufficient to release sexual behaviour. Type C homosexuality may be the result of the conditioning of a type A, or more probably, type B homosexual. The latter may modify its behaviour to type C, as the result of experiencing satisfaction from homosexual activity. Type D homosexuality may be caused, as shown in this paper, by the frustration of the reproductive behaviour in the presence of dominant males. Type E homosexuality may develop by conditioning from type D, in the same way in which type C may develop from type B. The occurrence of type E may also be the result of some physiological abnormality in an individual.

Types A, B and D homosexuality have all been shown to occur in male Sticklebacks. It has not always been possible to distinguish cases as being definitely of type A or B, but on the rare occasions when black males courted other black males, type B was certainly occurring, and in instances where at the nest a pseudofemale had taken over from a female, then it is almost certain to have been type A.

Types C and E homosexuality are, as far as I know, restricted to Primates. As already mentioned, the inversion of the sexual behaviour pattern may occur in heterosexual as well as homosexual situations, as it does for example, in the case of the female stickleback that fans.

Certain Authors (KINSEY, POMEROY and MARTIN, 1948 and FORD and BEACH, 1952) have discussed in detail the difficulties and confusions involved in the terminology of reports of homosexual behaviour. I feel that matters would be simplified if the term homosexual were retained and used in future for referring to all sexual relationships between two individuals of the same sex, and if all the synonyms for homosexual which have been used in the past were now abandoned. Use of the word homosexual as a general term along with the classification of the five basic types presented here, would perhaps prevent further ambiguity. It is realised that in studies of Human homosexual behaviour there will, naturally, be a need for the retention of a number of specialised terms such as, for example, Transvestitism (HIRSCHFIELD, 1940), which refers to individuals that obtain sexual pleasure from dressing as members of the opposite sex.

SUMMARY

The normal reproductive behaviour of the Ten-Spined Stickleback (*Pygosteus pungitius* L.) is described briefly, and the conditions that are found to produce abnormal courtship behaviour are reported.

The males of this species build nests in water-weeds, and the territory around each nest is defended by its owner. At a particular population density only some of a group of males are able to set up territories and build nests, and these individuals become dominant over the nestless males. If there is a scarcity of females, the dominant males behave towards the nestless males as if the latter were females. If there is a normal courtship in progress between a dominant male and a female, then the dominated males are often seen to compete with the female of the pair, and to exhibit the complete female courtship pattern. Such fish have been called "pseudofemales".

These observations reveal that the nervous mechanisms of the sexual behaviour pattern of the other sex are present, although they are not normally functional.

Pseudofemale behaviour is analysed further with model experiments. Its function appears to be the relief of frustration in this case, and it is noted that here where a whole pattern is being thwarted, the result is a displacement pattern. The speed of the inversion of the sexual role seems to exclude any possibility of hormonal change being responsible.

It is probable that the phemomenon has ecological significance, as an outcome of its occurrence is often the destruction of eggs.

Homosexuality in other species is briefly reviewed. Terminology is discussed, and a classification of types of homosexual behaviour is given.

LITERATURE

Baerends, G. P. and J. M. Baerends. (1950). An introduction to the study of the ethology of Cichlid Fishes. — Behaviour, Supplement I, p. 1-242.

Beach, F. A. (1938). Sex reversals in the mating pattern of the Rat. — Genet. Psychol. 53, p. 329-334.

—— (1948). Hormones and Behavior. — Hoeber, New York.

Blegvad, H. (1917). On the food of fish in the Danish waters within the Skaw. — Rep. Danish. Biol. Sta. 24, p. 19-72.

Ford, C. S. and Beach, F. A. (1952). Patterns of Sexual Behavior. — Eyre and Spottiswoode, London.

Hamilton, G. V. (1914). A study of the sexual tendencies in monkies and Baboons. — J. Anim. Behav. 4, p. 295-319.

Hirschfield, M. (1940). Sexual Pathology. — Emerson, New York.

Huxley, J. S. (1914). The courtship habits of the Great Crested Grebe (*Podiceps cristatus*). — Proc. Zool. Soc. London, p. 491-562.

Hynes, H. B. N. (1950). The food of freshwater sticklebacks. — J. Anim. Ecol. 19, p. 36-58.

Kinsey, A. C., W. B. Pomeroy and C. E. Martin (1948). Sexual Behavior in the Human Male. — Saunders, Philadelphia-London.

Leiner, M. (1929). Ökologische Studien an *Gasterosteus aculeatus* L. — Zs. Morphol. Ökol. Tiere 14, p. 360-400.

—— (1930). Fortsetzung der ökologischen Studien an *Gasterosteus aculeatus*. — Zs. Morphol. Ökol. Tiere 16, p. 499-540.

—— (1931a). Ökologisches von *Gasterosteus aculeatus* L. — Zool. Anz. 93, p. 317-333.

—— (1931b). Der Laich und Brutpflegeinstinkt des Zwerchstichlings, *Pygosteus pungitius* L. — Zs. Morphol. Ökol. Tiere 21, p. 765-788.

—— (1934). Die drei europäischen Stichlinge und ihre Kreuzungsprodukte. — Zs. Morphol. Ökol. Tiere 28, p. 107-154.

Marshall, F. H. A. (1922). The Physiology of Reproduction. — Longmans, New York.

Pelkwijk, J. J. ter, and N. Tinbergen. (1937). Eine reizbiologische Analyse einiger Verhaltensweisen von *Gasterosteus aculeatus* L. — Zs. Tierpsychol. 1, p. 201-218.

SELOUS, E. (1901). An observational diary of the habits of the Great Crested Grebe. — Zoologist 5, p. 180-182.
SEVENSTER, P. (1949). Modderbaarsjes. — De Levende Natuur 52, p. 160-168 and 184-189.
STONE, C. P. (1924). A note on feminine behaviour in adult male Rats. — Am. J. Physiol. 68, p. 39-41.
TINBERGEN, N. (1940). Die Übersprungbewegung. — Zs. Tierpsychol. 4, p. 1-40.
—— (1942). An objectivistic study of the innate behaviour of animals. — Biblioth. biotheor. 1, p. 39-98.
—— (1950). The hierarchical organisation of nervous mechanisms underlying instinctive behaviour. — Sympos. Soc. exp. Biol. 4, p. 305-312.
—— (1951). The Study of Instinct. — Oxford Univ. Press.
—— and J. J. A. VAN IERSEL. (1947). Displacement reactions in the Three-Spined Stickleback. — Behaviour 1, p. 56-63.
WHITMAN, CH. O. (1919). The Behavior of Pigeons. — Carnegie Inst. Wash. Publ. 257, p. 1-161.
WILLIAMS, E. G. (1944). Homosexuality: A biological anomaly. — J. Nerv. and Ment. Dis. 99, p. 65-73.
WUNDER, W. (1928). Experimentelle Untersuchungen an Stichlingen. — Zool. Anz. (Suppl.) 3. (Verh. Deutsch. Zool. Ges. 32). p. 115-127.
—— (1930). Experimentelle Untersuchungen am dreistachligen Stichling (*Gasterosteus aculeatus* L.) während der Laichzeit. — Zs. Morphol. Ökol. Tiere 16, p. 453-498.
ZUCKERMAN, S. (1932). The Social Life of Monkies and Apes. — Kegan Paul, London.

ZUSAMMENFASSUNG

Die Arbeit beschreibt kurz das normale Fortpflanzungsverhalten des zehnstachligen Stichlings und berichtet über Umstände, unter denen es abnorm wird.

Die Männchen besetzen und verteidigen ein Revier, in dem sie ein an Wasserpflanzen aufgehängtes Nest bauen. Sind zuviele Männchen in einem Behälter, so bauen nur wenige dominante Männchen; sie unterdrücken das Fortpflanzungsverhalten der übrigen. Bei Weibchenmangel behandeln die dominanten Männchen die anderen, als ob sie Weibchen wären. Wenn ein dominantes Männchen ein Weibchen zum Neste führt, können die unterdrückten Männchen mit diesem Weibchen konkurrieren, indem sie das vollständige weibliche Verhaltensinventar zeigen („Pseudoweibchen").

Demnach besitzt das Männchen die Verhaltensmechanismen auch des weiblichen Geschlechts, obwohl sie normalerweise nicht funktionieren. Umgekehrt zeigen trächtige Weibchen oft Übersprungfächeln, wenn es ihnen nicht gelingt, sich in ein Nest hineinzubohren. Nie jedoch fächelt ein Weibchen autochthon.

Wie Attrappenversuche lehrten, lässt sich durch den Anblick pseudoweibliches Verhalten sowohl eines balzenden Männchens wie eines balzenden Weibchens auslösen.

Pseudoweibliches Verhalten scheint dem unterdrückten Männchen eine gewisse Entspannung zu gewähren. Weiter ist es auch einem Fall von Übersprungverhalten ähnlich, wobei jedoch ein ganzer Verhaltenskomplex überspringt, der zudem normalerweise nie in Erscheinung tritt. Da solche Pseudoweibchen, die man in normalere Verhältnisse verbringt, sofort männliches Verhalten zeigen können, dürfte das Auftreten pseudoweiblichen Verhaltens nicht hormonal, sondern zentralnervös bedingt sein.

Es scheint nicht ausgeschlossen, dass die Erscheinung ökologische Bedeutung hat, in der freien Natur bei übergrosser Bevölkerungsdichte regelmässig auftritt und dann die Fortpflanzung herabzudrücken hilft.

Homosexualität bei anderen Tierarten wird kurz besprochen und eine Klassifikation der verschiedenen Typen vorgeschlagen.

THE CAUSATION OF PSEUDOFEMALE AND PSEUDOMALE BEHAVIOUR: A FURTHER COMMENT

by

DESMOND MORRIS [1])
(Department of Zoology and Comparative Anatomy, Oxford University)

(Rec. 12-X-1954)

INTRODUCTION

Pseudofemale behaviour has recently been reported for the Ten-spined Stickleback (*Pygosteus pungitius*) (MORRIS 1952) and the Zebra Finch (*Poephila guttata*) (MORRIS 1954). A male is said to be performing pseudofemale behaviour when it exhibits motor patterns belonging to the special sexual repertoire of the female. I refer here to the "special" sexual repertoire of the female because in many species mutual sexual displays (HUXLEY 1914) occur in which both sex partners perform together the same movements (*e.g.* Pre-copulatory head-tossing in Gulls: TINBERGEN 1953; and many examples in ARMSTRONG 1947, Ch. XI). In such cases, even if the male were behaving as a pseudofemale, it would be impossible to detect. The recognition of pseudofemale behaviour depends, therefore, on the existence of a distinct sexual "di-ethism" (*cf.* sexual dimorphism) in a particular species. Furthermore, it must be a positive di-ethism. By this I mean that there must be sexual behaviour differences between the sexes which do not consist solely of the male doing something and the female doing nothing. For example, the females of many species flee if they are sexually unresponsive, and simply remain still if they are responsive. "Staying still" acts as a sexual signal to the males concerned, but it cannot be used by ethologists in investigating pseudofemale behaviour. Those species in which both male and female perform special sexual actions are most valuable for such investigations.

Owing to the fact that males are, generally speaking, more elaborate in their sexual behaviour than females, it is much easier to study pseudomale

1) Acknowledgements are due to the following: Dr N. TINBERGEN for reading the MS and for his help and advice; Mr P. E. GUITON for many lengthy and valuable discussions; Mr U. WEIDMANN for preparing the German summary; and the Nature Conservancy for sponsoring this research.

than pseudofemale behaviour. BEACH (1948) was able to assemble more evidence of pseudomale behaviour than of pseudofemale behaviour, and he concluded that "the execution of feminine sexual patterns by genetic males occurs less frequently than does the display of male behaviour by the female". Although this is probably true, it is safer, at this stage, to substitute for "occurs less frequently", the words "has been observed less frequently", if a general statement is to be made. Both the Ten-spined Stickleback and the Zebra Finch exhibit positive di-ethism, and in both species I have now observed pseudofemale actions by males and pseudomale actions by females.

ADDITIONAL DATA

I have already reported the way in which male Ten-spined Sticklebacks have been observed to perform the entire mating pattern of the female (with the single exception of egg-laying!) under special circumstances. Since then I have observed females of this species performing part of the male sexual pattern, under the following circumstances [1]).

During the course of experiments in which ripe females are presented to isolated male sticklebacks, it is common practice to keep a number of females together in a tank in which they are fed liberally. Excessive feeding hastens the development of the eggs, and females are then taken from such a tank singly for experiments elsewhere when needed. In these "female tanks" the females become sexually motivated to a very high degree and, since no males are present, they may become intensely sexually thwarted. This is revealed by the fact that they may even take up the initial female courting posture in the absence of the appropriate releasing stimuli. Under these circumstances, I have seen, on several occasions, one ripe female perform the male courtship dance (a series of head-down jumps) to another ripe female. On each occasion, the latter followed behind the dancing female and was lead across the tank by it. Together the two female performed homosexual behaviour which was a replica of the first stages of the usual heterosexual courtship sequence of the species. As might be expected, these female homosexual courtships were comparatively brief, and were broken off, not by the following female, which was performing its usual motor patterns, but by the dancing pseudomale fish.

In the case of the Zebra Finch, male birds were seen to perform the female invitation-to-copulation display on many occasions. (This display consists of a rapid quivering of the tail while the body is held in a special

---

1) This has been briefly mentioned elsewhere (MORRIS 1954).

crouched posture.) More rarely, the females responded to this by approaching the soliciting male and gingerly placing one foot on its back, as if to mount it. In a few instances, females mounted their soliciting males and performed the full male copulatory pattern (beating wings, lowered and twisted tail, *etc.*). Females have not yet been seen to perform the male courtship dance in this species.

In a previous report of this behaviour of the Zebra Finch (MORRIS 1954), I pointed out that the strong arousal of the sexual tendency of the male and its subsequent thwarting by the absence of appropriate releasing stimuli from an unresponsive female, was an important causal factor in the pseudofemale display of this species. Since that time, I have recorded the exact behaviour sequences involved in a large number of courtships of this species, paying particular attention to the stage of the courtship sequence at which the pseudofemale display was given by the male. In twenty of the courtships, which were scored in this way, pseudofemale displays were seen, and most of these occurred at the same stage in the sequence and in the same type of courtship. Simplifying slightly, it can be said that there are four courtship situations in which these pseudofemale displays occurred:

1) In 17 out of the 20 cases, the following sequence was observed:
   a) Male performs full song-and-dance display.
   b) Female does not stay still, but hops about.
   c) Male abandons song-and-dance, and performs pseudofemale display.

2.) In 1 of the 20 cases, the following sequence was observed:
   a) Male performs full song-and-dance display.
   b) Female stays still, but does not solicit.
   c) Male abandons song-and-dance, and performs pseudofemale display.

3) In 1 of the 20 cases, the following sequence was observed:
   a) Male performs full song-and-dance display.
   b) Female stays still, but does not solicit.
   c) Male ignores absence of invitation display and mounts female.
   d) Female flees from under male, thus preventing copulation.
   e) Male performs pseudofemale display.

4) In 1 of the 20 cases, the following sequence was observed:
   a) Male performs the full song-and-dance display.
   b) Female stays still and solicits.
   c) Male mounts.
   d) Female flees from under male, thus preventing copulation.
   e) Male performs pseudofemale display.

From these observations it can be concluded that the male performs male sexual behaviour at any stage, when the female is responsive at that stage in the sequence. Also the male may, to some extent, "overlook" the unresponsiveness of the female and carry on to the next male stage in the sequence despite it. But the extent to which the male will do this is limited and there comes a point when the male abandons its male behaviour and switches to pseudofemale behaviour. The correlation between the sexual thwarting of the male and the appearence of the pseudofemale display seems very convincing.

Two points must be added here. Firstly, it must be stressed that after an apparently successful copulation, the male simply dismounts and then both birds usually shake and preen themselves, and there is no pseudofemale display. Secondly, it should be noted that the twenty cases described above were not selected in any special way. Many more instances of pseudofemale behaviour have been observed in this species, but these were the first twenty to be recorded in this particular series of observations. In conclusion, it should perhaps be mentioned that the data given here were obtained from four pairs of birds, each pair being isolated from the other three.

THE CAUSATION OF PSEUDOFEMALE AND PSEUDOMALE BEHAVIOUR

In the light of the further data given here and certain other facts, it is interesting to re-examine the causal aspect of the problem of pseudofemale and pseudomale behaviour. In the past, I have stressed this or that causal factor, according to the particular case I have been studying. I now wish to postulate, tentatively, the four causal factors which I consider form the fundamental causal basis of such behaviour, and which I think may well be found to apply at a very general level. They are as follows:

A. Some hormonal and/or structural abnormality of the sexual system.

B. Intraspecific submissiveness or subordination in males; intraspecific aggressiveness or dominance in females.

C. The arousal and subsequent thwarting of the sex drive.

D. The presence of the releasing stimuli for the sexual behaviour of the opposite sex.

I suggest that one, or some combination, of the above factors, will probably be found to account for the occurrence of any pseudofemale or pseudomale behaviour. Some examples of the type of evidence on which this suggestion is based must now be given.

C a u s a l f a c t o r A. must be included if a general statement is to be made, but it does not concern us here. BEACH (1948) has reviewed the evidence concerning this factor. The effects on sexual behaviour, of alterations in the gonadal and hormonal systems of animals, has been the subject of considerable experimentation in recent years. However, comparatively little experimental evidence exists concerning the purely neural causal factors B, C, and D. This is undoubtedly the direction which future research into the subject of the reversal of sexual roles should take.

Causal factor B. requires some explanation. I have already discussed in detail elsewhere, the way in which courtship can be thought of as a three-point conflict between the tendencies to flee, attack and mate (MORRIS 1954 and 1955). The exact balance between these three tendencies varies from species to species. Speaking loosely, the male of some species are rather aggrèssive towards their mates when courting, whereas others are rather "scared" of their mates. Clearly the mating tendency must ultimately become the strongest of the three, or sexual behaviour will not occur, but the relative strengths of attacking and fleeing can vary and the conflict between the stronger of these two and the mating tendency will then form the basis of the courtship pattern.

Not only is there a difference of this kind between species, but also between the male and female of each species. In many species it has been shown that the male takes the dominant role whilst the female is submissive. I shall deal with this type first. ZUCKERMANN (1932), writing of Baboons, has shown that in these animals, causal factor B is all important in producing pseudofemale and pseudomale behaviour. As he puts it (p. 289), "In a particular situation, the animal assumes the dominant or male sexual role, while a fellow assumes the reciprocal and submissive role of female. Such behaviour might be either homosexual or heterosexual. Mounting behaviour therefore depends fundamentally upon degrees of dominance, and at this level of analysis it seems purely accidental whether a particular response is homosexual or heterosexual". In other words, there is a particular balance of the fleeing, attacking and mating tendencies involved in the male sexual behaviour (which, for the sake of convenience can be described as an fAM-type [1]), and another balance in the female (FaM-type). When a male is unusually "socially inferior", subordinate, or submissive, its usual flee-attack balance (fA) will shift in the direction of that typical of the female (Fa) and its sexual behaviour will automatically and inevitably suffer a similar shift (fAM → FaM), and the male will show pseudofemale behaviour. If

1) F — flee; A — attack; M — mate; capital letter — strong activation; small letter — weak activation. For full discussion see MORRIS 1955.

a female is unusually "socially superior", dominant, or aggressive, then the reverse process to the above will result in her performing pseudomale behaviour. When, as often happens, these two shifts occur reciprocally, reversed mounting takes place.

The pseudofemale behaviour of the male Ten-spined Stickleback can be thought of in the same way. The male courtship is of the aggressive type (fAM) and intense pseudofemale responses were only obversed in males which had been beaten into a submissive, subordinate state.

Before going on to discuss the next causal factor, it must be stressed that active subordination is not always an inevitable prelude to pseudofemale behaviour. To illustrate this point, it is necessary to compare the Ten-spined Stickleback with the Zebra Finch. In the former species, the male is so aggressive in its usual courtship that it has to be severely beaten up and subordinated before it can be in a state which permits it to behave as a pseudofemale. But the male Zebra Finch, far from being aggressive to its female during courtship, is rather "scared" of her, and in this species the male behaves as a pseudofemale without previously being unusually subordinated. In other words, the very nature of the usual male sex pattern makes it easy for pseudofemale behaviour to occur, without the typical social reproductive organisation of this species being first thrown out of balance.

Causal factor C. must now be considered. I shall continue to use the Zebra Fnch as an example here. If the male Zebra Finch does not have to be intensely subordinated to behave as a pseudofemale, then what determines whether it shall react to a particular sexual situation with typical male, or with atypical pseudofemale responses? The answer to this has really already been given earlier in this paper, where it was clearly shown that the arousal and subsequent thwarting of the male sexual tendency leads to pseudofemale display. The same applies also to the male and female sticklebacks. In both cases, pseudofemale display occurred only when the fish were intensely sexually thwarted. In an earlier paper I have likened pseudofemale behaviour, in which this causal factor is important, to (an admittedly very aberrant form of) displacement activities, but I shall not pursue this comparison until further experimental data have become available.

Finally, Causal Factor D. completes the list and is illustrated by the female Zebra Finch. The bird may respond to a pseudofemale display of a male, by mounting it and making the typical male copulatory actions. Such a female is giving a pseudomale response and yet is neither sexually thwarted nor unduely aggressive or dominant. The main causal factor here appears to be the presence of the releasing stimuli for the sexual responses of the opposite sex. The male is giving the female invitation display to the female and the latter, despite its sex, responds "appropriately" to this, and accepts the invitation like a male.

Summarising the foregoing, it is now possible to examine just which causal factors are at work in the various examples of pseudofemale and pseudomale behaviour mentioned already, and in any other cases in which some causal basis has been proved or suggested. In doing this, factor A will be ignored completely, and only the purely neural cases will be dealt with.

1. Male Ten-spined Stickleback (MORRIS 1952). Factors B, C, and D are all important, especially C. I suspect that in very rare cases, pseudofemale behaviour occurs when B and D are only mildly involved.

2. Female Ten-spined Stickleback (in present paper). Factors C and D are important. Since there was no agonistic behaviour in the "female tanks", factor B could not be involved.

3. Male Zebra Finch (MORRIS 1954 and in present paper). Factor B passively involved: factor C most important; factor D never operating.

4. Female Zebra Finch (MORRIS 1954). Factor D apparently only the one involved.

5. Male Baboon (ZUCKERMAN 1932). Factor B most important; factors C and D not essential.

6. Female Baboon (ZUCKERMAN 1932). Factor B most important; factors C and D not essential.

7. Woodlark (NIETHAMMER 1937). It is reported that reversed mounting takes place after the male has made unsuccessful attempts to copulate with the female. This suggests that this species shows pseudofemale and pseudomale behaviour which is causally similar to that of the Zebra Finch.

8. Pigeon (WHITMAN 1919). An example of pseudomale behaviour which involves causal factor C is illustrated by this passage from WHITMAN: "Often the sexual impulses do not develop synchronously in a pair, and thus we may have a female dominated by passion mated with a male who is in a state of relative unreadiness. In such a situation, the female may make the advances and possibly assume the masculine role in courting and copulation." (p. 98).

9. Dog and 10. Rat (FORD and BEACH 1952, p. 143). Pseudomale behaviour involving factor C, in the same way as the Pigeon cited above. FORD and BEACH state: "It is not at all uncommon to observe the temporary display of masculine behaviour on the part of receptive females confronted with sexually sluggish males. Female dogs, rats, and other animals in heat may mount the male repeatedly if he is slow to assume the initiative in the sexual relationship.".

The above list is not intended to be a complete one. But it is sufficient already to show that, although the causation of sex role reversal differs from sex to sex and species to species, there are, nevertheless, always the same few basic factors involved. Sometimes a single factor alone appears to account for a particular instance. Elsewhere, this or that combination of factors appears to be involved.

CONCLUSION

Reversed mounting has been observed in a number of bird species other than those mentioned here. In some species, reversed mounting is said to be the usual sequel to ordinary mounting, and it is possible that in such species the reversed act has become a functional ceremony. But until more evidence is available, it must be borne in mind that it is often extremely difficult, with birds, to observe whether a copulation has been successful or not. Reversal sequels may, in fact, be following a t t e m p t e d copulations as in the case of the Zebra Finch.

In connection with reversal sequels, which are known to have followed unsuccessful copulation attempts, it may seem strange that the female, which was not prepared to accept the male in the usual way, was immediately prepared to respond to him with pseudomale behaviour when he then behaved as a pseudofemale. I suggest that the answer is that the approaching male, giving the male display, aroused the fleeing tendency of the female too much to permit the sexual tendency of the female to dominate her mood. When the male then stopped advancing towards the female and gave the invitation display, the fleeing tendency of the female was less stimulated, and sexual behaviour was permitted expression. But now the presence of the releasing stimuli for the sexual behaviour of the opposite sex acted on the female to switch her response from female to pseudomale. This suggestion is based on observations of the fleeing intention movements made by female Zebra Finches when performing such behaviour (see MORRIS 1954). The strength of the fleeing tendency appears to decline when the male Zebra Finch switches from male to pseudofemale behaviour, but even then, the female may show such strong tendencies to flee, that she will only gingerly place one foot on the back of the soliciting male. (It should also be mentioned that the occurrence of pseudomale behaviour is much rarer than that of pseudofemale behaviour in the Zebra Finch.)

Finally, a few points should be made concerning pseudofemale and pseudomale behaviour in our own species. Much has been written recently about homosexuality in human beings. Of particular interest are the contributions of KINSEY and his co-workers in America. Owing to the nature of their investigations, they were limited to making observations by interviewing techniques, but have done this on such a large scale, as to produce important results (see KINSEY, POMEROY and MARTIN 1948). In an earlier discussion of homosexuality, KINSEY (1941) had already dismissed the hormonal evidence that had been put forward to explain the general causation of homosexual behaviour in human beings, stating that "the circumstances of the first sexual experience, psychic conditioning, and social pressures are obvious factors in determining the pattern of behaviour. It would appear that no similar correlation has as yet been shown between hormones and homesexual activity.". More recently, he and his colleagues (KINSEY, POMEROY, MARTIN, and GEBHARD 1949) have emphasized the importance of the effects of experience in determining patterns of sexual behaviour: "In brief, the psychosexual pattern in the human animal originates in indiscriminate sexual responses which, as a product of conditioning and social pressures, become increasingly restricted in the direction of traditional interpretations of what is normal or abnormal in sexual behaviour.". This, I feel, is satisfactory as far as choice of a male or a female as a sexual partner is concerned, but once such a partnership exists, be it homosexual or heterosexual, it still remains for something to determine who shall play the masculine role, and who the femine role. I would tentatively suggest that factor C (submissiveness in the male; aggressiveness in the female), discussed earlier, is probably the most important causal factor in the production of pseudofemale or pseudomale behaviour in human beings (regardless of the s e x of the partner).

It is necessary to point out here the relationship between pseudofemale and pseudomale behaviour (inversion), on the one hand, and homosexuality, on the other. KINSEY, POMEROY and MARTIN (1948) have discussed this at length (p. 612-617) and complain that "It is unfortunate that students of animal behaviour have applied the term homosexual to a totally different sort of phenomenon...". They go on to say that pseudofemale and pseudomale behaviour is "what the students of animal behaviour have referred to as homosexuality". I have attempted to clarify this situation (MORRIS 1952) by introducing a classification of types of homosexuality, based on the idea that any sexual behaviour between two members of the same sex is homosexual behaviour (which is the way in which KINSEY and his colleagues use the term). Since the present paper is dealing with inversion as its central problem, it is useful to conclude with a tabulation of the relationship between inversion and homosexuality:

TABLE ONE

Motor Pattern of male	Male type	Partner	Relationship
Male	Masculine ♂	+ Feminine ♀	Heterosexual
Pseudofemale	Feminine ♂	+ Masculine ♀	Heterosexual
Pseudofemale	Feminine ♂	+ Masculine ♂	Homosexual
Male	Masculine ♂	+ Feminine ♂	Homosexual

The above table shows the four possible types of sexual behaviour available to the male. A similar table could be made for the female. It will be noticed that inversion is simply a matter of motor patterns, whereas homosexuality is simply a matter of partnership, and that although the two phenomena overlap, they can also be quite independent of one another.

SUMMARY

Pseudofemale behaviour is the performance by a male of motor patterns belonging to the special sexual repertoire of the female. Pseudomale behaviour refers to a similar inversion on the part of the female. The selection of suitable species for the study of this behaviour is discussed. Data (additional to those given in MORRIS 1952 and 1954) are presented for female Ten-spined Sticklebacks and for male Zebra Finches. Four basic causal factors are postulated for pseudofemale and pseudomale behaviour. They are: A. Hormonal or structural abnormality. B. Male submissiveness or female agressiveness. C. Arousal and thwarting of sex drive. D. Releasing stimuli for sex behaviour of opposite sex. Factor A is not considered in this paper. It is suggested that one, or some combination, of these factors will always be responsible for sexual behaviour inversions. Examples of the operation of various combinations of these factors are given and discussed. In conclusion, human pseudofemale and pseudomale behaviour is mentioned, and the relationship between inversion and homosexuality clarified.

LITERATURE

ARMSTRONG, E. A. (1947). Bird Display and Behaviour. — Lindsay Drummond Ltd., London.

BEACH, F. A. (1948). Hormones and Behaviour. — Hoeber, New York.

FORD, C. S. and F. A. BEACH. (1952). Patterns of Sexual Behaviour. — Eyre and Spottiswoode, London.

KINSEY, A. C. (1941) Homosexuality. Criteria for a hormonal explanation of the homosexual. — J. Clin. Endocrin. 1, p. 424-428.

——, W. B. POMEROY and C. E. MARTIN. (1948). Sexual Behaviour in the Human Male. — Saunders, Philadelphia-London.

——, ——, —— and P. H. GEBHARD. (1949). Concepts of normality and abnormality in sexual behaviour, p. 11-32, in Psychosexual Development in Health and Disease. — Grune and Stratton, New York.

MORRIS, D. (1952). Homosexuality in the Ten-spined Stickleback. — Behaviour 4, p. 233-261.

—— (1954). The reproductive behaviour of the Zebra Finch, (*Poephila guttata*), with special reference to pseudofemale behaviour and displacement activities. — Behaviour 6, p. 271-322.

—— (1955). The function and causation of courtship ceremonies. Fondation Singer Polignac — Colloque Internat. sur l'Instinct, June 1954. — In press.
NIETHAMMER, G. (1938). Handbuch der deutschen Vogelkunde. — Leipzig.
TINBERGEN, N. (1953). The Herring Gull's World. — Collins, London.
WHITMAN, C. O. (1919). The Behaviour of Pigeons. — Carnegie. Inst. Wash. Publ. 257, vol. 3, p. 1-161.
ZUCKERMANN, S. (1932). The Social Life of Monkeys and Apes. — Kegan Paul, London.

ZUSAMMENFASSUNG

Als pseudoweibliches Verhalten wird die Ausführung sexueller Verhaltensweisen, die dem spezifisch weiblichen Inventar zugehören, durch ein Männchen bezeichnet. Pseudomännliches Verhalten bezieht sich auf eine entsprechende Inversion im Verhalten des Weibchens. Die Wahl geeigneter Arten für das Studium solchen Verhaltens wird diskutiert. Neue Beobachtungen (zusätzlich zu jenen, über die in MORRIS 1952 und 1954 berichtet wurde) am Weibchen des Zehnstacheligen Stichlings und an Zebra-Finken-Männchen werden beschrieben. Vier grundsätzliche Kausalfaktoren werden für pseudoweibliches bezw. -männliches Verhalten postuliert: A. Hormonale oder morphologische Abnormalität. B. Unterwürfigkeit des Männchens oder Aggressivität des Weibchens. C. Aktivierung (arousal) und Frustration (thwarting) des Geschlechtsdranges. D. Anwesenheit auslösender Reize für geschlechtliches Verhalten des entgegengesetzten Geschlechtes. Faktor A wird nicht behandelt in dieser Arbeit. Es wird vermutet, dass ein einzelner oder eine Kombination dieser Faktoren immer verantwortlich sind bei Inversionen des geschlechtlichen Verhaltens. Beispiele für die Wirkungsweise verschiedener Kombinationen dieses Faktors werden angeführt und diskutiert. Abschliessend wird pseudoweibliches und pseudomännliches Verhalten beim Menschen erwähnt, und die Beziehung zwischen Inversion und Homosexualität klargestellt.

TERRITORY IN THE THREE-SPINED STICKLEBACK *GASTEROSTEUS ACULEATUS* L.

AN EXPERIMENTAL STUDY IN INTRA-SPECIFIC COMPETITION

BY

J. VAN DEN ASSEM

LEIDEN
E. J. BRILL
1967

I. INTRODUCTION

Individual animals of many kinds restrict all or part of their activities, especially during the reproductive phase of their life, to a topographical area of limited extent—limited in relation to their motor capacities. Such an area is not merely inhabited but virtually possessed by an animal, which does not tolerate the presence of conspecifics inside his holding. Such an area is called a territory.

To my knowledge, ALTUM (1868) [1]), a German zoology professor, seems to be one of the first who circumscribed the concept of territory in his book 'Der Vogel und sein Leben'. He noted that in many bird species breeding pairs have to space out at rather fixed mutual distances. He thought the size of a territory was dictated by the food requirements of the occupants and the food production of the locality.

Another important pioneer on territory matters is MOFFAT, an Irish gentleman, who read a remarkable paper before the Naturalists' Field Club in Dublin, 1903. This paper was published in the same year, but passed unnoticed for several decades. In MOFFAT's opinion the territory system plays an important role in checking numbers in a (bird) population: only those members of a population that succeed in claiming a 'parcel of land' are able to reproduce, while a considerable number is unsuccessful in this respect.

In 1920, HOWARD, who was unaware either of ALTUM's or MOFFAT's ideas, published a much quoted book: 'Territory in bird life'. This publication marks the beginning of an extensive literature on the subject, especially ornithological. Reviews of it were published by NICE, 1941 (with 399 references), and by HINDE, 1956 (with 188 references). Characteristic of these

---

1) It can hardly be doubted that at least some notions about territory phenomena are even older. In a criticism of HOWARD's theories, D. & L. LACK (Brit. Birds 27, p. 179-197, 1933) quoted GILBERT WHITE (The natural history of Selbourne, Letter XI to the Hon. DAINES BARRINGTON, dated Febr. 8. 1772, published 1789) as an early author. In the following issue of British Birds a letter to the Editor was published quoting the poet OLIVER GOLDSMITH ("The fact is, all these small birds mark out a territory to themselves which will permit none of their own species to remain in; they guard their dominion with the most watchful resentment and we seldom find two male tenants in the same hedgerow together", A History of the Earth and Animated Nature, Vol. V, p. 301, 1774). The correspondent adds, however, that, according to Dr JOHNSON, the author could hardly tell a horse from a cow; his contributions to biological science were probably derived from an older source.

According to MEIKLEJOHN (Ibis 98, p. 544, 1956) an allusion to territoriality among birds had already been made by the Emperor FRIEDRICH II VON HOHENSTAUFEN in his book *De arte venandi cum avibus*, written *ca* 1250.

large collections of papers is the almost complete lack of quantitative data. This is understandable because the major part of this literature is based on field observations of birds (364 out of the 399 references in Mrs NICE's paper deal with birds). Experimental work under more or less natural conditions with these animals is difficult to carry out in practice and accumulation of sufficient data is very time-consuming. The lack of quantitative data obviously did not prevent authors from putting forward hypotheses about the functions of territory. On the contrary, a multitude of suggestions was put forward, often on the basis of rather superficial observations on one species or another. In my opinion, measuring and counting should be primary requirements for any study on territory and its functions.

Aim of the study.

The aim of this study has been to present some objective evidence about the function or functions of territorial behaviour. When writing on the significance of territory in blackbirds, SNOW (1956, p. 446) commented that only "after very many years' observation it might be possible to analyse, for example, nest-success and the maintenance of the pair-bond in relation to such variables as size of territory and distance of the nest from other nests, and so to test objectively some of the suggested functions of territory, but available data so far are quite insufficient for this".

In contrast to blackbirds (and most other bird species) the Three-spined Stickleback seems to be a more favourable subject for quantitative work on territory problems: the animal is easily available, the males hold a well-defined territory, its reproductive behaviour is well-known, the animals can be studied in the laboratory under more or less natural conditions where various experiments can be carried out without severely interfering with the animal's activities. Lastly, the reproductive cycle is short enough to permit the collection of a reasonable amount of data in a reasonable time. Therefore, the present study was undertaken, partly based on previous work on this topic by VAN IERSEL (1958). Though no claim is made that all functions of stickleback territory are understood, some objective evidence has been found for at least some functions, as the reader may judge from the following chapters.

The topic under consideration has both ethological and ecological aspects. A point is made here that it is important to consider certain ecological concepts — competition, regulation of numbers, population structure, *etc.* — from an ethological point of view as well. Since any population is composed of living individuals, their behaviour must play an important role in producing the effects in which ecologists are most interested, and the study of relevant behaviour patterns may help in a better understanding.

In reviewing the literature I realized that my opinion is not an original one. TINBERGEN (1949) stressed a similar point. More recently, LACK (1966, p. 280) made a similar plea in stating that in the future experimental techniques, based on a sound knowledge of the 'natural ecology' of a species could be rewarding. I suspect that LACK implies that an understanding of a species' behaviour under natural conditions is a worthwhile part of ecology. PARK (1961), quoted by LACK (*l.c.*) phrases the same idea as "natural history is one of the prime sources of insight and knowledge for the modern ecologist". In fact, CHITTY (1960) and CHRISTIAN (1959) have long advocated this approach.

Definition of territory.

Because different kinds of territory have been found, different definitions have been proposed. A descriptive definition should contain the two elements indicated in the first paragraph: a) spatial restriction of (some) behaviour, and b) intolerance, leading to defence and resulting in isolation after the local removal or repulsion of conspecifics.

Among many formulations given, NOBLE's simple description (1939) — *a territory is any defended area* — seems satisfactory for the present paper since under appropriate conditions there is hardly ever any doubt as to whether or not a stickleback will defend an area.

In practice only the effects of defence may be manifest, *i.e.* the exclusion of conspecifics from an area by a (temporary) resident. The way in which this defence is accomplished may be difficult to assess; sometimes the mere presence of an owner inside his holding may suffice.

Most kinds of territories can be grouped as breeding territories. This applies also to the Three-spined Stickleback where territories are exclusively connected with breeding activities. It should be clear that a territory in itself is non-existent, it always is inseparably connected with appropriate behaviour of its owner. Territorial behaviour in the strict sense is such behaviour as serves to establish and maintain a territory.

The term territory as used in this paper, essentially following NOBLE, is described as follows: the territory of a Three-spined Stickleback is a topographical area of limited extent where a male-owner spends almost all of his time, and where conspecifics are excluded. Conspecifics are chased as soon as they trespass; the only exceptions are ripe females and, what I may call, sneaking males — both under specified conditions to be described later (see p. 6, 81, 100).

TABLE 73

Probability of entering nest in situations '♀ a.e. -1' for nests of solitary males and nests in rival situations

situation	♀ a.e. -1 total	♀ a.e.-1 and in
solitary	112	89 (80%)
rival	60	36 (56%)

$\chi^2 = 7.5$, $p < 0.01$.

ces in rival situations are not available, however. Qualitative observations certainly point to this effect. Overt aggression between rival males could almost continuously be observed during a courtship sequence, and a much longer time was required before leading of the female could be performed (cf. p. 91). This postponement of leading may have affected the female's willingness to enter (however, cf. data of Table 59). Moreover, — and this factor seems to be the more important — females were bitten as well during aggressive interactions between males. Such a treatment is very likely to have affected her willingness to enter a nest subsequently.

A direct interference with her entering was observed several times. In those cases a nest-owner, which showed the nest entrance to a female nearby, suddenly grasped the female's tail and dragged her away upon the intrusion of rival males which persisted in courtship and tried to lure the female away. Sneaking males (see next section) may also prevent a female from entering a nest.

SNEAKING

An important phenomenon relevant here is the occurrence of what I may call sneaking males. Sneakers may interfere directly with entering a nest.

Since sneaking was commonly observed in rival situations and, moreover, appeared to have important consequences for obtaining an offspring (*i.e.* it may lead to the fertilization of a clutch in a rival's nest; besides, it has important consequences for parental success), a description of sneaking behaviour is included here. An analysis of the underlying motivation will not be given; it is planned to treat this remarkable phenomenon more fully elsewhere.

Sneaking behaviour was exclusively observed in rival situations. It occurred most often during courtship but it was not absolutely linked with it; sneaking was also observed during the building- and sexual phases in the absence of a female. The actual presence of a rival's nest (or nest under construction) was probably essential in those situations: sneaking was frequently directed to the rival's nest, not to the rival himself (or to a female).

Males in parental condition were never observed to sneak.

This peculiar behaviour pattern is certainly no laboratory artifact; it was observed under natural conditions as well (SEVENSTER, *in litt.*).

Sneaking starts suddenly. A male that has taken part in courtship competition, or has seen courtship behaviour by another male in another territory may suddenly freeze; sometimes the change in behaviour is very dramatic. He then sinks to the bottom; no fin seems to be moved, only the dorsal spines are erected. In a very short time the male loses his bright nuptial colours (if he had them before) and adopts a cryptic pattern with dark bands across the back prevailing, much like the pattern of a female.

Most often a sneaker sinks to the bottom in the boundary area of his territory. He may lie motionless for some time, then he starts sneaking very slowly — virtually millimeter by millimeter so that one hardly notices his moving forward at all — in the direction of the rival's nest, while the rival is courting the female at that time. Moving forward seems to be effected by minute movements of the pectorals only; dorsal spines remain erected all the time.

A courting male may interrupt his courtship and chase one or more sneakers that have intruded into his area. A sneaker will almost always flee immediately to his own territory when attacked at this stage, and various behaviour patterns may follow, *e.g.* a new sneaking bout, or the sneaker may resume courtship behaviour.

Sometimes, however, a sneaker is not noticed. It was observed several times that a territory-owner passed a sneaker very closely obviously without noticing him, probably because of his immobility and cryptic colouration — a combination that seems to be quite effective in this respect. Often a sneaker appears to be fascinated by the rival's nest in such a way that he remains directed to it, and does not seem to notice that the female has passed to another territory, maybe his own.

A sneaker will make a sudden dash for the rival's nest as the last part of his invasive action. During courtship he will never dash for a nest in the absence of its owner (as may happen during the sexual- or building phase, see below, where sneaking serves another function and is probably caused by different factors); in a courtship context a sneaker will dash forward during a performance on the nest by its owner, *e.g.* boring, fanning, or creeping through. Especially creeping through makes a sneaker dash for the nest without exceptions (as does a female in a comparable situation, see p. 82). Another, most provocative stimulus for a sneaker to make the final dash is a female being shown the nest entrance, or a female entering the nest.

Following dashing forward several things may happen, presumably depending on the precise external situation and the sneaker's motivation.

In some cases the sneaker quivered unmistakably on the protruding tail of the nest-owner, who paused for some seconds inside his nest during his creeping through. Following quivering, a sneaker may make 'boring' movements on the nest (*N.B.* this kind of boring has to be distinguished from 'normal' boring as performed by a nest-owner), and subsequently he may take up a position in front of the nest entrance and remain motionless. In some cases the sneaker crept through the nest immediately after the nest-owner.

A sneaker that made a dash towards a boring or fanning nest-owner was only observed to make boring movements afterwards. These boring movements may be made in any place; frequently boring was performed on the roof part of a nest. This intensive boring appears to be a stimulus for other males — also for those in which no sneaking tendency was apparent before — to dash for the nest as well and join in boring (a phenomenon also observed in nature). During this boring erection of dorsal and ventral spines was observed in all cases.

Mass-boring may cause some disorder to a nest; it has to be distinguished, however, from boring with pulling material as may occur during earlier phases (see below, p. 108). In the present case nest material is never pulled away. The suffering nest-owner reacts to the general melee by biting furiously and dragging the intruders away from his nest by pulling their tail if he can get hold of it. A peculiar feature of the boring sneakers is that they are very stubborn in their action and difficult to chase. Moreover, they never react aggressively to the biting nest-owner. Besides biting, the owner of the raided nest may start bringing sand, spitting it on top of the roof part of his nest; sometimes such a quantity is brought that the entire nest gets more or less buried. All cases described so far will lead in the end to the sneaker(s) being chased by the nest-owner.

A male in sneaking position will always dash forward at the moment a nest-owner starts showing the entrance to a female. The sneaker may arrive at the nest opening just prior to the female, make some 'boring' movements and then take up a position in front of the nest while staying motionless. In this way the female may find the entrance blocked; she may make some efforts to enter the nest and then turn away — or is torn away by the, now aggressive, nest-owner that interrupts showing the entrance and starts biting. Courtship may then proceed in another territory.

No doubt such proceedings will explain part of the lower probability of entering by the female, as was recorded above (p. 94).

If a female does enter the nest, in spite of the sneakers already present, several behaviour patterns may follow again. Frequently some 'boring' was performed by the sneakers, in some cases they were seen to quiver on the female's tail. From qualitative observations I got the impression that tail-quivering was especially performed in those cases in which the nest-owner had turned away from the nest to collect sand. (Under normal circumstances a nest-owner does not bring sand when a female is inside his nest, it is induced by the presence of sneakers). Immediately following quivering the sneaker (or sneakers) will remain motionless, almost always with the snout at the nest entrance or on top of it.

From the observations I got the impression that in such a situation a nest-owner 'did not know how to handle the case'. To all appearances overt aggression by the nest-owner is inhibited if the rivals stay motionless in contact with his nest. The owner swims to and fro, sometimes delivering a very inhibited bite on a tail or some other part of a body. Staying motionless at some distance away from the nest will always result in some biting and in an attempt to chase the intruder. A nest-owner will at once bite furiously if a sneaker moves. Moving while in contact with the nest will also induce aggression, but it will stop if the sneaker is able to resume a motionless position. The way in which such sneakers are treated is rather reminiscent to how a male reacts to a not quite willing female which has assumed a position in front or on top of the nest instead of entering it.

If a sneaker manages to remain motionless at the nest entrance, the nest-owner will start quivering after some time; he will do this on the female's tail, but also, and in some cases almost exclusively, on the body of a sneaker which still remains motionless.

It need not surprise that the duration of the female's stay in the nest was much longer under such circumstances than was found in solitary situations (see p. 89; in the latter situation it was of about the same duration in all territory sizes, also the sub-minimal ones). In rival situations the overall average was 91 seconds (121 ×). If the observations are separated into cases with sneakers present and sneakers absent, the average of the latter group (64 seconds) is not significantly different from solitary situations; the average of the former (182 seconds) differs from that of solitary males (durations solitary *vs* rival situations, Mann-Whitney U test, $p<0.001$). Since the situation 'sneakers present' could not be standardized, the durations of staying inside the nest were extremely variable; a maximum duration of 560 seconds was recorded.

It was never observed that a female did not spawn in these situations;

sneakers clearly delay spawning but never seem to prevent it once the female is inside the nest.

Once a male has succeeded to induce a female to spawn inside his nest he passes through a most important moment of his existence because, when fertilizing a freshly laid clutch, he contributes to the next generation. At this very moment, however, the competition with other males has not yet come to an end. After the female has spawned and left the nest, a sneaker will try to enter at once and fertilize the clutch himself. In about half of all observed cases he was able to do so before the legitimate owner did.

That the sneaker ejaculated indeed could be deduced from those cases in which the nest-owner, probably due to his very high aggression towards the sneaker which left his nest, did not fertilize the clutch himself. In such cases eggs developed normally. (From experience we know that eggs can only be fertilized during a limited period after spawning, maximally about 30 minutes). In one case eggs were removed after a sneaker had passed through the nest and before the owner had a chance to do so: all eggs appeared to have been fertilized and developed in a normal way when ventilated artificially.

It is unknown whether ejaculating a few seconds ahead of another male results in a higher success in fertilization, but it may be expected that the male being first will have the major share in fathering offspring. A sneaker almost always attempts to rob part of the eggs he has fertilized. A description of this behaviour, and its effect, will be dealt with in the next chapter, p. 122.

The description of sneaking as given above does not deviate much from the one given by Morris (1952) in a paper entitled 'Homosexuality in the Ten-spined Stickleback', in which a similar behaviour of a related species, *Pygosteus pungitius*, is described. The author mentions behaviour of the Three-spined Stickleback as much the same in this respect, with which I completely agree. I disagree, however, with Morris' interpretations.

Morris calls sneaking 'pseudo-female behaviour' because he thinks the pattern is essentially similar to the sexual behaviour pattern of the female, the reason why he denotes it as homosexual ('homosexual being any sexual performance involving members of the same sex'). This question will be dealt with more fully elsewhere; below a few arguments will be presented against this view.

Morris' main arguments for the homosexual character of sneaking in the Ten-spineds are the following: sneakers (pseudo-females in his terminology) will follow a zigzagging male towards his nest; they will compete with the real female for a place at the opening; they do not take a male's position if a female model is presented in front of the nest and the nest-owner has been removed, but take up a position in front of the nest as well, and may enter, pausing inside the nest as a real female does when spawning.

Furthermore, he states that occurrence of sneaking is limited to such males that are unable to claim a territory and build a nest themselves because of crowded conditions in the tank, and this, he supposes results in sexual frustration ('being highly motivated sexually but being prevented from fully expressing themselves sexually by the presence of dominant males').

From my own observations I conclude first of all that sneaking behaviour is particularly suited to make an invader as inconspicuous as possible. A female, on the contrary, makes herself as conspicuous as possible once she has entered a territory.

A sneaker never follows another male in a female fashion, although he may dash forward as if following a leading nest-owner. Clearly, the nest guides him; he certainly does not go there because he is led by the nest-owner. From MORRIS' description it is also apparent that sneaking is activated by the presence of a female being courted, or by certain behaviour a nest-owner performs on his nest; his interpretation that the nest-owner leads the sneaker goes too far. Further, sneakers do not attempt to enter the nest as a female does when the owner shows the entrance. They may, however, perform boring movements — not necessarily in the opening only — probably in search of fresh eggs. They do not take up therefore the male's position as if showing the entrance in MORRIS' experiments with a female model, but they take up another one because the situation 'stickleback with snout in nest opening' releases boring, not showing. This is the more convincing since, once the female is in the nest, a sneaker may in fact take up the nest-owner's role: in the situation 'tail protruding from the opening' quivering may be performed. I suggest that it is quite possible that this would have occurred in MORRIS' experiments as well (in the Ten-spined version) if the model had been placed further inside the nest.

Finally, pausing inside the nest (as MORRIS calls it) may be normal creeping through in the absence of a clutch which is almost always of a much longer duration than passing through the nest with ejaculation (as occurs after spawning). Fertilizing eggs is not a female characteristic, and it can hardly be thought of as being just an accidental by-product if the whole context in which it occurs is taken into account.

Because of these descriptive arguments MORRIS' interpretation of sneaking behaviour is discarded.

Regarding MORRIS' claim that sneaking can only be performed by inferior males (males without a territory or nest themselves), it can be stated that sneaking is very common in territory-owners as well (as in fact it also is in the Ten-spined Stickleback). Further, sneaking is not the outcome of a long term sexual frustration of males in MORRIS' sense; it may follow a courtship immediately. Moreover, it is not restricted to high densities as MORRIS claims; it was also observed in a simple rival situation in a 600 × 100 cm² ditch. (Sneaking at low densities was also observed in the Ten-spined Stickleback, SEVENSTER, pers. comm.). That sneaking is not restricted to inferior males, and has in fact nothing to do with inferiority, nor with long term sexual frustration, can be derived from a quantitative analysis of the probability of sneaking in relation to territory size rank numbers.

Probability of sneaking and territory size.

In complicated rival situations it was found that the tendency to sneak could not be correlated with the rank number of performing males; the probability of sneaking was about equal for males of any rank number. Relevant data are represented in Tables 74, 75.

TABLE 74

Sneaking and dashing towards a nest in 31 cases of 'female at the nest entrance' in complicated rival situations in 150 × 100 cm²

rank number attacking male	rank number of attacked nest (1)	(2)	(3)	(4)	(5)	(6)	(7)	total in rows
(1)	×	4	5	0	5	0	0	14
(2)	22	×	7	5	6	0	0	40
(3)	14	4	×	0	4	0	0	22
(4)	24	6	8	×	2	0	0	40
(5)	4	5	6	0	×	0	0	15
(6)	2	0	5	0	0	×	0	7
(7)	0	0	0	0	0	0	×	0

When a female is at the entrance of a nest all males present, except the owner of this nest, can be counted as potential sneakers. In Table 75 a calculation is given for the rank number frequency of these potential sneakers. The expected frequency of attacks has been calculated on the assumption that there is no preference for sneaking in relation to rank number. Because

TABLE 75

Frequency of potential sneakers related to rank numbers

rank number nest	(1)	(2)	(3)	(4)	(5)	(6)	(7)
no of males present	31	31	31	29	10	4	1
no of sit. '♀ a.e.'	19	4	6	1	1	0	0
freq. pot. sneakers	12	27	25	28	9	4	1
In total 106 potential sneakers which performed 138 attacks							
attack freq. observed	14	40	22	40	15	7	0
attack freq. expected	15.6	35.1	32.5	36.4	11.7	5.2	1.3

$\chi^2 = 5.5$, df $= 5$, $0.3 < p < 0.5$.

observed and expected frequencies agree rather well, I conclude that, given the situation 'female near a nest', all males have an equal chance of performing sneaking with dashing to a nest.

The hazard of a fertilization of a clutch by a rival was found to differ between experiments: in simple situations in 60 × 30 cm² — the rival always being at a short distance away — it was seen in 44% of all cases (28 out of 64 courtships); in complicated situations in larger areas in 24% (24 out of 99 courtships). Small territories in complicated situations may suffer more often from rivals in this respect than larger territories do, but the differences found in 150 and 300 × 100 cm² are not significant (Table 76).

TABLE 76

Fertilization of a clutch by a rival in territories of different size

♀ spawns in nest (1)	52×: nest-owner fertilizes eggs exclusively 38×, 73%	other males fertilize eggs as well 14×, 27%
♀ spawns elsewhere	27×: nest-owner fertilizes eggs exclusively 17×; 63%	other males fertilize eggs as well 10×, 37%

$\chi^2 = 0.45$, $p = 0.50$.

Functional aspects of sneaking behaviour.

In MORRIS' experiments with males without a nest, robbing of eggs was followed by egg consumption. Under similar circumstances Three-spined Sticklebacks will do the same; consumption will occur after the robber has swum around with a mouthful of eggs, as if it were in search of a nest to bring them home. Nest-owners, however, both in the Three - and in the Ten-spined Stickleback, always bring the stolen eggs to their own nest and take care for them in the usual way (see below, p. 122, 123).

This leads to a consideration of the possible function of sneaking. MORRIS indicates two possible functions: pseudo-female behaviour has been selected for as an elaborate egg-eating device in crowded situations where food is otherwise scarce; it is suggested to act in this way as a density dependent population regulating device bringing down high densities to a lower level again where sneaking will not occur any longer; and, second possibility, it is a "displacement outlet relieving sexual frustration", with egg-eating as accidental outcome.

Apart from the arguments presented above which render these functional explanations questionable, it seems hard to visualize how this behaviour could have been selected for if only non-territorial individuals, which have been excluded from reproduction, perform this behaviour.

In my opinion sneaking serves a different function. Therefore, I suggest a third possibility. Sneaking behaviour may be performed by any male stickleback (sexually motivated). This behaviour pattern has the clear advantage for its performer — and was thus selected for — that it provides a possibility of fathering an extra number of offspring by fertilizing a clutch outside the own nest (an advantage obtained at the expense of a rival, and thus extra effective). Sneaking may be seen as a device for stealing a fertilization, and certainly not as a device for obtaining some extra food. After a sneaker has passed through a nest, he always makes an attempt to rob eggs. A description of robbing, and the functions it probably serves, will be dealt with in the next chapter, see p. 122, 129.

Sneaking in a non-courtship context.

A few remarks will be made here on sneaking in a non-courtship context, as can be observed during nest-building, or after the nest has been finished. The motor pattern involved in this kind of sneaking is identical to that during courtship. A difference might be that the sneaker pauses more often in places that provide some cover. It is quite clear from observations that this way of moving forward is intended to remain unnoticed is a hostile environment. Again, this moving forward is unquestionably directed to a rival's nest or nest-beginning. At a certain moment the sneaker makes a dash for this nest, but here no attempt seems to be made to enter. On the contrary, the sneaker starts pulling at the opening or at the roof until part of it gives way. Thereupon he carries this nest material to his own nest. Since in all experiments plenty of material was present all over the tanks, there was certainly no need for material to be collected at a rival's nest. In cases in which mass-pulling occurred — other males appear to be stimulated to dash for a nest that is being thus treated — serious damage to a nest could be the result; sometimes a nest seemed to be destroyed quite effectively. No quantitative data are available for this phenomenon. A high nest density is no requisite, nest-pulling was observed in simple rival situations in a 600 × 100 cm^2 area as well.

Dashing for a rival's nest need not occur in the presence of the rival being active at his nest at that time (as was observed during courtship). From observations one gets the impression that the position of a rival's nest was perfectly known by the other males present. Sneaking might serve the purpose of moving a potential competitor into a disadvantageous position by damaging his nest.

Same characteristics as noted above, *i.e.* a tendency to wreck a rival's construction, a perfect knowledge of the local topography, a secretive way of moving in, and the absence of overt aggression in a marauder have been described by MARSHALL (1954, p. 44) for the (blue) males of the Satin Bower-bird (*Ptilinorhynchus violaceus*). These males may wreck each other's bower and steal coloured display objects. MARSHALL's description is very reminiscent of the state of affairs in sticklebacks in this respect: "A wrecking rival will hop in stealthily, rather than fly boldly through open timber. He comes by the most direct route from his position at the time he observes that the bower is undefended. At the bower (he) works silently and tears down beakfuls of the walls and strews them about in disorder. (If disturbed by the returning owner) he usually snatches up a beakful of blue feathers or glass as he flees. He never stays to fight".

Territory size, courtship-success, and success in fertilization.

In complicated rival situations it has been shown that males, in competition for a female, have by no means equal chances to induce a female to enter their nest: males with large-sized holdings are the most successful in this respect. The function of territory dimensions cannot, however, be described in such simple terms if the chance of fathering offspring is taken into account. The state of affairs is then greatly complicated because a number of additional factors appear to play a role.

The territory hierarchy, based on size of property, which favours the owners of larger holdings, is disrupted by sneaking: males with smaller territories may fertilize more clutches (in a rival's nest) than those they are able to collect by courtship in their own nest. The same factor, however, favours the owners of larger territories as well — as long as they have no eggs in their own nest. Because chances are that the owners of large territories will collect a clutch earlier than other males do, the former will stop

(It may be noted in this respect that in complicated 1-1 situations egg yield of the largest territories was more closely in agreement with the expected yield than was found in other territories: largest territories 98.7%, smallest 93.7% and others 108.1%).

Stealing of eggs.

Stealing of eggs is no doubt the principal factor causing egg losses in 1-0 rival situations. Stealing was rather frequently observed, its proceedings will be described immediately below. In fact this description is more or less a continuation of the description of sneaking (see p. 100).

Sneakers which succeed in fertilizing a clutch always attempt to steal eggs later on. After leaving the nest through the exit they invariably turn around and try to get hold of a mouthful of eggs, in spite of the furious attacks by the nest-owner. A successful robber dashes off and puts the stolen eggs inside his own nest, where he is going to take care for them in the usual way. After the first raid a second, and even a third, may follow.

Not only sneakers which have just fertilized may make an attempt to rob eggs. Frequently the violent 'boring' movements involved attract other males as well, which hurl themselves at the nest and join in the 'boring' (probably without first making an attempt to fertilize the clutch). During this 'boring' both dorsal and ventral spines are erected without an exception; the general colouration of these males resembles the pattern of sneakers.

Egg-robbers seem to be attracted almost irresistibly by a nest containing fresh eggs, in spite of the furious defence by its owner. A most conspicuous characteristic is the absolute absence of any sign of overt aggression in the robbing males. This is the more conspicuous because the robbing tendency will disappear suddenly once overt aggression becomes manifest again. This may happen if a male, that has put some stolen eggs into his own nest, is attacked on his way for a second raid at his territory boundary. If he reacts by biting or threatening, no further egg-robbing will occur. This seems to be the reason that mostly only part of a clutch is robbed. Rarely is a nest plundered completely (this may happen especially in small territories, see p. 121).

A possible mechanism preventing robbers from a complete plunder might be the following. It is known (VAN IERSEL, 1953; SEVENSTER-BOL, 1962) that the situation 'fresh eggs in nest' *per se* (or, 'robbed eggs in nest' in case of egg-stealing) stimulates aggression in the nest-owner. Judging from qualitative observations it appears that as soon as a male becomes aggressive his tendency to rob eggs disappears. Since aggression seems to be incompatible with robbing, this change might set a limit to robbing. The possible function of this limitation will be discussed later (see p. 131).

A deficit of egg yield was always found for those nests in which stealing had been observed. As was stated above, a correlation was found between deficit and territory rank number (*i.e.* territory size). If robbing is the main factor causing these deficits it has to be expected that not all nests are equally affected by stealing. This appeared to be the case indeed. In 1-0 complicated rival situations, in 31 courtship sequences followed by spawning, 19 times a clutch was laid inside nest (1), and 12 times inside some other nest. The number of raids including robbing of eggs observed in these cases were 8 times on nest (1), and 28 times on a nest of lower rank, which means no doubt a much more favourable outcome for the (1) nests.

Parental-success with stolen eggs.

In the 1-0 situations the thief himself is not very well off either. Most often the stolen eggs were found to have disappeared 5 days later. Only 9 males out of 35 (27%) were found with exclusively stolen eggs in their nests on day 5; 24 ended with an empty nest.

A cause of this failure might be an insufficient number of eggs. For parental behaviour to develop properly a minimum number of eggs is required. In many cases the number of eggs obtained by robbing might be smaller than this minimum required to start or to maintain parental care. In such a case the clump of eggs will no longer be treated as an object for parental care after a certain time, but instead will be removed from the nest and may be consumed. Egg-consumption was observed in several cases.

The same reasoning applies to a nest that was severely raided; by stealing the number of eggs may be reduced to less than the necessary minimum.

An attempt was made to get an impression of the number of eggs stolen per raid. A calculation was made (for nests in 60 × 30 and 150 × 100 cm^2 tanks) from the deficits found in nests in which robbing was seen, divided by the number of successful raids observed. In 11 nests that were robbed 32 times a total deficit of 1834 eggs was estimated; the average number of eggs stolen per raid thus amounts to about 57 eggs. This figure agrees rather well with one direct count made for two successive raids: 125 eggs, or 52 eggs per raid.

Furthermore, the observed yields obtained by robbing, and the deficits estimated for nests in which robbing was seen, do not have a uniform distribution, but one from which it can be deduced that portions of a certain size (about 50 eggs), or some multiple, prevailed. The unit of robbing, a mouthful, probably has a rather constant value.

The question arises whether the number of robbed eggs alone can be the primary cause for the failure of many 'robbing cycles'. To collect some information on this point a number of solitary males were presented with 50 eggs (after spawning and fertilization, the clutch was reduced to this size and put back in the nest). Five males thus treated out of 6 completed a parental cycle; one male ate his eggs on the third day.

Solitary males which were offered the same number of eggs without having fertilized them (the clutch had been fertilized by another male in another tank) interrupted a parental cycle more frequently (2 males performed parental behaviour until the end of the 5th day, 5 interrupted the cycle). The difference between both groups is not quite significant (Fisher's Exact Probability test, $p = 0.07$), but the tendency seems to be clear enough: fertilization of a clutch seems to be a factor of importance for completing a parental cycle (see also VAN IERSEL, 1953), at least if small numbers of eggs are involved.

I conclude from these data that a failure of parental care due to small numbers of eggs can be expected for those robbers which did not take part in fertilizing eggs.

In a rival situation the proportion of males which complete a parental cycle after they have fertilized a clutch and subsequently have made one successful raid (*i.e.* have stolen about 50 eggs) is significantly smaller than the corresponding proportion in a solitary situation (2 out of 14 — with 43 and 23 eggs respectively on day 5 —, and 5 out of 6, see above; Fisher's Exact Probability test, $p<0.01$). It appears that among rivals other factors, apart from fertilization, and probably inherent in the rival situation — the frequent aggressive interactions may be important as such — have an adverse influence on parental success of robber-males. It can be expected that the effects of these factors add to the effect of the small number of eggs in males which did not fertilize a clutch. In three of such males none had any eggs left on day 5; in a solitary situation 2 out of 7 had. (One robber male retained his booty although he had not fertilized it, but this case was aberrant in so far that the robber got hold of the entire clutch, which was spawned in front of a nest instead of inside, and brought it home in one single raid. This case provides an additional indication of the importance of egg numbers, here 256 eggs).

It might be expected that the chance of retention of eggs increases with the numbers present. The yield of males which made one raid was compared with that of males which attacked more often. Out of 14 males robbing once, 2 (14.3%) still had eggs on day 5, as against 6 (37.4%) out of 16 which robbed more often (for 3 robbers that retained eggs in their nest the number of raids was unknown). The difference between both groups is not significant, however. (Fisher's Exact Probability test, $p = 0.18$).

In rival situations the egg yield of a nest, in which a female had spawned and which was raided subsequently, was either more than 50 eggs (usually far more), or no eggs at all. In the latter cases the owners had presumably removed what was left by the robbers: an amount too small to induce parental care under rival conditions.

If the robber had an opportunity to collect an additional clutch later on (obtained by courtship and spawned in his own nest), the egg yield observed at the end of his parental cycle was greater than expected from the weight loss of his female alone; apparently the stolen eggs had been safe here. (For 5 such males expected: 2422 eggs and found: 2794 eggs, a surplus of 15.5%).

From the results described above I conclude that several factors have an influence on parental-success with stolen eggs. a). Fertilization of eggs preceding robbing; fertilizing eggs has a positive effect. b). Numbers of eggs stolen; success is positively correlated with egg numbers. (The same factor applies to males being robbed; too few eggs left will end in a parental failure). c). Obtaining a full clutch in addition to a number of stolen eggs has a positive effect. (However, in these cases not only an additional number of eggs was involved but an additional fertilization as well). d). Factors inherent in the rival situation, probably the frequent aggressive interactions, have a negative effect (see also p. 132).

Robbing of eggs is not exclusively limited to the period immediately following spawning. In some cases eggs were found in nests where the owner had not been observed to rob them at that moment. In one case robbing of eggs was observed from a nest on the third day of the parental cycle of the owner.

Probably the share of these cases of robbing is relatively small. It may not surpass 25% of all cases in 1-0 situations, as it can be calculated from the following data. In 26 nests in a rival situation which received a clutch of eggs, egg yield was significantly below expectation in 16 cases. In 12 of these nests robbing had been observed (75%), in 4 nests (25%) no robbing had been recorded. Robbing had not been seen either in the remaining 10 nests where yields and numbers expected agreed. An equal proportion of not-observed robbing followed from the egg-yield of robbers. Stealing was observed and eggs were present in 9 cases, stealing observed but no eggs left at the end of the cycle in 24 cases. In 3 nests eggs were present where the owner had not been observed to steal them. Assuming that the chances of retaining eggs for not-observed robbers is similar to that of known robbers, the total number of not-observed robbers can be estimated at about 11, or 25% of the total number of all robbers.

Synchronization of parental cycles.

Summarizing the data on deficits of egg yield in rival situations, it can be stated that they are primarily caused by stealing. Robbing males removed eggs from nests in which a female spawned. The majority of the stolen eggs got lost (especially so if the robber did not collect an additional clutch himself by courting a female, as was the case in 1-0 experiments). The owners of raided nests may contribute to the general deficit if they consume what was left of the original clutch.

In a synchronized breeding population, in which all males receive a clutch during the same period of time, egg losses appear to be at a minimum — though stealing of eggs was also observed in such a situation. However, stealing does not seem to affect the total egg yield in those cases (cf. Tables 82, 84).

Synchronization of a population may be realized if all settled males are willing to court a female, and if the number of females available suffices to provide all males with at least one clutch within a short period (in the experiments usually within 24 hours).

Even with a sex ratio of 1 : 1, a surplus of females may indeed be available because a number of males will be unsuccessful in claiming a territory. Moreover, a female is able to produce a clutch of eggs every third day, provided temperature and feeding conditions are optimal. In nature nests containing 2 or 3 clutches were found on several occasions (once 4 clutches); old and empty nests were never found. (No systematic investigations have been made on this point, however. It will be an issue for further research).

Synchronization of males may be the outcome of a selection for maximal reproductive success of individual males. In settling and building in rival situations (in nature nesting in a solitary situation will probably not occur) it was found that a male's chances of success in settling are greatest the earlier he is able to do so. There must be a limit to this ability, that will be approached as closely as possible by a selection for an early start. Within a population this will lead to a simultaneous settling, which will tend to synchronize all subsequent breeding behaviour (SEVENSTER's observation on settling of a population supports this view). VAN MULLEM *et col.* (1964) who studied stickleback populations in the wild, concluded that reproduction was largely synchronized. Net-catches indicated that changes in numbers, owing to the appearance of young, occurred during certain periods of a rather restricted duration. The results of the experiments and observations

It appeared that the proportion of well-developed eggs in stolen 'clutches' was smaller than in spawned clutches (cf. Table 86). In the former group the proportion of clutches with 50% or more of all eggs present affected was distinctly larger than in the latter ($\chi^2 = 4.58$, $p<0.05$).

Development of stolen eggs was not differently affected if another full clutch was added subsequently (as occurred in 1-1 situations, cf. Table 86).

Mortality, retardation, and territory rank number.

Trials in complicated rival situations 1-0 with known territory sizes, provide some indication that development of eggs may differ between territories of different size. Data are represented in Table 88; they have been

TABLE 88

Mortality, retardation and development of eggs in territories (1) and elsewhere

situation	no of clutches	total dead	total very retarded	total retarded	total well-developed
nest (1)	9	19 (1.0%)	68 (3.4%)	831 (41.0%)	1172 (58.0%)
other nests	14	121 (3.3%)	995 (26.6%)	2425 (64.8%)	1193 (31.9%)

grouped into two categories, territories (1) and other territories, because the small sample does not warrant a further splitting.

Judged from the overall rate of well-developed eggs, conditions for egg-development were more favourable in (1) nests than elsewhere. A comparison of proportions of well-developed eggs in individual clutches of both groups failed to indicate a significant difference, however (Mann-Whitney U test, p = 0.18, two-tailed).

Is stealing of eggs functional?

The ability to sneak and subsequently fertilize a clutch in a rival's nest has the clear advantage for its performer that it will increase his chances of fathering offspring, especially in those cases in which he is the first male to pass through the nest.

The function of stealing eggs is not so obvious. Stealing always follows a fertilization of a rival's clutch, though many attempts may be unsuccessful. By fertilizing eggs elsewhere a male seems to be already in a most favourable position. Several likely arguments can be put forward *not* to steal eggs.

a. Nest-owners that take part in fertilizing the clutch (and the majority will do so, in spite of sneaking males that may have preceded) will start a parental cycle and will ventilate the eggs appropriately.

b. Even if a nest-owner does not take part in fertilizing, he will nevertheless start a parental cycle, and cease behaving sexually, if the number of eggs present in his nest suffices to produce this effect. Stealing eggs will have a negative effect in this respect.

c. A sneaker will be able to continue behaving sexually if he does not collect eggs; in the absence of parental behaviour inhibiting effects of parental care (induced by the presence of eggs) on sex are of course non-existent.

d. It was found that moving eggs resulted in a total loss in many cases, and in an inferior development of those eggs in others.

e. Once a sneaker has put stolen eggs into his own nest, it will reduce his chances of leading a female himself, albeit for a short period only. Such a refractory period is absent if no eggs are present in his nest (SEVENSTER-BOL, 1962).

Egg-losses may just be an experimental artifact. In nature the supply of females could certainly be such that all males are able to collect one or more clutches by courting; egg-losses from stealing would not occur under those circumstances. This, however, does not inform us of the possible function of stealing.

There is certainly no need that eggs of one clutch be distributed over different nests to improve development by a better ventilation. A single male is able to ventilate an undivided clutch (and even a number of clutches) in such a way that results are maximal; no doubt clutch size and fanning capacity are well-adapted features.

One hypothesis is that robbing is merely a by-product of the situation in which it occurs. It is well-known that a cluster of eggs outside the nest is immediately put inside. Eggs may get outside a nest by different causes; sometimes they were spawned in front of the entrance. Since eggs left outside will soon perish, the tendency to push them into the nest has a clear function. Possibly then, robbing is comparable to retrieving eggs. In the experiments described earlier, robbing was often found to have apparent negative results, but possibly the reaction has been retained because its net result (retrieving eggs in all possible situations) is still positive.

Judged from observations, however, this hypothesis seems to be unlikely. The motivation of robbing males does not seem to fit the situation 'eggs found outside the nest which have to be retrieved', but far more points to a specific development fitting a specific situation, and serving a specific function.

Another hypothesis is that egg-stealing and bringing home of eggs leads to fewer interactions between rivals in the period following spawning. This may result in a better arrangement of eggs inside the nest, and subsequently in a better ventilation. This explanation does not seem to be very probable either. Following stealing the robbed nest-owner often remains aggressive for a rather long period. Moreover, during robbing, or directly following it, sand may be spat on the roof part of the nest — probably an adequate defence against further stealing, but hardly a means to facilitate ventilation.

Still another hypothesis is that egg-robbing prevents interactions, or at least ameliorates the effects thereof, during the parental cycle. It was found that rival situations 1-1 had a greater yield of well-developed eggs than 1-0 situations (see Table 86). Further, stealing by males that collected a complete clutch by courtship was never observed. From these data it could be reasoned that egg-stealing might be an advantageous device for arriving at a 1-1 situation under those circumstances where the female supply might fall short of providing every male with a full clutch within a short period of time. If robbers could collect eggs on several occasions from different nests they would be able to collect a sufficient number for starting a full parental cycle. Besides, if the supply of females does not fall short, stealing only one mouthful will bring a sneaker already in a weak parental condition (in case a fertilization has preceded stealing), and the addition of a normal clutch afterwards will prevent the stolen eggs from getting lost.

In conclusion, the most probable hypothesis on the function of egg-stealing seems to be that robbing has been selected as a means to synchronize the parental cycles in a population of settled males; fertilizing eggs elsewhere and stealing part of them gives a male a better chance of offspring than without stealing.

One objection, that seems contrary to this conclusion, should be noted at the end. Robbing was not always linked with a preceding fertilization (attempt); it was observed at least once at a nest in which the eggs had been spawned three days before. The function of robbing in such a case is not clear; the robber will start a parental cycle in favour of another male's offspring, while cutting off a further possibility to father his own during the same cycle.

From males used in the laboratory we sometimes got the impression that those which had performed parental behaviour earlier in the season remained in a condition fit for reproduction over a longer period subsequently than males that had not. With this possibility in mind, it is suggested that stealing of eggs fertilized by a rival might be favourable to keep a male in a better physiological condition, so that he is more likely to start a new reproductive cycle again.

FANNING DURING THE PARENTAL CYCLE

Regular ventilation of the eggs is necessary for development of the embryos. The nest-owner provides for this by fanning (p. 112), *i.e.* causing a water current to pass through the nest that brings in sufficient oxygen and removes waste products. Fanning characteristics, such as the total time spent in fanning, the number of fanning bouts, and the duration of individual fanning bouts, were assessed daily during the parental cycle in 30-minute periods.

It seems obvious to compare parental-success and data for fanning of males in solitary and rival situations in order to find a correlation. It may be expected that the smaller proportion of well-developed eggs as found for rival males — especially in a complicated situation — goes along with insufficient or inefficient fanning, or both.

Observations of complicated situations make this assumption very probable. One frequently observes a male that seems to interrupt his fanning activities on the approach of a rival at the boundary. The impression is that such fanning would have lasted longer if the rival had been elsewhere at that moment, and no aggressive interaction had occurred subsequently. Indeed, if one separates all fanning bouts into bouts followed immediately by an aggressive interaction, and bouts without aggression following, it appears that the average bout length of the former is distinctly shorter than the latter. Therefore, bouts followed by an aggressive interaction will be called interrupted.

The average durations of fanning bouts of either category on consecutive days of the parental cycle are represented in Table 89. These data were obtained from a sample of males in simple 1-0 and simple 1-1 situations. From Table 89 it follows that on all days of the cycle interrupted bouts appear to be of a shorter duration than normal bouts.

TABLE 89

Average durations of interrupted and not interrupted fanning bouts for different days of the parental cycle

situation	category of fanning bout	day of the parental cycle D 1	D 2	D 3	D 4	D 5
simple 1-1	+ aggr.	5.8 (38)	8.2 (35)	7.9 (63)	7.3 (80)	8.0 (54)
	— aggr.	8.3 (104)	11.2 (124)	11.7 (187)	12.5 (246)	16.5 (222)
simple 1-0	+ aggr.	5.5 (28)	9.4 (49)	6.6 (83)	9.2 (65)	10.0 (102)
	— aggr.	10.8 (63)	10.3 (152)	9.6 (133)	12.2 (97)	15.2 (290)

(In brackets number of fanning bouths)

IX. DISCUSSION

No territory, be it topographically fixed or not, comes into being without relevant behaviour of an owner. Relevant behaviour has to be understood as any behaviour which results in an animal claiming and maintaining a holding. This actual claiming and maintaining of a holding will be called territorial behaviour. For a limited period of time a territory may continue to exist even in the absence of the owner, because his neighbours have learned about it. A continuous reinforcement of the processes involved remains necessary, however, since the removal of an owner eventually results in a change in the *status quo*.

Strict territorial behaviour comprises agonistic behaviour — frequently intra-specific only — such as attack, or attack intentions, or threat. In fact, we may include in the term 'territorial behaviour' anything that can be shown to reduce a competitor's aggression. Territorial behaviour in this sense is restricted, however, to a social situation. A solitary male will not perform such behaviour. (He may perform aggressively motivated appetitive behaviour — patrolling his area in search of other males — but this type of behaviour cannot be distinguished unambiguously from behaviour that is differently motivated). Nevertheless, it certainly makes sense to consider the area held by a solitary male a territory — at least in the Three-spined Stickleback — because a solitary male will show overt aggression upon experimental confrontation with a conspecific.

NOBLE's (1939) definition of a territory — *any defended area* — provides for most of these considerations, but should perhaps be modified to include solitary situations. It could be rephrased as *any area that will be defended* — if an appropriate stimulus is offered.

The term 'defence' should be used in a functional sense. As it was already pointed out, it will often be difficult to prove defence. The results of an owner's defence, however, can often be determined without much difficulty. Therefore, it would perhaps be preferable to redefine a territory as *any area where the mere presence or the behaviour of an owner (or owners) excludes or would exclude the simultaneous presence of conspecifics.*

On theoretical grounds, HUXLEY (1934) supposed that there has to be a minimum size for any territory. If a territory boundary is pushed back in the direction of the territory centre, defence will become so intense at a certain point that it resists further compression. A territory seems to be compressible only up to a certain point; HUXLEY used the well-known analogy of rubber discs as an illustration of this model (p. 64).

In experiments with the Three-spined Stickleback the existence of a minimum size has been proven. The density of territories did not rise above a certain value and no additional males could be accommodated (p. 54). In solitary situations a functionally minimum size was found which affected success in reproduction (*i.e.* 30 × 30 cm^2). A solitary male may defend an area of smaller dimensions than this critical minimum — and therefore such a sub-minimal holding should still be called a territory — but it is no longer functioning properly. In free rival situations territory dimensions hardly ever dropped below this critical value; territory dimensions as such usually did not set a limit to reproduction in these situations. In solitary males success in nest-building and courtship appeared, within a considerable range of dimensions, to be independent of territory size. If the dimensions dropped below the critical value, nest-building was largely inhibited (p. 26). In the same situation success in courtship is significantly reduced for two reasons. First, a much longer time is needed before the female can be led to the nest (p. 84), or, in other words, duration of pair formation — which is in sticklebacks a very temporary affair — is considerably increased. Secondly, the chances of the female entering the nest, once she has arrived at the entrance, is significantly decreased (p. 86). Thus, the chances of obtaining a clutch, and hence, of fathering offspring, is reduced.

This reduction of success is probably due to the male being made aggressive by the female's sudden and continual presence near the nest, which is the aggression centre of the territory (p. 89). In such a situation a prolonged time seems to be needed to shift the owner's aggression-sex balance to a relatively sexual state. Therefore, it seems to be not just the presence of a female, but *a female at a certain distance from the nest* that represents an important factor in pair formation, at least at the beginning of courtship.

The percentage of females entering the nest was much smaller in sub-minimal territories than in those of larger size (p. 86). Presumably, the aggression-sex balance required for a successful courtship is more unstable in sub-minimal territories, because it may switch back immediately to a relatively aggressive state if the female, following her arrival at the entrance to the nest, does not enter at once. An alternative explanation, however, may be that in sub-minimal territories the male's greater aggressiveness during courtship may have induced in the female a correspondingly greater tendency to flee. This in turn may have affected her willingness to enter the nest, even when she had been courted up to this point for the same length of time as in otherwise similar situations that had led to successful entering — see Table 59.

Once a female had entered a nest, the males were exactly similar in their ability to make the female spawn (p. 76).

Theoretically there is also a maximum size of a territory. An unlimited space is by definition no territory; an area from which other animals are excluded, implies that it has to be of some limited extent. The maximum size of a stickleback's territory has not been investigated, but it is worth considering the issues involved.

A territory-holding male frequently visits his nest, even though it is empty; in the largest available area (600 × 100 cm^2) he visited it on the average once every two minutes. To all appearances the time between leaving the nest and starting to return has a relatively stable value, independent of actual territory size. This time is available for the outward trip, and thus sets a limit to the radius of the territory. Size seems to be important only in so far as it has a direct bearing upon the duration of the return trip: the larger the territory the longer it takes to get back (p. 41). Although the behaviour of the male obviously determines to a large extent the size of a territory in general, it would be quite wrong to suppose that this is the only factor. There are many factors, among which we might mention clearness of the water, and heterogeneity of the habitat, which will also have a marked effect upon territory size.

Territory-owners do not act indifferently towards each other; on the contrary, they interact continuously. Such interactions between candidates for the same territory start during the settling phase, and will lead to a rather regular spacing-out of the participating males (p. 63). The effects of spacing-out were found to be still maximal in the largest area used (600 × 100 cm^2), since if we start our experiment with one male at the end of the tank, and introduce the second one 24 hours later, the latter always built at the greatest possible distance away. Additional males, introduced later still, settled and built according to the same pattern, constructing their nests at a site almost equidistant and maximally away from the two nearest neighbours. A description of the pattern of nesting, and a possible explanation of how it is brought about, is given on p. 59, 65.

An area of limited dimensions cannot accommodate an unlimited number of males. Once a certain density has been reached the numbers remain constant. Such an area is said to be fully occupied (p. 54). The density of a fully occupied area does not have a strict numerical value; like all parameters which depend on biological values it is variable, and in stickleback, it seems to depend on the way in which the area is settled. When an area is settled simultaneously, its territorial density is considerably higher than it is when there is a succession in settling. This result follows from it being easier for a male to obtain a territory if he has to compete only with males that are also looking for territories than with those that are already settled. In the former case such a male has to divide his aggression among several opponents. Moreover, aggression is not yet maximal at this stage; it is after the nest has been built that aggression reaches its peak level (VAN IERSEL, pers. comm.).

The highest densities observed in a fully occupied area — in HUXLEY's model the point where a further compression of territories is no longer possible — agrees rather well with the minimum size for a territory as established for success in building and courtship (p. 55).

Next to a spacing-out mechanism that prevents over-crowding, another phenomenon, called attraction, is frequently reported in the literature. KALELA (1952, 1954) quotes a number of authors (*e.g.* KULLENBERG for *Fringilla coelebs*, PYNNÖNEN for *Dryobates major*, RINKEL for *Vanellus vanellus*), and discusses in this context the possible functions of the supposed centripetal effects of territorial behaviour (see below). Though qualitative data provide some evidence that males are attracted to one another there is no experimental evidence, to my knowledge. Mutual attraction certainly might cause animals to settle in a group. However, clustering may be caused quite differently, and less direct. In sticklebacks at a certain stage settling and nest-building behaviour were activated by interactions with conspecifics (p. 51). Therefore, even in an otherwise homogeneous habitat territories may become clustered as a result of such a mechanism, *i.e.* for reasons other than a strict behavioural attraction. Another possibility is that animals require some special feature in the environment, apart from the presence of conspecifics, to settle down. Clusters founded in this way have, of course, nothing to do with an attraction. An attraction mechanism may be effective in the Three-spined Stickleback. Preliminary experiments did not yield conclusive evidence. It is an issue for further research.

Breeding in a territory-cluster may lead to greater reproductive success, in spite of the proven disadvantages (see below). In the Three-spined Stickleback nest-building may be stimulated by conspecifics under certain conditions (p. 51). In other experiments, not here reported, there was a difference in the sex-aggression ratios between the social and the solitary situation (a shift towards a relatively more sexual state in the former situation). Besides, females may find a group of males more attractive than they find a solitary male.

Stickleback males settled in a rival situation possess territories of variable size (p. 95). Such males act as competitors; the individual success in reproduction can be affected significantly by their interactions.

When a ripe female was introduced into an experimental rival situation, the duration of a successful courtship (measured as the time lapse between introduction and entering) was considerably prolonged by aggressive interactions between males. At the same time the chances of nest-entering had decreased, compared with those in solitary controls (p. 99). Clearly, the aggressive interactions between males interfered with the necessary change in the balance between the sexual and aggressive tendencies, resulting in a state of affairs very much like that in sub-minimal territories in solitary situations.

Males in rival situations did not have equal chances of leading a female to a nest and fertilizing the clutch. Owners of the largest territories had a distinct advantage (p. 97). A correlation was found between the total surface of territories and the numbers of females that entered the nest (p. 98). The notion of a distribution of females according to a mere probability seems obvious here: the probability of staying in a certain area will be correlated with its size. However, owing to continuous interactions between males, and between males and females, courtship procedures may be far from simple in complicated situations. Females seldom stuck to one courting male but on the contrary, changed partners frequently. The duration of her stay in a certain area did not seem correlated with the area's surface in a simple way, but unfortunately sufficient detailed records are not available.

In the experiments reported on, one female was introduced at a time. The degree of interactions, and the effects of introducing more females at a time are therefore unknown. In nature such a situation may be quite usual.

In the case of those males that were unable to lead a female to their own nest, the ability to sneak is another factor adding to the chances of fertilizing eggs (p. 100). The tendency to sneak could not be correlated with the rank number of the performing males; sneaking was about equally probable for males of any rank number. Males owning small territories in complicated rival situations are more likely to have their eggs fertilized by a sneaker than owners of large territories are (p. 107).

Having fertilized a clutch of eggs, a male enters a refractory period and is unable to lead a female to his nest again. This period lasts about half an hour and depends on the presence of eggs inside the nest (SEVENSTER-

BOL, 1962). During this period those neighbours that did not fertilize eggs will have a better chance of success in courtship. Moreover, one male does not collect an unlimited number of clutches; 5 seems to be about the maximum possible. The conclusion that the owner of the largest territory is most successful in reproduction is too simple, additional factors greatly complicate the state of affairs.

Mutual interactions between males also affect individual reproductive success during the last phase of the cycle (parental care). Three factors were identified that reduce success in complicated rival situations: stealing of eggs, interruption of fanning bouts, and insufficient development of the fanning tendency proper. Interruption of fanning — which results in inefficient ventilation of the clutch — and insufficient fanning cause an increased mortality and retard the development of the embryos. In addition to this, slow development may even lead to higher mortality in later life (see p. 128).

Losses due to stealing of eggs, egg mortality, and retardation were least pronounced in large territories; in small territories many eggs may be lost and development may be severely retarded (p. 121, 129).

Within a group of territorial rivals, synchronization of their reproductive behaviour appears to be of great importance for ultimate success in reproduction. In experimental situations where there is any male present without a clutch of eggs in his nest (1-0 situations, p. 118), many more eggs were lost and development was much more retarded than was the case when all males had received eggs (1-1 situations). Possibly a 1-0 situation is only an experimental artifact, and 1-1 situations are the rule under natural conditions, either because there are enough ripe females for each male to receive at least one clutch, or, because those that do not get females manage to steal enough eggs (p. 131). Males whose behaviour was synchronized (1-1 situations) did interact but the effects were far less severe (p. 119, 127).

Synchronization in a population of sticklebacks probably originates right at the start of the reproductive cycle — mutual stimulation of nest-building occurs in males which are in contact with conspecifics (p. 44). If there are enough females, synchronization will be maintained up to the conclusion of the parental cycle.

A state of affairs under natural conditions and directly comparable to a 1-0 situation in sticklebacks, was described by TOMPA (1964), who studied a population of song sparrows (*Melospiza melodia*) on Mandarte Island, B.C. Male song sparrows trying to settle among established territory-owners interfered with their breeding. The same was true for territorial males that remained single. These unmated individuals remained aggressive throughout the breeding season and caused continual disturbance. A year in which total density was average, an unusually large number of unmated, territory-holding, males resulted in a low reproductive success of breeding pairs. According to TOMPA this was caused by breaking of pair bonds and temporary or permanent inhibition of breeding in many pairs. In another year, when the total density was high, a relatively low percentage of eggs hatched, most probably caused by lack of proper incubation due to interference by non-territorial males.

In a number of cases the adverse effects of aggressive interactions on the ultimate breeding success may be far more complicated, and less direct. JENKINS' (1961) findings in partridge, where the amount of interactions between pairs during late winter was thought to determine the survival of chicks later in the season, may be an example. Possibly some delayed mortality among young may be found in sticklebacks as well. Mortality after hatching may be correlated with the amount of interactions the males experienced during their parental phase, but no data are available to check this point.

Competition for a territory occurred in all experiments in which more than one male was used. Some only of the males present were able to settle, although the rest were perfectly capable of settling as shown by their success once conditions became more favourable (p. 52).

In nature it is very likely that there are more males present than there is space for them to occupy territories. The densities observed by SEVENSTER (see p. 55), and certainly by BERTIN (p. 55) suggest that, at least locally, the available area was fully occupied.

Competition for an area, and the existence of a minimum size for an appropriate territory make territorial behaviour a potential agent for limiting the number of breeding individuals, as already suggested by MOFFAT (1903), and later authors such as KLUYVER & TINBERGEN, 1953; TOMPA, 1964; SNOW, 1958; DELIUS, 1965; JENKINS *et al.*, 1963. The experimental results with sticklebacks support this hypothesis.

LACK (1966) on the other hand doubts that numbers are limited in this way. In a population of Great Tits studied over an extensive period near Oxford, the numbers of pairs breeding in nest boxes fluctuated between restricted limits during 19 years. In one year, however, a sudden and very considerable increase was observed (amounting to a 69% increase over the preceding year). LACK reasons that no such high numbers would be possible if territorial behaviour had indeed limited numbers in previous years. He derived a similar argument from TOMPA's work, where also numbers were very much higher in one year than they had been in the others.

Although settling in song birds will proceed in a more complicated way than in sticklebacks — especially because of the mixing of generations and faithfulness to a territory held in a previous year — I suggest that fluctuations in numbers as found could result from the way settling of candidates took place, with simultaneous and successive settling as extremes of a scale of possibilities. TOMPA's work may give some support to this view: a peak density was reached after a catastrophic year when many settled, older males succumbed in an unusually severe and late snowstorm. Subsequently the area was resettled by younger males that probably all settled just at about the same time, and so arrived at an exceptional density.

The occurrence of apparently very diverse densities of breeding pairs in different habitats is another argument by LACK against the limiting effect of territorial behaviour. Blackbirds in the Oxford Botanical Gardens breed at a density about 10 times higher than that in Wytham wood; the breeding density of song sparrows on Mandarte Island was likewise about 10 times higher than that in certain areas in Ohio (NICE, 1937). LACK's argument does not seem to hold. Characteristics of the habitat will certainly affect the actual size of territories. In sticklebacks it is possible to accommodate many more males in a tank with screens and a well-spaced vegetation than in an entirely bare tank (*e.g.* cf. WUNDER, 1930, Figs. 11 and 13).

Another of LACK's arguments (1966, p. 279) is that there is no need to account for a density limiting effect of territorial behaviour where territory already has a significance in pair formation. This reasoning does not hold either; the second possibility does not logically exclude the first, on the contrary, the two may very well be connected as in fact it seems to be the case in sticklebacks. Success in courtship was found to be greater in the larger territories, and the larger territories also had an effect on density.

KALELA (1952, 1954) and WYNNE-EDWARDS (1962) think of the territory system as an important device for regulating population density. Regulation of breeding density through territoriality can only be thought of if territorial behaviour not only prevents over-crowding by limiting numbers breeding, but also ensures in one way or another that density does not drop below a certain minimum. Such a regulatory effect, which is connected with the notion of an optimal density, will be brought about, according to KALELA and WYNNE-EDWARDS, by inter-group selection. WYNNE-EDWARDS even goes so far to suggest inter-group selection as being far more important than selection at the individual level (natural selection), stating that where the interests of a group are opposed to those of the individual, it is the latter that will be sacrificed. In my opinion, however, WYNNE-EDWARDS does not present objective evidence in favour of his view; I agree with the criticism of LACK (1966), and prefer natural selection as the simpler hypothesis.

Size of an individual territory is an important factor in determining the ultimate reproductive success of the owner. The question can then be raised whether it is some individual predisposition that makes a male fitter than his rivals to claim a larger area in the competition for a territory. Or, to go one step further, are genetical factors involved and is territory size a character selected for?

Some experiments were done to test whether any consistency in territory size is apparent among settling males. An affirmative result would permit one to conclude that a predisposition, due either to ontogenetical or to genetically controlled factors was present indeed. The data, however, do not lead to an unambiguous conclusion (p. 76). The owner of the largest territory was sometimes once again among the top-ranking males if the experiment was appropriately repeated, but sometimes he was not. A second, and much more important step, will be to investigate the chances that offspring of a male with the largest territory will claim a large territory when competing with the offspring of a male that has claimed only a small area. This point is an issue for further research.

If it were found that the ability to claim a large holding is indeed a factor that is (partly) controlled genetically, the existence of a factor which limits territory size is a necessary consequence of this. A high level of the aggressive tendency, necessary to hold a very large territory, might well be disadvantageous in a courtship situation and reduce his chances of obtaining offspring (see also TINBERGEN, 1957). Interference by rivals in a complicated situation appears to be of such a kind that a certain territory owner, that has to exhibit agonistic behaviour in defending his property against trespassers, is at the same time refraining from other behaviour patterns that would positively affect his reproductive success. For example: interactions with rivals during a courtship sequence postpone leading of the female to the nest and decrease the chances of her entering it. Agonistic behaviour towards sneaking males sometimes result in a failure to fertilize eggs. Interactions during the parental phase may result in an increased mortality or a severe retardation of the eggs. Probably some compromise is realized in such a way that a male's aggression is sufficient for defending a large enough area, yet, is low enough to allow the female at the nest within a minimum period of courtship. Possibly experiments with the offspring of males selected for different levels of aggression will give evidence on this problem.

No positive evidence is available that territory size in the stickleback is affected by the availability of food. Food for the adult might just be another of these environmental factors which affect the male directly, but it does not seem to be very probable. The territory is not likely to act as a food reserve for the young because the male chases them away soon after they have become free-swimming.

Spacing-out by the formation of territories may reduce predation, but such a factor is not very probable either in the present case — though evidence is lacking. Adult sticklebacks may have only few aquatic predators, as their spiny armament has been proven quite effective (HOOGLAND. MORRIS & TINBERGEN. 1956). SCHENK (pers. comm.) observed that sticklebacks were brought in as food by terns (*Sterna hirundo*). The density observed in nesting sticklebacks suggests that a reduction of predation by spacing-out in not very likely in this case.

It is likely that in the Three-spined Stickleback the significance of its territory, given the organization of reproductive behaviour as it has been selected for in evolution, has to be found in a reduction of interference by conspecifics during reproductive behaviour.

In my opinion this seems to be the main function for territorial behaviour. As an incidental consequence of this, territorial behaviour appears to function as a regulatory mechanism. Regulation of numbers is a by-product of selection for something else of advantage at the individual level — contrary to WYNNE-EDWARDS who states that this regulatory mechanism is selected *because* it regulates numbers.

Retrospective: The behaviour of the male three-spined stickleback in a natural situation: a quantitative description

Article published in *Behaviour* volume 41 (1972), p. 232–241

R.J. Wootton
(IBERS, Edward Llwyd Building, Aberystwyth University, Aberystwyth, SY23 3DA, Wales, UK)

The genesis of this study (Wootton, 1972) lay in a camping holiday at Long Beach on the west coast of Vancouver Island in the mid-1960's. Bob Black and I were both post-graduates in the Department of Zoology, University of British Columbia studying the behaviour of male three-spined sticklebacks, *Gasterosteus aculeatus*. By chance, we found a set of splash pools in a rocky outcrop on Long Beach, in which male *G. aculeatus* were breeding. Although time was limited, we made qualitative and some quantitative observations on the males. When we compared our observations with those of other field studies of the behaviour of male *G. aculeatus*, we were surprised to find that little had been published. Bob Black wrote a short note presenting our results that demonstrated that quantitative data could be collected from natural populations (Black & Wootton, 1970).

After completing my PhD in late 1968, I took up a post-doctoral fellowship in the Department of Psychology in Durham University. I wanted, eventually, to follow up my PhD research on the measurement of aggression in male *G. aculeatus* with studies on the neural and endocrine control of aggression. The move to Durham was to gain experience in the use of neurophysiological techniques in the research group of Dr. J. Delius. My post-doctoral research used an operant technique to map the sites in the brain of domestic pigeons that would support electrical self-stimulation (Delius et al., 1976).

In the summer of 1969, someone told me that stickleback could be seen in the River Wear in Durham City. At lunch time, I walked down to the river and found a 30 m section where I could see stickleback in shallow water along the river edge. After this serendipitous find, I spent suitable lunchtimes over the next nine weeks making observations at this stretch, while continuing the pigeon research in the mornings and afternoons. I recorded the positions of the nests of the male *G. aculeatus* and the distance between them, and I quantified the behaviour of selected males. The study (Wootton, 1972) had two main aims. The first was to confirm that the male behaviour observed in the laboratory was also observed under natural conditions. The second was to test a hypothesis, suggested by my PhD research, that the territory size of a male *G. aculeatus* would initially contract and then expand over the course of a parental cycle. My observations confirmed that the behaviour of the males in the natural population was similar to that of males held in laboratory conditions. There was limited evidence that the territory size of a parental male decreased as the male spent more time fanning the nest, but I did not observe any subsequent increase in territory size when the time spent fanning decreased as the parental cycle progressed. An unexpected observation was the frequency of nest raiding by males in this population (Wootton, 1971).

As an account of the behaviour of *G. aculeatus* in a natural population, my research (Wootton, 1971, 1972) was rapidly superseded by a more detailed account of the behaviour in a lacustrine population (Kynard, 1978). Kynard was making his field observations in 1969 and 1970, so our data collection overlapped. Subsequently, reports on behaviour in a natural population became more common. There were several notable studies. FitzGerald and his research group observed anadromous *G. aculeatus* in salt marsh pools at Isle Verte in the St. Lawrence Estuary (reviewed in Whoriskey & FitzGerald, 1994). This research programme augmented field observations with field experiments. Mori (1993) described the behaviour of *G. aculeatus* in springs and streams in Japan. Foster and her co-workers exploited the numerous divergent populations found in lakes of the Cook Inlet Basin of Alaska to describe the inter-population variations in behaviour of *G. aculeatus* (Foster et al., 2008). In Europe, Bakker's research team used field observations to identify the correlates of reproductive success in male *G. aculeatus* (e.g., Kraak et al., 1999).

After I moved from Durham to the University College of Wales Aberystwyth in 1969, my research interests switched to more ecological questions, including studies on feeding, growth, reproductive traits and demography of *G. aculeatus* and other fish species. This change reflected my teaching responsibilities at Aberystwyth, which were in the areas of population and community ecology. The culmination of my early years of research on the stickleback was the publication of a monograph on the biology of sticklebacks (Wootton, 1976). As with most careers in scientific research, chance played a role. Chance had taken me to the University of British Columbia at a time when there was a research interest in the biology of the stickleback (Hagen, 1967). Chance also led to the discovery of sites where I could observe the behaviour of the stickleback under natural conditions.

References

Black, R. & Wootton, R.J. (1970). Dispersion in a natural population of three-spined sticklebacks. — Can. J. Zool. 48: 1133-1135.

Delius, J.D., Williams, A.S. & Wootton, R.J. (1976). Motivation dependence of brain self-stimulation in the pigeon. — Behav. Process. 1: 15-27.

Foster, S.A., Shaw, K.A., Robert, R.K.L. & Baker, J.A. (2008). Benthic, limnetic and oceanic threespine stickleback: profiles of reproductive behaviour. — Behaviour 145: 485-508.

Hagen, D.W. (1967). Isolating mechanisms in a threespine stickleback (*Gasterosteus*). — J. Fish. Res. Bd. Can. 24: 1637-1692.

Kraak, S.B.M., Bakker, T.C.M. & Mundwiler, B. (1999). Sexual selection in sticklebacks in the field: correlates of reproductive, mating and paternal success. — Behav. Ecol. 10: 696-706.

Kynard, B.E. (1978). Breeding behaviour of a lacustrine population of threespine sticklebacks (*Gasterosteus aculeatus* L.). — Behaviour 67: 178-207.

Mori, S. (1993). The breeding system of the three-spined stickleback *Gasterosteus aculeatus* (Forma Leiura) with reference to spatial and temporal patterns of nesting activity. — Behaviour 126: 97-124.

Whoriskey, F.G. & FitzGerald, G.J. (1994). Ecology of the threespine stickleback on the breeding grounds. — In: The evolutionary biology of the threespine stickleback (Bell, M.A. & Foster, S.A., eds). Oxford University Press, Oxford, p. 188-206.

Wootton, R.J. (1971). A note on the nest-raiding behaviour of male sticklebacks. — Can. J. Zool. 49: 960-962.

Wootton, R.J. (1972). The behaviour of the male three-spined stickleback in a natural situation: a quantitative description. — Behaviour 41: 232-241.

Wootton, R.J. (1976). The biology of the sticklebacks. — Academic Press, London.

THE BEHAVIOUR OF THE MALE THREE-SPINED STICKLE-BACK IN A NATURAL SITUATION: A QUANTITATIVE DESCRIPTION

by

R. J. WOOTTON [1] [2]
(Department of Psychology, Durham University, U.K.)

(With 4 Figures)
(Rec. 7-IV-1971)

INTRODUCTION

Through the work of the Leiden and Oxford ethologists, the reproductive behaviour of the male Three-spined Stickleback has become one of the classical subjects for detailed analysis (TER PELKWIJK & TINBERGEN, 1937; VAN IERSEL, 1953; GUITON, 1960; SEVENSTER, 1961; and VAN DEN ASSEM, 1967). It is, therefore, surprising that little has been published on the behaviour of the stickleback in a natural situation, especially as one of the strengths of the ethological approach to animal behaviour has been the emphasis on the behaviour of the animal in its natural surroundings.

This paper presents the results of some sixty hours of field observations on sticklebacks. It thus provides a preliminary, but quantitative basis for a comparison of field observations with data collected in laboratory studies.

MATERIALS AND METHODS

The study area was a 30 m stretch of the River Wear in Durham City, England (O.S. Sheet 85, grid reference 272421). The sticklebacks could be seen in the shallow water along the river edge.

The positions of the nests of the male sticklebacks were measured to the nearest 5 cm. At selected nests the behaviour of the resident male was recorded for a 20 min period. Each behaviour pattern was given an alphabetic symbol, and the sequence of behaviour in the 20 min period was written down as it

1) This study was carried out while I held an S.R.C. Post-doctoral Fellowship in the Department of Psychology at Durham. I would like to thank Professor F. V. SMITH for his hospitality.

Dr T. B. POOLE commented on an early version of this manuscript.

2) Present address: Department of Zoology, University College of Wales, Aberystwyth, Wales, U.K.

DISCUSSION

The behaviour of the male stickleback in a natural situation is similar to that shown by males held under laboratory conditions. The fanning cycles recorded in this study were similar to the cycles described in detail by VAN IERSEL (1953), VAN DEN ASSEM (1967) and WOOTTON (1968). The changes which followed creeping-through were also similar to those described by SEVENSTER (1961) and analysed in detail by NELSON (1965). In this case the only difference was that the increase in vacuum zig-zagging described by SEVENSTER and NELSON was not seen during the field observations.

VAN DEN ASSEM (1967) has described nest raiding behaviour identical in all respects to that described in this paper, although his was a laboratory study. Clearly such behaviour is an important component in the interactions between territorial males, but as yet no convincing explanation of its function has been given (WOOTTON, in preparation). A raiding male seemed to take either material or eggs from a raided nest, and would travel upto as much as 300 cm to make a raid.

Changes in the aggression of males during the breeding cycle have been measured in the laboratory by SEVENSTER (1961) and in more detail by WOOTTON (1968 and in prep.). Aggression decreases over the first seven or eight days of a twelve day parental period. After fry become visible the aggression of the male increases considerably. On the basis of these findings and the studies of MORRIS (1958) and BLACK (1969), WOOTTON suggested that during the parental period, territory size would first tend to decrease but then would increase again as the fry began to move further from the nest. The field observations support the first part of this hypothesis. As the fanning of the parental male increased, the area defended often decreased. But there was no evidence of an increase in territory size after the fry had hatched. The one male observed until its fry had dispersed, then abandoned its territory. Clearly the relationship between aggression as measured in laboratory studies and changes in territory size when a number of males are present requires further study. SCHOENER (1968) has reviewed a number of studies on territorial birds whose territories also show changes in size during the breeding cycle.

Although the classical accounts of the courtship behaviour of the male stickleback emphasize the zig-zag, lead and show components of courtship (TINBERGEN, 1953), the studies of SEVENSTER (1961), NELSON (1965) and WILZ (1967, 1970), and these field observations have shown that in male-female interactions the circling, dorsal-pricking and visits to the nest that include bouts of fanning are also prominent. Circling and dorsal-pricking by the male seems to inhibit the female both from following the male when the male returns to the nest and from leaving the area. WILZ has interpreted circling and dorsal-pricking followed by a return to the nest that includes fanning as a necessary interlude in the courtship sequence, an interlude in which the male switches from a primarily aggressive motivation to a primarily sexual motivation and during which the female tends not to leave the area. This interpretation is supported by the field observations.

The field observations were, of necessity, restricted in both time and scope. They do, however, suggest that the descriptions and interpretations of the behaviour of sticklebacks that are obtained from laboratory studies are relevant to the behaviour of sticklebacks in natural situations.

SUMMARY

The behaviour of male Three-spined Sticklebacks in a stretch of the River Wear in Durham City is described. Quantitative data on parental behaviour and the effects of creeping through agreed closely with data collected under laboratory conditions. A decrease in territory size during the parental phase was noted. Nest raiding and egg stealing were also observed. Circling and dorsal-pricking were important components of courtship behaviour.

REFERENCES

ASSEM, J. VAN DEN (1967). Territory in the Three-spined Stickleback (*Gasterosteus aculeatus*), an experimental study in intra-specific competition. — Behaviour Suppl. 16, p. 1-164.

BLACK, W. R. (1969). Aggressive behaviour, territoriality and parental success in Three-spined Sticklebacks. — M.Sc. Thesis. University of British Columbia, p. 1-68.

GUITON, P. (1960). On the control of behaviour during the reproductive cycle of *Gasterosteus aculeatus*. — Behaviour 15, p. 163-184.

IERSEL, J. J. A. VAN (1953). An experimental analysis of the parental behaviour of the male Three-spined Stickleback (*Gasterosteus aculeatus*). — Behaviour Suppl. 3, p. 1-159.

LEINER, M. (1930). Fortsetzung der ökologischen Studien an *Gasterosteus aculeatus*. — Z. Morph. u. Ökol. Tiere 16, p. 499-540.

MORRIS, D. (1958). The reproductive behaviour of the Ten-spined Stickleback (*Pygosteus pungitius* L.). — Behaviour Suppl. 6, p. 1-154.

NELSON, K. (1965). After effects of courtship in the Three-spined Stickleback. — Z. vergl. Physiol. 50, p. 569-597.

PELKWIJK, J. J. & TINBERGEN, N. (1937). Eine reizbiologische Analyse einiger Verhaltensweisen von *Gasterosteus aculeatus* L. — Z. Tierpsychol. 1, p. 193-200.

SCHOENER, T. W. (1968). Sizes of feeding territories among birds. — Ecology 49, p. 123-141.

SEVENSTER, P. (1961). A causal study of a displacement activity (fanning in *Gasterosteus aculeatus*). — Behaviour Suppl. 9, p. 1-170.

TINBERGEN, N. (1953). Social Behaviour in Animals. — Methuen, London, p. 1-150.

WILZ, K. J. (1967). The organization of courtship behaviour in sticklebacks. — D.Phil. Thesis, Oxford, p. 1-162.

—— (1970). Causal and functional analysis of dorsal pricking and nest activity

Retrospective: Breeding behavior of a lacustrine population of threespine stickleback (*Gasterosteus aculeatus* L.)

Article published in *Behaviour* volume 67 (1978), p. 178–206

Boyd Kynard
(BK-Riverfish, LLC., 28 Echo Hill Rd., Amherst, MA 01002, USA)

By the time of my field study during 1969–1970, Tinbergen's (1951) laboratory studies on the breeding behaviour of threespine stickleback were already considered to be classics because they laid much of the foundation for the study of ethology. However, studies on the behaviour of wild sticklebacks during spawning were lacking. In the spring of 1969, I was a PhD student who had been searching for more than a year for an ideal site in which to study threespine stickleback breeding in the wild. A field study that examined breeding in a wild population could place the extensive laboratory observations into a natural context, and perhaps discover new ecological aspects of breeding. When I visited Wapato Lake in April, 1969 to help a fellow graduate student with his project on nearby Lake Chelan, I saw hundreds of threespine stickleback breeding in the shoals of the lake and declared loudly "this is my PhD" — I never looked back.

In the late 1960s and early 1970s, Washington state and British Columbia were sites of intense research by Don Hagan and Don McPhail and their students on the ecology and evolution of threespine stickleback. I was a student of Don Hagan and my study (Kynard, 1978) on the behavioural ecology of breeding in one stickleback population was only a small part of the diverse research on sticklebacks in the Northwest at that time. Research by our stickleback group revealed many aspects of reproductive biology, predation and foraging, evolution of lateral plates and spines, and ecological genetics (reviewed by Wootton, 1976, 1984; Bell & Foster, 1994).

Wapato Lake provided a natural laboratory in which to observe and quantify many aspects of breeding by stickleback and to test hypotheses only previously tested in the laboratory. The conceptual behavioural-ecology model of reproduction from my study included mate choice (such as the discovery of females courting males), nest raiding by territorial males and by groups of females, habitat selection by males, effects of breeding time and habitat selection on reproductive success of males, and aggression by breeding males toward other males and toward a predator. Concurrent with my study, a fellow graduate student at the University of Washington, D. Semler (1971), found sexual selection for red-throated Wapato Lake stickleback males and Moodie (1972) studied nesting ecology and reproductive success in a stickleback population in the Queen Charlotte Islands of British Columbia. Thus, the Canadian-USA effort on sticklebacks was special, broad, and deep.

My study (Kynard, 1978) emphasized reproductive success of males relative to spawning timing and nesting ecology. This work served as a foundation for later research on stickleback diversionary displays (e.g., Foster, 1988, 1994), egg kidnapping (e.g., Ridley & Rechten, 1981) and other studies of stickleback behaviour in a natural context. The Wapato population declined during 1970, and I gathered the most extensive data to date on the likely adverse affects of introduced fish (young-of-the-year sunfish) on stickleback (Kynard, 1979). This work was one of the first to raise concern about stickleback conservation (reviewed by Foster et al., 2003; von Hippel, 2008), and illustrates the need for researchers studying wild populations to be able to quickly adapt a research programme to unexpected and rare events, particularly when those events concern environmental degradation.

References

Bell, M.A. & Foster, S.A. (1994). Introduction to the evolutionary biology of the threespine stickleback. — In: The evolutionary biology of the threespine stickleback (Bell, M.A. & Foster, S.A., eds). Oxford University Press, Oxford, p. 1-27.

Foster, S.A. (1988). Diversionary displays of paternal sticklebacks: defenses against cannibalistic groups. — Behav. Ecol. Sociobiol. 22: 335-340.

Foster, S.A. (1994). Inference of evolutionary pattern: diversionary displays of threespine stickleback. — Behav. Ecol. 5: 114-121.

Foster, S.A., Baker, J.A. & Bell, M.A. (2003). The case for conserving threespine stickleback populations: protecting an adaptive radiation. — Fisheries 28: 10-18.

von Hippel, F.A. (2008). Conservation of threespine and ninespine stickleback radiations in the Cook Inlet Basin, Alaska. — Behaviour 145: 693-724.

Kynard, B.E. (1978). Breeding behavior of a lacustrine population of threespine sticklebacks (*Gasterosteus aculeatus* L.). — Behaviour 67: 178-206.

Kynard, B.E. (1979). Population decline and change in frequencies of lateral plates in threespine sticklebacks (*Gasterosteus aculeatus*). — Copeia 1979: 635-638.

Moodie, C.E.E. (1972). Morphology, life history, and ecology of an unusual stickleback (*Gasterosteus aculeatus*) in the Queen Charlotte Islands, Canada. — Can. J. Zool. 50: 721-732.

Ridley, M. & Rechten, C. (1981). Female sticklebacks prefer to spawn with males whose nests contain eggs. — Behaviour 76: 152-161.

Semler, D.E. (1971). Some aspects of adaptation in a polymorphism for breeding colors in the threespine stickleback (*Gasterosteus aculeatus* L.). — J. Zool. Lond. 165: 291-302.

Tinbergen, N. (1951). The study of instinct. — Clarendon Press, Oxford.

Wootton, R.A. (1976). The biology of the sticklebacks. — Academic Press, London.

Wootton, R.A. (1984). A functional biology of sticklebacks. — University of California Press, Berkeley, CA.

Behaviour, LXVII, 3-4

BREEDING BEHAVIOR OF A LACUSTRINE POPULATION OF THREESPINE STICKLEBACKS (*GASTEROSTEUS ACULEATUS* L.)

by

BOYD E. KYNARD[1] [2]

(College of Fisheries, University of Washington, Seattle, Wash., U.S.A.)

(With 9 Figures)
(Acc. 29-II-1978)

INTRODUCTION

The freshwater population of threespine sticklebacks in Wapato Lake, Washington, U.S.A., has been examined for general life history and lateral plate frequency (HAGEN & GILBERTSON, 1972), genetics of the 3 lateral plate morphs (HAGEN & GILBERTSON, 1973b), inheritance of number of lateral plates (HAGEN, 1973), trout predation on lateral plate phenotypes (HAGEN & GILBERTSON, 1973a), and sexual selection of throat color in breeding males (SEMLER, 1971). In the present investigation, their breeding behavior was studied using an ethological approach. During 1969 and 1970, about 1000 hours were spent conducting observations and experiments in the field. Emphasis was placed on examining individual nesting males for (1) boldness to predators, (2) changes in level of aggression during the reproductive cycle, (3) location of nests relative to water depth, amount of shelter around the nest, and distance to their nearest neighbor, (4) predation of rival's offspring, (5) success in egg collection and parental care, and (6) homing ability. Females were investigated for clutch size and for their courtship of males. Population phenomena examined were timing and length of the breeding season, differential mortality of the sexes during post-reproductive mortality, and raiding of nests by groups of non-breeding male and female adults.

1) Present address: Massachusetts Coop. Fish. Res. Unit., Univ. of Massachusetts, Amherst, Mass. 01003, U.S.A.

2) I wish to thank D. W. HAGEN for advice and support during the early stages of the work which was part of a Doctoral Dissertation at the College of Fisheries. Many useful discussions were held with D. MCPHAIL, M. JOHNSON, G. E. E. MOODIE, D. SEMLER, K. LARSON, B. ALLEE, and especially with L. G. GILBERTSON. Thanks are also due to R. WYDOSKI, G. W. BROWN, and R. COCKRAN for advice in the final stages of the study. Dr J. J. A. van IERSEL provided a valuable review of the manuscript. E. AYERS provided manuscript assistance.

fungused eggs. Since males in dense cover tended to have fewer fungused eggs and they were also closest to other males, increased interference by neighboring males did not seem to cause an increase in the number of fungused eggs. This is supported by VAN DEN ASSEM (1967) who compared males in solitary and rival situations for breeding success and found rival situations produced retarded egg development but not increased mortality of eggs. In Wapato males, no correlation was found between number of fungused eggs and total number of eggs which would indicate a simple explanation, *i.e.*, the greater the number of nest eggs, the greater the egg mortality due to problems of providing adequate ventilation to the eggs. Also, examination of fungused eggs in nests did not indicate that eggs on the bottom of the nest or in the interior of the egg mass were particularly susceptible to fungus. No data on fungused eggs was collected on males in 1970.

With a reproductive cycle of about 2 weeks, a male could nest consecutively during the peak breeding months of May and June and produce 4 broods of young. However, close-consecutive nesting did not appear to be prevalent among males. Of the 23 males in area A that were marked before their eggs hatched and were followed during their care of fry, only 4 (17%) were later observed with another nest. One male immediately built a nest 43 cm from his original site and completed another cycle. The other three males declared new territories and built nests in area A within two weeks. The 19 other males were never seen again. They could have nested at another location, but this seemed unlikely since males seemed to prefer a "home" area in field experiments.

Examination of the stomachs of territorial males revealed extensive egg predation in all lake areas. Males in every phase of reproduction were observed sneaking from their territory into a rival's territory to steal eggs. Evidence that males were not eating their own nest eggs was that (1) sneaking behavior of males which resulted in stealing eggs from a rival's nest was observed many times, (2) a negative correlation between stage of nest eggs and stage of cannibalised eggs ($F_{1/450} = 8.74$, $p<0.01$), and (3) the stomachs of 34.7% of the predatory males contained eggs of a different stage than their nest eggs. The large number (65.3%) of males eating eggs that were the same stage as their own was expected since the timing of breeding in many males was synchronized. In spite of this, regression analysis indicated there was still a significant trend for males with early eggs to steal eggs from older nests and for males with EE or fry to steal eggs from younger nests (see evidence #2).

Males from the rock areas did not have as high an incidence of egg predation as those in area A ($p<0.05$) (Table 4) and they also ate fewer eggs

TABLE 4

Comparison of stomach contents of males from Area A and Rock areas

	Stomach contents			
	Empty	Insects	Insects & eggs	Eggs
Area A	5 (5.8)*	22 (25.6)	39 (45.3)	20 (23.3)
Rocks	7 (8.6)	33 (40.8)	20 (24.7)	21 (25.9)
p	<0.05			

* ( ) is percent.

although the difference was not significant. Predatory males from area A ate a mean of 14.6 eggs (n = 60) while those from the Rocks areas ate 11.60 (n = 42) (t_{100} = 1.71, $p>0.05$). The incidence of egg predation in the lake was relatively unchanged during 1969 and 1970, respectively, 59.9% (100 out of 167 ate eggs) and 64.3% (401 out of 624 ate eggs) ($p>0.05$). There was no difference in the mean number of eggs in the stomachs of these predatory males, *i.e.*, 12.1 (s = 12.4) for 1969 and 11.3 (s = 11.2) for 1970 (t_{789} = 0.74, $p>0.05$).

Males ate eggs during every phase of their cycle but there was a significant difference between males in different reproductive stages for the number eating eggs (Table 5). The greatest frequency of predatory males was found

TABLE 5

Comparison of males in various phases of reproduction for the number cannibalizing eggs

	Reproductive phase			
	EN	PEE	EE	F
Predatory males	46 (43.0)*	115 (85.2)	76 (92.7)	23 (41.1)
Non-predatory males	61 (57.0)	20 (14.8)	6 (7.3)	33 (58.9)
p	<0.005			

* ( ) is percent. Stages of males were empty nest (EN), pre-eyed egg care (PEE), eyed-egg care (EE), and fry care (F).

in those caring for eggs, with 85% of those caring for PEE and 92% of those caring for EE also stealing eggs. It was unexpected that such a large percentage (41%) of the males caring for F would contain stolen eggs. Since actual digestion rate of eggs wasn't determined, some males that were collected while caring for fry might have stolen their last eggs just before their

EE hatched into F. Therefore, the 41% is probably an overestimate, but surely many males did steal eggs while caring for F. An examination of all males with stomach eggs revealed that those caring for EE stole significantly more eggs on the average than EN, PEE and F males which ate similar numbers of eggs ($t_{256} = 2.7$, $p<0.01$) and (Fig. 8). The average number of eggs EE males stole (17.3) was about twice that of males in the other stages, *i.e.*, EN (8.1), PEE (9.1), and F (7.0). Males caring for F were either the worst at stealing eggs or more probably, they spent the least amount of time stealing eggs, because they only had an average of 7 eggs in their stomachs.

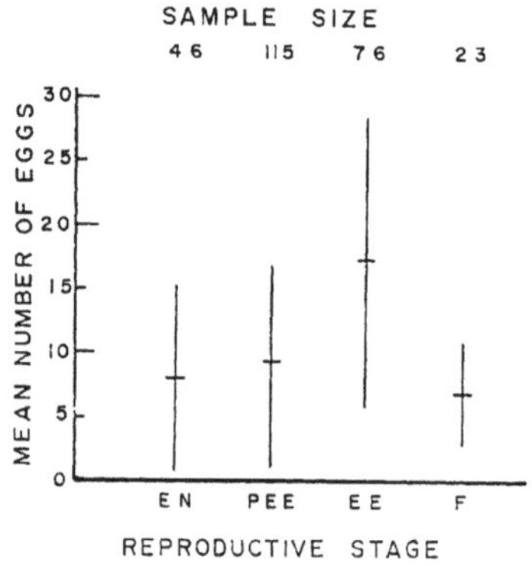

Fig. 8. Comparison of males in various stages of reproduction for number of cannibalised eggs. Stage of males was empty nest (EN), pre-eyed egg care (PEE), eyed-egg care (EE), and fry care (F). Length of vertical bar is one S.D.

Regression analysis also indicated the diet of a nesting male changed during the breeding cycle. Number of eggs in the stomach was positively correlated with reproductive phase of the male ($F_{1/498} = 4.78$, $p<0.05$) and size of male ($F_{1/465} = 9.36$, $p<0.01$). Correlations with the number and stage of nest eggs were reported earlier.

Predation was greatest on PEE and only 25% of all predatory males contained EE or F (Fig. 9). Nest eggs that were less than about 2 days old but older than undeveloped, *i.e.*, in the pre-orbital orb stage, were especially susceptible to cannibalism. The reason for the low number males that stole undeveloped eggs may be that, during the more or less synchronized phase of fresh undeveloped eggs, most males were still mating and seemed to spend most of their time patroling their territories looking for females and intruder males. This is supported by the high number of males (57%) in the EN stage that had not stolen any eggs (Table 5). Of the 499 males that contained stickleback progeny, only 5 (1%) contained F. Preliminary experiments

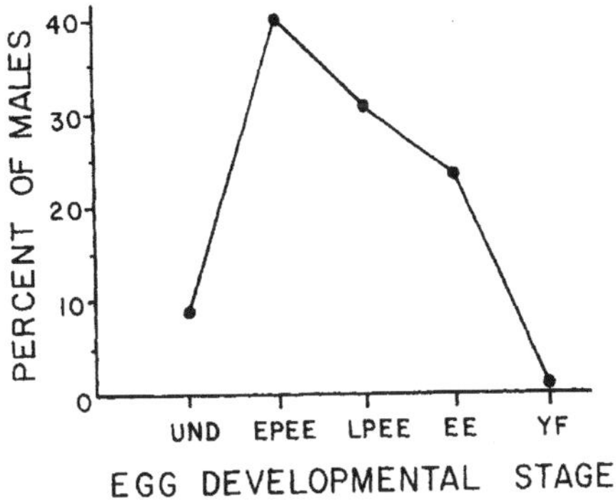

Fig. 9. Time of predation on developing nest eggs as indicated by the percent of predatory stickleback males eating undeveloped eggs (UND), early pre-eyed eggs (EPEE), late pre-eyed eggs (LPEE), eyed-eggs (EE), and yolk-sac fry (YF).

indicated that yolk-sac F did not digest in the stomach very rapidly at 13°C and that the eyes and pelvic girdle were identifiable 24 hr after being eaten. If F were being consumed in large quantities by males, it seems likely that the large sample of males observed in this study would contain more of them. This seems particularly true since yolk-F in the open nest were significantly more susceptible to predation than PEE which were concealed inside the nest, respectively, 3.45 min and 19.29 min average time between removal of parental male until first predation on his progeny by a neighboring male ($t_{(19)} = 4.84$, $p<0.01$). Since several of the PEE tests were terminated at 30 min without predation having taken place, the actual average time of predation on PEE would be considerably longer than 19.29 min. Possibly the quickness with which unguarded nests with F were raided was due to their conspicuous attempts at swimming while in the opened nest. In the 14 cases where actual predation was observed, the predator was a neighboring male. Once the intruder discovered yolk-F in the nest they were easily captured. After a few minutes, other neighboring males usually joined in to devour all F. Thus, the low incidence of predation of F would not seem to be due to the difficulty of catching yolk-F in the nest but rather to the presence of a parental male.

FEMALE FECUNDITY AND BEHAVIOR

Clutch size in females was positively correlated with their standard length and the least squares equation that described the regression line ($Y = -132.8 + 5.4$, $n = 150$, where Y = number of eggs and X = fish length) was similar to those studied by Moodie (1973) and Hagen (1967). Females collected in May and July differed significantly in their mean clutch size

(t_{148} = 2.67, $p<0.01$). Those captured in May had a mean standard length of 42.4 mm (n = 25) and a mean clutch size of 95.8 eggs (s = 36.8). In contrast, the mean size of females in July was 53.9 mm (n = 125) and their mean fecundity was 159.6 eggs (s = 54.5). This represented an increase of 63.8 eggs/female or a 67% increase in the mean number of eggs/female in about 2 months. Since males that nested in May had an average clutch size of 1000 eggs, they would have to mate with about 10 females to collect that number of eggs. In July, males collected only about 550 eggs and, thus, they mated with only 3-4 females.

Female courtship of males was commonly observed in Wapato Lake. Female movements were similar to the typical "heads up" position of acceptance that a female gives a male. This was done hovering in front of the male with an occasional slight forward motion of the body that moved the female toward the male. If not receptive, the male would either ignore this behavior and charge the female, or in extreme cases, grasp the female by the caudal peduncle and carry her out of his territory. If chased or carried away, the female would often return to the nest faster than the male and resume courting postures. If receptive to the advances of the female, the male immediately led the female to the nest and, with his body upright just off the bottom, he pointed to the nest entrance. If she entered, spawning occurred normally and the spent female left the territory.

The incidence of females courting males increased dramatically after the peak breeding months of May and June, *i.e.*, May—26/192 males (13.5% with female courting), June—18/176 males (10.2% with female courting), July— 49/66 males (74.2% with female courting), and August—69/82 males (84.2% with female courting) ($p<0.001$). By the end of the breeding season, any male with a nest was seldom found without several females constantly courting him. Even males caring for EE were courted and despite repeated vicious attacks, females rushed to court them every few minutes.

GROUP NEST RAIDING AND POST-REPRODUCTIVE MORTALITY

Adult sticklebacks in loosely schooled groups of several hundred were often observed moving over the breeding areas of nesting males. These groups were particularely abundant in area A, where captures with dip nets indicated they were composed primarily but not entirely of females, *i.e.*, (67% females; 33% males, n = 28). These groups swam slowly over nesting males, and when they stopped, 10 to 30 fish descended to the bottom and began poking and biting at the bottom. Resident males vigorously defended their area but all were not successful. Eventually a nest with eggs was discovered by one or more of the raiders, which then began to "bore" into the nest and tear it to pieces. The raiders and particularly boring fish did not react aggressively to the resident male but they were very difficult to chase away from the territory. Immediately after boring began, many of the group descended on the nest and soon individual raiders with large clumps of eggs darted from the nest. Other fish followed and the eggs were quickly devoured. Males with red throats appeared to be raided more often than other males, *i.e.*, red-throated males were raided in 7 out of 12 cases and based on their frequency in the breeding area, they should have been raided only 2 out of 12 times. Fisher's exact test probability was $p>0.05$ and, although it was not quite significant at the 0.05 level, the results indicated the need for further study. The raider packs were extremely wary of the observer and this was responsible for the small number of observations on raided males. Raiding of nests by "raider packs" represents one method that females and non-territorial males steal eggs and, also, a possible selection pressure against red-throated males which, in Wapato Lake, seem to have a mating advantage over non-red-throated males from sexual selection (SEMLER, 1971).

In June of 1970, the sex ratio of adult sticklebacks contained a significant excess of females, *i.e.*, 55.5% female: 44.4% male, n = 4661 (HAGEN & GILBERTSON, 1973). A comparison of this ratio (2075 males: 2586 females) with the eighty-five dead fish collected after post-reproductive mortality began in mid-July indicated that females and males died in expected frequencies in the early mortality period, respectively, 38 males (44.7%): 47 females (55.3% ($p>0.05$). This was quite different from the August sex ratio where dead females outnumbered males 12 : 1. There was a significant difference between the sex ratios of early and late post-reproductive fish, *i.e.*, for mid-July, males (38) : females (47) and for mid-August, males (11) : females (136), $p<0.001$.

DISCUSSION

The major period of reproduction on the Wapato Lake sticklebacks was about 100 to 114 days, *i.e.*, during April, May, June and July. Comparison of the termination dates of breeding for 1969 and 1970 suggests a similar but somewhat earlier cycle for 1969. The 100 to 114 day breeding season of this inland lacustrine population, which is located at 48° N latitude, was shorter than would be expected if the population was located in a maritime climate. HAGEN (1967) recorded a 124 day season for a maritime stream *leiurus* population at 49° and MOODIE (1972) found a 90 day season in a more northern maritime lake population at 53°. During a 100 day breeding season, a male could possibly complete 6 or 7 cycles if he nested consecutively. Since only one male out of the 23 that were observed nested immediately after completing a cycle and only 4 males renested at all, the number of cycles most males completed was probably considerably less than the maximum estimate.

Although territorial males were found from May through September of 1970, the most successful breeding was limited to the earliest 60 days. Males that nested late in the breeding season, *i.e.*, after July, collected fewer eggs in their nests, reared fewer offspring to EE, and, also fewer were able to complete a parental care cycle. The 3 peaks of breeding on May 1, June 15, and August 1, indicated that some, but certainly not all, breeding was synchronized. Males had a 13 to 14 day cycle in early June. Even allowing for normal deviation in developing time of eggs due to cooler or warmer temperatures of other times, the period of time between breeding peaks does not support the idea at close consecutive nesting of males. Synchronized breeding was also found in a stream population of *leiurus* in the Netherlands by VAN MULLEM (1967) who found 2 micro-peaks of breeding within each Spring and Fall breeding period.

LENGTH OF REPRODUCTIVE CYCLE

MOODIE (1972) estimated Mayer Lake sticklebacks had an 18 day reproductive cycle. At maximum daytime temperatures of 21° C, the average Wapato male had a 13 to 14 day cycle that consisted of 1 to 2 days for nest building and courtship, 5 to 6 days parental care of eggs, and about 6 days care of F. The cooler water temperatures found in Mayer Lake undoubtedly contributed somewhat to the difference in length of cycles. The great variability among Wapato males for the duration of F care was unexpected since males that deserted F <4 or 5 days old left offspring that still swam rather feebly. Although a windstorm was responsible for the departure of two males with F that left on days 1 and 2, 8 other males left before their F were 5 days old. Why these males left wasn't obvious. Since they were marked on their dorsal spine for identification, marking could have effected their behavior, but this seemed unlikely since other similarly marked fish did not abandon their nests.

REPRODUCTIVE ECOLOGY

Males were most abundant in sheltered nesting habitat. Experimental tests, which gave Wapato Lake males a choice between shelter and open nesting habitat, indicated that they preferred shelter habitat, and that competition for territories in shelter was important in explaining the distribution of males relative to the amount of available shelter (KYNARD, 1979a). The evolution of this preference may be related to the reproductive advantages associated with nesting in shelter habitat that were found in this study and by MOODIE (1972), *i.e.*, more males in shelter habitat reared their offspring to fry, and also, nests in open habitat tended to have more fungused eggs than nests in cover.

Depth of water over the nest was also important in selection of a nest site. Experimental males from Wapato and nearby Roses Lake exhibited a strong habitat preference for water >45 cm deep (KYNARD, 1979a). A genetic basis for this preference was indicated in males from Roses Lake. These males did not have their choice affected by experience because they never nested in the wild prior to laboratory experiments. Since there apparently is a genetic predisposition to choose deep water, there should be selective advantage(s) associated with the choice. In both of these lakes, one selective factor against males that nested in shallow water was the frequent wind storms which caused nest destruction and disturbance of males. The declining water level found in Wapato Lake during spawning could also cause differential nest survival relative to water depth. During observations on males breeding in a stream, WOOTTON (1972) noted that one male in shallow water lost his nest because of the declining water level. Wapato and Roses Lake males also spawn during a gradual decline of the water level which in 1969 and 1970 began in mid-May and reached a maximum of −1.4 m by July 1. It seems probable that nesting in deep water would be adaptive in either ecological situation, *i.e.*, in areas of strong winds or in areas where the water level declined. In Wapato and Roses Lakes, both are probably important selective factors responsible for the strong deep-water habitat preference of males. Since the number of eggs in the nest of Wapato Lake males was positively correlated with increasing depth of water over the nest, males nesting in deep water may have other reproductive advantages, *i.e.*, perhaps females prefer to remain in deep water to feed, *etc.*, and/or females may prefer the courtship or breeding color of deep water males. The last possibility is suggested by the results of SEMLER (1971), who found red-throated males were preferred by females in Wapato Lake, and they also nest in deeper water than other males (SEMLER, 1971; KYNARD, unpublished data).

The spatial dispersion of territorial males was not random but was more uniform than expected. Since the area was very heterogeneous relative to amount of cover present and males were known to space themselves out with respect to density of cover through territorial aggression, the results from the internest distance analysis were not unexpected. The 41 cm average internest distance did not allow very much space between males. The small internest distance of most males is a good reflection of the tendency of males to nest in sheltered locations. Territorial males of *Gasterosteus aculeatus* form *trachurus* that were breeding in natural splash pools also appeared to be dispersed uniformly (Black & Wootton, 1970). Intermale aggression seemed responsible for the uniform spacing of Wapato Lake sticklebacks during breeding. Intermale aggression has been shown in laboratory observations to be positively correlated with increased territory size (van den Assem, 1967) and the present positive regression of internest distance and level of aggression of males in nature supported this correlation. Since territory size is not necessarily correlated in every case with internest distance, further measurements of actual territory size and aggression in males are needed.

MALE BEHAVIOR ON NESTS

The level of aggression of breeding sticklebacks has often been determined in laboratory studies by counting the number of bites given a test male in a tube, *i.e.*, van Iersel (1958), Sevenster (1961), Segaar (1961), Sevenster-Bol (1962), Wilz (1972), and Wootton (1971). These workers found a high level of aggression in the EN stage rather than the low level found in Wapato males. It could be that the EN stage is so short in Wapato fish, that a high level of aggression in the EN stage could not be reached.

Results of the present field experiments gave confirmation and functional explanation for some of the results of laboratory workers. In laboratory tests, Wootton (1971), Black (1971), and Huntingford (1976) reported that biting frequency changed during a male's reproductive cycle with the greatest intensity and frequency of biting occurring during the F phase. In addition, recently a correlation was found to exist between the level of a male's territorial aggression and his boldness toward predators (Huntingford 1976). In the present tests involving predator exposure and level of aggression of nesting males, those caring for F were both bolder to the trout and more aggressive toward the tester male than other males. Since offspring were found to be most vulnerable to predation while in the F stage, the function of the high level of aggression and boldness in males with F appeared to be related to providing maximum protection for these young.

Changes in body color also took place during a male's cycle with the greatest frequency of black and black-blue sided males occurring during EN, when a male was courting females, and later when he was caring for F. Wootton (1970, 1976) and van Iersel (1953) noted a darkening of a male's body (dark back and vertical bars) at the end of a parental cycle, but not during the sexual phase, and this seemed related to making males more cryptic. A more cryptic body color seems favored in sticklebacks under predation. Male sticklebacks breeding in dark, tea colored waters of Chehalis River, Washington, were black bodied and this seemed related to making males more cryptic in the presence of western mudminnow (*Novumbra hubbsi*) that preyed on stickleback eggs and young (McPHAIL, 1969). Also, trout predation favoring non-red throated males seems responsible for the low percentage of conspicuous red-throated males in Wapato Lake (SEMLER, 1971) and Mayer Lake, British Columbia (MOODIE, 1972a, b). Whether black males have a selective disadvantage from trout predation isn't known. Black males in Wapato Lake, which were most abundant during mating and F care stages, were not cryptic but were very conspicuous to this observer. Since mating and F phases are two periods when a male needs no interference from other males, the correlation suggests a black body may have intimidation value for other males and perhaps in attracting females (like red-throat color). Like changes in aggression and boldness that occurred during a male's cycle, the changes in body color seemed related, at least partially, to increasing intimidation of intruders. A similar hypothesis to account for the changes in aggression of nesting males to both conspecifics and a variety of cohabiting species was suggested recently by HUNTINGFORD (1977).

In 1970, 58% of all males that were caring for F had black bodies when they were collected. Whether all of the other 42% could become black isn't known but during observations many males did change their mid- and ventro-lateral body color within a few minutes. Males were also flexible for body color during the sexual phase. This rapid change in chromatic response of the melanophores on a stickleback's sides was demonstrated recently by BURTON (1975), who found the ventral part of their sides could become darker or lighter in 1-2 min whereas the dorsal surface took 3-5 min. Since males were flexible for body color even when caring for F, there are probably other selective factors operating on body color even during this critical period of parental care. Of those males caring for F, many also contained stolen eggs. Observations on many egg stealing males revealed that none had black sides or were aggressive toward the resident male. WOOTTON (1976) and VAN DEN ASSEM (1967) also noted that sneaking males were not aggressive and VAN DEN ASSEM noted sneakers were cryptic. Perhaps retaining the ability to become drab and inconspicuous within a few minutes is related to continuing their sneaking and stealing of eggs. By remaining flexible for body color and level of aggression, males caring for F could become black and intimidate intruders or lose their intimidation color and aggressive behavior, become drab, and for brief periods steal eggs. Any nest raiding done by these males must be of short duration because experiments indicated they probably had <4 min to return before their fry would be discovered and preyed upon by a neighboring male. In addition, the low predation rate on fry in spite of their vulnerability to predation (Fig. 9), the low number of eggs eaten by males guarding fry (Fig. 8), and the low incidence of egg predation by males caring for fry (Table 5), all indicated that fry were given much better parental care by males than eggs.

Information on the number of eggs collected by male sticklebacks in the wild is limited. HAGEN (1967) found a clutch size of 50 to 60 in stream *leiurus*, and MOODIE (1972) found a mean of 773 mid-developed eggs (n = 133) in the nests of the large Black Stickleback from Mayer Lake, B.C. The clutch size of Wapato males varied according to their breeding location in the lake, *i.e.*, males in the vegetated area collected about 1000 eggs while males in the rocky areas collected <400 eggs. Since fewer females were also observed in the rocky areas, the difference in clutch size may indicate that females prefer the vegetated area for feeding and swimming or, possibly, that they prefer mates that nested in the vegetated littoral area. This also presumes that females in the Rocks and area A were of equal size, a factor not investigated in this study. Although experiments indicated a definite preference of males for vegetated over non-vegetated nesting habitat (KYNARD, 1979a), nothing is known about female habitat preference. The present study indicated that males had superior reproductive success if they spawned early in the season in a vegetated, littoral area of the lake and selected nest sites in concealment and deep water. MOODIE (1972) also found reproductive advantages associated with nesting in different lake habitats and close to shelter.

Several workers have reported egg cannibalism in natural populations of breeding threespine sticklebacks (WOOTTON, 1970, 1972; SEMLER, 1971; MOODIE, 1972; and review by HYNES, 1950). Males in Wapato Lake lost eggs to neighboring males who sneaked into their territory to steal, and, also, to the raider packs. The movements and behavior of sneaking Wapato males, many of which were in the parental phase, appeared similar to the sneaking behavior of the non-parental males that was described in detail in the laboratory studies of VAN DEN ASSEM (1967). Wapato Lake males raided nests most heavily while in the parental phase (Fig. 8). This was quite different from the conclusion reached by WOOTTON (1976) that parental males did not make nest raids.

Sneaking of individual males into a neighbor's territory seemed clearly related to egg cannibalism in Wapato Lake fish. Sneaking that led to a stolen fertilization was never observed in Wapato Lake. This differed from the extensive laboratory observations of VAN DEN ASSEM (1967) who concluded sneaking and egg stealing could serve more than one function, *i.e.*, to steal a fertilization in a rival's nest (if the female was present in the nest), or to transfer some eggs back to his own nest or to destroy another nest. Stealing to eat eggs wasn't observed, possibly because the males were well fed. The stealing situation observed by VAN DEN ASSEM would most likely exist in nature where breeding females were scarce, a circumstance entirely foreign to the Wapato Lake population.

Although the stealing of eggs by "raider packs" of non-breeding males and females has not been reported in the literature before, the work of HYNES (1950) on Birket, a small stream in Cheshire, England, established that non-breeding *leiurus* adults cannibalised eggs. Since non-breeding adults from such greatly dissimilar areas as Birket and Wapato Lake ate eggs, the occurrence of the raider packs may be widespread. Female predation of eggs at Wapato Lake was probably an important factor in the mortality of nest eggs because (1) the raider packs were abundant with 10 to 15 groups seen daily in area A, (2) large numbers of females participated in each raider pack, and (3) SEMLER (1971) found 39% of the females he collected at Wapato Lake contained a mean of 7.15 eggs in their stomachs. Females and non-breeding males probably obtain eggs only with the raider packs, since no individual theft by a female or non-breeding male was ever observed during the two summers of observations at Wapato Lake. At this time, no satisfactory explanation can be offered for the occurrence of the raider packs. The author knows of no other vertebrate where groups of non-breeding male and female adults make similar cannibalistic raids on breeding conspecifics.

Nesting males stole eggs that were predominantly EPEE, *i.e.*, less than about 2 days old, although progeny were eaten in every stage of development prior to hatching. The predatory females examined by SEMLER (1971) and HYNES (1950) also ate PEE although we can't be certain if most predation occurred in the earliest 2 days. These field results differed from the laboratory results of BLACK (1971) where most nests were raided just before the eggs hatched. After the completion of mating, Wapato Lake males covered their nests with debris and their nests remained concealed for about 2 days after egg collection, *i.e.*, during the period of most egg losses to neighboring males and probably also to the raider packs. A similar period of concealment was reported in a stream population of *leiurus* in the River Wear, England (WOOTTON, 1972). After 2 days, during which there was no fanning of eggs, and very little patroling by males, males began to fan and soon a conspicuous fanning hole developed in the nest at the entrance hole. During the period of nest concealment, this observer, who was a visual predator with a search image for nests, could only locate nests after lengthy observations of a male's behavior, *i.e.*, eventually they would spit additional debris onto their nest. Once the fanning hole developed, the nest was easily located. The concealment of the nest during the period of most egg losses suggests that this behavior functions to reduce discovery and losses to egg predators. It also allows a male to be absent from the nest to forage for insects and eggs for a long time since concealed nests with PEE weren't usually found by a predatory neighbor until 19+ min.

What peculiarity of nesting males allowed 81% of the eggs eaten by females and 78% of those eaten by nesting males to be stolen while the victim's eggs were PEE's? The level of aggression of males caring for PEE wouldn't seem to be the answer, since males caring for PEE and EE had similar levels of aggression (Fig. 6). Males did seem to be absent from their nests most during this period, particularly during the 2 day period of nest concealment. WOOTTON (1972) also noted that the time males spent at their nests increased during the reproductive cycle. Perhaps leaving the nest unguarded with a chance for egg loss to predators is the gamble a male takes to obtain necessary food resources for himself and, ultimately, for the protection of his progeny, which will demand more of his time and energy for parental care toward the end of the cycle.

Since egg stealing by males was so prevalent that they even left their unguarded progeny in conspicuous circumstances, *i.e.*, in nests with ventilation holes or totally opened nest in the case of yolk-F, there should be a significant selective advantage related to egg stealing. The most obvious advantage is obtaining additional food resources. Egg predation by nesting males seemed to be related to a shortage of food resources in the small territories of males where energy expenditure in territorial defense and rearing of progeny for about 2 weeks in the warm waters of Wapato Lake appeared enormous. Insect food resources appeared limited in area A as only 34.8% of the nesting males' stomachs contained only insects. Among the 45.8% of the males that were eating insects plus eggs, the amount of insect material in their stomachs was greatly dwarfed by the large volume of eggs. In area A, males nested close to one another and the mean internest distance was only 41.1 cm. The small size of his territory and his parental duties restrict a male's space and time for foraging and undoubtedly affect his feeding strategy during the reproductive cycle. The strategy seemed to be to steal eggs through the entire breeding cycle, concealing the nest when possible, and to only relax this activity when mating or when caring for F.

FEMALE BEHAVIOR

Another aspect of mating behaviour that was different from the classical description of male-female relationships presented by TINBERGEN (1951, 1953) was the common occurrence of females courting males and scarcity of male zig-zag courtship. Females actively courted males by traveling from territory to territory and repeatedly giving head-up displays to resident males, even those not in the sexual phase. Also, zig-zag courtship dance of males was usually not used prior to mating. When a female entered a male's territory, the male usually moved straight toward her, then she gave head-up display, and the male would either lead her directly to the nest and mate or attack and try to chase her away. Although male courtship appeared unnecessary during most of the breeding season, males used zig-zag courtship during early weeks of reproduction. Perhaps changes in relative abundance of females and males that were willing to mate were responsible for changes in occurrence of male zig-zag courtship and female courtship of males during the summer. Males with nests that were ready to mate seemed much more abundant than females during early weeks of breeding, thus, early males competed for the limited numbers of females using zig-zag courtship. As willing females became more numerous, zig-zag courtship by males was unnecessary to attract a mate and, eventually willing females became so numerous relative to number of males, that females actively courted males. Thus, during the breeding season, courtship changed from predominantly male courtship of females to predominantly female courtship of males.

An increase in the excess of females during the late summer was indicated by the increased incidence of observations in July and August where females were courting males, and by the mortality data in which males and females died in equal proportions early but where females outnumbered males 12 : 1 in the later period. The excess of females in the sex ratio of the population becomes established early in juveniles by 2.5 mo of age and remains throughout their one year of life until post-reproductive mortality (HAGEN & GILBERTSON, 1973a; and present study). The selective factors that effect the sex ratio have not been studied in detail but trout predation on Wapato Lake sticklebacks did not favor females (HAGEN & GILBERTSON, 1973a).

In general, the breeding biology of Wapato Lake sticklebacks was similar to other populations reported in the literature. Several aspects of their behavior appeared unique but this may only be due to the scarcity of detailed field studies. The ethological approach using a combination of field observations and experiments gave verification and, occasionally, a functional explanation to the results of some laboratory studies. By simultaneously examining many aspects of breeding behavior, many interrelationships were revealed which permitted a more intimate understanding of the total breeding system. It seems that in addition to being a good subject for detailed laboratory studies of reproductive behavior, *Gasterosteus aculeatus* is also an excellent subject for examination of breeding behavior in the field. Only there can the adaptive significance of many aspects of their behavior be fully appreciated and understood.

FEMALE STICKLEBACKS PREFER TO SPAWN WITH MALES WHOSE NESTS CONTAIN EGGS

by

MARK RIDLEY and CATHERINE RECHTEN[1][2]
(Animal Behaviour Research Group, Department of Zoology, Oxford, England)

(Acc. 30-IX-1980)

INTRODUCTION

The male threespined stickleback (*Gasterosteus aculeatus*) is territorial and builds a nest. After spawning with one female, he will spawn with further females during the next two to three days. He then looks after the eggs and fry for two to three weeks. There is no maternal care. The female stickleback produces several clutches of eggs at intervals of about four to eleven days, spawning with several males. Each clutch contains about 50-150 eggs. (For references see WOOTTON, 1976).

Male sticklebacks fight amongst each other for the possession of territories. When a female enters a terri'ory, the male tries to attract her to his nest by courting. This is exactly the kind of mating system in which we would expect female choice to occur. Natural selection should favour females that preferentially lay their eggs in such a way that they will have a relatively high survival chance.

One factor that will affect egg survival is the male's ability to defend them. It is known that eggs survive better if the male's territory is larger (VAN DEN ASSEM, 1967; BLACK, 1971), and in fact females have been shown to prefer males with larger territories (VAN DEN ASSEM, 1967). Also, males with red throats are at an advantage in territorial disputes, and for a polymorphic population SEMLER (1971) has shown that females prefer males with red throats.

Another factor that might affect egg survival is the number of eggs in the nest. The number of eggs could influence the survival chance of each egg in several ways. Here are three of the more convincing arguments.

---

1) We thank Marian and Richard DAWKINS for reading the manuscript, and Alan GRAFEN for many conversations on the subject. M. R. is supported by an E. P. A. Cephalosporin Research Studentship, C. R. by the Deutsche Forschungsgemeinschaft.

2) C. R. present address: MPIV, D-8131 Seewiesen, West Germany.

(1) ROHWER (1978) suggested that during the parental phase, male sticklebacks might occassionally eat some of their own offspring. The more eggs there are, the lower the chance that a particular egg will be eaten.

(2) A similar argument to ROHWER's can be made concerning predators other than the parental male. If the eggs are exposed to any predators that eat only a small part of the clutch, the risk of predation to any one egg decreases with the number of eggs in the nest.

(3) After receiving his first clutch, a male will spawn with further females for only the next two to three days. It seems likely that the risk of predation on the eggs is especially high during the courtship phase: only during this phase does a male allow other fish near his nest. The time from laying until the beginning of the parental phase is shorter for later clutches. The risk of predation should therefore be lower for later clutches.

These arguments give good reason to suppose that the number of eggs in a nest affects the survival chance of the individual egg. If this is so, females should prefer to spawn in nests that already contain eggs, as ROHWER realized. However, it is also possible to imagine arguments why females should prefer to spawn in empty nests, or should prefer the number of clutches in the nest to be below some upper limit. For instance, the efficiency of aeration by the male might decrease as clutches are added.

The experiments described below were designed to test whether the probability of a female's spawning in a nest depends on the presence or absence of eggs.

METHODS

Threespined sticklebacks were collected in February and May 1980 from Radley Pond, near Oxford. The were kept on a 16/8 h L/D schedule at 18-20°C and fed live *Tubifex*. The early batch required about two weeks of this treatment to come into breeding condition. Experiments were conducted from March to July. The females were kept singly or in pairs in small tanks, males in a large communal tank. Experimental males were moved singly to the experimental tanks, which were 30 cm high and has a base of 48 × 60 cm. The males usually built a nest and were ready to court females within 24 h of transfer. For an experiment a female was either introduced directly into a male's tank, or placed in a neighbouring tank of equal size, which was then opened to the male. The two procedures produced no obvious difference in outcome.

In our experiments, we had two classes of males: those with, and those without eggs. A male was transferred to the 'with eggs' class by presenting him with successive females until one spawned. A male with eggs was transferred to the 'without eggs' class by destroying the nest, after which the male would build again.

DEFINITIONS

The sexual behaviour of the threespined stickleback is so well known (Tinbergen, 1951; Wootton, 1976) that only the more relevant points are mentioned below.

A female stickleback courted by a male will either flee, or respond with a head-up posture and eventually follow him to his nest. We recorded *no follow* if the female had not responded to the male's courting within three minutes of being noticed by the male, or (if she did respond) if she had not followed him to the nest entrance within ten minutes. Otherwise the result was scored as *follow*, and a note was made of the time interval between the start of courtship and the female's arrival at the nest entrance. Occasionally a female followed a male down to the nest, but did not find the nest entrance or orientate herself correctly towards it. Such females were allowed to try again, but scored as *no follow* if they had not managed a correct approach within the set time limit. A female that had reached the nest entrance correctly put her snout into the nest. She could then either enter the nest and *spawn*, or back out again immediately and leave the nest — a *refuse*. Very rarely a female entered a nest completely or partially, remained motionless for a few minutes, and then left it without spawning (as did some females observed by Wootton, 1974, p. 851). This we called a *sit*.

RESULTS

The female's first decision: to follow or not?

One way in which a female could exert a preference for males with eggs is by being more ready to follow them. We did not think this very likely, because a female decides whether to follow when she is at some distance from the nest, where she can hardly be imagined to see or smell the eggs. The only plausible potential cue is the behaviour of the male; males with eggs might, for some reason, court differently from those without.

We tested whether the probability of following is influenced by the presence of eggs in the nest. We present the data from 51 females and 45 males. Females were introduced to two classes of males, those with and those without eggs, and we counted the frequencies of following and not following in both cases. Each male could be used at most once in each class, and each female at most once per spawning cycle (females spawned every three to six days). Males in the 'with eggs' class may have had one or more than one clutch. Experiments were carried out on the first or second day after the male acquired his first clutch.

Of 38 times that females were presented to males without eggs, following was recorded 25 times (a proportion of .66). Of 31 times that females were presented to males with eggs, following was recorded 20 times (a proportion of .65). The proportions were almost identical in the two cases, and the data gives no reason to conclude that a female's decision whether to follow the male is influenced by his possessing eggs.

The female's second decision: to spawn or refuse?

A much more likely time for a female to exert a preference for males with eggs is after she has followed the male to his nest, when she sticks her snout in its entrance. The female can easily be imagined to sense at this stage whether the nest contains eggs, and any preference for nests with eggs would manifest itself in her probabilities of refusing and spawning. So we tested whether these probabilities were influenced by the presence of eggs in the nest.

We used the same animals as in the previous section. Again there were two classes of males, those with and those without eggs, but this time we ignored cases in which the female did not follow. We counted the frequencies of refusing and spawning to the two classes of males. The same criteria were used for inclusion of results as in the previous section. The proportion of nest inspections that led to a 'sit' was too low (7%) to be included in the analysis.

Of 34 times that females followed males without eggs, 14 resulted in spawning (a proportion of .41), the rest resulted in refusing. Of 41 times that females followed males with eggs, 21 resulted in spawning (a proportion of .51), the rest in refusing.

The null hypothesis that eggs do not increase the probability that a female will spawn could be rejected at the significance level of $p = .13$ (Fisher exact probability test). On the data presented so far, therefore, there is only meagre evidence for female choice of nests with eggs. But it points to the possibility that among the females there are some that are more willing to spawn with a male that has eggs than with a male that has no eggs. The sequential choice experiments described in the following section test for these females.

Sequential choice experiments: Are there females that are more likely to refuse a male without eggs than a male with eggs?

If some of the females that refused did so just because the male lacked eggs, they should readily spawn with a male that does have eggs. A sequential choice design is the most obvious way of revealing these females.

Females that had refused one male became eligible for running in a sequential choice experiment. They were presented to a second male (male B) within one hour of refusing the first (male A). There are four possible designs: A-/B+ (first male without eggs, second male with eggs), A-/B-, A+/B+ and A+/B-. We must distinguish between effects due to females choosing males with eggs and effects due to a male being second.

For example, a female might be more likely to spawn with the second male, regardless of whether he has eggs, simply because she has been courted more. The A-/B+ experiment must be controlled by the A-/B-experiment to show any effect of females choosing males with eggs. The A+/B+ and the A+/B- designs provide additional controls. In each of the four experimental desings, each male could appear no more than once as first and once as second male, and each female no more than once per reproductive cycle.

Table 1 gives the results. A remarkably high proportion (.65) of the females that had refused a male without eggs went on to spawn when presented to a male with eggs. Females were significantly more likely to spawn with the second male in the A-/B+ design than in the A-/B- control (Fisher exact probability test, p = .028). In the two designs where the female had refused a male *with* eggs, the proportion of spawns with the second male, whether or not he had eggs, was very low. The result of the sequential choice experiments seem to us to be conclusive evidence that female choice of males with eggs exists.

How long do females take to reach the nest?

The time interval between the beginning of courtship and the female's arrival at the nest might be thought to depend both on the willingness of the female to spawn and on the male's courtship. If the courtship of males with eggs differs from that of males without eggs, we might expect the female's following times to reflect this difference. If the female's decision to spawn or refuse is influenced by factors other than the presence of eggs, we would expect the following times before refusals to differ from those before spawnings.

Four classes of following times were measured: the female could follow a male with or without eggs, and in either case the follow could result in a refusal or a spawning. Twenty-two independent measurements were made in each class. Median latencies of following to a nest without eggs

TABLE 1

Female responses to second male after refusing the first one

Male A	Male B	Total N	Spawn	Refuse	Proportion of spawns
—	+	17	11	6	.65
—	—	10	2	8	.20
+	+	10	0	10	.00
+	—	10	1	9	.10

were 52 s for females that spawned, and 100 s for females that refused. Median latencies of following to a nest with eggs were 32 s for females that spawned, and 127 s for females that refused. We performed a two-way analysis of variance on the logarithmically transformed times, the two factors being the presence or absence of eggs, and whether a refusal or spawning resulted. Nearly all the variance attributable to the factors was explained by the difference between times to a refusal and times to a spawning ($F_{1,84} = 35.6$, $p < 0.01$); the effects due to presence of absence of eggs ($F_{1,84} = 2.24$, n.s.) and the interaction ($F_{1,84} = 2.79$, n.s.) being negligible. The times to a refusal are significantly longer than those to a spawning. A knowledge of how long a female has taken to reach the nest, therefore, gives information on whether she will refuse or spawn. No significant effect of the presence of absence of eggs was shown, again tending to confirm our idea that the effectiveness of male courtship does not differ in the two cases.

Our explanation of the results of the sequential choice experiment was that some of the females who refuse a male without eggs will spawn with a male with eggs. For the females in the A-/B+ design we measured the times to refuse the first male. We compared these times for those females that spawned with the second male with those for females that refused the second male. If the time to follow is some measure of the female's readiness to spawn, the times to refuse the *first* male would be shorter for those that subsequently spawned. However, the median time to refuse the first male for those females who spawned with the second was 1 min 17 s; the median time for those that refused the second was 1 min 37 s. The difference between the two distributions was statistically insignificant (Mann-Whitney $U_{10,13} = 58.5$, n.s.). So simplistic an understanding of the times to follow is evidently not valid.

DISCUSSION

We concluded from our results that the presence of eggs in a male's nest increases the chance that a female will spawn with him. We also conclude that this preference takes effect after the female has followed the male to his nest. The female looks in the nest,and is more likely to spawn in it if it contains eggs.

Our results lead us to the following account of stickleback courtship. The female first decides whether to follow. This decision is probably influenced by the female's readiness to spawn (Wootton, 1974), the territory size (van den Assem, 1967), and the male's appearance (Semler, 1971) and courting (Gross, Franck & Eikenberg, 1980). It is probably not influenced by whether the male has eggs: in our experiments males with eggs were no more likely to be followed than those without, and there was no difference in the following times. Females are apparently unable to distinguish between males with and without eggs by their courtship. This makes functional sense because any males that signalled in any way that they had no eggs would be selected against. Once the female has reached the nest she can either spawn or refuse. This decision is influenced by the presence of absence of eggs (Table 1), and by other factors. The fact that following times before refusals differ from those before spawnings implicates factors other than presence of eggs: when they reach the nest females must already vary in their probability of spawning.

Our experiments do not prove that it is the presence or absence of eggs that directly determines the probabilities of refusals and spawnings. The most obvious factor that might be associated with the presence or absence of eggs is the quality of the nest. After receiving a clutch of eggs the male extends his nest (Leiner, 1929; van Iersel, 1953). Our result, that females choose males with eggs, might have been caused by female's choosing large nests. We see these as mere alternative mechanisms: *de facto* female choice of males with eggs would occur in either case. Anyway, from functional considerations, we do not think it likely that females are using nest quality as a cue to choose males. If they were we would expect males to extend even empty nests, which they do not. If females used the presence of eggs as a cue to choose males, males with empty nests should kidnap eggs, which they do, as further discussed below.

Another factor that could be associated with the presence of absence of eggs is the quality of the male himself. Males could differ in some property that influenced their chance of spawning. Good quality males would then be more likely to be in the 'with eggs' class. Males in this class might then be more likely to be spawned with, not because they have eggs but because of their high quality. This is a serious objection, but we doubt its validity. In our experiments we picked in advance which male to transfer to the 'with eggs' class. By successively presenting lots of females to the chosen male, he could reliably be transferred to the 'with eggs' class. Furthermore, we examined our data for variation in the spawning frequencies of the males. We found no indication of any differnces between males, certainly none great enough to account for our results.

So the most likely explanation of our results is that females prefer to spawn with males that have eggs. This result suggests new functional explanations for several components of stickleback courtship.

Function of female refusal.

Female refusal has received little attention, and the only previous explanation we have found is a motivational one given by WOOTTON: 'If the female is not quite ready to spawn, she may merely nose part of the way into the nest and then retreat' (1976, p. 190). However, such a simple motivational explanation is incompatible with out result that the probability of refusal varies with the presence or absence of eggs. Our results suggest that female refusal has a function: females inspect the nest and are more likely to refuse if it would be disadvantageous to spawn in it, for instance because it lacks eggs.

The motivational explanation is not completely wrong. The fact that following times differ between females that subsequently spawn and refuse shows that females already vary in their probability of spawning before they inspect the nest. We would expect females to vary in their tendency to spawn because this tendency increases for each female through the spawning cycle. At an extreme stage of egg maturation a female is physiologically forced to spawn even if no male is present (WOOTTON, 1976, p. 141; our observations). As a female approaches this stage she should become less and less choosy.

Function of egg kidnapping.

Territorial male sticklebacks may kidnap eggs from neighbouring nests and rear them. ROHWER (1978) argued that egg kidnapping would be adaptive if females preferred to spawn with males that already possessed eggs (LI & OWINGS, 1978, citing KODRIC-BROWN, pers. comm., incidentally mentioned a similar argument). Our demonstration of such female choice, therefore, supports ROHWER's hypothesis. Once female choice of males with eggs has become prevalent, egg kidnapping by males will necessarily be adaptive, regardless of why female choice was originally selected for. In our Introduction we gave some reasons why females might be selected to prefer to spawn with males that possessed eggs.

Function of displacement fanning.

Fanning during stickleback courtship has been the exemplar of a functionless displacement activity. TINBERGEN (1951, p. 153), for example, wrote: "Displacement 'fanning' acts as an outlet of sexual motivation, but it has no meaning as a social releaser." More recently, a functional explanation has been suggested by MCFARLAND (1974) and COHEN & MCFARLAND (1979). They thought that nest maintenance is necessary during courtship: '... it seems to me that successful courtship involves not only inducing a female to lay eggs in the nest, but also maintaining the nest in good condition to receive the eggs' (McFarland, 1974, p. 222).

Rohwer (1978) has suggested another function for fanning during courtship, which is supported by our results. We have shown that females prefer to spawn in nests with eggs. What better way for a male to show a female that he has eggs than by fanning them? The following evolutionary sequence could be imagined. Originally only males with eggs fanned. Females could have then chosen males with eggs by only following those that fanned. Males without eggs would then be selected to fan when a female appeared, and displacement fanning would become established as a deceptive social releaser.

SUMMARY

Does the presence of absence of eggs in a male stickleback's nest affect the chance that a female will spawn with him? Female threespined sticklebacks (*Gasterosteus aculeatus*) were presented alternatively to males with or to males without eggs.

Our results show that the chance that a female will follow a male to his nest is unaffected by whether he has eggs. Once a female has reached the nest, she can either enter it and spawn, or back out and refuse. There was a tendency for females to be more likely to spawn than to refuse if the nest contained eggs.

In a sequential choice experiment females that had refused a male without eggs were then presented to a second male, either with eggs or, as a control, without eggs. Females were significantly more likely to spawn with the second male if he possessed eggs.

The finding that females prefer to spawn with males with eggs suggests functional explanations for female refusal, male egg kidnapping, and male 'displacement fanning'.

REFERENCES

Assem, J. van den (1967). Territory in the three-spined stickleback, *Gasterosteus aculeatus* L. An experimental study in intral-specific competition. Behaviour, suppl. 16, p. 1-164.

Black, R. (1971). Hatching success in the three-spined stickleback (*Gasterosteus aculeatus*) in relation to changes in behaviour during the parental phase. — Anim. Behav. 19, p. 532-541.

Cohen, S. & McFarland, D. J. (1979). Time-sharing as a mechanism for the control of behaviour sequences during the courtship of the three-spined stickleback (*Gasterosteus aculeatus*). — Anim. Behav. 27, p. 270-283.

Gross, H. P., Franck, D. & Eikenberg, G. (1980). Courtship success in the three-spined stickleback: behavioural determinants. — Z. Tierpsychol.

Iersel, J. J. A. van (1953). An analysis of the parental behaviour of the three-spined stickleback (*Gasterosteus aculeatus* L.). Behaviour, suppl. 3, p. 1-159.

Leiner, M. (1929). Ökologische Studien an *Gasterosteus aculeatus*. — Z. Morph. Ökol. Tiere 16, p. 499-540.

Li, S. K. & Owings, D. H. (1978). Sexual selection in the three-spined stickleback: II. Nest raiding during the courtship phase. — Behaviour 64, p. 298-304.

McFarland, D. J. (1974). Time-sharing as a behavioural phenomenon. — Adv. stud. behav. 5, p. 201-225.

Rohwer, S. (1978). Parent cannibalism of offspring and egg raiding as a courtship strategy. — Amer. Nat. 112, p. 429-440.

SEMLER, D. E. (1971). Some aspects of adaptation in a polymorphism for breeding colours in the threespine stickleback (*Gasterosteus aculeatus*). — J. Zool. 165, p. 291-302.
TINBERGEN, N. (1951). The study of instinct. — Clarendon Press, Oxford.
WOOTTON, R. J. (1974). Changes in the courtship behaviour of female three-spined sticklebacks between spawnings. — Anim. Behav. 22, p. 850-855.
—— (1976). The biology of sticklebacks. — Academic Press, London.

ZUSAMMENFASSUNG

Laicht ein Stichlingsweibchen bevorzugt mit einem Männchen, dessen Nest schon Eier beenthält? Um dies zu untersuchen setzten wir laichbereite Weibchen zu Männchen mit Eiern und ohne Eier.

Die Wahrscheinlichkeit, daß ein Weibchen dem Männchen zum Nest folgen wird, ist unabhängig davon, ob er Eier besitzt. Wenn ein Weibchen das Nest erreicht hat, kann sie entweder hineinkriechen und ablaichen oder verweigern und davonschwimmen. Die Weibchen zeigten eine Tendenz, mit größerer Wahrscheinlichkeit abzulaichen, wenn das Nest Eier enthielt.

In einem Sukzessiv-wahlversuch wurden Weibchen, die bei einem Männchen ohne Eier verweigert hatten, zu einem zweiten Männchen gegeben. Die Weibchen laichten signifkant häufiger mit dem zweiten Männchen, wenn er Eier besaß.

Der Befund, daß Weibchen bevorzugt mit Männchen ablaichen, die schon Eier besitzen, führt zu funktionellen Erklärungen für das Verweigern der Weibchen, das Eierrauben der Männchen und 'Übersprungsfächeln'.

PARENTAL REACTIONS TO YOUNG IN STICKLEBACKS (GASTEROSTEUS ACULEATUS L.)

by

E. FEUTH-DE BRUIJN and P. SEVENSTER
(Div. Ethology, Dept. of Zoology, University of Leiden, The Netherlands)

(Acc. 1-IX-1982)

1. Introduction

It would seem that the very first requirement for a parenthal behaviour-system to develop is the prevention of noxious behaviour towards the offspring. Yet, especially in the case of predators, infanticide is not an uncommon event. So it becomes an interesting question how in such animals the destruction of their own eggs and young is avoided and replaced by favourable reactions. In the three-spined stickleback exactly this situation is found. Sticklebacks are predators and are in fact preying upon animals of small size including conspecific eggs and young (a.o. LEINER, 1929; HYNES, 1950; BLACK & WOOTTON, 1970; SEMLER 1971).

In the parental phase of its breeding cycle, however, the male is confronted with his own eggs and young and the tendency to devour them has to be suppressed or cancelled. The parental behaviour of a male towards its own eggs has been the subject of an extensive study by VAN IERSEL (1953), who analysed the interplay of causal factors underlying the activity of fanning, which is predominant in the parental phase (ably summarised by WOOTTON, 1976). Here it is relevant that under certain conditions this parental behaviour was found to break down and the eggs were then in fact eaten by the father. This suggests, that normally eating the eggs is somehow suppressed by factors inducing parental behaviour. In the present paper the behaviour of a male stickleback towards newly-hatched young is considered. The following short description may serve as an introduction to the experiments on which we want to report. Under the laboratory conditions used in our research (16 hours light, and 18-20°C) eggs hatch on the 7th or 8th day after fertilisation (in code D_7 or D_8 after D_0). The young are born with a large yolk-sac and are comparatively passive and immobile at first. From time to time the tail is moved to and fro in bouts at a high frequency, in a fluttering sort of motion, and sometimes jerky movements are made consisting of a few powerful tailbeats. Already after a few hours this may occasionally lead to considerable displacements of the young of the order of several cm, which may carry them to the rim of the nest or even some distance away from the nest. Such straying young are picked up by the father, carried back to the nest and spit into the nest, which by this time has become a rather loose and hollow structure. This activity of the male henceforward will be called "retrieving". In this stage the male becomes extremely attentive to any small moving particles, and constantly searches the area around the nest to collect any young that may have gone astray.

However, hourly more young will have hatched and more of them become increasingly mobile. Already on the day of hatching (usually D_7, seldom D_8 in our experiments) many young start "jumping": beating their tails vigourously, they rise up to 10 cm above the nest or the sand bottom before they passively sink down again. During these jumps the male may interfere, and retrieve them. As the number of young and their mobility increase the male soon spends a great deal of his time on searching for wandering young and retrieving them. In the course of the next day (usually D_8) there is a further increase in these activities of both male and young. The latter are now capable of higher jumps and in the afternoon (*i.e.* some 24 hrs after hatching) a number may even reach the surface of the tank (in our observations 25 or 40 cm above the sand).

At the surface they gulp a little bubble of air to fill their airbladder (which is necessary to start subsequent gasformation: the sticklebacks belong to the physoclysts). Often the airbubble is actually visible as a silvery speck in the young, when after reaching the surface they sink down again. With the newly-gained buoyancy they are now capable of floating or hovering at any depth without constantly swimming upwards as before. In addition the yolk-sac is quickly shrinking and disappears entirely on the second day (D_9), which also facilitates mobility. On this second day (roughly 48 hrs after hatching) a large portion of the young manage to reach the surface and gulp air. It is certain that at least some young make a trip to the surface more than once, but it has not been established whether they take in air on each of these occasions. Many attempts to get to the surface are counteracted by the father who is continuously active in trying to retrieve the young both on their upward journey and when they are coming down. With the number of attempts increasing and the young presumably becoming more agile and skilled swimmers, the male more and more frequently fails to prevent the young from reaching the surface. On the third day (72 hrs after hatching) the number of attempts decreases considerably and practically all young have air in their bladders.

Towards the end of that day and on the next only a few more attempts are observed. The young now either disperse or assemble in a dense cloud, usually above or near the nest. In the latter case it is not quite clear whether this is brough about exclusively by the father's retrieving or whether in addition the young react to one another, as they definitely do at a later stage, when they swim about in schools. The account just given is based on semiquantitative observations, and certainly variations on the main theme do occur; the whole process deserves to be scrutinised more thoroughly. Perhaps the most fascinating feature of the episode described is the "antagonism" between the father and the young, which reminds one of the "parent-offspring conflict" in the sociobiological litterature (Trivers, 1974): whereas the young are striving hard towards the surface, the male is doing his utmost to prevent them from getting there. Occasionally a few young are found much later which most of the time stay close to the bottom and when moving have to swim upwards with vigorous tailbeating. It seems that these did not succeed in filling their airbladder, and consequently have no buoyancy. Are they victims of their father's stubborn retrieving behaviour? On the other hand provisional observations in a more natural setting, *i.e.* with a second territorial male in the tank, seemed to indicate that the young are much less exposed to predation by the rival male because of their father's retrieving.

When the father was taken away, the young seemed to disperse more quickly and farther from the nest, which must entail a greater risk of straying into neighbouring territories and being eaten by the owners. In this respect it might be disadvantageous for a young to gulp air prematurely and thus become more mobile at an earlier age, and this might explain the male's attempts to keep the young from reaching the surface, until they are strong and skilled swimmers capable of escaping him. Then, of course, it remains a question, why the young do not delay their trip to the surface themselves, in other words why their motivation to go and gulp air is premature. Leiner (1929) implicitly suggests that the male by chasing the young is virtually training them to escape from adult sticklebacks.

At present we must leave it at that. In this paper we will be exclusively concerned with the male's behaviour towards 1-day old young. It is quite obvious both from the litterature and from our own observations that newly-hatched young (and even much older ones) are eaten by adult or halfgrown sticklebacks and, what is especially relevant, by territorial, nest-owning males. Already Leiner (1929) reports: "Wiederholt konnte ich beobachten, dasz fremde Männchen, die im gleichen Aquarium ein

Nest, aber noch keine eigenen Jungen hatten, die herumschwimmenden Kleinen auffraszen, während es der Vater nicht tat. Offenbar besitzt dieser während der Zeit der Jungenbewachung instinktmäszige Hemmungen, die ihn hindern, die Jungen (auch fremden) zu fressen. Diese Hemmungen müssen aber graduell sehr verschieden sein. Es wurden nämlich Männchen beobachtet, die allem Anschein nach ihre eigenen Kinder auffraszen."

It is with this tendency of the male stickleback to eat his young and the inhibition, usually preventing it, that we will be concerned.

Pilot experiments had indicated that three factors might be important in determining the male stickleback's behaviour towards newly-hatched young.

1. The *degree of satiation* of the male determined by the feeding schedule.
2. The *stage of the breeding cycle* the male has reached.
3. The *number of clutches* in the male's nest.

So we decided to investigate the influence of these three factors in a more standardised way, by presenting 1 day-old young on a dish near the nest.

2. Material and methods

Fish were used from two populations in The Netherlands: most were caught on the island of Tholen, where a coastal migratory population is found, and only a few were obtained from a trout-nursery in a brook near Vaassen, where the population is non-migratory.

No differences between the populations have been established so far with respect to the behaviour under study, and we will therefore not refer to the origin of the individual fishes, one of the tables excepted. The fish were caught during their first autumn or winter, and almost certainly had no breeding experience. Once in the laboratory they were kept in large concrete tanks at ± 10°C and 8 hrs light, *i.e.* winter conditions. When needed, groups were transferred to aquaria of 125 × 40 × 50 cm, with sand and plants, and at ± 20°C and 16 hrs light. Under these summer conditions they become reproductive in a week or so. As soon as their sex could be recognised, males and females were separated. Males in nesting condition were finally transferred to the experimental tanks of 60 × 40 × 40 cm, with a sand bottom and a row of *Vallisneria* or similar plants along the hind wall, and some algae to serve as material for nestbuilding. All animals were fed with *Tubifex* worms, chironomid larvae and live *Artemia* of various size from a laboratory culture.

Some males were kept for the purpose of fertilising clutches and supplying us with young for the experiments. The clutches were removed from their nest 1 or 2 hours after fertilisation. They were then artificially ventilated either in plastic cups with an airstone or in a special hatching apparatus, in which clutches were placed in small cups each under a jet of running well-aerated tapwater of 20°C. The hatching success was practically 100%. In all experiments young of 1 day old were used. As described, such young have a large yolk-sac and are still immobile most of the time, except for some bouts or wriggling and jumping. At this stage they could easily be presented on large "watchglasses", *i.e.* shallow glass dishes of 10 cm diameter. To increase the visibility of the young the dishes were painted white. In a standard test 10 young of 1 day old were presented to the experimental

TABLE 6. Serial presentatisons of ten-spined young as compared to three-spined, to hungry and satiated males

Satiated	3C	Ten-spined	±	±	+	+
		Three-spined	+	+	+	+
Hungry	3C	Ten-spined	–	–	–	–
		Three-spined	+	+	+	+
			D_4	D_5	D_6	D_7

TABLE 7. Serial presentations of young to old Tholen-males

Satiated	3C	+	+	+	+		+
	1C	+	+		+		
Hungry	3C	+	+	+	+		+
	1C	+	+	+	+		+
		D_1	D_2	D_3	D_4	D_5	D_6

C = number of clutches; D_n = n^{th} day of the parental phase; ○ = ignoring; ± = testing; – = eating; + = retrieving.

These preliminary experiments seem to indicate that retrieving is released by fairly specific stimuli.

9. Aberrant reactions in a group of males

A group of males from Tholen had been kept in the laboratory for almost a year (most of the time under winter conditions). When 18 of these males were given single presentations on various days in the parental phase, we found to our surprise that they always retrieved no matter on what day: Table 7. For this phenomenon we have no explanation: the fish had not gone through a breeding cycle in the laboratory, and the only exceptional feature we can think of is the age at which they were tested, and the long winter period to which they had been exposed.

10. Discussion

Three-spined stickleback males show intensive "retrieving" behaviour on the day of hatching and for some days afterwards: they take young that have wandered from the nest in their mouth, carry them back and deposit them on the nest. From our serial presentations of strange 1 day old young at a distance of 10 cm from the nest it appears that the tendency to retrieve emerges some days before hatching (Table 3). A few males already start retrieving on D_2, some only on D_5 or D_6, but most on D_3 or D_4. Before that day young will be ignored, tested or eaten. The tendency to eat them depends for a major part on the nutritional state of the male concerned: in our serial presentations all hungry males, quite a number of normally fed males, and only a few satiated males ate the young presented to them before they started to retrieve. Males quickly learn to hunt for young and eat them and once they have accepted them as prey this may interfere with the tendency to retrieve them: in that case all our hungry males (and one normally fed) went on eating the young presented until the day of hatching. Normally fed males, however, that had taken to eating young, nevertheless switched to retrieving before the day of hatching (except one). Further it appears from our single presentations (Table 4) that even hungry males retrieve before the day of hatching if they come across young for the first time on D_3 or later in the parental phase. Similarly in serial presentations the hungry males would stop eating the young after a "sudden satiation" (with young or other food), and switch to retrieving instead. In these cases the satiation apparently reveals a latent tendency to retrieve. Therefore the most likely explanation of the data is that there is a competition between eating and retrieving young in the advancing parental phase. Only if hunger is severe and young have become accepted as food by previous experiences, will eating prevail over retrieving in this period. Whether the tendency to retrieve in an individual male gradually increases in strength or whether it emerges suddenly in full strength in an all-or-nothing fashion, cannot be decided on our data so far.

Table 3 seems to show that there is a gradual increase in the number of males switching to retrieving over the days, from D_2 onwards. Also: those normally fed males that had consistently eaten young in the earlier tests tended to switch to retrieving rather later than others (nrs 12, 14, 21, 23, 25). Assuming that those were the more hungry males in their category, this suggests that retrieving is only gradually gaining precedence over eating and the process in the hungrier males takes longer. Especially, since we may take it for granted that hunger is also increasing over the days. So a gradually increasing tendency to retrieve seems the most likely explanation.

The data do not reveal any clear correlation with the number of clutches, which has been found to be a most important factor in determining the amount of fanning (VAN IERSEL, 1953). The age of the clutches has not been varied independently of the stage of the parental phase, and the relation with other parental activities (*e.g.* opening up the nest) or the possibility of common factors in the system remain to be explored.

The outcome of the sudden satiation experiments has raised the interesting question whether the hungry males, used to eating young, on D7 first satiate themselves by eating part of their offspring and then retrieve the strange young in the test. In view of our experiments this seems quite possible, and is supported by the two cases, where the males retrieved on D8, instead of D7, in parallel to the young hatching on D8. On the other hand we have seen that satiation of a hungry male, used to eating young, need not immediately lead to retrieving: it takes some hours to become effective. The tests on D7, however, often must have followed closely on the appearance of the first young, and yet the males promptly retrieved. It remains possible therefore that on the day of hatching the tendency to retrieve (*e.g.* because of the appearance of young *on* the nest!) is so much stronger that the competing hunger is outdone.

The question might also be raised in this connection why the hungry males should not satiate themselves on their eggs much earlier. LEINER (1929) has pointed out that the stickleback male in his breeding period, and especially in the parental phase, is restricted in his feeding opportunities because of his confinement to the territory and the limited time left by his intensive parental care for the nest. Consequently, LEINER suggests, a breeding male must be fasting most of the time. ROHWER (1980) in an interesting discussion of breeding strategies of sticklebacks and other fishes, has taken up this point again.[1]) He suggests that the male stickleback may in fact use part of his own eggs as a food store or emergency ration, thereby exploiting the female's greater freedom of movement and consequent ability to collect food which for the greater part is stored in her eggs. Certainly we should pay more attention to such possibilities in the future, and more data on the situation in nature should become available. It does seem unlikely, however, that males in nature would often find themselves in a predicament like that of our experimental hungry males. And even those hungry males seemed to produce reasonable numbers of young, though admittedly we did not count them. Awaiting further evidence, we feel that it is unlikely that they have eaten large numbers of eggs. What is found occasionally is that the male breaks

[1]) The article contains other interesting ideas and suggestions although, regrettably, the author sometimes does not seem to be aware of the fundamental difference between a *causal* (proximate) explanation and a *functional* (ultimate) one, as in his discussion of "displacement fanning" and of the "waning of the parental drive".

off his parental behaviour and devours all the eggs (see also VAN IERSEL, 1953). For more normally fed males (which is not the same as naturally fed!) VAN DEN ASSEM (1967) did not find losses of eggs for which he could not account by other factors, like stealing or mortality from moulds. In any case, from our present data it seems unlikely that males would ever eat their own young unless they have, by hunger and previous exposure to strange young, learnt to accept them as food. How often this is likely to occur in the natural situation remains to be seen.

The premature tendency to retrieve is another subject for speculation. In the period between say D_2 and D_5 the male is sure to meet with strange young only. For a constant temperature of 24°C (which seems an unlikely climate for most sticklebacks in nature) VAN IERSEL (1953) still finds about 5 days for the development of eggs. At present it is difficult, functionally speaking, to see why within those 5 days sticklebacks already should retrieve young if they find some. It will be important to study the influence of the temperature and the developmental stage of the eggs on the emergence of the tendency to retrieve.

Finally we want to touch briefly on the specificity of the retrieval reaction which follows from our presentation of ten-spined young to three-spined males in the parental phase. As appears, from our description of the males' behaviour during these presentations, an excessive amount of testing is found in the tests concerned, accompanied by prolonged "mumbling", before the young are spit out again or, possibly, retrieved in the end (see above). In presentations of three-spined young such frequent testing and such persistent "mumbling" bouts never occur. Neither do they occur in the presentations of young perches, which were promptly eaten. Presumably then, this testing and mumbling has to do with the close resemblance of ten-spined young to three-spined young which makes it difficult for the male to decide between eating, retrieving or testing. The long bouts of "mumbling" strongly suggest that the stimuli which release retrieving are of a chemical nature and that by "mumbling" the males are actually sampling the young gustatorily.

Summary

Three-spined sticklebacks readily eat conspecific young. Only when males have one or more clutches of eggs in their nest, do they develop a tendency to retrieve young and carry them to the nest. This tendency becomes evident between the second and the sixth day after egglaying and fertilisation, but most commonly on the third or fourth day. Even quite hungry males retrieve young in this period of their breeding cycle. However, if such males have previously learnt to accept young as food, they eat young also in this period. Sudden satiation of such males in the period concerned shows that the tendency to retrieve

FILIAL EGG-CANNIBALISM IN THE BROOK STICKLEBACK, CULAEA INCONSTANS (KIRTLAND)

by

IRENE G. SALFERT and G. ERIC E. MOODIE[1][2]

(Department of Zoology, University of Manitoba, Winnipeg, Manitoba, Canada, R3T 2N2 and Department of Biology, University of Winnipeg, Winnipeg, Manitoba, Canada, R3B 2E9, respectively)

(With 5 Figures)
(Acc. 7-IX-1984)

Introduction

Male parental care is widespread in boney fishes, occurring in at least 57 families (Blumer, 1979). Males may guard the eggs, construct nests, fan the eggs and provide various other services for the eggs or fry. Rohwer (1978) suggested males which exhibit significant parental care may resort to eating some of their own eggs when the level of invested care is high and food is limiting. Sacrificing a portion of the current brood would ultimately benefit the male by maximizing its survival and future reproductive output. Rohwer described this behaviour as "filial cannibalism", in contrast to "heterocannibalism", the eating of unrelated conspecifics. The widespread occurrence of heterocannibalism (Black, 1971; Black & Wootton, 1970; DeMartini, 1976; Hynes, 1950; Jewell, 1968; Kynard, 1979; Semler, 1971) makes the detection of filial cannibalism somewhat difficult.

Rohwer (1978) also predicted filial cannibalism would affect courtship strategies. For example, a female should select males already possessing eggs in order to minimize the number of its eggs being cannibalized by the male. Ridley & Rechten (1981) showed female threespine stickleback did prefer males whose nests contained eggs. However, females could choose to lay their clutches in nests containing eggs for other reasons.

[1]) We are grateful to Dr J. G. Eales and Mrs S. Shostak for their generous support. This study was supported by a Natural Sciences and Engineering Research Council of Canada Operating Grant to G.E.E.M. and Postgraduate Scholarships to I.G.S.

[2]) Requests for reprints should be sent to G.E.E.M.

Though persuasive in theory, little evidence exists to prove that filial cannibalism is a normal parental strategy in territorial species. Although the hypothesis is widely accepted (CARLISLE, 1982; KODRIC-BROWN, 1983; POLIS, 1981; THORMAN & WIEDERHOLM, 1983), much of the evidence supporting filial cannibalism is based on circumstantial evidence such as stomach content analysis. ROHWER's arguments were largely based on the extensive literature on the biology of the threespine stickleback, *Gasterosteus aculeatus* (reviewed by WOOTTON, 1976).

Our study, based on laboratory experiments and field observations, shows filial cannibalism is probably a normal feature of the reproductive behaviour of the brook stickleback, *Culaea inconstans*. The reproductive biology of *C. inconstans* is like that of *G. aculeatus* (WOOTTON, 1976).

Methods

In the laboratory eggs were labelled with radioactive iodine (^{125}I). Iodine has been shown to accumulate in ovaries of reproductive *Ictalurus* species (EALES, pers. comm.; TARRANT, 1971). Detection of egg cannibalism by males was then possible.

In the field, egg cannibalism was studied in stickleback known to be parental males. The importance of eggs in the parental male diet was determined, as well as the developmental stage of cannibalized eggs. Either hetero- or filial cannibalism was inferred from the stage of eggs in neighbouring nests.

Radiotracer study.

The tracer experiment consisted of four trials. Fish of trials I-III were from Lake 200 near Erickson, Manitoba. Fish of trial I were caught on 27 May, 1982, and those of trials II and III in September, 1982. Trial IV utilized fish from Delta Marsh, Lake Manitoba, obtained 25-26 May, 1983, they were caught late in their breeding cycle and large numbers died in the laboratory. Few results were obtained from this group.

Laboratory rearing.

Stickleback were maintained at temperatures ranging between 4 and 20°C, on a 8 h light 16 h dark cycle. Fish were fed daily on frozen adult brine shrimp (*Artemia*).

To bring fish into reproductive condition, about 15-20 individuals were placed in 76 l tanks at room temperature with a 16-18 h light 8-6 h dark cycle. These tanks contained about 2 cm of sand, plastic plants, and filamentous algae. Males displaying nuptial colouration and aggressive behaviour were removed and placed in 76-l tanks (3-5 males per tank); these were the test males.

Test tanks.

Experimental tanks (Fig. 1) were located in a controlled environment chamber at 14-16°C with a 16 h light 8 h dark cycle. The tanks contained about 2.5 cm of sand and two clusters of plastic plants positioned in opposite corners. Filamentous green algae and moss were provided as nesting material. Once two males began nest-building, other fish were removed from the tank. Males were separated by a removable plastic opaque partition containing a screened opening (14 cm diameter) while labelled females were presented. A third control male was placed in a suspended plastic cage (Fig. 1).

males strongly suggested the occurrence of filial cannibalism. The eggs in the stomachs of at least seven parental males were in the same stage as the eggs in their nest and different from the eggs of their nearest neighbour.

Discussion

Despite the fact that ROHWER's paper is regularly cited in contexts implying the existence of filial cannibalism (CARLISLE, 1982; KODRIC-BROWN, 1983; POLIS, 1981; THORMAN & WIEDERHOLM, 1983), little evidence exists to support this hypothesis.

The difficulty in demonstrating filial cannibalism lies in distinguishing it from heterocannibalism which is common in sticklebacks (BLACK, 1971; BLACK & WOOTTON, 1970; HYNES, 1950; KYNARD, 1979; ROGERS, 1968; SEMLER, 1971), and other groups (DEMARTINI, 1976; JEWELL, 1968).

ROHWER offered the following five predictions which he believed would permit the differentiation of filial and heterocannibalism. Unfortunately, some of the required data are difficult to obtain and confirmation of some of the predictions would still only infer the existence of filial cannibalism.

Prediction a.

If filial cannibalism occurs, isolated parental male stomachs will contain eggs nearly as often as males having egg-guarding neighbours. The dugout population provided data on the egg-predation of parental males with neighbours. These data strongly infer that filial cannibalism occurred in at least seven of 13 males containing eggs. Of course, there is also significant heterocannibalism by parental males. Almost 35% of parental male *G. aculeatus* may eat eggs of a different stage than their own (KYNARD, 1979).

Prediction b.

ROHWER (1978) suggested that if females prefer males with eggs already present, there will be courtship related egg stealing wherein the stolen eggs will not be eaten. This theft should occur before the male has obtained eggs via courtship. In contrast, if only heterocannibalism occurs, theft of eggs should be more common near the end of the cycle when males have fasted longer. This complex prediction is probably invalid. Some data were misinterpreted, others overlooked and much was unpublished at the time the prediction was formulated.

ROHWER supported this prediction with references to VAN DEN ASSEM (1967) and WOOTTON (1971) who observed raiding by males without eggs. However, VAN DEN ASSEM's (1967) study was confined to the laboratory and he indicated (p. 126) why robbery in nature could be less common (see below). WOOTTON observed raiding in two natural populations. He did not state whether the eggs were eaten so perhaps either heterocannibalism or courtship-motivated theft was involved. MCKENZIE (1974) observed raiding in *Culaea* and noted most eggs were lost during the melee. BLACK (1971), in a laboratory study, reported raided eggs "are rarely adopted, but usually eaten".

In many cases where raided eggs were not immediately eaten, the raider fertilized the stolen eggs while they were in the victim's nest (VAN DEN ASSEM, 1967; MCINERNEY, 1969; SARGENT & GEBLER, 1980). In other cases, ROHWER's cuckoldry explanation appears valid: LI & OWINGS (1978a) and ROWLAND (1979) never observed the raider removing eggs after having fertilized them in another male's nest.

The only evidence supporting the concept of courtship related theft is VAN DEN ASSEM's (1967) observation of five males (p. 125) which appeared not to eat their stolen eggs. The indirect method by which non-consumption was determined was statistical and thus open to uncertainty.

The importance of courtship-related theft rests on the assumption that female sticklebacks prefer males with clutches. Recent studies cast doubt on the relative importance of female choice in sticklebacks. VAN DEN ASSEM (1967) noted several reasons why female numbers may effectively exceed males: a) not all males will obtain territories; b) females produce clutches every three days (*Culaea* is the same, MOODIE, unpublished) whereas males require 12 days to complete a nesting cycle; and c) males mate with several females. LI & OWINGS (1978a) also argue that the importance of female choice may have been over-estimated. In a field study, KYNARD (1979) described female courtship of non-courting males as common. MOODIE (unpublished) and HAGEN (pers. comm.) have also observed females' courtship of unreceptive males in *Gasterosteus* on the Pacific and Atlantic coasts, respectively. Rather than representing a courtship strategy, it is possible that theft and non-consumption of eggs fertilized by another male may be just a mistake (COYNE & SOHN, 1978).

Against the scant evidence for non-consumptive egg-stealing is a large body of data demonstrating heterocannibalism later in the brood cycle (BLACK, 1971; BLACK & WOOTTON, 1970; KYNARD, 1979; present study) as well as studies of seined or trapped males which suggest the existence of heterocannibalism (HYNES, 1950; MOODIE, 1972; SEMLER, 1971; WORGAN & FITZGERALD, 1981). However, there is unfortunately no way of assessing the relative frequency of early egg theft for courtship purposes and later egg theft involving heterocannibalism, as the prediction requires.

The prediction is also confounded by the likelihood that heterocannibalism may increase fitness not only as stated by ROHWER (because eggs are a source of energy) but also because it must represent an effective form of interference competition (LI & OWINGS, 1978b). For this reason and contrary to the prediction, heterocannibalism (and raiding in general) should take place anytime in the brood cycle.

In summary, prediction b is weakly supported, often contradicted by empirical evidence, and may also be unsupportable in theory.

Prediction c.

If filial cannibalism is occurring, parental males should eat eggs more often than females (because their foraging opportunities are limited while their own eggs provide an available food supply) (ROHWER, 1978). Data from the dugout strongly supports this prediction. Two thirds of the parental males ate eggs, significantly more often than males and females caught with traps over the same time span. Eggs were a more important part of the diet of parental males than of trap-caught fish. This may suggest parental males were filial cannibals as well as heterocannibals.

In general then, this prediction is well supported by the data from parental males and also by the less conclusive data from trap catches or other capture methods (HYNES, 1950; SALFERT, 1984; SEMLER, 1971).

Prediction d.

If filial cannibalism takes place, it should result in males eating the last few eggs or larvae remaining in the nest at the end of the brood cycle. Such behaviour has been observed (FEUTH-DE BRUIJN & SEVENSTER, 1983; VAN IERSEL, 1953; MCKENZIE, 1974). In the present study, parental males in the laboratory may have cannibalized fry 2-4 days after a hatch. This could, however, be a laboratory artifact resulting from the intermittent feeding cycle typically applied to captive fish or the confined nature of aquaria.

Field studies on the other hand provide little evidence for cannibalism on larvae: KYNARD (1979) found fry in stomachs of only five of 499 cannibals. Similarly, fry were observed in 10 of 200 brook stickleback stomachs containing offspring (SALFERT, 1984). WORGAN & FITZGERALD (1981) examined three Gasterosteid species and reported low dietary importance indices for stickleback larvae, though the males of two species ate many eggs.

There are differences in the stage of eggs eaten by parental males occupying different environments. In a more favourable habitat (abundant food and low risk of fry predation), males assumed to be parental ate more eggs of a late stage of development, thus supporting this prediction (SALFERT, 1984). These males could have been more ready to undertake another breeding cycle and thus more prone to eat the last remaining eggs of an asynchronous hatch. In the dugout, where food was limited and fry were cannibalized (SALFERT, 1984), eggs of an earlier stage were eaten.

There was no apparent tendency for males in the dugout population to eat eggs remaining from asynchronous hatches. Males did not abandon free-swimming fry, and often "herded" fry within the territory. Similarly, MOODIE (1972) and KYNARD (1979) describe some males remaining with the fry for 8-9 days.

Prediction e.

In filial cannibals with several brood cycles per season, fewer eggs should be eaten from the last brood of the season than from earlier broods. ROHWER (1978) lacked data to test this prediction and no relevant data have appeared to date.

In summary, though two predictions are unsubstantiated (b and e), there is evidence from the field supporting predictions a, c, and d. In addition to the field data, the tracer study clearly shows that filial cannibalism occurs in *C. inconstans.* Filial cannibalism is also described or alluded to in laboratory studies by VAN IERSEL (1953), MORRIS (1958), VAN DEN ASSEM (1967), BEUNE (1974), and FEUTH-DE BRUIJN & SEVENSTER (1983). The present study also shows that filial cannibalism occurs on a scale that exceeds the simple consumption of diseased or dead eggs.

Given the existence of filial cannibalism, it remains to determine its function. ROHWER (1978) argued it serves nutritional needs. Certainly there is some laboratory evidence that filial cannibalism can be released by hunger (FEUTH-DE BRUIJN & SEVENSTER, 1983) and field evidence that cannibalism in general is correlated with food availability and stomach fullness (FOX, 1975; KYNARD, 1979; SALFERT, 1984).

There are other data, however, which suggest that in sticklebacks at least, filial cannibalism may have the important reproductive function of increasing a male's chances of obtaining a maximum number of eggs per brood (— similar to ROHWER's courtship prediction c, page 435, though he assumed such cannibalism is related to female choice of mate).

Several authors have noted filial cannibalism seems to be triggered by small numbers of eggs in the nests of males which have fertilized only one clutch (VAN DEN ASSEM, 1967; BEUNE, 1974; VAN IERSEL, 1953; MORRIS, 1958). Given that males regularly fertilize several females per brood in nature and appear capable of successfully caring for very large numbers of eggs (>4,100 in the present study; >3,000 in BEUNE, 1974) and that the male cannot accept new clutches indefinitely (VAN IERSEL, 1953), filial cannibalism may be the best means of terminating small broods when there is the possibility of obtaining a significantly larger brood in the immediate future. Such a function could account for the persistence of filial cannibalism in the tracer study in the face of an apparent surplus of alternate food and a scarcity of gravid females. Of course, while the ultimate function of filial cannibalism may be reproductive as described here, it would provide a subsidiary nutritive benefit as well. At present, it is impossible to say whether nutrition or brood maximization is the ultimate explanation. Perhaps either might dominate depending on the relative abundance of alternate food and gravid females. Additional studies are needed which vary food levels and the number of clutches (*i.e.* eggs) a male fertilizes.

Summary

Filial cannibalism was directly verified in the laboratory. All test males (n = 43) ate some or all of their own ^{125}I labelled eggs. Eight males successfully reared some eggs to hatching. This success rate is comparable to those in the literature for other Gasterosteid species.

Parental males and their nests were removed individually from a dugout. Differences in the importance of eggs in diet and the frequency of egg-consumption between trap-caught fish and parental males offered indirect evidence for the occurrence of filial cannibalism. Of 22 egg-guarding males which cannibalized eggs, 21 contained eggs at the same stage as eggs in their nest. In cases where the nearest neighbouring nest contained eggs of a different age, filial cannibalism (rather than egg-robbery) could be assumed in at least seven of 13 parental males which ate eggs.

References

ASSEM, J. VAN DEN (1967). Territory in the three-spined stickleback, *Gasterosteus aculeatus* L., an experimental study in intra-specific competition. — Behaviour suppl. 16, p. 1-164.

BEUNE, H. M. (1974). Hatching success in the three-spined stickleback (*Gasterosteus aculeatus* L.): a causal and functional analysis on a Pacific North American population of form *trachurus*. — B.Sc. thesis. University of British Columbia.

UNDERSTANDING THE EVOLUTION OF BEHAVIOR IN THREESPINE STICKLEBACK: THE VALUE OF GEOGRAPHIC VARIATION

SUSAN A. FOSTER[1,2])

(Department of Biological Sciences, University of Arkansas, Fayetteville, AR 72701, U.S.A.)

(With 4 Figures)

(Acc. 18-VI-1995)

Summary

The endemic radiation of the threespine stickleback in North America possesses features that will permit us to gain unique insights into evolutionary pattern and process. Already, comparison of populations has demonstrated a remarkable level of adaptive differentiation in reproductive behavior across populations in both southern British Columbia, Canada, and the Cook Inlet Region of Alaska, U.S.A. In small, shallow lakes, threespine stickleback feed predominantly on benthic invertebrates in the shallow littoral zone. They form large groups that attack nests defended by males, and consume all of the young within. In contrast, in larger, more oligotrophic lakes, adult stickleback feed primarily on plankton in large open-water foraging groups. They do not form groups that cannibalize young in nests. In these non-cannibalistic populations, males perform more conspicuous courtship behavior than do those in cannibalistic populations, apparently because risk of attack by visually hunting conspecifics is low. Courtship in these lakes more often includes the conspicuous zig-zag dance, and less often, meandering and dorsal pricking. Anadromous populations tend to be cannibalistic, but also are often behaviorally intermediate. They incorporate the zig-zag dance in courtship more often than do males in cannibalistic freshwater populations, but also usually incorporate the less conspicuous dorsal pricking behavior. These results suggest that the ancestral condition was intermediate and that both cannibalistic and non-cannibalistic freshwater populations display some derived elements in their reproductive repertoires.

This kind of comparison among populations thought to have been independently derived from the marine form can provide insights into the patterns and causes of adaptive diversification. For populations such as these, carefully chosen from disparate geographic locations so as to preclude common evolutionary history after colonization of freshwater,

[1]) Present address: Department of Biology, Clark University, 950 Main Street, Worcester, MA 01610-1477, U.S.A.; e-mail: sfoster@vax.clarku.edu

[2]) I wish to thank Piet Sevenster and Theo Bakker for tremendous patience and John Baker and Sydney Cameron for drawing figures. Because this paper is the culmination of work done in Michael A. Bell's laboratory and because his thinking and comments have added so much to my way of thinking about evolutionary pattern and process, I must especially thank him for his seminal role in this work. This research was supported by a NIMH National Research Service Award, by NSF grant DEB-9108132 and by NSF Presidential Faculty Fellowship DEB-9253718.

 Behaviour 132 (15-16)

an unresolved polytomy is probably the best phylogenetic model. For closely-spaced lakes within drainages, branching trees, reflecting gene flow after colonization, are probably more realistic models. Resolution of such phylogeographic relationships requires the acquisition of appropriate molecular data. Both approaches, with their implicitely historical view of character evolution, provide windows that will permit us an unusually clear view of historical changes in behavioral phenotypes and of their causes.

> *This insignificant fish has a ceremonious sex life. To attract females the male builds a house, changes color and does a kind of dance, a ritual singularly useful to the study of instinct.*
> Niko TINBERGEN (1952)

Introduction

With this deceptively simple passage, Niko TINBERGEN began a popular account of the seminal research on the ethology of the threespine stickleback (*Gasterosteus aculeatus*) which he had performed with J. VAN IERSEL. At the time, ethologists assumed that many stereotyped behavioral displays, including those involved in the courtship ritual of this stickleback, were invariant components of the behavioral repertoires of all of the members of a given species (HINDE & TINBERGEN, 1958; TINBERGEN, 1964; LORENZ, 1970; DEWSBURY, 1990). That TINBERGEN held the view that display elements in the stickleback courtship ritual were examples of such 'species-typical' behavior patterns is apparent in both this popular account (TINBERGEN, 1952) and his scientific writings (*e.g.* TINBERGEN, 1951). Ironically, the species with which TINBERGEN had played as a child, and studied so effectively in later years, is proving to be among the most variable species ever studied. This diminutive fish shows tremendous geographic variation not only in behavior, but also in life history and morphology (see papers in BELL & FOSTER, 1994b).

The first indication of geographic variation in display behavior in the threespine stickleback was published by Kenneth WILZ in 1973. In this paper WILZ described quantitative differences in the behavior of males from two different populations. One was from the brackish waters of Naragansett Bay, Rhode Island, USA and the other from streams and canals in and around Oxford, United Kingdom (WILZ, 1970, 1973). His observations on the two populations, made under similar laboratory conditions, indicated that the 'American males' were far more aggressive than were males of the 'European variety' and that they incorporated more dorsal pricking and nest-directed activity in their courtship than did their European counterparts. Overall, courtship in the Naragansett Bay population was a comparatively protracted and inefficient affair. In contrast, courtship in the British population incorporated the zig-zag dance more frequently and was shorter in duration (WILZ, 1973).

Because WILZ studied only two populations, each on a different continent, and because WOOTTON (as cited by WILZ, 1973) had observed behavior similar to that of the Naragansett Bay population in threespine stickleback nesting in tidepools in northwestern North America, WILZ reasonably interpreted the differences as reflecting varietal differences between the two continents. Today we know that equivalent variation in aggressive (HUNTINGFORD, 1982; BAKKER, 1986, 1994), anti-predator (GILES & HUNTINGFORD, 1984; FOSTER & PLOCH, 1990; HUNTINGFORD *et al.*, 1994), and reproductive behavior (FOSTER, 1988, 1994a, b) often occurs on a much smaller spatial scale. Indeed such differences can be apparent among populations inhabiting different lakes in a single drainage. As is the case with morphology, the variation often reflects differences in ecological conditions among habitats, and is apparently adaptive (see papers in BELL & FOSTER, 1994).

In this paper, I will extend WILZ's observations, providing quantitative evidence of differences among populations on a local scale that parallel the differences in reproductive behavior described by WILZ (1973). I will also provide evidence of associations between the behavioral differences and particular ecological conditions that help to explain their adaptive value and will argue that the same selective forces have molded courtship behavior in disparate geographic regions. Finally, the direction of evolutionary change will be examined by comparison with ancestral marine populations. The value of population comparison in evolutionary inference will be discussed in closing.

Historical zoogeography and rationale

The threespine stickleback fish, *Gasterosteus aculeatus*, has undergone a remarkable endemic radiation in freshwater habitats in recently deglaciated regions of northwestern North America, Scotland, and other less intensively studied regions in the temperate zone of the Northern Hemisphere. Restricted primarily to coastal freshwater habitats and marine waters in temperate and arctic regions, this small fish must have been absent from much of the northern part of its range at the time of the last glacial maximum (Bell & Foster, 1994b, for review). As glaciers began to recede 22,000 years ago (less in some areas; Foster & Bell, 1994; McPhail, 1994), marine and/or anadromous stickleback colonized the newly-formed freshwater habitats, giving rise not only to phenotypically diverse freshwater populations, but also to new species (McPhail, 1984, 1994).

The radiation of threespine stickleback in recently deglaciated freshwater habitats is particularly intriguing because plesiomorphic (ancestral) character states can be ascertained with unusual assurance. This is, in part, because the morphology of marine and anadromous threespine stickleback has changed little in the last 10 million or more years as indicated by fossil remains (Bell, 1977, 1994) and because both are remarkably uniform throughout the range of the species complex (Bell & Foster, 1994b). Thus, the morphological character states expressed in marine and anadromous populations are probably plesiomorphic.

Recent phylogenetic reconstructions among populations of threespine stickleback using allozyme (Haglund *et al.*, 1992; Buth & Haglund, 1994) and mtDNA sequence data (Ortí *et al.*, 1994) provide evidence that the marine and anadromous forms are ancient. A set of Asian populations forms the sister group to all other *G. aculeatus*, which in turn, are divided into separate monophyletic groups in the Atlantic and Pacific basins. The fossil record shows that the threespine stickleback, now with a discontinuous distribution between the Atlantic and Pacific basins, has been in both oceans for at least 1.9 million years (Bell, 1994). Thus, character states shared between Pacific and Atlantic marine populations are presumably primitive relative to the postglacial diversification in either basin (Foster & Bell, 1994). In this instance, we have data beyond the usual outgroup comparison to support our hypotheses of ancestral states.

A remarkable attribute of the postglacial radiation of freshwater stickleback is that many of the population differences have proven adaptive (papers in Bell & Foster, 1994a for reviews). The prominent role of natural selection in molding population phenotypes is suggested in that similar freshwater habitats in geographically distant areas typically are inhabited by populations with similar, apparently adaptive, morphological phenotypes. These habitats are thought to have been colonized independently by marine stickleback, which are in turn, thought to have given rise independently to similar, adaptive phenotypes (reviewed in Bell &

FOSTER, 1994b). If populations are selected from different drainages, independent origin of similar, derived phenotypes would be inferred. Within drainages, however, phylogenetic reconstruction of population relationships awaits appropriate molecular data for analysis (WITHLER & MCPHAIL, 1985; BELL & FOSTER, 1994b).

One of the most dramatic examples of ecotypic differentiation in the freshwater radiation involves trophic divergence between populations in shallow lakes with extensive littoral zones and those in deep, steep-sided, oligotrophic lakes. In southwestern British Columbia, where the divergence has been best studied, many populations in deep lakes have become specialized for feeding predominantly on plankton in open water (limnetic populations, *sensu* MCPHAIL, 1984). In small, shallow lakes, populations are instead specialized for foraging on large benthic invertebrates in the littoral zone (benthic populations; MCPHAIL, 1984, 1994; LAVIN & MCPHAIL, 1985, 1986; SCHLUTER & MCPHAIL, 1992). In limnetic fish, the mouth is comparatively narrow and the snout long (enhancing sucking action), the eye is larger, and the fish have long, closely-spaced gill rakers. Benthic fish are deeper bodied, have larger mouths and smaller eyes, and have short, widely-spaced gill rakers that prevent escape of prey but do not impede water movement. These suites of characters are thought to adapt the ecotypes to their respective environments (HAGEN & GILBERTSON, 1972; MCPHAIL, 1984; LAVIN & MCPHAIL, 1986; SCHLUTER & MCPHAIL, 1992).

Benthic populations differ from limnetic populations with respect to several aspects of behavior as well. One of the differences involves the cannibalistic proclivities of the two ecotypes (FOSTER, 1988, 1994a, b; RIDGWAY & MCPHAIL, 1988). In British Columbia, stickleback in benthic populations form large groups which forage on benthic invertebrates. The groups attack nests guarded by parental males whenever they are detected. If the nest contains embryos or yolk sac fry, they are consumed. Cannibalistic attacks resulted in the failure of 68% of nests under observation during one breeding season in a British Columbia lake (FOSTER, 1988). Although fish in limnetic populations form planktivorous foraging groups, cannibalism of young in nests has not been observed in any population (FOSTER, 1994a, b).

In the next sections of this paper I will provide quantitative data documenting this association between cannibalism and foraging mode in lacustrine populations in both British Columbia, Canada, and the Cook Inlet region of Alaska. In addition, I will provide evidence of divergence in courtship behavior between the ecotypes in Alaska that parallels that previously described in British Columbia (FOSTER, 1994a) and quantitative data will be presented for populations in both regions to facilitate comparisons. To permit inference of the direction of evolutionary change upon invasion of freshwater I provide a similar data set for one Alaskan anadromous population and more limited data for two others.

Study sites

Six of the lacustrine populations are in the Strait of Georgia region of British Columbia, Canada. This region was ice covered at the height of the Fraser Glaciation and existing coastal lakes and streams were formed during isostatic rebound (increase in land elevation after ice melt) over the last 13,000 yr (MATHEWS *et al.*, 1970; HOWES, 1983). Sproat, Cowichan, and Crystal lakes are located on Vancouver Island in disjunct river drainages. Hotel, Garden Bay and North lakes are on the Sechelt Peninsula across the Strait of Georgia from Vancouver Island. Hotel and Garden Bay lakes are connected by a stream approximately 0.3 km in length but are of divergent phenotypes. They both are connected to North Lake through creeks and lakes draining a higher elevation lake (Klein Lake) in two directions. North Lake was probably independently colonized through a much shorter direct connection to the Agamemnon Channel to the north. All of these lakes are between 48 and 50 degrees north latitiude. They range from 8 - 3,800 ha in surface area.

The five Alaskan lakes are located within the MatSu Valley north of the Knik Arm of Cook Inlet, between 61 and 62 degrees north latitude. Like much of coastal British Columbia, this region was covered by ice during the recent Fraser Glaciation, so that the study populations are only about 10,000 yr old (REGER & UPDIKE, 1983). Willow and Lynne lakes are located near the confluence of Willow Creek and the Susitna River ca. 65 km inland from the river's junction with Cook Inlet. Willow Lake is connected to Willow Creek via a short, direct tributary. Lynne Lake is a relatively isolated lake located in the upper portion of a different Willow Creek tributary. At least 25 stream km and seven intervening lakes separate these populations. Big Beaver Lake is located on a tributary to the Meadow Creek system ca. 22 km (straight-line) southeast of Lynne Lake. Meadow Creek is part of the Big Lake - Fish Creek system, which enters Cook Inlet separately from the Susitna River. Stephan Lake is situated about 10 km southwest of Big Beaver Lake and is near the headwaters of Goose Creek, also a tributary of Fish Creek. However, at least 40 stream km and two intervening lakes separate these populations. Bruce Lake is an isolated water body located at the upper end of the Little Meadow Creek watershed, about 15 straight-line km east of Big Beaver Lake. About 20 stream km and six lakes separate Bruce Lake from Big Beaver Lake, the closest study site. The five lakes range from 8 and 113 ha in surface area. The anadromous population was studied on its spawning grounds, a tidal marsh system just upstream from the mouth of the Anchor River on the Kenai Peninsula. The anadromous population is at least, 190 straight-line km south southwest of the nearest lacustrine study population, at about 60 degrees north latitude.

Stickleback breeding areas were similar at all of the study sites. Nests were situated in the littoral zone in less than 3.5 m of water, in areas littered with organic debris. Cover afforded by vegetation varied from 0 to 50% in the breeding area, and twigs, stones, logs, and fallen trees provided additional cover. Only in Sproat Lake did a substantial proportion (10-20%) of males nest in vegetation.

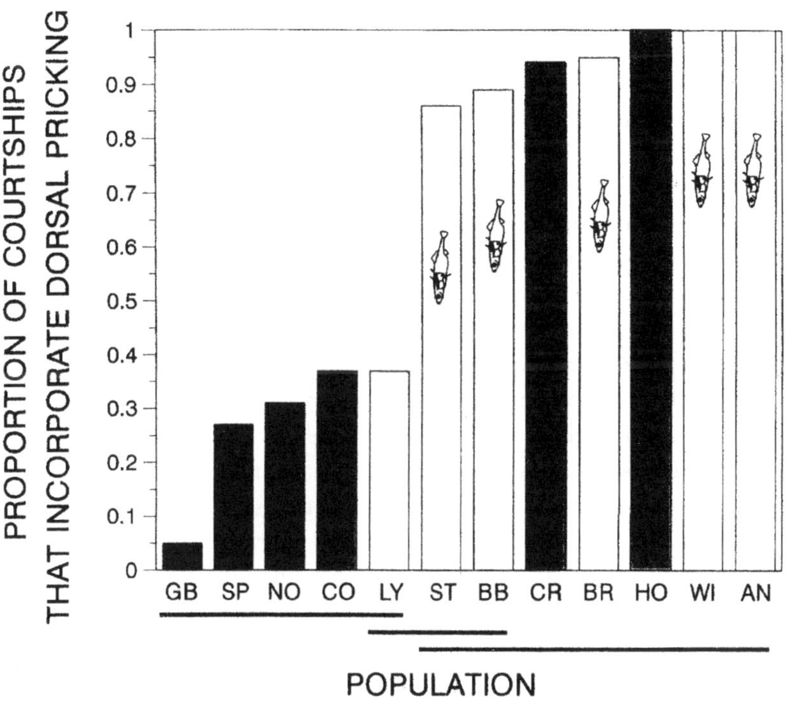

Fig. 2. The proportion of courtships in each of 12 populations of threespine stickleback that incorporated dorsal pricking. Bars with stickleback in or above them represent populations in which group cannibalism was observed. Site names as in Fig. 1. Lines below the X-axis connect populations that did not differ significantly at the 0.05 level (G statistic).

clivity (Tables 2 and 3). Undoubtedly, this was partly due to our inability to discriminate aggressive swims from straight swim solicitation (the functions could be inferred only by female response which may not have been an accurate reflector of male intent), and partly to inability to discriminate a rise from the nest from a straight approach to a waiting female.

Discussion

The behavior of males in cannibalistic and non-cannibalistic populations differs predictably, and similarly, in British Columbia and Alaska. In cannibalistic populations, courtship tends to be less conspicuous than in non-cannibalistic populations in that the zig-zag dance is missing or is a relatively minor element of courtship, and dorsal pricking is the most common form of courtship. The difference is exacerbated by differences in the expression of the zig-zag dance. In non-cannibalistic populations, the zig-zag is often initiated when a female is first noticed by a male. She may be several meters away still, and he will zig-zag the full distance to her with pronounced side to side jumps. In cannibalistic populations, on the other hand, zig-zag dancing tends to appear later in the courtship ritual, is typically directed to a female who is much nearer because courtship has already been initiated, and is less pronounced, possibly because of the nearness of the female.

The value of inconspicuous courtship in cannibalistic populations is probably that foraging stickleback detect the nests they attack on the basis of nest-directed behavior patterns of the male and others with whom he interacts. Several lines of evidence suggest that stickleback do not detect nests visually on the basis of their structural characteristics. For example, males often cannot find their nests when moved unless a distinctive object is also moved (WARD & FITZGERALD, 1987), groups of stickleback often feed within a few cm of nests without discovering and attacking them, and females must be shown the entrance of the nest in order to gain access unless they have watched males work on their nests (personal observations).

Dorsal pricking may be advantageous in cannibalistic populations not only because it is inconspicuous, but also because it enables males to survey their surroundings before leading females to their nests (SARGENT, 1982). They are therefore able to avoid leading a female to the nest when groups are nearby. In contrast, the zig-zag dance is probably favored in non-cannibalistic populations because males that perform the display are more likely to be discovered by females than are those that do not. When cannibalism is absent from a population, males can court elaborately without risk to their nests or to embryos already contained within, and the effects of sexual selection can be expressed more fully.

The differences in courtship behavior may well reflect ecotypic differentiation. Not only are the behavior types associated with heritable differences in trophic morphology (HAGEN, 1973; BENTZEN & MCPHAIL, 1984; FOSTER *et al.*, 1992), laboratory observations also suggest that there may be heritable differences among populations in the tendency to perform the two components of the courtship dance (WILZ, 1973; RIDGWAY & McPHAIL, 1984; unpublished data). Males in all populations that have been studied are able to perform both forms of courtship; only laboratory rearing experiments will permit unambiguous assessment of possible genetic bases for the differences in expression of the behavior patterns in nature and in the laboratory.

Comparison of the marine populations (ancestral) with derived, freshwater populations, suggests that the ancestral condition is one in which benthic foraging and group cannibalism are common and courtship is relatively inconspicuous. The Anchor River population is particularly interesting in this context because there were no benthic foraging groups in the ponds at the time most of the behavioral observations were made. Large groups were seen at other times during the breeding season, suggesting that the risk of group cannibalism is temporally unpredictable in tide pools, and that inconspicuous courtship is probably advantageous at some times. This tendency is apparently retained even when groups are absent, as indicated by the uniform inclusion of dorsal pricking in courtship and the intermediate frequency of the zig-zag dance in courtships.

In this examination of local and regional patterns of differentiation of courtship behavior, the assumption has been made that each derived freshwater population can be treated as an independent data point because each has evolved the characteristics it posesses independently. Our knowledge of the zoogeographic history of the threespine stickleback strongly supports this interpretation when the populations are geographically separated from one another on a spatial scale likely to preclude gene flow subseqent to colonization of fresh water (BELL & FOSTER, 1994b). Certainly, the freshwater populations possessing derived characters that are phenotypically similar, must have evolved them independently when they are from geographic regions as disparate as the Cook Inlet region of Alaska and southern British Columbia. Within regions, different drainages are also very likely to have been colonized independently.

The appropriate model of relationships among populations chosen in this way is an unresolved polytomy (Fig. 3). This is because all of the populations can be assumed to have been independently drawn from a relatively uniform marine ancestor and to have had no subsequent common evolutionary history. We now know this model is oversimplified for

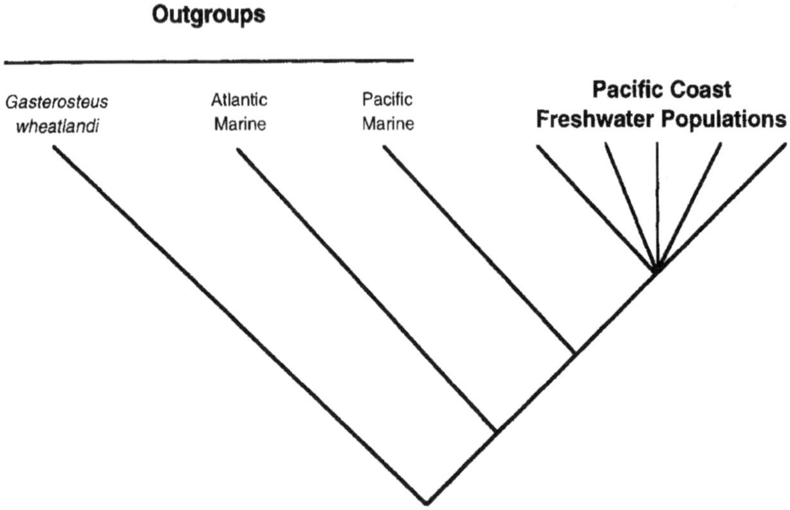

Fig. 3. A simplified phylogeny of the threespine stickleback showing outgroups (excluding the Asian clade for simplicity) for the freshwater radiation of threespine sickleback in northwestern North America.

the Pacific region, although an equivalent model would be appropriate for Atlantic populations. In the Pacific, there exist two ancient clades that are phenotypically cryptic but that can be readily distinguished with allozyme, RFLP, and sequence data (Haglund *et al.*, 1992; Buth & Haglund, 1994; Ortí *et al.*, 1994). From southern British Columbia south, the model in Fig. 3 would be appropriate if the Asian-Pacific clade was inserted between *Gasterosteus wheatlandi* and the Atlantic clade. The appropriate model is less clear in regions of overlap between the two Pacific clades. If the clades are reproductively isolated, two separate trees could be constructed. If not, a reticulate model would be appropriate.

Understanding relationships among the clades is essential if character state polarity (direction of evolutionary change in character states) is to be correctly inferred because these groups serve as evolutionary outgroups for the Pacific radiation. An outgroup, or sister group, is that which is most closely related to the group under study (the ingroup). In practice, outgroups are not always easy to identify because more than one possible sister group may exist (Donoghue & Cantino, 1984; Maddison *et al.*, 1984). However, if a single sister group can be identified, the simplest rule for assessing polarity is to assume that, for a character that varies across members of the ingroup, the primitive (plesiomorphic) character state is that also found in the outgroup. Thus, the derived (apomorphic) character states must be those found only in the ingroup.

The discussion thus far, has assumed that the three anadromous populations serve as an appropriate sampling of the marine outgroup for the freshwater radiation. Character states found in all three are assumed to be ancestral. However, there are only limited data for marine populations and the clade(s) to which the three anadromous study populations belong are unknown. Nevertheless, the inferences derived from this population comparison are probably correct for reasons outlined below.

Population comparisons, like those made here, would be more robust if there existed cladograms, based on molecular data, expressing relationships between populations within and among regions. At present, the methods for resolving such relationships within the stickleback species complex have not been developed, but there is reason to expect that they soon will be (FOSTER & BELL, 1994). Once such cladograms are available, we will be able to overlay behavioral character states from the different populations onto the cladograms to examine patterns of evolutionary change among populations on different spatial scales (see AVISE *et al.*, 1987; AVISE, 1994 for a discussion of phylogeographic approaches to the study of evolutionary change).

To illustrate this approach, I will use the hypothetical example shown in Fig. 4 which presents relationships among four populations in two river drainages. The populations in the same river drainages are portrayed as having more recent common evolutionary histories than those in different drainages, a reasonable expectation. I have overlayed the characters on the tree that have been discussed in this paper plus one additional character; diversionary display behavior. This is a behavior pattern used by males in cannibalistic populations when defending their nests against conspecific groups (FOSTER, 1988, 1994a, b; RIDGWAY & MCPHAIL, 1988). In these populations males exhibit a display that ends in rooting, and that appears to mimic feeding (WHORISKEY & FITZGERALD, 1985; FOSTER, 1988, 1994a, b; RIDGWAY & MCPHAIL, 1988; WHORISKEY, 1991). This behavior has been called diversionary display, following the avian literature (ARMSTRONG, 1949a, b; SKUTCH, 1955), as it often diverts groups from the trajectory that would take them into the male's territory, causing them instead to follow the male and attempt to feed. In limnetic populations, males do not perform diversionary displays at the approach of a group. Instead they court gravid females or attack individuals in the groups directly (FOSTER, 1988, 1994a, b).

Primitive character states can be inferred from outgroup analysis. Although relatively few marine populations have been studied to date, representatives of both the Atlantic and Pacific North American clades forage on benthic invertebrates during the breeding season, cannibalize young in nests, and perform diversionary displays (WHORISKEY & FITZGERALD, 1985, 1994; FOSTER, 1994a, b; unpublished data). Thus, these are likely to be the primitive character states as depicted in Fig. 4, with the derived character states being, respectively, planktivory, loss of cannibalistic behavior, and loss of diversionary display behavior.

If, upon overlaying the character states for each population on our hypothetical cladogram, we obtained the distribution of characters portrayed in Fig. 4, the most parsimonious interpretation would be that the suites of characters defining a limnetic population had evolved independently on two occasions, once in each drainage. If instead, all of the limnetic populations were clustered on the same branch of the tree, a single transition, followed by dispersal, would be the most parsimonious explanation.

The role of natural selection in promoting the evolution of the limnetic form could also be assessed by mapping the lake type onto the cladogram and testing for associations between habitat (lake) type and the two suites of behavioral characters. If there were only two deep, steep sided lakes, and they contained the limnetic populations, an appropriate inference would be that limnetic populations evolved from the marine ancestor twice in response to local selection.

Knowledge of the direction of evolutionary change in homologous behavioral character states (character polarity), can provide insights into evolutionary pattern in ways not possible otherwise. For example, one can (1) trace the transformational sequence by which one behavioral state gave rise to another within a group (*e.g.* HINDE & TINBERGEN, 1958; LAUDER, 1986), (2) determine whether exposure to a particular environmental regime has preceded the evolution of a similar, homoplasious behavior more than once, suggesting adaptation (*e.g.* BAUM & LARSON, 1991), or (3) clarify which character states are ancestral among variable

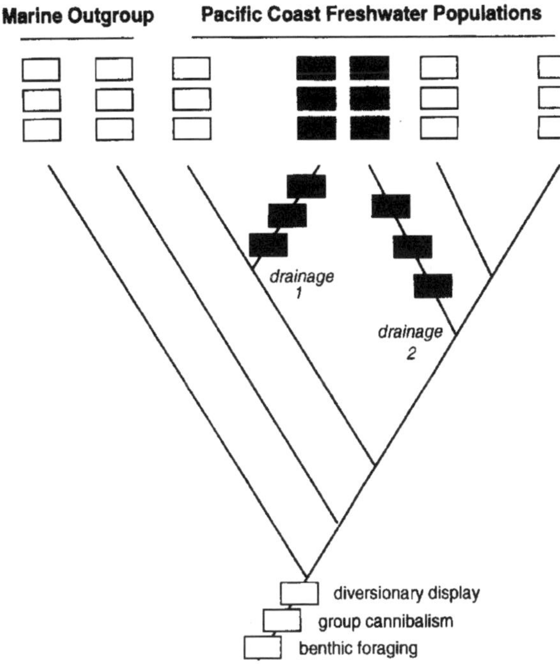

Fig. 4. A hypothetical phylogeny of five lacustrine populations of threespine stickleback in two river drainages. Unshaded boxes denote ancestral character states and shaded, derived. Character states are described in the text.

populations (*e.g.* AVISE *et al.*, 1987; AVISE, 1994; FOSTER & CAMERON, 1995). The postglacial radiation of the threespine stickleback provides an extremely rare opportunity to make such comparisons among populations.

Comparisons among populations that display ecotypic differentiation can be particularly valuable for inferring the causes of differentiation (ARNOLD, 1992; FOSTER *et al.*, 1992). Populations are likely still to reside in the habitats that have selected for the differences, and there tend to be fewer differences among populations than among species, reducing the possibility of misinterpretation due to the existence of covarying characters. Population comparisons are also uniquely suited to the study of microevolutionary pattern. Comparative studies such as these must be interpreted with caution however, and should be seen as a method of generating evolutionary hypotheses which can be tested experimentally or using carefully designed measurements of fitness of the phenotypes across populations (ENDLER, 1986; LAUDER *et al.*, 1993; LEROI *et al.*, 1994; FOSTER & CAMERON, 1995). All of these approaches are feasible with threespine stickleback.

The data presented here demonstrate that specific ecological conditions are often associated with behavioral phenotypes in ways that can be readily interpreted in an adaptive context. Indeed, as in the case of morphological phenotypes (BELL & FOSTER, 1994b; MCPHAIL, 1994), there seems to be a remarkably strong influence of natural selection on at least some elements of the the reproductive behavior of threespine stickleback. Other aspects are less apparently adaptive. Population differences in sneaking behavior (GOLDSCHMIDT *et al.*, 1992, unpublished data) and in forms of diversionary display behavior (FOSTER, 1988, 1994a,b) are less readily interpreted in an adaptive context. Thus, genetic drift (including founder effects) and gene flow may also have played a role in molding population phenotypes. Comparison of threespine stickleback populations provides a unique opportunity to explore the relative roles of these forces in molding behavioral phenotypes in this remarkable evolutionary radiation.

References

ARMSTRONG, E.A. (1949a). Diversionary display. Part 1. Connotation and terminology. — Ibis 91, p. 88-97.
—— (1949b). Diversionary display. Part 2. The nature and origin of distraction display. — Ibis 91, p. 179-188.
ARNOLD, S.J. (1992). Behavioural variation in natural populations. VI. Prey responses by two species of garter snakes in three regions of sympatry. — Anim. Behav. 44, p. 705-719.
AVISE, J.C. (1994). Molecular markers, natural history and evolution. — Chapman and Hall, New York.
—, ARNOLD, J., BALL, R.M., BERMINGHAM, E., LAMB, T., NEIGEL, J.E. & SAUNDERS, N.C. (1987). Intraspecific phylogeography: the mitochondrial DNA bridge between population genetics and systematics. — Ann. Rev. Ecol. Syst. 18, p. 489-522.
BAKKER, T.C.M. (1986). Aggressiveness in sticklebacks (*Gasterosteus aculeatus* L.: a behaviour genetic study. — Behaviour 98, p. 1-144.
—— (1994). Evolution of aggressive behavior in threespine stickleback. — In: The evolutionary biology of the threespine stickleback (M.A. BELL & S.A. FOSTER, eds). Oxford University Press, Oxford, p. 345-380.
BAUM, D.A. & LARSON, A. (1991). Adaptation reviewed: a phylogenetic methodology for studying character macroevolution. — Syst. Zool. 40, p. 1-18.
BELL, M.A. (1977). A late Miocene marine stickleback, *Gasterosteus aculeatus aculeatus*, and its zoogeographic and evolutionary significance. — Copeia 1977, p. 277-282.

Predators & Parasites

Frank A. von Hippel

> I remember some years ago keeping a small Pike in the same vessel with a number of Sticklebacks, and I shall never forget the persistence with which first one and then another of these fish attacked the Pike's tail. Occasionally the larger fish would retaliate and try to swallow one of its tormentors, but the sauce piquant of those formidable spines always proved too potent. After a few violent shakings of the head, the Stickleback was forcibly ejected from the cavernous jaws of its would-be devourer. The result of these repeated attacks on the Pike's tail was a gradual diminution of that organ, and I, feeling sorry for the victim of these cowardly attacks on the rear, took him out of the aquarium and turned him into a pool of water. — The Reverend W. Houghton (1879)

Ecology as a discipline developed much of its theory from the study of density-dependent mortality factors, chief among them predation, parasitism and competition. The study of competition in stickleback came early, due to the competitive interaction between males that initiates territory formation and nest building (e.g., van Iersel, 1953; Morris, 1958; van den Assem, 1967). The study of predation also came early, due to curiosity about the formidable spines and robust lateral plates of the threespine stickleback (Hoogland et al., 1957). But these studies were carried out by pioneering ethologists before the critical interactions between behaviour and ecology were understood – before the advent of behavioural ecology (Kruuk, 2003). The full ecological implications of their work would come later, as ecologists and evolutionary biologists recognized the population level consequences of arms races between predators and prey, between competitors, and between parasites and hosts. Therefore ecological and evolutionary theory that employed the stickleback model relied on earlier ethological work on the causation of behaviour, often carried out by Niko Tinbergen and his students.

 Tinbergen's Legacy in Behaviour, 347-352

The threespine stickleback's body armour attracted the curiosity of Tinbergen's group, who were interested in the interplay of form and function. In the first paper of this section, Hoogland, Morris and Tinbergen (1957) wrote, "Since sticklebacks, when seized by a Pike, respond by raising all their spines, and keeping them rigidly erected as long as they are in the Pike's mouth, it was naturally supposed that the conditioning process in the Pike had something to do with the presence of the spines." Clearly the behaviour is a critical part of the stickleback's defence, and without it the "weapon" is almost useless (Fig. 1). "The series with the de-spined sticklebacks show that it is actually the spines that afford this protection. The sticklebacks' behaviour, which makes them raise their spines exactly when needed, and the structure of the spines and their erection mechanisms... are both beautifully adapted to this function." In fact, Hoogland et al. (1957) noted that "the overwhelming majority of sticklebacks survived after having been snapped up, and many of them survived even after very severe mauling."

Hoogland et al. (1957) observed that the spines of the threespine stickleback are especially effective against predators, and threespine stickleback are much less timid than ninespine (ten-spined) stickleback, which have smaller spines. This may be a key feature that led to the development of the threespine stickleback as a model species, as "it is partly due to *Gasterosteus*' boldness that it is so well suited for laboratory studies of its behaviour... It would seem as if *Gasterosteus* can afford to be 'tame' (and thus waste less time by fleeing) because of the protection offered by its spines." These differences also account for the outcome of interspecific interactions between sympatric stickleback species and dominance relations between them. The morphology and behaviour interact to determine landscape distribution patterns of stickleback species.

Another density-dependent mortality factor profoundly impacts individual stickleback and the populations to which they belong — parasitism, which affects competitive interactions and vulnerability to predators. While robust spines may protect a stickleback from a small pike, they do not provide protection from parasites. Sticklebacks are host to a large diversity of parasites, but the one that has received the most attention is the remarkable tapeworm *Schistocephalus solidus*. *Schistocephalus* manages to manipulate the behaviour and appearance of its stickleback host when it has reached the ideal transmission size in order to increase the chances that the stickleback will be consumed by avian predators, the next host in the life cycle of the

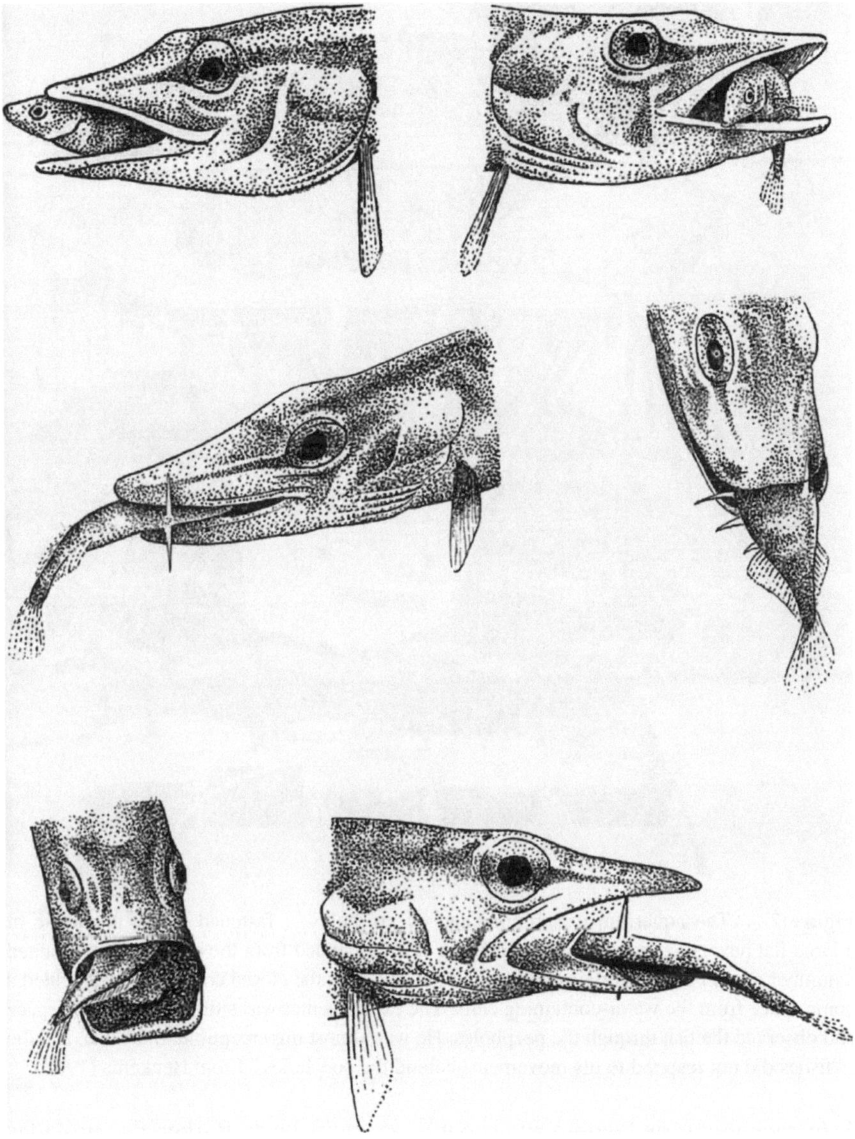

Figure 1. "Drawings taken from 16 mm film showing the variety of positions of *Gasterosteus* in the mouth of the Pike." From Hoogland et al. (1957).

worm (Barber et al., 2004). In the second paper of this section, Iain Barber and Felicity Huntingford (1995) reviewed experiments on the effects of

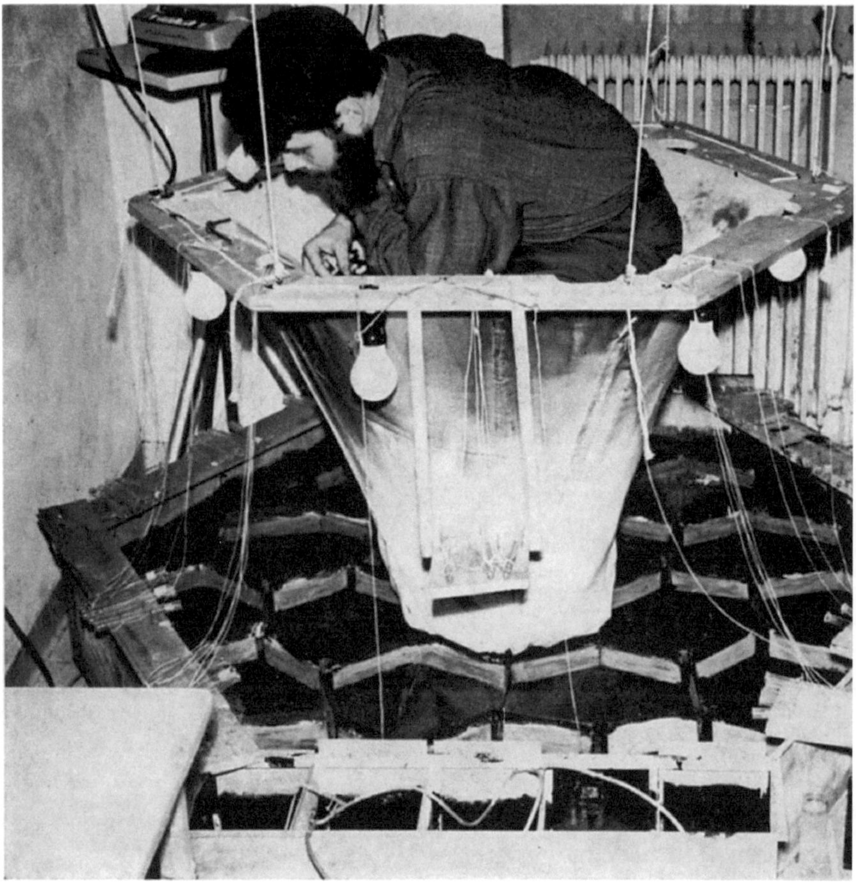

Figure 2. "The aquarium was illuminated by six bulbs... fastened to the underside of a large flat hexagonal wooden ring. This ring was suspended from the ceiling and contained a number of peepholes. Curtains connecting the ring with the closed compartment shielded a conic space from the water-containing cells. The experimenter was sitting in this conic space and observed the fish through the peepholes. He was almost imperceptible to the fish, and the fishes did not respond to his movements behind the peepholes." From Beukema (1968).

Schistocephalus on feeding and shoaling in stickleback. Barber also provides a retrospective of this work here.

The effects of hunger on stickleback feeding behaviour have been well studied (e.g., see Beukema, 1968 for an analysis of hunger and experience on stickleback feeding behaviour, and for a unique experimental maze and observational blind; Fig. 2). As Barber and Huntingford (1995) showed, *Schistocephalus* both constricts the stickleback's stomach and competes with it for nutrients. This has profound consequences for the stickleback's behaviour, including dietary preference and shoaling decisions. The risks that the stickleback takes to obtain food in its nutritionally-deprived state, and the loss of the protection of the shoal, increase the risk of predation. That risk was already elevated due to manipulation by the worm of the stickleback's behaviour and appearance, thereby leading to increased transmission to piscivorous birds. Due to the complexities and experimental tractability of the stickleback-*Schistocephalus* system, it has developed into a fantastic model in parasitology.

Predation by piscivorous birds, fish and insects impose different selection pressures, which may account for population variability in defensive traits. In the third paper of this section, Reimchen (1995 excerpts; accompanied by a retrospective) explored the functional context of intrapopulation variability of lateral plates and found that trout predation selected for an increase in plate number while bird predation selected for a decrease. Because the importance of various predators changes seasonally (and dramatically), variability in lateral plates and other traits within a population is maintained. Stickleback with different lateral plate phenotypes may differ in their behaviour (habitat selection, response to predators, etc.) so the effects of predators may be indirectly mediated by behaviour and/or a direct consequence of selection by predators of particular phenotypes (Reimchen, 1995). Regardless, Reimchen argued that the stickleback model supports the view that phenotypic variation both within and between populations is due primarily to selection, rather than "a non-adaptive product of history."

Not all lateral plates are equal. In the final paper of this section, Reimchen (2000) demonstrated that the posterior plates provide protection from being swallowed by a fish predator by disrupting the predator's jaw movements. This then increases the probability that the stickleback will escape after being captured. Interestingly, Reimchen's (2000) results also suggest that some of the variance in escape rate found by Hoogland et al. (1957) was due to protection afforded to the threespine stickleback by its lateral plates, rather than only due to the spines.

Geographic variation of stickleback armour traits has provided some of the most compelling evidence of fine-scaled adaptation to environmental pressures. Sticklebacks readily and rapidly adapt to changes in their environment, making them an excellent model system for studying the evolutionary consequences of a changing planet. For example, the interacting forces of climate change, invasive species, pollution and habitat degradation can be explored by measuring evolutionary responses of disparate stickleback populations. The relative impact of the components of global change vary over geography, and so perhaps the stickleback, because of its adaptability and widespread occurrence, will become a key model in research efforts to understand and mitigate the diverse forces of environmental degradation. The stickleback is already an important model organism for studies of one form of environmental degradation, that of contaminants (see section Physiology & Behaviour). Following decades of work on their body armour and defensive behaviour, and on parasites and host manipulations, the threespine stickleback is now poised for a career testing the evolutionary outcome of human-induced global change.

References

van den Assem, J. (1967). Territory in the three-spined stickleback, *Gasterosteus aculeatus* L. An experimental study in intra-specific competition. — Behaviour, Supplement XVI: 1-164.

Barber, I. & Huntingford, F.A. (1995). The effect of *Schistocephalus solidus* (Cestoda: Pseudophyllidea) on the foraging and shoaling behaviour of three-spined sticklebacks, *Gasterosteus aculeatus*. — Behaviour 132: 1223-1240.

Barber, I., Walker, P. & Svensson, P.A. (2004). Behavioural responses to simulated avian predation in female three-spined sticklebacks: the effect of experimental *Schistocephalus solidus* infections. — Behaviour 141: 1425-1440.

Beukema, J.J. (1968). Predation by the three-spined stickleback (*Gasterosteus aculeatus* L.): the influence of hunger and experience. — Behaviour 31: 1-125.

Hoogland, R., Morris, D. & Tinbergen, N. (1957). The spines of sticklebacks (*Gasterosteus* and *Pygosteus*) as means of defence against predators (*Perca* and *Esox*). — Behaviour 10: 205-236.

Houghton, W. (1879). British fresh-water fishes. — W. Mackenzie, London.

van Iersel, J.J.A. (1953). An analysis of the parental behavior of the male three-spined stickleback (*Gasterosteus aculeatus* L.). — Behaviour, Supplement III: 1-159.

Kruuk, H. (2003). Niko's nature; a life of Niko Tinbergen and his science of animal behaviour. — Oxford University Press, Oxford.

Morris, D. (1958). The reproductive behaviour of the ten-spined stickleback (*Pygosteus pungitius* L.). — Behaviour, Supplement VI: 1-154.

Reimchen, T.E. (1995). Predator-induced cyclical changes in lateral plate frequencies of *Gasterosteus*. — Behaviour 132: 1079-1094.

Reimchen, T.E. (2000). Predator handling failures of lateral plate morphs in *Gasterosteus aculeatus*: functional implications for the ancestral plate condition. — Behaviour 137: 1081-1096.

THE SPINES OF STICKLEBACKS (*GASTEROSTEUS* AND *PYGOSTEUS*) AS MEANS OF DEFENCE AGAINST PREDATORS (*PERCA* AND *ESOX*)

by

R. HOOGLAND, D. MORRIS and N. TINBERGEN

(Department of Zoology, University of Leiden, Department of Zoology and Comparative Anatomy, University of Oxford and Department of Zoology and Comparative Anatomy, University of Oxford)

(With 15 Figs.)

(Rec. 27-I-1956)

1. INTRODUCTION

Many species of fish have evolved sharp spines which are often alleged to serve as defensive weapons against predators. Most frequently these spines are modified fin-rays. The sticklebacks possess this type, and it was the purpose of this study to analyse the effect of these weapons upon two predators, the Perch (*Perca fluviatilis* L.) and the Pike (*Esox lucius* L.). The species of stickleback studied were the Three-spined Stickleback (*Gasterosteus aculeatus* L.) and the Ten-spined Stickleback (*Pygosteus pungitius* L.), the only two members of the family which occur in European freshwaters. The reactions of the predators to these fish were compared with those to various non-spined freshwater fish, viz., the Minnow (*Phoxinus phoxinus* (L.)), the Roach (*Rutilus rutilus* (L.)), the Rudd (*Scardinius*

 Tinbergen's Legacy in Behaviour, 353-359

erythrophthalmus (L.)), and the Crucian Carp (*Carassius carassius* (L.)) [1].

These two predators and five prey species are all known to occur together in the wild, and HARTLEY (1947) has found sticklebacks, Minnows and Roach amongst the stomach contents of the predators.

Some preliminary tests by TINBERGEN in 1939 appeared to show that young Pike of approximately 15 cm long had considerable difficulty in swallowing adult Three-spined Sticklebacks, and also that they soon learned to leave sticklebacks alone, while continuing to eat non-spined fish of the same and even considerably larger size. Since sticklebacks, when seized by a Pike, respond by raising all their spines, and keeping them rigidly erected as long as they are in the Pike's mouth, it was naturally supposed that the conditioning process in the Pike had something to do with the presence of the spines. This problem was tackled experimentally by HOOGLAND in 1948. Later, in 1952, MORRIS, engaged in a study of the behaviour of *Pygosteus*, extended HOOGLAND's work studying Perch as well as Pike, and *Pygosteus* as well as *Gasterosteus*. The observations done so far throw light not only on the function of the sticklebacks' spines but enabled us at the same time to analyse the predator's behaviour to a certain extent. Also, striking differences were found between the two species of sticklebacks, and these could be correlated with differences in their reproductive behaviour and other aspects of their ecology generally.

A film of the predatory behaviour of Pike towards sticklebacks and non-spined fish was taken by TINBERGEN, and copies are in the Department of Zoology at Leiden and in the Department of Zoology and Comparative Anatomy at Oxford.

2. MATERIAL AND METHODS

Five Pike and several hundreds of *Gasterosteus*, Rudd, Roach and Crucian Carp were collected near Leiden in Holland; three Pike and five Perch, as well as all sticklebacks and Minnows used in MORRIS' experiments were collected from ponds and streams in Wiltshire and Oxfordshire in England. The prey specimens offered were adult sticklebacks of both species and non-spined fish of approximately the same size as adult Three-spined Sticklebacks, or slightly larger (Fig. 1). The Pike and Perch varied from 10 to 25 cm; naturally, their size increased during the course of the tests.

---

1) Apart from Roach and Rudd we may occasionally have used other, similar species. Since we used young fish of approximately stickleback size which are not easily determined, and since the responses of the predators to all these non-spined fish were always the same, we did not attempt to determine the exact species. The majority of these fish were taken from waters in which larger Roach and Rudd were abundant.

when the stickleback was moving within view without even evoking a response of level 1) were not noted, and in fact it would be difficult to design a scoring system for them.

The occasional snapping responses to the stickleback, as recorded on days 2, 6, 7, 10, 13, 15, 17, 19, 29, 30, 31, and 32 were for the greater part due to a disturbing factor which we did not try to eliminate: the Pike became positively conditioned to the experimenter's movements when introducing a fish into the tank, and sometimes shot forward at the prey immediately on its introduction. The "fives" scored on these 12 days were all of this type (open triangles) except 2 (those on day 10 and day 32), which are indicated by solid triangles.

8. THE RESPONSES OF PERCH AND PIKE TO DE-SPINED STICKLEBACKS

In a number of tests, sticklebacks were offered to Perch an Pike after their spines had been cut off. In order not to injure the sticklebacks, the bases of the spines were left intact, and the result was that „de-spined" fish were not as smooth as the non-spined fish used; yet the main obstacles preventing the predators from swallowing sticklebacks were gone, and the remaining stumps were less sharp than the intact spines. In short series with Perch and Pike, involving about 150 *Gasterosteus,* such de-spined fish were almost invariably eaten. The following figures apply to two series done with Pike A and C and de-spined *Gasterosteus.* Of 131 L. offered, 121 were eaten and 5 refused. Of 63 de-spined *Gasterosteus* offered, 55 were eaten and 8 refused. The slightly larger proportion of *Gasterosteus* refused may be partly due to their being still coarser than the non-spined species, partly to the fact that the sticklebacks were always offered after the L. Even with this handicap, it is obvious from a comparison of these series with the foregoing ones that the protection enjoyed by sticklebacks is mainly due to their spines.

9. DISCUSSION

It seems clear that both *Gasterosteus* and *Pygosteus* are, to Perch and Pike of the sizes used in these tests (10-25 cm) less attractive prey than four non-spined species of fish, and also, that the larger-spined *Gasterosteus* is much less vulnerable to such Perch and Pike than *Pygosteus*; the difference between *Pygosteus* and non-spined fish is however still demonstrable.

The series with the de-spined sticklebacks show that it is actually the spines that afford this protection. The sticklebacks' behaviour, which makes them raise their spines exactly when needed, and the structure of the spines and their erection mechanisms (studied by HOOGLAND, 1951) are both beautifully adapted to this function.

The rejection of sticklebacks by these predators is really based on two distinct processes. Many of the rejections were due to a non-conditioned response to the strong mechanical stimuli received after the predator had seized the stickleback. This response alone accounts for a sufficient number of rejections to prove the survival value of the spines, particularly since the overwhelming majority of sticklebacks survived after having been snapped up, and many of them survived even after very severe mauling. In addition to this, however, both Perch and Pike learn to respond to visual stimuli by avoiding sticklebacks before they have even touched them. All records scored as "Avoidance" in Figs. 6, 13, and 14 refer to this conditioned response.

The break in the reaction chain may occur either at the end of stalking, or already much earlier. Of the total of 162 observations recorded in series E (Fig. 15) in which the chain was interrupted before contact was made, 14 were broken off after the leap, 18 after stalking, 81 after turning the body towards the stickleback, and 49 already after the Pike had merely turned its eyes towards the stickleback. It would certainly be interesting to know which the visual stimulus situation is to which the Pike's avoidance becomes conditioned. From some preliminary tests it seems that the main characteristic is the way a stickleback moves; which of course is quite characteristic.

An ecological interpretation of our findings meets with some unexpected difficulties. We saw already that a starved Pike, instead of learning to avoid sticklebacks, learns to eat them. We have no indications however that such a Pike would actually prefer sticklebacks to non-spined fish, and in fact this seems unlikely.

A more serious problem is whether Pike and Perch of larger size would avoid sticklebacks. Our tests, all done with predators less than 25 cm long, naturally cannot answer this. However FROST (1954) has made a thorough study of the stomach contents of Pike of various size caught in Windermere, and her results seem in some respects contradictory to ours. FROST reports that Minnows are much more abundant in the lake than *Gasterosteus*; that Pike up to about 40 cm eat more Minnows than *Gasterosteus*; and that larger Pike eate more *Gasterosteus* than Minnows. The percentage of Minnows declines with increasing size of the pike; this may be explained by assuming that the optimum size of prey grows with increasing length of the Pike. The percentage of *Gasterosteus* increases with increasing size of the Pike up to the size-class 50-59 cm; this would mean that as the Pike grows the spines bother it less. After that there is a drop, which again may have to do with the shift in optimum prey-size. The puzzling feature is however that the percentage of *Gasterosteus* actually mounts above that of the Minnows [1]).

The only conclusion we can draw is that in Windermere *Gasterosteus* is on the average less well protected from Pike than the Minnow. In this connection it may be worth pointing out that, although we used Minnows as one of four non-spined fish, our conclusion is not that stickebacks are generally better protected than Minnows, but that they are better protected than de-spined sticklebacks. The Minnow may have anti-predator adaptations which work even better than spines, at least against large Pike.

In this context it may also be worth speculating whether the conditions under which *Gasterosteus* live in Windermere are typical of the species. Since FROST reports that "sticklebacks, although at no time of great importance as food, are eaten mainly in May and June, at which time they congregate for breeding" (p. 351), her findings may mean that the stickleback's habit of migrating to shallow waters to breed has evolved as an adaptation which enables it to get away from large predators. Perhaps the Windermere population has too few shallow streams at its disposal and thus is subject to more severe predation in spring than other populations.

Whatever the solution may be, we believe that we are entitled to conclude that 1) sticklebacks derive a certain protection from their spines, and 2) *Gasterosteus* is much better protected than *Pygosteus*. The second point is worth elaborating. MORRIS (1954) has shown that there are considerable differences in behaviour between the two species, and we suggest that our present study enables us to relate some of these differences with those in the effectiveness of the spines as anti-predator devices. *Pygosteus* is much more timid than *Gasterosteus*, and it is partly due to *Gasterosteus*' boldness that it is so well suited for laboratory studies of its behaviour. Any *Gasterosteus* brought into the laboratory in spring can be observed without special precautions within a day, even in tanks with very little cover. *Pygosteus* requires much cover, and even so remains shy for weeks on end, and often retires to the least accessible corners for nest building, so that mirrors, observation screens, *etc.*, have often to be used in order to see its complete behaviour. It would seem as if *Gasterosteus* can afford to be

1) We must call attention to a misprint in FROST's paper which seems to contradict this statement. On p. 357 she mentions that the average numbers eaten are 0.16 for sticklebacks, 0.88 for Minnows. From her table III to which this statement refers it is clear that the figure for the Minnows is actually 0.08.

"tame" (and thus waste less time by fleeing) because of the protection offered by its spines.

Another difference between the two is in the type of nesting habitat they select. *Gasterosteus,* while preferring some vegetation, chooses a relatively open type of habitat; and often a few vertical weeds (offered in or outside the tank), or even crude visual imitation of weeds will induce it to settle and build. *Pygosteus* builds in dense weeds, and will only accept an open habitat after long habituation to the observer.

The difference in behaviour of the females is equally striking. MORRIS (*loc. cit.*) has shown that female *Pygosteus* do not school in spring, but settle on territories of their own round those of the males, where they stay in cover most of the time. None of us has ever seen large schools of female *Pygosteus* roaming about in the spawning season, whereas the females of *Gasterosteus* are always easily seen, travelling about in schools up to 59 individuals strong, out of which the males pick those that respond to their courtship.

Finally, it seems not impossible that the difference in the nuptial colours of the males has something to do with the different need for procryptic coloration. Both species are usually camouflaged, showing general colour resemblance with the background, countershading, and disruptive cross-barring on the back. In the breeding season, *Gasterosteus* males turn a vivid red underneath; *Pygosteus* males become jet black. There seems little doubt that the red of *Gasterosteus* makes it much more conspicuous than the black *Pygosteus* males ever are. Moreover, during the actual sexual phase of the cycle, the back of *Gasterosteus* becomes a bluish white, which reverses the usual countershading, and makes the fish very conspicuous indeed, both when seen from above and when seen from the side. *Pygosteus* males also reverse their countershading, but the effect is far less spectacular.

All this leads us to conclude that *Pygosteus* has so to speak compensated for its inferiority as regards spine effectiveness by means of other anti-predator devices such as better procryptic coloration, greater timidity, and a stronger tendency to keep in cover.

A number of authors have argued that *Gasterosteus* must have evolved from a *Pygosteus*-like ancestor. BERTIN (1925) gives anatomical evidence in support of this. A comparative study of nest building behaviour tends to support this conclusion, for *Gasterosteus* is the only species of stickleback that does not build its nest suspended in water weeds. In view of the evidence given in this paper it seems reasonable therefore to consider *Gasterosteus* as a species which has specialised in reduction of number and increase of size of spines, which has enabled it to reduce the time to be spent in fleeing and hiding in cover. This has resulted in various peculiarities of its reproductive behaviour, such as the selection of more open habitats, and the schooling and wandering behaviour of the females.

We believe that this is an example of the way in which diverse and seemingly independent systems such as anti-predator defence (involving morphological and physiological-cum-ethological components) and the reproductive system are in fact interrelated, and do not evolve quite independently. The two species, derived from common stock, have specialised in different types of anti-predator defence, and this has had repercussions in a wide field of other characteristics.

9. SUMMARY

Feeding experiments with Pike (*Esox lucius* L.) and Perch (*Perca fluviatilis* L.) as predators, and Three-spined Sticklebacks (*Gasterosteus aculeatus* L.), Ten-spined Sticklebacks (*Pygosteus pungitius* L.), Minnows (*Phoxinus phoxinus* (L.)), Roach (*Rutilus rutilus* (L.)), Rudd (*Scardinius erythrophthalmus* (L.)) and Crucian Carp (*Carassius carassius* (L.)) as prey species show that both species of stickleback enjoy a demonstrable degree of protection from these predators. This protection is much better for *Gasterosteus* than for *Pygosteus*; tests with de-spined sticklebacks show that it is mainly due to the spines.

A description of the predatory behaviour of Perch and Pike is given; analysis of the experiments shows that (1) sticklebacks are rejected when, after being snapped up, their spines hurt the predator's mouth; (2) after very few experiences both Perch and Pike become negatively conditioned to the sight of sticklebacks and avoid them before they have made contact. As a result of this conditioning, the last links of the predators' feeding chain drop out; in Pike, mere fixation with two eyes was often sufficient to recognise the prey as unwanted.

The possession of few and large spines in *Gasterosteus* must be regarded as more specialised than that of many small spines as found in various other species of stickleback. There is a correlation between the extreme development of this anti-predator device in *Gasterosteus* and (1) its boldness, (2) its tendency to select a more open nesting habitat, (3) the schooling and wandering tendencies of the females in the spawning season, and (4) its relatively conspicuous nuptial colours; these correlations suggest an interrelationship between anti-predator devices and reproductive behaviour and structures; the evolution of these two systems must have occurred in conjunction with each other.

LITERATURE

BERTIN, L. (1925). Recherches bionomiques, biométriques et systématiques sur les Épinoches (Gastérostéidés). — Ann. Inst. Ocean. Monaco. 2, p. 1-204.

BREDER, C. M. (1926). The locomotion of fishes. — Zoologica (N.Y.) 4, p. 159-297.

DEELDER, C. L. (1951). A contribution to the knowledge of the stunted growth of Perch (*Perca fluviatilis*) in Holland. — Hydrobiol. 3, p. 357-378.

FRISCH, K. v. (1941). Über einen Schreckstoff der Fischhaut und seine biologische Bedeutung. — Z. vergl. Physiol. 29, p. 46-145.

FROST, W. E. (1954). The food of the Pike, *Esox lucius* L., in Windermere. — J. Anim. Ecol. 23, p. 339-360.

Retrospective: The effect of *Schistocephalus solidus* (Cestoda: Pseudophyllidea) on the foraging and shoaling behaviour of three-spined sticklebacks, *Gasterosteus aculeatus*

Article published in *Behaviour* volume 132 (1995), p. 1223–1240

Iain Barber
(Department of Biology, University of Leicester, UK)

Plerocercoid larvae of the bird tapeworm *Schistocephalus solidus* are common parasites of three-spined sticklebacks in many still water populations. Stickleback serve as the second intermediate host of the parasite, becoming infected after ingesting parasitized copepods. The parasites utilise host-derived energy to grow to a spectacularly large size in the body cavity of host sticklebacks, grossly swelling the host fish's abdomen as they do so (Fig. 1). The nutritional demand imposed by growing parasites typically has devastating consequences for the energetic status of stickleback hosts, and infections are typically associated with dramatically reduced gonad development in females and reduced body condition in general (Arme & Owen, 1967; Pennycuick, 1971; McPhail & Peacock, 1983; Tierney et al., 1996), though Heins & Baker (2008) recently reviewed cases where infection is associated with reduced impact.

The energetic costs of infection also have a number of interesting consequences for host behaviour. Elegant experiments by Manfred Milinski and Nick Giles demonstrated that sticklebacks infected with *Schistocephalus* plerocercoids were more likely to engage in risky foraging opportunities, including feeding in the presence of potential predators (Giles, 1983, 1987; Milinski, 1984, 1985). The fact that the fish's normal anti-predator behaviour could be affected by a nutritionally demanding parasite such as *Schistocephalus* is particularly interesting because the parasite can only attain sexual maturity in the intestine of a predatory bird. Hence, the parasite's reproductive fitness depends absolutely on the fate of its stickleback host, inviting the

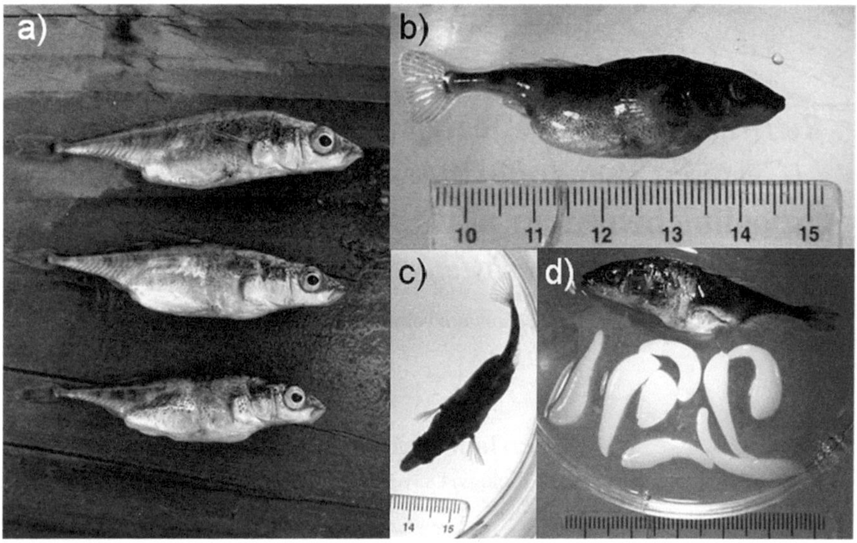

Figure 1. *Schistocephalus solidus* infected three-spined stickleback. (a) Three *Schistocephalus* infected stickleback from Walby Lake, Alaska, showing a range of infection-associated distension of the body cavity. (b)–(d) A heavily infected male stickleback from a park pond in Leicestershire, UK, showing (b) lateral and (c) dorsal profiles; (d) the same fish post-dissection, showing eight plerocercoids recovered from the body cavity.

question as to whether the 'manipulation' of normal host behaviours may have arisen as a parasite adaptation (Poulin, 1994).

Schistocephalus infections also have the potential to affect other antipredator behaviours, including the formation of shoals (loose aggregations), which is one of the primary antipredator behaviours of small fish (Pitcher & Parrish, 1993). Shoaling behaviour provides antipredator benefits through 'safety in numbers'-type mechanisms ('many eyes,' 'dilution effect,' 'attack abatement,' etc.); however, being a member of a shoal also incurs costs, including increased competition for food (Krause & Ruxton, 2002). Hence, the extent to which individuals engage in shoaling behaviour should reflect the outcome of a cost-benefit trade-off.

Schistocephalus infection has many effects on host fish that potentially alter the costs and benefits of shoal membership. Previous work showed that *Schistocephalus*-infected stickleback handled prey less efficiently than non-infected conspecifics (Cunningham et al., 1994), suggesting they may be poor competitors for food, and hence suffer increased costs of shoal membership. Furthermore, infected fish have altered physical appearance, potentially reducing the benefits of shoaling, such as predator confusion, that rely on phenotypic homogeneity among group members.

In a previous study I demonstrated that parasitized individuals spent more time away from a stimulus shoal (composed of non-infected conspecifics) and were more likely to engage in 'risky' lone foraging in a maze (Barber et al., 1995). Subsequently, I attempted to explain these results by adopting a cost/benefit approach to the question of shoaling (Barber & Huntingford, 1995). The aims of the paper were to review a number of contemporary studies that had examined the foraging behaviour of *Schistocephalus*-infected stickleback, synthesise their results (and attempt to reconcile some apparent discrepancies), and develop a simple model that could explain the observed deviant shoaling behaviour of *Schistocephalus*-parasitized stickleback.

One of the reasons for the popularity of the stickleback-*Schistocephalus* system as a model for testing the behavioural effects of parasites stems from the fact that infected fish are visually distinguishable, making it possible to compare the behaviour of naturally-infected fish with those of non-infected fish from a single population. This was the approach taken in all of the studies reviewed in our paper (Barber & Huntingford, 1995). However, though the use of naturally infected fish is highly convenient, it is also problematic, as it is impossible to rule out other explanations for the phenotypic correlation between behaviour and infection status. Other explanations include the possibility that pre-existing behavioural variation could generate differential exposure to parasites; for example, fish differing in competitive ability might experience different levels of exposure to parasites if low quality food is more likely to be infected. Alternatively, underlying genetic variation could, equally plausibly, influence both the parasite resistance and behaviour of fish. The use of wild-caught, naturally-infected fish also raises the potential problem of unquantified co-infections. Consequently, although it was clear that parasitized fish behaved very differently from non-parasitized conspecifics in the studies reviewed in the paper, supporting the hypothesis that *Schistocephalus* altered host behaviour, further insight could only be gained by undertaking experimental infection studies.

Fortunately, stickleback are bred and reared relatively easily in the laboratory, and even more fortuitously, pioneering work by the parasite physiologist J.D. Smyth in the 1950s had developed protocols for the *in vitro* culture of adult helminths, including *Schistocephalus* (summarized in Smyth, 1990). It is therefore possible to generate infective stages of *Schistocephalus* and feed them to lab-bred stickleback (via experimentally infected copepods), and hence compare the behaviour of fish that had been experimentally infected with the parasite with those that had been sham-exposed. Furthermore, because the size of the parasite can be measured non-invasively using photographic image analysis techniques (Barber, 1997), it is possible to link host behaviour changes to parasite mass, and hence to the parasite's infectivity to the definitive host (Tierney & Crompton, 1992).

In Barber et al. (2004) we adopted this approach to track changes in the antipredator behaviour of individual experimentally-parasitized and non-parasitized stickleback, and link changes in the behaviour of infected fish to parasite mass. Our results showed that reduced antipredator behaviour was only observed in experimentally infected fish once their plerocercoids had reached a mass (50 mg) that had previously been shown to be infective to the definitive host. This provided the first experimental demonstration that *Schistocephalus* was responsible for observed changes in behaviour, and the results corroborated observational studies of the behaviour of naturally infected stickleback (Giles, 1983; Tierney et al., 1993; Ness & Foster, 1999).

Further studies are needed to investigate the physiological mechanisms that underpin behavioural change in this system. For example, Øverli et al. (2001) found elevated concentrations of the serotonin metabolite, 5-hydroxyindoleacetic acid (5-HIAA) in the brainstem of naturally-infected stickleback when compared with non-infected fish. As a consequence, the ratio of 5-HIAA:serotonin, an indicator of stress in fish, was elevated in both the hypothalamus and brainstem regions of infected stickleback. This may indicate chronic stress among infected fish; however, it is also possible that host behaviour change is mediated through neurochemical manipulation, as in other systems (Thomas et al., 2005). Experimental infection studies that link monoamine levels to host behaviour in the stickleback-*Schistocephalus* system are now required.

The stickleback-*Schistocephalus* experimental system also permits detailed investigations into the physiological impacts of infection. In Barber et al. (2008) we reviewed the results of a number of studies, undertaken primarily by our group, that examined the impact of infection on host growth, energetics and development. These studies were undertaken under a range of social situations and ration levels, giving us the opportunity to examine how rearing conditions influenced the parasite's impact on the host. Our analysis suggested that infection phenotypes that most closely resembled those found in the majority of natural populations were most likely to arise when hosts were reared under competition for limited food resources, or were fed a restricted, or temporally variable, diet. In contrast, when access to food was unrestricted, infected fish were able to sustain high growth rates and lay down energy reserves. These results supported earlier work by Candolin & Voigt (2001), who showed that the impacts of *Schistocephalus* infection on male reproduction in 'benign' laboratory conditions overestimated the performance of infected males in the field, and emphasised the need for further studies examining how infection status and environmental factors interact to determine the fitness of parasitized individuals in natural populations.

In summary, the experimental stickleback-*Schistocephalus* system provides a remarkably amenable model for examining a highly diverse range of research questions in behaviour, parasitology and evolutionary ecology in both the laboratory and field. The recent sequencing of the stickleback genome only increases the utility of the system, allowing the detailed study of molecular genetic interactions between hosts and parasites.

References

Arme, C. & Owen, R.W. (1967). Infections of three-spined stickleback *Gasterosteus aculeatus* L. with plerocercoid larvae of *Schistocephalus solidus* (Müller 1776) with special reference to pathological effects. — Parasitology 57: 301-304.

Barber, I. (1997). A non-invasive morphometric technique for estimating cestode plerocercoid burden in small freshwater fish. — J. Fish Biol. 51: 654-658.

Barber, I. & Huntingford, F.A. (1995). The effect of *Schistocephalus solidus* (Cestoda: Pseudophyllidea) on the foraging and shoaling behaviour of three-spined sticklebacks, *Gasterosteus aculeatus*. — Behaviour 132: 1223-1240.

Barber, I., Huntingford, F.A. & Crompton, D.W.T. (1995). The effect of hunger and cestode parasitism on the shoaling decisions of small freshwater fish. — J. Fish Biol. 47: 524-536.

Barber, I., Walker, P. & Svensson, P.A. (2004). Behavioural responses to simulated avian predation in female three-spined sticklebacks: the effect of experimental *Schistocephalus solidus* infections. — Behaviour 141: 1425-1440.

Barber, I., Wright, H.A., Arnott, S.A. & Wootton, R.J. (2008). Growth and energetics in the stickleback-*Schistocephalus* host-parasite system: a review of experimental infection studies. — Behaviour 145: 647-668.

Candolin, U. & Voigt, H.R. (2001). No effect of a parasite on reproduction in stickleback males: a laboratory artefact? — Parasitology 122: 457-464.

Cunningham, E.J., Tierney, J.F. & Huntingford, F.A. (1994). Effects of the cestode *Schistocephalus solidus* on food intake and foraging decisions in the three-spined stickleback *Gasterosteus aculeatus*. — Ethology 9: 65-75.

Giles, N. (1983). Behavioral effects of the parasite *Schistocephalus solidus* (Cestoda) on an intermediate host, the three-spined stickleback, *Gasterosteus aculeatus* L. — Anim. Behav. 31: 1192-1194.
Giles, N. (1987). Predation risk and reduced foraging activity in fish — experiments with parasitized and non-parasitized three-spined sticklebacks, *Gasterosteus aculeatus* L. — J. Fish Biol. 31: 37-44.
Heins, D.C. & Baker, J.A. (2008). The stickleback-*Schistocephalus* host-parasite system as a model for understanding the effect of a macroparasite on host reproduction. — Behaviour 145: 625-645.
Krause, J. & Ruxton, G.D. (2002). Living in groups. — Oxford University Press, Oxford.
McPhail, J.D. & Peacock, S.D. (1983). Some effects of the cestode (*Schistocephalus solidus*) on reproduction in the threespine stickleback (*Gasterosteus aculeatus*): evolutionary aspects of a host-parasite interaction. — Can. J. Zool. 61: 901-908.
Milinski, M. (1984). Parasites determine a predator's optimal feeding strategy. — Behav. Ecol. Sociobiol. 15: 35-37.
Milinski, M. (1985). Risk of predation of parasitized sticklebacks (*Gasterosteus aculeatus* L.) under competition for food. — Behaviour 93: 203-215.
Ness, J.H. & Foster, S.A. (1999). Parasite-associated phenotype modifications in threespine stickleback. — Oikos 85: 127-134.
Øverli, Ø., Páll, M., Borg, B., Jobling, M. & Winberg, S. (2001). Effects of *Schistocephalus solidus* infection on brain monoaminergic activity in female three-spined sticklebacks *Gasterosteus aculeatus*. — Proc. Roy. Soc. B Biol. Sci. 268: 1411-1415.
Pennycuick, L. (1971). Quantitative effects of three species of parasites on a population of three-spined sticklebacks, *Gasterosteus aculeatus*. — J. Zool. 165: 143-162.
Pitcher, T.J. & Parrish, J.K. (1993). Functions of shoaling behaviour in teleosts. — In: Behaviour of teleost fishes (Pitcher, T.J., ed.). Chapman & Hall, London, p. 363-439.
Poulin, R. (1994). The evolution of parasite manipulation of host behavior — a theoretical analysis. — Parasitology 109: S109-S118.
Smyth, J.D. (1990). *In vitro* cultivation of parasitic helminths. — CRC Press, Boca Raton, FL.
Thomas, F., Adamo, S. & Moore, J. (2005). Parasitic manipulation: where are we and where should we go? — Behav. Proc. 68: 185-199.
Tierney, J.F. & Crompton, D.W.T. (1992). Infectivity of plerocercoids of *Schistocephalus solidus* (Cestoda, Ligulidae) and fecundity of the adults in an experimental definitive host, *Gallus gallus*. — J. Parasitol. 78: 1049-1054.
Tierney, J.F., Huntingford, F.A. & Crompton, D.W.T. (1993). The relationship between infectivity of *Schistocephalus solidus* (Cestoda) and antipredator behavior of its intermediate host, the three-spined stickleback, *Gasterosteus aculeatus*. — Anim. Behav. 46: 603-605.
Tierney, J.F., Huntingford, F.A. & Crompton, D.W.T. (1996). Body condition and reproductive status in sticklebacks exposed to a single wave of *Schistocephalus solidus* infection. — J. Fish Biol. 49: 483-493.

THE EFFECT OF SCHISTOCEPHALUS SOLIDUS (CESTODA: PSEUDOPHYLLIDEA) ON THE FORAGING AND SHOALING BEHAVIOUR OF THREE-SPINED STICKLEBACKS, GASTEROSTEUS ACULEATUS

by

IAIN BARBER[1]) and FELICITY A. HUNTINGFORD[2])

(Fish Behaviour and Ecology Group, Division of Environmental and Evolutionary Biology, University of Glasgow, Glasgow G12 8QQ, Scotland, U.K.)

(With 6 Figures)
(Acc. 24-V-1995)

Summary

In this paper we review recent experimental work on the effects of the parasite *Schistocephalus solidus* (Cestoda: Pseudophyllidea) on the feeding behaviour of three-spined sticklebacks (*Gasterosteus aculeatus* L.). We also discuss how increased feeding motivation and subsequent altered foraging behaviour may be a mechanism for parasite-associated changes in the shoaling behaviour of infected sticklebacks. The presence of *S. solidus* plerocercoids in the body cavity constricts the stomach, increases the handling time for large prey and consequently reduces the profitability of such prey for infected fish. This is reflected in a switch in dietary preference from large to small prey in the laboratory and in altered stomach contents and impaired nutrient reserves in the wild. By altering their hosts' nutritional state by direct competition for nutrients from digested food (and possibly indirectly by altering diet and reducing competitive ability) and also by altering the fishes' appearance, such parasites have the potential to alter the costs and benefits involved in joining a shoal of conspecifics. Experimental work on the shoaling decisions of *S. solidus*-infected sticklebacks supports this hypothesis, and such behavioural modification is discussed in the context of the manipulation hypothesis of parasite transmission.

Introduction: biology of the parasite

Plerocercoids of *Schistocephalus solidus* MÜLLER 1776 (Cestoda: Pseudophyllidea) are frequent parasites of the three-spined stickleback (*Gasterosteus aculeatus*) throughout its range in the Northern hemisphere. The

[1]) Author to whom correspondence should be addressed; e-mail: I.Barber@udcf.gla.ac.uk
[2]) We are grateful to Theo BAKKER, Claus WEDEKIND and Piet SEVENSTER for critical comments on an earlier version of this manuscript, and to all who attended the 2nd International Symposium on Stickleback behaviour held in Leiden, 1994, for stimulating ideas and conversation. IB is supported by a U.K. N.E.R.C. studentship.

stickleback is the second intermediate host in the parasite's life cycle; fish become infected after eating infected copepods (the first intermediate host) and the parasite normally only reaches sexual maturity in the gut of a piscivorous bird (the definitive host) following the ingestion of infected sticklebacks (Wootton, 1976). The parasite grows rapidly in the body cavity of the fish host with increases in dry weight of up to 70% per week having been recorded (Meakins & Walkey, 1973). During this time, *S. solidus* places a considerable demand on the energy reserves of its host, and infected fish need to eat more food to maintain body weight (Pascoe & Mattey, 1977). Because of the added energetic demands of the parasite, the resting metabolic rate of infected sticklebacks is increased, and swimming for them is energetically more costly, presumably because of their reduced streamlined shape (Lester, 1971).

S. solidus and stickleback foraging behaviour

Physical restrictions on foraging.

Arme & Owen (1967) describe the gross pathology of *S. solidus* infection in the stickleback host, which includes displacement of the host viscera as the plerocercoid grows in the body cavity of infected fish. Since, in heavy infections, the parasite can weigh at least as much as the host (Arme & Owen, 1967; pers. obs.), this is likely to place a severe limitation on space availability within the body cavity. After periods of intensive feeding, the greatly distended stomach of uninfected sticklebacks normally expands to fill most of the available space in the body cavity (Hart & Gill, 1992). It is therefore likely that the stomachs of infected fish have much less room available for distension, reducing the volume of prey which could be consumed during one feeding bout. Experimental evidence supports this hypothesis, and *S. solidus*-infected sticklebacks have been shown to consume fewer prey items than size-matched uninfected conspecifics by various workers (Milinski, 1985; Cunningham *et al.*, 1994), strongly suggesting that the 'appetite' – the number of prey items taken before voluntary cessation of feeding (Wootton, 1990) – of infected fish is greatly reduced by the physical presence of plerocercoids in the body cavity.

The effect of *S. solidus* infection on prey handling time and competitive ability.

In an experiment designed to test the effect of *S. solidus* infection on the foraging efficiency of sticklebacks, CUNNINGHAM *et al.* (1994) investigated the behaviour of infected and uninfected sticklebacks preying individually on sequentially-presented live bloodworms (Fig. 1a). Infected fish were found to make more prey fixations towards and more manipulations of each prey item before eventual consumption than uninfected individuals, and they exhibited a correspondingly longer handling time per item. This resulted in infected fish taking longer to reach satiation (Table 1), which, when combined with known detrimental effect of the parasite on energy reserves (PASCOE & MATTEY, 1973) and manoeuvrability (ARME & OWEN, 1967), strongly suggests that infected fish in shoals of otherwise uninfected individuals should be physically poor competitors for food.

As MILINSKI (1990) points out, poor physical competitors need to be 'clever' to obtain sufficient food for vital functions, and parasitised sticklebacks appear to mediate their survival through alterations to their foraging strategy. They may achieve this by either foraging in habitats which offer low levels of competition, or by exhibiting a change in their prey selection criteria to concentrate on less contested prey. Experimental evidence for the first proposal is provided by GILES (1983), who demonstrated that sticklebacks infected with *S. solidus* would return to food patches more quickly than uninfected conspecifics after a simulated avian attack. Presumably the infected fish were prepared to accept a perceived elevated risk of predation in return for the possibility of increasing their food intake rate by foraging in a competitor-free environment. Later experiments by MILINSKI (1985) showed that infected sticklebacks would feed in close proximity to a live predatory cichlid in order to maintain a constant rate of food intake, whereas uninfected fish fed only on the prey items most distant from the predator, and even then with a reduced intake rate. Evidently, under risk of predation the apparent 'boldness' of infected fish may be a mechanism enabling them to compensate for any competitive disadvantage the parasite may convey. A review of how uninfected and *S. solidus* infected sticklebacks forage under the threat of predation is provided by MILINSKI (1990).

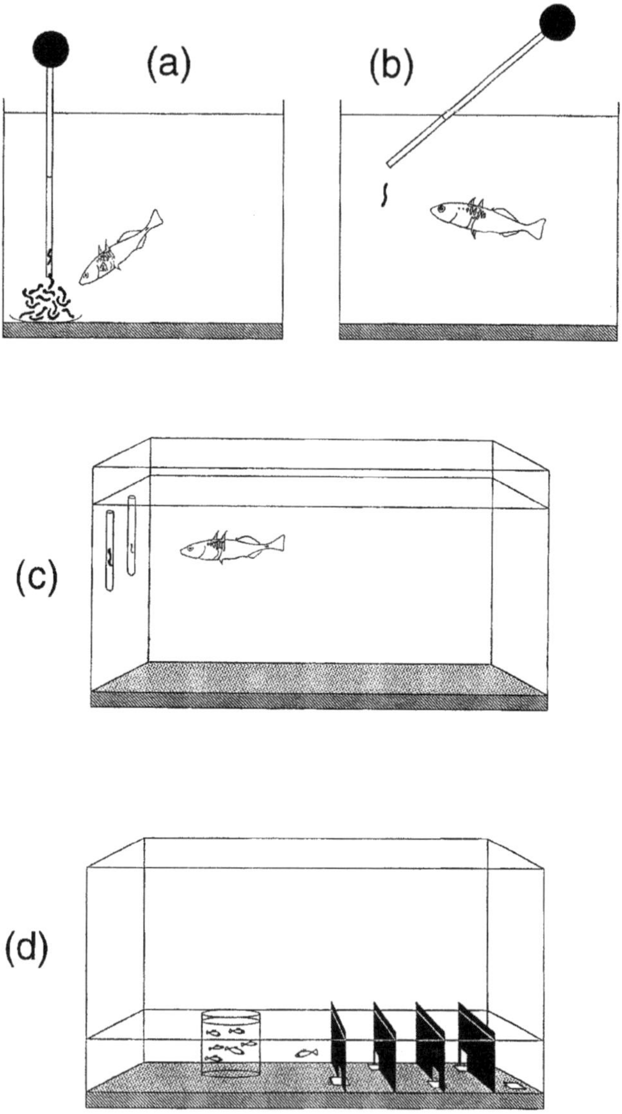

Fig. 1. The design of various experiments described in the text. (a) Prey handling time, (b) foraging efficiency and (c) prey size preference, as described by CUNNINGHAM *et al.* (1994); (d) the shoaling decisions experiment (BARBER *et al.*, in press)

TABLE 1. The feeding efficiency of uninfected (N = 18) and *S. solidus*-infected (N = 21) sticklebacks

Behaviour	Uninfected fish	Infected fish	Statistical significance
No. of prey manipulations	1 (0 - 5)	19 (5 - 52)	p << 0.01
No. of prey fixations	5 (1 - 8)	27 (9 - 75)	p << 0.01
Time taken to reach satiation (s)	94 (51 - 148)	293 (150 - 542)	p << 0.01
Handling time / item (s)	5.8 (4.7 - 7.9)	30.8 (28.1 - 42.3)	p << 0.01

Values given are medians with interquartile ranges. Statistical differences between the two categories of fish were determined by Wilcoxon-Mann-Whitney tests. Data from CUNNINGHAM *et al.* (1994)

The modified prey choice of *S. solidus*-infected sticklebacks.

In the absence of significant immediate predation pressure, however, it is likely that infected fish maximise their food intake rate by showing some change in prey selection. When feeding on two size classes of *Daphnia*, in direct competition with unparasitised fish in the absence of any predatory threat, *S. solidus* infected sticklebacks were found to preferentially select the smaller prey items (MILINSKI, 1984). This strategy allowed infected fish to avoid energetically-expensive competitions for the larger prey, which were preferred by the unparasitised fish. RANTA (1995) also reported differences in the food intake rate and prey size selection of infected and uninfected sticklebacks feeding on two sizes of *Daphnia*. However, in his study although infected fish had a lower food intake rate than uninfected fish, infected fish were found to consume a relatively greater proportion of the larger prey. By selecting the larger prey, infected fish were able to achieve equivalent short-term energy intake rates to uninfected fish, which fed at a higher rate and preferred the smaller individuals. RANTA (1995) accounts for the discordance between his and MILINSKI's (1984) results by pointing out that the sticklebacks used by MILINSKI had been starved prior to experimentation, whereas his own study fish had been allowed to feed *ad libitum* for 4 hours prior to the trials. Using a simple model, it is relatively simple to explain these apparently disparate results graphically (see Fig. 2 and legend for explanation).

It would appear in both of the above studies that when competition is high, infected fish should 'make the best of a bad situation' and select the smaller, apparently less profitable food items. However, foraging on these small food items may not always be sub-optimal for infected sticklebacks.

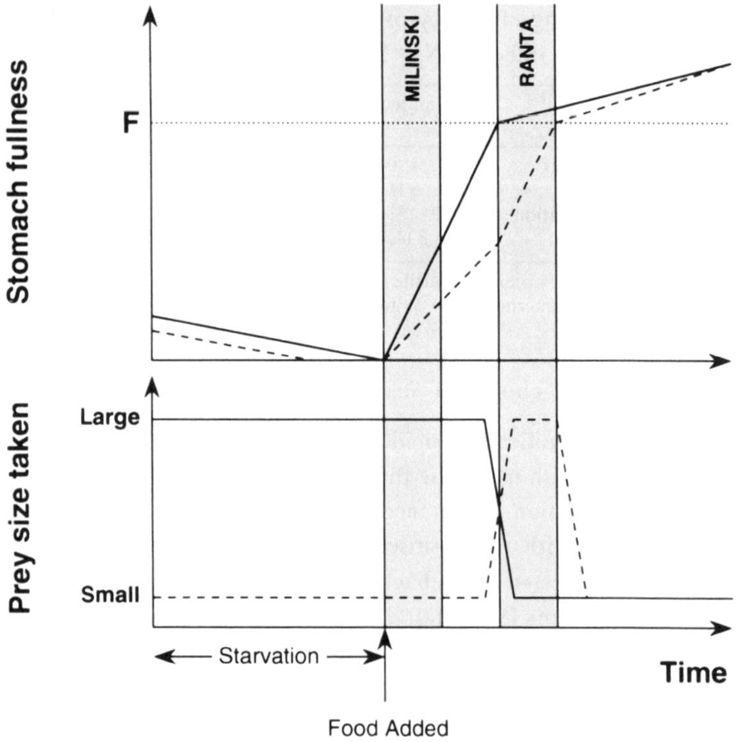

Fig. 2. Diagrammatic interpretation of the results of the experiments of MILINSKI (1984) and RANTA (1995). When the two prey sizes of *Daphnia* are initially presented to the fish, infected fish (dashed line) are probably outcompeted for the larger more profitable individuals (see text), and instead feed on the smaller prey. Since prey profitability is known to change with stomach fullness (HART & GILL, 1993), it is likely that there is a threshold level of stomach fullness ('F') above which it becomes energetically more profitable to feed on small prey. When uninfected fish (solid line) reach this threshold their prey choice switches and larger *Daphnia* become available to the infected fish, allowing them to increase their food intake. At some point later, infected fish will reach their stomach fullness threshold and switch back to small prey. It is likely that the apparently incompatible results obtained from the two experiments were due to the fact that MILINSKI's fish were tested immediately following a period of starvation, whereas RANTA's fish were allowed to feed *ad libitum* before the trial.

The increased prey handling time of *S. solidus* infected sticklebacks described by CUNNINGHAM *et al.* (1994) results in the same sized prey being differentially profitable to fish of different infection status, and so fish infected with *S. solidus* may be expected to exhibit a different prey choice preference to their uninfected conspecifics. In an experimental test of this hypothesis, CUNNINGHAM *et al.* (1994) studied the short-term foraging efficiency of infected and uninfected sticklebacks with either completely empty or partially-full stomachs feeding on single live bloodworms of one of two size categories (large / small) (Fig. 1b). Previously-determined energy values for both sizes of prey were then used to calculate the energy gain, in Joules, per second ($J.s^{-1}$) spent handling the item for each stomach fullness and prey size combination for both classes of fish. For uninfected sticklebacks, large prey were found to always be more profitable on a $J.s^{-1}$ basis, irrespective of stomach fullness. The profitability of all prey was much reduced for infected fish, for which both size classes yielded similar $J.s^{-1}$ values when the stomach was empty. However, with a partially-full stomach smaller prey became energetically more profitable than large prey (Fig. 3a). Subsequent behavioural experiments determined the actual prey size preference of infected and uninfected sticklebacks with either empty or partially-full stomachs (Fig. 1c). When presented with a choice between the two size classes of prey, all fish tested selected the most energetically profitable individual item, which meant that infected fish with partially full stomachs preferred the smaller prey items (Fig. 3b). These results demonstrate that the prey choice of both infected and uninfected sticklebacks reflects prey profitability, which is in turn dependent on both infection status and the level of stomach fullness of the individual concerned.

Effect of *S. solidus* infection on diet composition in natural habitats.

Do these altered prey preferences demonstrated in the laboratory relate to the field situation, where the variety of prey encountered is inevitably more varied? In a study of the natural diet of three-spined sticklebacks in an urban pond in Edinburgh, Scotland, TIERNEY (1994) used stomach analysis to quantify differences in both the foraging success (stomach fullness) and prey selection (diet composition) of *S. solidus*-infected and uninfected fish. Sticklebacks were sampled monthly and the dietary data collected were analysed by season (spring: March-May; summer: June-August; autumn: September-November; winter: December-February). Using the points system of HYNES (1950), distributions of stomach fullness categories were found to be similar for uninfected and infected fish during the autumn and spring, but uninfected fish had significantly fuller stom-

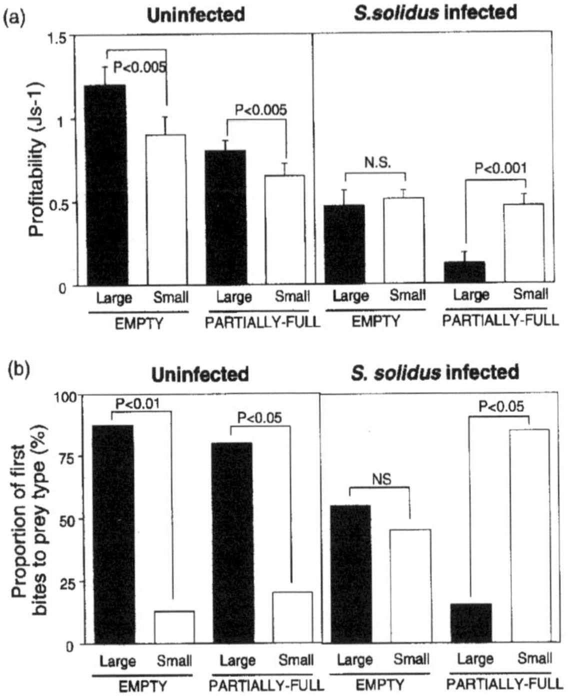

Fig. 3. (a) The profitability (in J.s^{-1}) of large and small bloodworms to uninfected (N = 18) and *S. solidus* infected (N = 21) sticklebacks. The profitability of each prey size to fish with both empty and partially-full stomachs is shown. Median values with interquartile ranges are given. Wilcoxon-Mann-Whitney tests were used to calculate statistical differences. (b) The initial prey choices of uninfected (N = 18) and *S. solidus* infected (N = 21) sticklebacks, measured as the proportion of first bites made towards test tubes containing small or large individual prey items. The prey choices of fish with both empty and partially-full stomachs are shown. χ^2 tests were used to calculate statistical differences between categories. (Redrawn from Cunningham *et al.*, 1994).

achs during the winter and summer months. The diet composition of infected and uninfected fish also differed at certain times of the year, suggesting that during periods of high intraspecific competition infected fish show a switch in diet. Earlier work by Jakobsen *et al.* (1988) also detected a difference in the diet of uninfected and infected fish at certain times of the year when food competition was likely to be most severe. *S. solidus* parasitised females were found to take a larger proportion of benthic prey than uninfected females, which appeared to prefer feeding in the limnetic zone on large cladocerans. This suggests that the altered diet was a consequence of altered habitat selection by infected fish, but whether this was a result of increased feeding competition, or whether some other factor restricted them to the benthic zone, is unclear.

It appears that the physical and energetic restrictions imposed by *S. solidus* change the foraging decisions taken by stickleback hosts by increasing their nutritional requirements whilst a) reducing their ability to compete successfully with others for food, b) decreasing the amount of food which can be consumed in one feeding bout, and c) having a detrimental effect on their foraging efficiency. It is likely that in the field these changes in foraging behaviour may have broader implications for the ecology of parasitised hosts than simply altering diet selection. For instance, by altering the foraging efficiency of infected sticklebacks, *S. solidus* may have an effect other facets of its host's behaviour. Social foraging in many species of fish, including the three-spine stickleback (Keenleyside, 1955; Wootton, 1984) is known to be common, and the formation of shoals is an evolutionary successful adaptation to the combined pressures of predator evasion and the location of patchily-distributed food resources. By altering the foraging behaviour of infected fish, *S. solidus* could theoretically change the value of being a shoal member.

S. solidus and shoaling behaviour

The costs and benefits of shoaling.

The two main benefits accrued by an individual fish by joining a shoal of conspecifics are protection against predation (see Pitcher & Parrish (1993) for a recent review) and enhanced food detection (Pitcher *et al.*, 1982; Street & Hart, 1985). Recent work carried out on *G. aculeatus* fry has shown that when food is unlimited, individuals which live in shoals have a higher specific growth rate than those which are raised in isolation (Peuhkuri *et al.*, 1995), a possible result of the increased efficiency of social foraging. However, the major cost of group living – competition for resources (Bertram, 1978), and most frequently, food – limits the extent to which fish form shoals. Accepting that organisms have evolved to maximise their fitness (Hamilton, 1964), we assume the decisions of animals to join, remain with or leave a group to be optimal under any particular set of circumstances, and such decisions are likely to be dependent on the outcome of trade-offs between the costs and benefits involved (Fig. 4). Certain environmental, physiological and morphological factors determine the values of the costs and benefits involved with shoal membership, and as they change we may expect the subsequent shoaling behaviour of individuals to change also. Fig. 5 shows some of the factors which may affect the costs and benefits involved, with factors which may be altered by *S. solidus* infection being shaded in the diagram. For instance, the increased level of food competition experienced when joining a shoal is likely to be more important to a nutritionally-stressed poor competitor (*e.g.* a *S. solidus*-infected stickleback) than to a well-fed dominant fish. Many of the antipredator benefits of shoaling depend on all shoal members having identical behaviour and appearance, and individuals which are distinguishable from the rest of a group are known to suffer a higher risk of predation (Ohguchi, 1981; Landeau & Terborgh, 1986), so the similarity of an individual to other shoal members is likely to be an important consideration when deciding whether to join or not. *S. solidus* causes marked swelling of the abdomen (Arme & Owen, 1967) and in some populations is associated with a deterioration of skin pigment (LoBue & Bell, 1993), both of which render infected fish visually distinguishable from uninfected conspecifics. Such changes may therefore alter the antipredator benefits accrued by being in a shoal, as increased predator protection through the confusion effect may not available to infected fish.

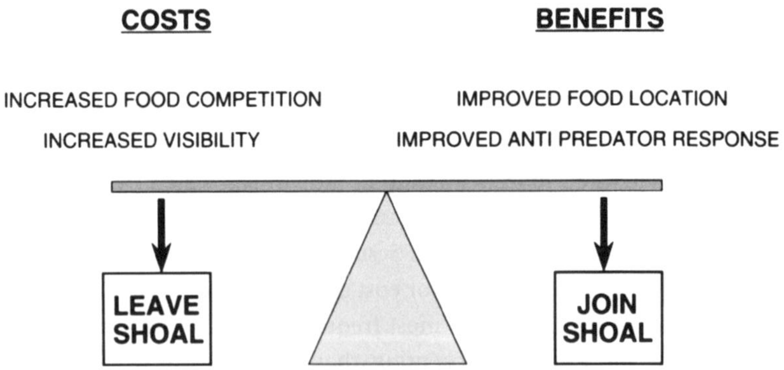

Fig. 4. A diagrammatic explanation of the costs and benefits of shoal membership, and how changes in their importance may alter the shoaling decisions taken by individual fish.

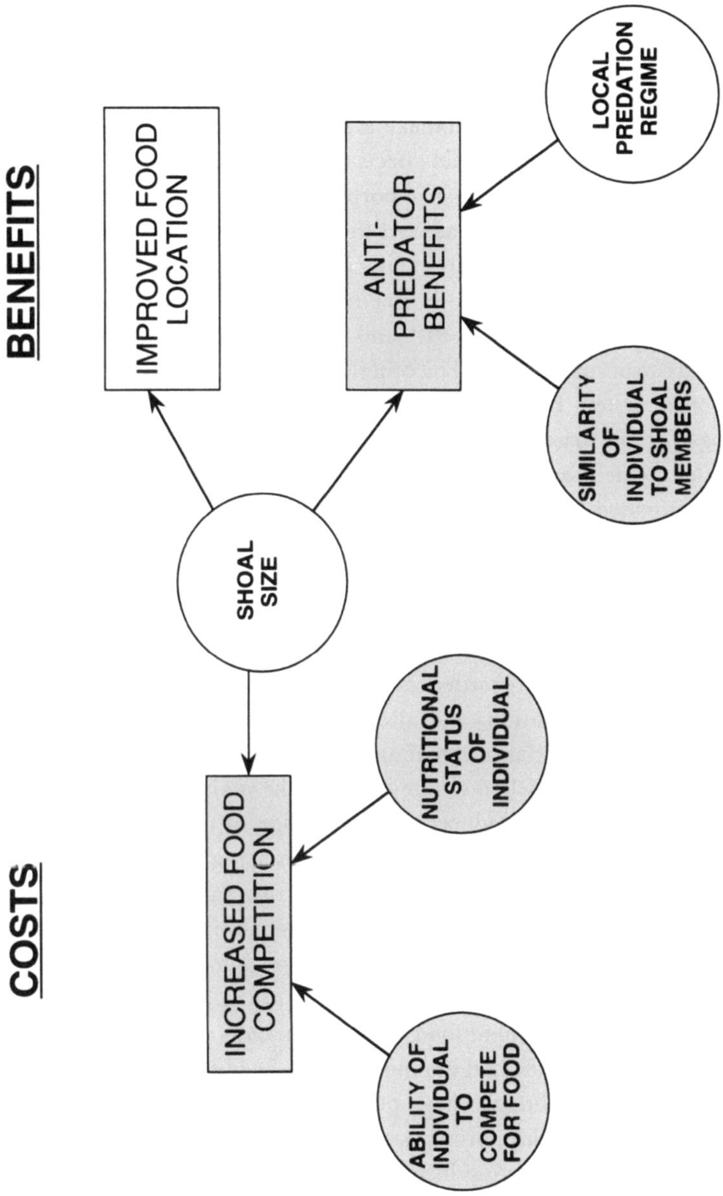

Fig. 5. Some of the factors (circles) affecting the costs and benefits (rectangles) associated with shoal membership. Factors which may be affected by *S. solidus* infection are shaded.

The effect of *S. solidus* infection on host shoaling decisions.

The hypothesis that *S. solidus*-infected fish should exhibit altered shoaling behaviour because the parasite changes the costs and benefits involved was tested experimentally by BARBER *et al.* (in press). An experiment was designed to study the differential effects of food deprivation on the trade-off between individual foraging opportunity and antipredator shoaling behaviour of infected and uninfected three-spined sticklebacks (Fig. 1d). In the centre of one half of a 60 cm aquarium, in a perforated transparent plastic cylinder, were housed six size-matched uninfected sticklebacks. These fish formed a captive 'shoal', and were allowed to settle for several days before experimentation. The other half of the aquarium was divided into a maze system by opaque plastic partitions fitted with small offset openings. At the base of each of these openings was located a plastic feeder onto which four live bloodworms were tethered; this ensured that the prey provided a moving stimulus. On release from a settling chamber close to the shoal, individual test fish had a choice of joining the captive shoal (taking up a position close the cylinder) or feeding. Test fish then could return to the captive shoal, or continue on to the next feeder, out of visual contact with the other fish. The behaviour of test fish in the experimental tank was recorded over a 30 minute period immediately after being fed to satiation, and after 24 and 48 hour periods of food deprivation. Uninfected fish were found to spend the largest proportion of time close to the shoal when satiated, and as the period of food deprivation was increased, they reduced the proportion of time spent with the shoal (Fig. 6a), spending instead more time foraging alone in the maze system (Fig. 6b). This reflects the shift in the costs and benefits of shoaling; as the fish became more hungry, they were more willing to leave the safety of the shoal to forage alone in the maze. Infected fish, however, spent much less time within one body length of the captive shoal when satiated (the condition where food competition costs should be minimal, and therefore where they had the most to gain by joining the shoal) (Fig. 6a). However, even though they spent less time with the shoal when satiated, infected fish did not increase the amount of time they spent in the maze (Fig. 6b), preferring instead to take up positions outside of the shoal whilst still remaining within visual contact of shoal members.

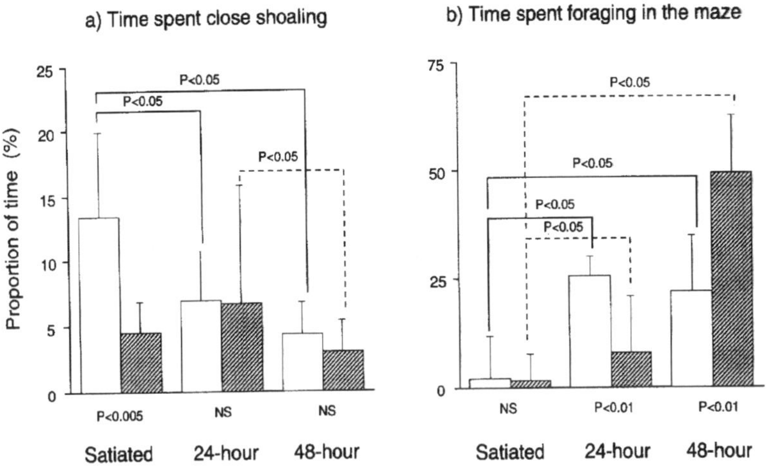

Fig. 6. The behaviour of uninfected (open bars, N = 13) and *S. solidus*-infected (hatched bars, N = 14) sticklebacks in the shoaling decisions experiment. (a) The proportion of time spent close shoaling (within one body length of the captive shoal). (b) The proportion of time spent foraging in the maze system, out of visual contact with the shoal. Median values are shown. Friedman ANOVA with *a posteriori* testing was used to calculate statistical differences between different satiation treatments of uninfected and infected fish, and Mann-Whitney U tests were used to show statistical differences between uninfected and infected fish within any particular satiation treatment (Redrawn from Barber *et al.*, in press).

S. solidus and host antipredator behaviour.

Previous workers have suggested that *S. solidus*-infected fish may be exposed to higher levels of avian predation since they appear to swim higher in the water column and are less effectively countershaded than uninfected conspecifics. Lester (1971) suggested that parasite-induced oxygen stress would force infected fish into shallow water which would be largely devoid of piscivorous fish but where avian predators would be more common. The strategy of moving away from the shoal may allow infected sticklebacks to make the most of any individual foraging opportunities by reducing local levels of food competition; alternatively, their lack of homogeneity with other shoal members may make it undesirable to remain with a shoal where, once the shoal was attacked, they would be visually-distinguishable to a predator. In any case, such manipulation of shoaling behaviour is likely to alter the predation regime of infected fish, and therefore could have consequences for the transmission of the parasite. It is clear that by altering shoaling behaviour, one of the primary antipredator defences of small prey fish such as the stickleback, the effects of *S. solidus* infection on the predator-prey relationships of their hosts may be more complex than first imagined.

Discussion

The foraging decisions of *S. solidus* infected sticklebacks.

Infection with *S. solidus* is associated with changes both in the foraging behaviour (appetite; feeding efficiency; prey choice) observed in the laboratory, and in the natural diet composition and foraging success of three-spined sticklebacks in the wild. The reduced efficiency with which infected fish handle and consume prey is likely to make them poor competitors when food availability is a limiting factor in the natural environment. Such a reduction in competitive ability may be responsible for the observed lower energy reserves in wild-caught infected fish during the summer (where the population reaches its highest density) and the winter (when food is likely to be most scarce) (TIERNEY *et al.*, in press). The observed differences in the diet composition of infected fish when compared with healthy conspecifics may be a result of this reduced competitive ability, whereby they are restricted to feeding on smaller, less attractive lower-energy foods such as ostracods (ALLEN & WOOTTON, 1984).

Fish that have a poor competitive ability should be under pressure to move into regions where food competition is less severe. This usually means compromising some other aspect of habitat or patch suitability, such as the threat of predation (GILES, 1983; MILINSKI, 1985). Under risk of predation the apparent 'boldness' of infected fish may compensate for any competitive disadvantage. In regions where fish are not subject to heavy predation pressure, other competition-reducing strategies are available to infected fish, such as the change in prey size and type preference.

The shoaling decisions of *S. solidus* infected sticklebacks.

Hunger level, demonstrated as being important in determining whether fish join shoals, has also been shown to be important in determining the positions taken up by fish in polarised schools. KRAUSE (1993) studied the position preferences of experimentally food-deprived roach (*Rutilus rutilus*) and chub (*Leuciscus cephalus*) in the field, and found that hungry fish moved initially to the front of the school, where higher feeding rates (O'CONNELL, 1972; KRAUSE *et al.*, 1992) may be obtained in return for a possible increased risk of predation (but see PARRISH, 1989).

Infected fish have a reduced stomach capacity and it is known that they have an increased metabolic demand (LESTER, 1971). It therefore seems reasonable to assume that at any point in time, infected fish are likely to be more hungry than uninfected conspecifics. It has been shown that in the laboratory, *S. solidus* infected sticklebacks were more willing than uninfected conspecifics to leave the apparent safety of a shoal to forage alone when hungry. However, they also spent less time close shoaling when it was predicted they had most to gain by joining the shoal. It may be that the effect of oddity prevents infected fish joining the shoal under these conditions, and this is currently being investigated.

Ecological consequences of altered behaviour in *S. solidus* infected sticklebacks.

The life cycle of *S. solidus* requires transmission between three hosts. Infected piscivorous birds pass out eggs into the water along with the faeces; these eggs hatch and the first stage larvae are taken up by cyclopoid copepods. If infected copepods are consumed by three-spined sticklebacks, the parasite larvae burrow through the gut wall of the fish and grow in the body cavity until they reach infectivity (at a size of 50 mg; TIERNEY & CROMPTON, 1990). If they are then eaten by a piscivorous bird, the parasite rapidly matures in the intestine of the bird and reproduces. If parasite manipulation is to play a part in improving the chances of successful transmission from stickleback to definitive host predator, then the parasite must act to increase the probability of its host being consumed by a suitable definitive host (a 'target' predator).

There is a considerable selection pressure on such parasites to be transmitted, as those that fail to infect a definitive host die without reproducing. Therefore, any manipulation of the host by its parasite must be specific — it would not be sufficient, for instance, to affect general antipredator behaviour if this also made hosts more susceptible to non-target predators, unless target predators were significantly more common. In a recent study of the effects of parasites on the schooling positions of fish, KRAUSE & GODIN (in press) show that killifish (*Fundulus diaphanus*) infected with the trematode parasite *Crassiaphiala bulboglossa* (the definitive host of which is a piscivorous bird) take up peripheral positions. It is claimed that by taking up such positions, parasite transmission may be enhanced by increasing predation pressure on their hosts. This, however, is dependent on the predation strategy of the most common avian predator at any particular site — some birds may preferentially attack the centre of the shoal, whilst piscivorous fish (non-target predators) may feed on peripheral fish. Present studies have shown that *S. solidus*-infected sticklebacks are more likely to be forced away from the shoal, probably brought about by parasite-mediated alterations to the costs and benefits involved with shoal membership. If this occurs in natural habitats as well as in the laboratory, then it seems probable that parasitised fish would be found less often as shoal members, and more often as lone individuals. Such a change in behaviour would undoubtedly alter the susceptibility of parasitised fish to the various target and non-target predators present in any particular habitat. However, before increased parasite transmission can be advocated, it will be essential to understand the prey selection mechanisms of both target and non-target predators.

References

ALLEN, J.R. & WOOTTON, R.J. (1984). Temporal patterns of diet and rate of food consumption of the three-spined stickleback (*Gasterosteus aculeatus* L.) in Llyn Frongoch, an upland Welsh lake. — Freshwater Biol. 14, p. 335-346.

ARME, C. & OWEN, R.W. (1967). Infections of the three-spined stickleback, *Gasterosteus aculeatus* L., with the plerocercoid larvae of *Schistocephalus solidus* with special reference to pathological effects. — Parasitology 57, p. 301-304.

BARBER, I., HUNTINGFORD, F.A. & CROMPTON, D.W.T. (in press). The effect of hunger and cestode parasitism on the shoaling decisions of small freshwater fish. — J. Fish Biol.

BERTRAM, B.C.R. (1978). Living in groups: predators and prey. — In: Behavioural ecology: an evolutionary approach (J.R. KREBS & N.B. DAVIES, eds). Blackwell, Oxford, p. 64-97.

CUNNINGHAM, E.J., TIERNEY, J.F. & HUNTINGFORD, F.A. (1994). Effects of the cestode *Schistocephalus solidus* on food intake and foraging decisions in the three-spined stickleback, *Gasterosteus aculeatus*. — Ethology 97, p. 65-75.

GILES, N. (1983). Behavioural effects of the parasite *Schistocephalus solidus* (Cestoda) on an intermediate host, the three-spined stickleback, *Gasterosteus aculeatus* L. — Anim. Behav. 31, p. 1192-1194.

—— (1987). Predation risk and reduced foraging activity in fish: experiments with parasitised and non-parasitised three-spined sticklebacks, *Gasterosteus aculeatus* L. — J. Fish Biol. 31, p. 37-44.

HAMILTON, W.D.J. (1964). The genetical evolution of social behaviour. — J. theoret. Biol. 7, p. 1-53.

HART, P.J.B. & GILL, A.B. (1992). Constraints on prey size selection by the three-spine stickleback: Energy requirements and the capacity and fullness of the gut. — J. Fish Biol. 40, p. 205-208.

Hynes, H.B.N. (1950). The food of freshwater sticklebacks (*Gasterosteus aculeatus* and *Pygosteus pungitius*) with a review of the methods used in studies of the food of fishes. — J. Anim. Ecol. 19, p. 36-58.

Jakobsen, P.J., Johnsen, G.H. & Larsson, P. (1988). Effects of predation risk and parasitism on the feeding ecology, habitat use and abundance of lacustrine threespine stickleback (*Gasterosteus aculeatus*). — Can. J. Fish. Aq. Sci. 45, p. 426-431.

Keenleyside, M.H.A. (1955). Some aspects of the schooling behaviour of fish. — Behaviour 8, p. 83-248.

Krause, J. (1993). The relationship between foraging and shoal position in a mixed shoal of roach (*Rutilus rutilus*) and chub (*Leuciscus cephalus*): a field study. — Oecologia 93, p. 356-359.

——, Bumann, D. & Todt, D. (1992). Relationships between the position preference and nutritional state of individuals in schools of juvenile roach (*Rutilus rutilus*). — Behav. Ecol. Socobiol. 30, p. 177-180.

—— & Godin, J-G.J. (in press). Influence of parasitism on the shoaling behaviour of banded killifish, *Fundulus diaphanus*. — Can. J. Zool.

Landeau, L. & Terborgh, J. (1986). Oddity and the 'confusion effect' in predation. — Anim. Behav. 34, p. 1372-1380.

Lester, R.J.G. (1971). The influence of *Schistocephalus plerocercoids* on the respiration of *Gasterosteus* and a possible resulting effect on the behaviour of the fish. — Can. J. Zool. 49, p. 361-366.

LoBue, C.P. & Bell, M.A. (1993). Phenotypic manipulation by the cestode parasite *Schistocephalus solidus* of its intermediate host, *Gasterosteus aculeatus*, the threespine stickleback. — Am. Nat. 142, p. 725-735.

Meakins, R.H. & Walkey, M. (1973). Aspects of the *in vivo* growth of the plerocercoid stage of *Schistocephalus solidus*. — Parasitology 67, p. 133-141.

Milinski, M. (1984). Parasites determine a predator's optimal feeding strategy. — Behav. Ecol. Sociobiol. 15, p35-37.

—— (1985). Risk of predation of parasitised sticklebacks (*Gasterosteus aculeatus* L.) under competition for food. — Behaviour 93, p. 203-216.

—— (1990). Parasites and host decision-making. — In: Parasitism and host behaviour (C.J. Barnard & J.M. Behnke, eds). Taylor & Francis, London, p. 95-116.

O'Connell, C.P. (1972). The interrelationship of biting and filter feeding activity of the northern anchovy (*Engraulis mordax*). — J. Fish. Res. Bd. Can. 29, p. 285-293.

Ohguchi, O. (1981). Prey density and selection against oddity by threespined sticklebacks. — Adv. Ethol. 23, p. 1-79.

Parrish, J.K. (1989). Re-examining the selfish herd: are central fish safer? — Anim. Behav. 38, p. 1048-1053.

Pascoe, D. & Mattey, D. (1977). Dietary stress in parasitised and non-parasitised *Gasterosteus aculeatus*. — Z. Parasitenkunde 51, p. 179-186.

Peuhkuri, N., Ranta, E., Juvonen, S-K. & Lindströom, K. (1995). Schooling affects growth in the three-spined stickleback, *Gasterosteus aculeatus*. — J. Fish Biol. 46, p. 221-226.

Pitcher, T.J., Magurran, A.E & Winfield, I.J. (1982). Fish in larger shoals find food faster. — Behav. Ecol. Sociobiol. 10, p. 149-151.

—— & Parrish, J.K. (1993). Functions of shoaling behaviour in teleosts. — In: Behaviour of teleost fishes, 2nd edn. (T.J. Pitcher, ed.). Chapman & Hall, London, p. 363-439.

Ranta, E. (1995). *Schistocephalus* infestation improves prey-size selection by three-spined sticklebacks, *Gasterosteus aculeatus*. — J. Fish Biol. 46, p. 156-158.

Siegel, S. & Castellan, N.J. (1988). Nonparametric statistics for the behavioural sciences. — McGraw-Hill, New York.

Street, N.G. & Hart, P.J.B. (1985). Group size and patch location by the stoneloach, *Noemacheilus barbatulus*, a non-visually foraging predator. — J. Fish Biol. 27, p. 785-792.

1240 BARBER & HUNTINGFORD

TIERNEY, J.F. (1994). Effects of *Schistocephalus solidus* (Cestoda) on the food intake and diet of the three-spined stickleback, *Gasterosteus aculeatus*. — J. Fish Biol. 44, p. 731-735.

—— & CROMPTON, D.W.T. (1992). Infectivity of plerocercoids of *Schistocephalus solidus* (Cestoda: Ligulidae) and fecundity of the adults in an experimental definitive host, *Gallus gallus*. — J. Parasitol. 78, p. 1049-1054.

——, HUNTINGFORD, F.A. & CROMPTON, D.W.T. (in press). The impact of *Schistocephalus solidus* (Cestoda) on body condition and reproduction in an annual population of sticklebacks, *Gasterosteus aculeatus*. — J. Fish. Biol.

WOOTTON, R.J. (1976). The biology of the sticklebacks. — Academic Press, London.

—— (1984). A functional biology of sticklebacks. — Croom Helm, London.

—— (1990). The ecology of teleost fishes. — Chapman and Hall, London.

Retrospective: Living in the ecological theatre

Articles published in *Behaviour* volume 132 (1995), p. 1079–1094; volume 137 (2000), p. 1081–1096

T.E. Reimchen

(Department of Biology, University of Victoria, P.O. Box 3020, Victoria, B.C., Canada, V8W 3N5)

The major processes underlying evolutionary change seemed quite well-established to this Canadian undergraduate in 1967. Although few questioned the primacy of form and function for character differences among species, the significance of variation within a species was in dispute, and one of the most prolific researchers of the 1970's and 1980's noted that '... some geographical variation within a species is clearly adaptive, but much is a non-adaptive product of history' (Gould, 1984). Variation within populations was generally considered developmental noise around an optimal phenotype preserved by normalizing and canalizing selection (Mayr, 1963; Futuyma, 1979), or in the case of discontinuous phenotypes, such as balanced polymorphisms, an adaptive response to ecological or physiological pressures (Ford, 1964).

My first exposure to intra-population variability was as a summer assistant to Ric Moodie, who was doing his PhD on a giant black stickleback from Haida Gwaii, an archipelago off the west coast of British Columbia. Ric identified differences in timidity and microhabitat of phenotypes which correlated with their number of bony lateral plates along their trunk (Moodie, 1972). The experience of studying these individual behavioural differences with Ric impacted my professional trajectory, including my choice of graduate school. I attended the University of Liverpool and worked under Arthur Cain and Philip Sheppard, both former students of E.B. Ford, who had carried out some of the seminal studies on the ecological genetics of natural populations of lepidopterans and molluscs (Ford, 1964).

My thesis research focused on shell colour variation in intertidal snails. I found a distinctive spatial geometry between the substrate, juvenile snails and predatory fish that provided evidence for fine-tuned adaptation at several levels including ontogenetic, intrapopulation and macrogeographical variation (Reimchen, 1979, 1989). These data in combination with the apparent limited role of any obvious founder effects, pleiotropy or constraints that are often employed to account for large and small scale geographical variation led me towards pan-selectionism. My post-doctoral plans were to use similar approaches to examine the extent of ecological influence on continuous/quantitative traits. Van Valen (1965) had already written an important paper on morphological variability and niche breadth in bird populations proposing that intraspecific competition could favour the evolution of specialists when broad niche space was available (see also Levene, 1953). Although Van Valen focused on trophic structures, I was interested in variability in stickleback predator defences. Evidence already existed that among stickleback populations, mean number of bony lateral plates was associated with the presence of predatory fish and that in the absence of these predators, stickleback tended to have fewer plates (Hagen & McPhail, 1970; Hagen & Gilbertson, 1972; Moodie & Reimchen, 1976; Gross, 1978).

With this background and funds from Joseph Nelson, I made my way in 1976 to a cabin on Drizzle Lake in a remote bog on Haida Gwaii. There, my partner and I resided for the next decade, investigating variability in the endemic populations of stickleback. This was probably not the best period to do a detailed study of selection and variability within a population, given some of the shrill critiques at the time on pan-selectionism and adaptationism (e.g., Gould & Lewontin, 1979). Although the lake was off the electric grid, I installed photovoltaic panels on nearby treetops and a small wind generator on the lake shore, and was able to produce sufficient power for lights as well as various electronics such as a 'portable' computer, printer and photographic darkroom.

Residency at the lake throughout the year substantially modified my conceptual interpretation of selection in the wild. Each week would often reveal a shift in the abiotic and biotic conditions that I would not see again until the next year at the same time (e.g., Reimchen & Douglas, 1980). The predation regime in July was very different from that of October which was distinctive from that of April. It seemed to me that it would be impossible to meaningfully interpret variation without this temporal detail of life history interactions, and that the apparent widespread evidence for non-functional variability in populations (Gould, 1984) was more likely a simple consequence of a limited understanding of the actual selective landscapes.

While stickleback researchers had focused on predatory fish as agents of selection (reviewed by Wootton, 1984), some of the first useful observations at our field site occurred when I saw avian piscivores capturing stickleback. Initially, these infrequent bouts of predation seemed of marginal demographic relevance compared with the impact of resident trout, but with daily surveys throughout the year, 16 species of avian piscivores, including grebes and loons, were observed each over a relatively narrow but different time window (Reimchen & Douglas, 1984). I established a three dimensional sampling grid for stickleback throughout the lake and made replicated transects every two months for three years and maintained records on the diel and seasonal foraging behaviour of trout and avian piscivores. The 19 transects allowed 11,000 stickleback to be scored for a diversity of traits. I found that variation in the frequency distribution of lateral plates among subadults, but not adults, was strongly concordant with seasonal and yearly shifts in the relative proportion of stickleback consumed by avian versus salmonid predators. As with concurrent studies of quantitative trophic characters on Galapagos finches (Grant et al., 1976) and more recent studies on trends in stickleback spine phenotypes from an adjacent lake (Reimchen & Nosil, 2002), the data suggested fine-tuned adaptations, with rapid population responses of subadults to seasonal periodicity in the selective landscape, described in one of the two papers reproduced here (Reimchen, 1995). Yet, the adult size classes, which are subject to only minor predation, showed no changes among seasons or years, indicating pseudostasis in morphology. Without having seen the cyclical trends in the subadults, this could have been interpreted as a developmental constraint to morphological change and limitation to the precision of adaptation (e.g., Gould & Eldredge, 1977).

Part of the morphological assessment of stickleback that I included in my field study was the extent of bilateral asymmetry of defence structures. An expanding literature at the time suggested that slight asymmetries reflected developmental instability from various combinations of inbreeding, physiological stress, parasitism and disruption of co-adapted gene complexes (Soulé, 1972; Parsons, 1990; Møller & Swaddle, 1997; Lemberget & McCormick, 2009; however, see Wiener & Rago, 1987). Consistent with fitness reduction, Ric Moodie and I had previously shown from a survey of some 20 Haida Gwaii lakes that plate asymmetries in stickleback were uncommon in populations with abundant predatory fish (Moodie & Reimchen, 1976). However, during my seasonal transects at Drizzle Lake, I thought something was amiss, as about one-half of the stickleback were asymmetric for plates even though they experienced extensive predation. I had expanded stickleback collections from all of the unsampled lakes throughout the archipelago and this provided an ideal opportunity to evaluate baseline levels of asymmetry in over 100 of the undisturbed populations. My PhD student Carolyn Bergstrom undertook these analyses in the late 1990's. Results to emerge from some of this research showed that in most populations, 30% to 60% of all individuals were asymmetric (Bergstrom & Reimchen, 2000). As a proxy for developmental stress and instability, these frequencies seemed unexpectedly high for pristine populations that appeared to lack any obvious source of physiological stress. We expected that at least there would be widespread examples for reduction in frequency of asymmetrics across age classes, particularly as these populations were subject to a diverse predation regime (Reimchen, 1994). Yet we did not observe any such evidence, implying that fitness between symmetrics and asymmetrics was similar.

This paper, as well as more recent data, indicate opposing processes on plate asymmetries, including their increased susceptibility to parasitism (Reimchen & Nosil, 2001), increased reproductive benefits relative to symmetric fish (Moodie & Moodie, 1996) as well as isotopic differences between left and right biased lateral plate asymmetries, possibly due to microhabitat differences in trophic niche (Reimchen et al., 2008; Reimchen & Bergstrom, 2009). The latter may be mediated through the lateralized capture behaviour of vertebrate predators (Bisazza et al., 1998; Vallortigara & Rogers, 2005). If these data are more broadly applicable, they suggest that fluctuating asymmetries are far more ecologically interesting than the mere by-product of developmental instability and may yield as yet hidden insights into individual niche space and fitness in the ecological landscape.

The second paper (Reimchen, 2000) reproduced here analyzes predator-handling failures of *Gasterosteus* differing in lateral plate and spine expression and offers a functional interpretation for the geological persistence and geographical distribution of the completely-plated morph, the ancestral phenotype in this taxon (Bell, 1981). Although the bony lateral plates covering the trunk of the threespine stickleback would seem to be an obvious predator defence, the widespread loss of posterior trunk plates in numerous freshwater populations subject to extensive predation weaken such an assumption (reviewed by Bell, 1976; Wootton, 1976, 1984). Their persistence in marine waters from the Miocene (Bell, 1977) had been correlated with physiological adaptations to salinity and cold temperatures (Heuts, 1947) while their occurrence in some northern fresh waters could be a historical signature of post-glacial migration routes (Münzing, 1963), introgression from marine stickleback (Miller & Hubbs, 1969; but see Bell retrospective in this volume) as well as epistatic linkages to low water temperatures or extreme winter conditions (Hagen & Moodie, 1982). The reluctance of researchers to consider direct functionality to the complete row of lateral plates was probably due to the earlier and influential work of R. Hoogland, Desmond Morris and Niko Tinbergen, who in 1957 provided evidence that the long spines on stickleback reduced feeding success of predatory fish relative to that for short-spined and unspined species. The reluctance may also have been accentuated by the prevalence of morphological stasis in marine habitats and inferred developmental constraints (Cronin, 1985).

These constraints or the physiological correlations seemed improbable to me given the large taxonomic diversity of predators taking stickleback (Reimchen, 1994), but it was the occurrence of pursuit and manipulation failures by these predators that gave me insights to the functionality of the fully-plated condition. In a review of predator-prey literature, Vermeij (1982) argued that while successful predation produced demographic consequences to the prey, unsuccessful or inefficient predators were the engine of defensive adaptations. Emerging from my geographical surveys of stickleback on the islands, I started to suspect a relationship between water clarity, predator inefficiencies and stickleback defences. Low-plated stickleback prevailed throughout the archipelago and the only three populations with completely-plated stickleback occurred in mountain lakes which were distinctive because of their exceptional water clarity combined with a steep shore profile and lack of any littoral zone. These conditions were spectrally and structurally similar to the open clear water habitats of marine stickleback and I inferred that this would potentially facilitate increased probability of capture by predatory fish with the subsequent advantage to post-capture defensive armour. This led to a series of experiments on trout-handling behaviour (Reimchen, 1992), similar to those of Hoogland et al. (1957), but where both spines and plates were evaluated. I showed that the completely plated morph had higher escape probabilities than low-plated stickleback, even when the latter had much larger and more robust spines than the former. The higher escape was probably due to abrasive interference between the plates and the pharyngeal jaws of predatory fish. This hypothesis received recent support in a natural experiment in Lake Washington where the completely-plated morph increased relative to the low-plated morph following recent improvements in clarity of the lake (Kitano et al., 2008) and provides a functional context to broader evolutionary trends (i.e. Bell, 2001; Peichel, 2005).

At the present, sitting at my laptop in Victoria, 1000 km from Drizzle Lake, and even further from the ecological theatre that structured natural selection within this remote population, I think that the substantive issues that emerged during our many years of observation on the lake were the predictable spatial and temporal variability in sources of mortality, variability in the efficiencies of predators, and the differences in these processes even among adjacent populations. These mortality factors are fundamental for interpreting intra- and inter-population phenotypic variability, yet are rarely characterized for most species. Determining the extent of fine-tuned adaptive variation and efficacy of selection in natural populations still has tremendous research potential one and a half centuries after the general framework of natural selection was established.

References

Bell, M.A. (1976). Evolution of phenotypic diversity in *Gasterosteus aculeatus* superspecies on the Pacific Coast of North America. — Syst. Zool. 25: 211-227.

Bell, M.A. (1977). A late Miocene marine threespine stickleback, *Gasterosteus aculeatus aculeatus*, and its zoogeographic and evolutionary significance. — Copeia 1977: 277-282.

Bell, M.A. (1981). Lateral plate polymorphism and ontogeny of the complete plate morph of threespine sticklebacks (*Gasterosteus aculeatus*). — Evolution 35: 67-74.

Bell, M.A. (2001). Lateral plate evolution in the threespine stickleback: getting nowhere fast. — Genetica 112: 445-461.

Bergstrom, C. & Reimchen, T.E. (2000). Functional implications of fluctuating asymmetry among endemic populations of *Gasterosteus aculeatus*. — Behaviour 137: 1097-1112.

Bisazza, A., Rogers, L.J. & Vallortigara, G. (1998). The origins of cerebral asymmetry: a review of evidence of behavioral and brain lateralization in fishes, reptiles and amphibians. — Neurosci. Biobehav. Rev. 22: 411-426.

Cronin, T.M. (1985). Speciation and stasis in marine Ostracoda — climatic modulation of evolution. — Science 227: 60-63.

Ford, E.B. (1964). Ecological genetics. — Wiley, New York.

Futuyma, D. (1979). Evolutionary biology. — Sinauer, Sunderland, Mass.

Gould, S.J. (1984). Covariance sets and ordered geographic-variation in *Cerion* from Aruba, Bonaire and Curacao — a way of studying nonadaptation. — Syst. Zool. 33: 217-237.

Gould, S.J. & Eldredge, N. (1977). Punctuated equilibrium: the tempo and mode of evolution reconsidered. — Paleobiology 3: 115-151.
Gould, S.J. & Lewontin, R.C. (1979). Spandrels of San-Marco and the Panglossian paradigm — a critique of the adaptationist program. — P. Roy. Soc. Lond. B Bio. 205: 581-598.
Grant, P.R., Grant, B.R. & Smith J.N. (1976). Darwin's finches — population variation and natural selection. — Proc. Nat. Acad. Sci. 73: 257-261.
Gross, H.P. (1978). Natural-selection by predators on defensive apparatus of three-spined stickleback, *Gasterosteus-aculeatus* L. — Can. J. Zool. 56: 398-413.
Hagen, D.W. & Gilbertson, L.G. (1972). Geographic variation and environmental selection in *Gasterosteus aculeatus* L. in the Pacific northwest, America. — Evolution 26: 32-51.
Hagen, D.W. & McPhail, J.D. (1970). The species problem with *Gasterosteus aculeatus* on the Pacific coast of North America. — J. Fish. Res. Board Can. 27: 147-155.
Hagen, D.W. & Moodie, G.E.E. (1982). Polymorphism for plate morphs in *Gasterosteus aculeatus* on the east coast of Canada and an hypothesis for their global distribution. — Can. J. Zool. 60: 1032-1042.
Heuts, M.J. (1947). Experimental studies on adaptive evolution in *Gasterosteus aculeatus* L. — Evolution 1: 89-102.
Hoogland, R., Morris, D. & Tinbergen, N. (1957). The spines of sticklebacks (*Gasterosteus* and *Pygosteus*) as a means of defence against predators (*Perca* and *Esox*). — Behaviour 10: 205-236.
Kitano, J., Bolnick, D.I., Beauchamp, D.A., Mazur, M.M., Mori, S., Nakano, T. & Peichel, C.L. (2008). Reverse evolution of armor plates in the threespine stickleback. — Curr. Biol. 18: 769-774.
Lemberget, T. & McCormick, M.I. (2009). Replenishment success linked to fluctuating asymmetry in larval fish. — Oecologia 159: 83-93.
Levene, H. (1953). Genetic equilibrium when more than one ecological niche is available. — Am. Nat. 87: 331-333.
Mayr, E. (1963). Animal species and evolution. — Belknap Press of Harvard University Press, Cambridge, Mass.
Miller, R.R. & Hubbs, C.L. (1969). Systematics of *Gasterosteus aculeatus*, with particular reference to intergradation and introgression along the Pacific coast of North America: a commentary on a recent contribution. — Copeia 1969: 52-69.
Møller, A.P. & Swaddle, J.P. (1997). Asymmetry, developmental stability, and evolution. — Oxford University Press, Oxford.
Moodie, G.E.E. (1972). Morphology, life history and ecology of an unusual stickleback (*Gasterosteus aculeatus*) in the Queen Charlotte Islands, Canada. — Can. J. Zool. 50: 721-732.
Moodie, G.E.E. & Moodie, P.F. (1996). Do asymmetric sticklebacks make better fathers? — P. Roy. Soc. Lond. B Bio. 263: 535-539.
Moodie, G.E.E. & Reimchen, T.E. (1976). Phenetic variation and habitat differences in *Gasterosteus* populations of the Queen Charlotte Islands. — Syst. Zool. 25: 49-61.
Münzing, J. (1963). The evolution of variation and distributional patterns in European populations of the three-spined stickleback, *Gasterosteus aculeatus*. — Evolution 17: 320-332.
Parsons, P.A. (1990). Fluctuating asymmetry and stress intensity. — Trends Ecol. Evol. 5: 97-98.
Peichel, C.L. (2005). Fishing for the secrets of vertebrate evolution in threespine sticklebacks. — Dev. Dynam. 234: 815-823.

Reimchen, T.E. (1979). Substrate heterogeneity, crypsis, and colour polymorphism in an intertidal snail (*Littorina mariae*). — Can. J. Zool. 57: 1070-1085.

Reimchen, T.E. (1989). Shell colour ontogeny and tubeworm mimicry in a marine gastropod *Littorina mariae*. — Biol. J. Linn. Soc. 36: 97-109.

Reimchen, T.E. (1992). Injuries on stickleback from attacks by a toothed predator (*Oncorhynchus*) and implications for the evolution of lateral plates. — Evolution 46: 1224-1230.

Reimchen, T.E. (1994). Predators and morphological evolution in threespine stickleback. — In: The evolutionary biology of the threespine stickleback (Bell, M.A. & Foster, S.A., eds). Oxford University Press, Oxford, p. 240-276.

Reimchen, T.E. (1995). Predator-induced cyclical changes in lateral plate frequencies of *Gasterosteus*. — Behaviour 132: 1079-1094.

Reimchen, T.E. (2000). Predator handling failures of lateral plate morphs in *Gasterosteus aculeatus*: functional implications for the ancestral plate condition. — Behaviour 137: 1081-1096.

Reimchen, T.E. & Bergstrom, C.A. (2009). The ecology of asymmetry in stickleback defense structures. — Evolution 63: 115-126.

Reimchen, T.E. & Douglas, S.D. (1980). Observations of loons (*Gavia immer* and *G. stellata*) at a bog lake on the Queen Charlotte Islands. — Can. Field Nat. 94: 398-404.

Reimchen, T.E. & Douglas, S. (1984). Seasonal and diurnal abundance of aquatic birds on the Drizzle Lake Reserve, Queen Charlotte Islands, British Columbia. — Can. Field Nat. 98: 22-28.

Reimchen, T.E., Ingram, T. & Hansen, S.C. (2008). Assessing niche differences of sex, armour and asymmetry phenotypes using stable isotope analyses in Haida Gwaii sticklebacks. — Behaviour 145: 561-577.

Reimchen, T.E. & Nosil, P. (2001). Lateral plate asymmetry, diet, and parasitism in threespine stickleback. — J. Evol. Biol. 14: 632-645.

Reimchen, T.E. & Nosil, P. (2002). Temporal cycles in divergent selection on spine number in threespine stickleback. — Evolution 56: 2472-2483.

Soulé, M. (1972). Phenetics of natural populations. 3. Variation in insular populations of a lizard. — Am. Nat. 106: 429-446.

Vallortigara, G. & Rogers, L.J. (2005). Survival with an asymmetrical brain: advantages and disadvantages of cerebral lateralization. — Beh. Brain Sci. 28: 575-589.

Van Valen, L. (1965). Morphological variation and width of ecological niche. — Am. Nat. 99: 377-389.

Vermeij, G.J. (1982). Unsuccessful predation and evolution. — Am. Nat. 120: 701-720.

Wiener, J.G. & Rago, P.J. (1987). A test of fluctuating asymmetry in bluegills (*Lepomis macrochirus rafinesque*) as a measure of pH-related stress. — Environ. Pollut. 44: 27-36.

Wootton, R.J. (1976). The biology of the sticklebacks. — Academic Press, London.

Wootton, R.A. (1984). A functional biology of sticklebacks. — University of California Press, Berkeley, CA.

PREDATOR-INDUCED CYCLICAL CHANGES IN LATERAL PLATE FREQUENCIES OF GASTEROSTEUS

by

T.E. REIMCHEN[1,2])

(Department of Biology, University of Victoria, P.O. Box 1700, Victoria, B.C., Canada)

(With 6 Figures)
(Acc. 12-VI-1995)

Summary

This study investigates ecological aspects to intrapopulation variability in the lateral plates of threespine stickleback (*Gasterosteus aculeatus*) at a coastal lake in western Canada. The frequency distribution of the plates, which was slightly skewed to the right and leptokurtic, varied spatially and seasonally. Cyclical frequency changes in plates were most pronounced on sub-adult size classes and these closely tracked (multiple r = 0.77) seasonal differences in the relative proportion of the two major predator groups, trout and diving birds. Trout predation was correlated with an increase in plate number while bird predation was correlated with a reduction in plate number. Such divergent selection pressures, which are probably common in natural populations, would account for the presence of multiple phenotypes and allow for rapid shifts in defensive morphology among different populations where predation regime differs.

Introduction

There has been major empirical and theoretical progress in understanding evolutionary processes in natural populations (reviews in FUTUYMA, 1979; ENDLER, 1986). One of the areas which has received limited attention is evaluating the extent to which intrapopulation variability in continuously distributed traits is functional. Investigations of trophic structures show that variability is advantageous where there is habitat diversity as different phenotypes can exploit slightly different niches (VAN VALEN, 1965; SCHOENER, 1968; POWELL & TAYLOR, 1979; GRANT & PRICE, 1981;

1) E-mail address: REIM@UVVM.UVIC.CA
2) I am grateful to J. ENDLER for comments on the manuscript, to the numerous colleagues including S.D. DOUGLAS and J. MILLER for field assistance and to J.B. FOSTER for original support of the project. The research was initially funded by the Ecological Reserves Unit, Government of British Columbia and subsequently by the Natural Sciences and Engineering Research Council of Canada (grants A5457 to J.S. NELSON and A2354 to the author).

PATTERSON, 1983; HEDRICK, 1985; GRANT & GRANT, 1989; WILSON, 1989; SMITH, 1990). However, direct evidence for such habitat specialization has not been extensively reported in the literature and this has led some workers (*i.e.* FUTUYMA & MORENO, 1988) to suspect that such genotype specialization is probably not important in natural populations.

I have undertaken a study of a meristic trait associated with predator defence to determine, as with previous studies of trophic characters, whether there is a functional context to intrapopulation variability. The bony lateral plates on threespine stickleback (*Gasterosteus aculeatus*), which are associated with predator defense (HAGEN & GILBERTSON, 1972; MOODIE, 1972; MOODIE & REIMCHEN, 1976; GROSS, 1978; REIMCHEN, 1983; BAŃBURA *et al.*, 1989; reviews in WOOTTON, 1984, REIMCHEN, 1994), are highly variable in number and position among individuals and populations. Plate numbers are heritable and controlled at multiple loci (HAGEN & GILBERTSON, 1973a) and they usually exhibit a bell-shaped frequency distribution. Over a four year period I monitored activity of piscivores at a remote lake off the west coast of Canada and intensively tracked both spatial and temporal variability in the lateral plate phenotype frequencies on an endemic population of stickleback. The results provide insight into the functional attributes of lateral plate variability and allow comment on the general importance of microevolutionary processes in population differentiation.

Study area and life history

Drizzle Lake (110 ha) occurs in a low elevation *Sphagnum* bog on the Queen Charlotte Islands off the west coast of Canada. Lake perimeter and bathymetry are simple and there are no physical impediments to fish movement within the lake (Fig. 1). The stickleback are large-bodied forms which are reproductively isolated from small-bodied phenotypes found in adjacent streams (REIMCHEN *et al.*, 1985). Molecular analyses indicate that the population is derived from marine stickleback, probably in post-glacial periods (GACH & REIMCHEN, 1989; O'REILLY *et al.*, 1993). During spring and summer, stickleback occupy diverse habitats in the lake (littoral/limnetic: benthic/pelagic) but during autumn and winter, fish are more abundant offshore than inshore. First reproduction occurs at approximately three years and lifespan can reach at least 8 years in the lake (REIMCHEN, 1992b).

Approximately one-half of the yearly predation on stickleback (N = 562 000) is caused by trout (*Oncorhynchus clarki*) and one-half is caused by avian piscivores, principally loons (*Gavia immer*) and grebes (*Podiceps grisegena*). Trout take primarily small fish (< 50 mm), secondarily sub-adults (50-70 mm) and low numbers of adults (> 70 mm). Birds take primarily sub-adults and adults. Both groups of piscivores are found over much of the lake but trout tend to be more common near the shore and birds are more prevalent in open water (REIMCHEN, 1990, 1994). Proportion of these two predator groups change substantially within and among seasons.

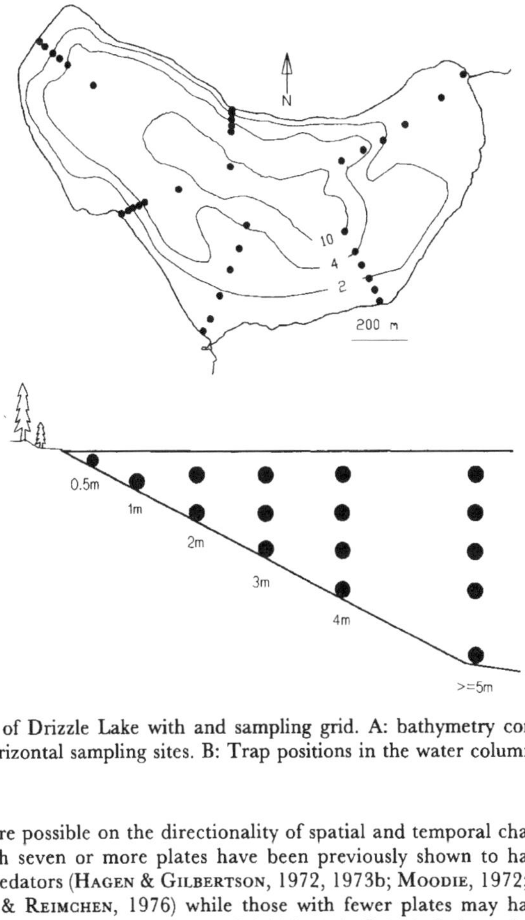

Fig. 1. Outline of Drizzle Lake with and sampling grid. A: bathymetry contours (m) and horizontal sampling sites. B: Trap positions in the water column.

Predictions are possible on the directionality of spatial and temporal changes in plates. Stickleback with seven or more plates have been previously shown to have advantages against trout predators (Hagen & Gilbertson, 1972, 1973b; Moodie, 1972; Moodie *et al.*, 1973; Moodie & Reimchen, 1976) while those with fewer plates may have advantages against bird predators (Reimchen, 1994). Stickleback in Drizzle Lake typically have from two to eight plates and if the range in plates is an adaptation to trout and bird predators, the frequency distributions should be shifted to higher counts with increased trout predation and to lower counts with increased bird predation. In contrast, uniformity of phenotype frequencies over these spatial and temporal ecological gradients would indicate that the plate variability was not an adaptation to the divergent predation regime.

Methods

Nineteen sampling transects were undertaken from 1980 to 1983 spaced four to ten weeks apart. In each transect, I set 96 wire-mesh funnel traps around the lake to encompass six different geographical sectors, six horizontal sites from shoreline to open water, and one or more sites in the water column (Fig. 1). Horizontal positions corresponded to depth contours of 0.5, 1, 2, 3, 4 and > = 5 m. These were baited (1 cc old cheddar cheese) and left for 48 hours. On retrieval, all adult fish, which comprised about 90% of each sample, were

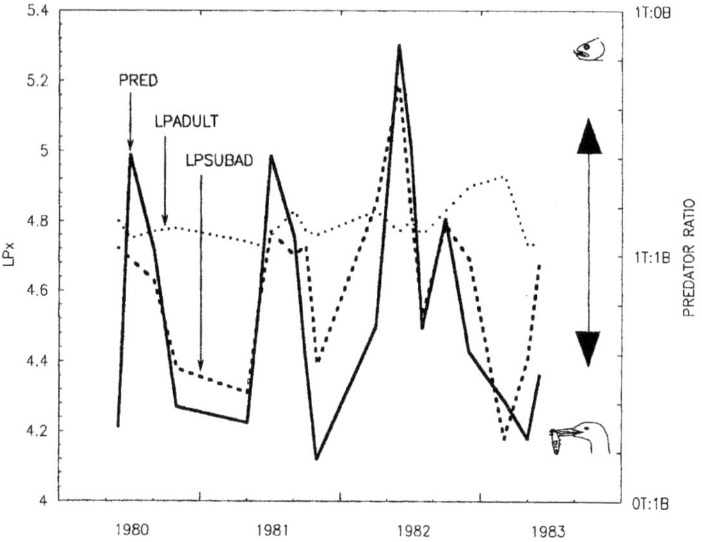

Fig. 6. Summary plot of mean number of lateral plates for adult (LPADULT) and sub-adult (LPSUBAD) stickleback against predation regime (PRED) over the study period. Right axis shows relative predation by the major groups (trout - T and diving birds - B).

Rather, it is due to the cyclical tracking between RATIO and LP (Fig. 6). The three major periods with increased bird predation over the 4 years were in each case followed by a sharp reduction in LPx.

Directional selection differentials were calculated for mean LP between sequential samples among the 19 transects (Table 1). For adults, absolute values for directional selection (i, sign ignored) ranged from 0.01 to 0.25 ($\bar{x} = 0.05$) of which a single comparison (Trans #18) reached statistical significance ($p < 0.02$). For sub-adults, i (sign ignored) ranged from 0.02 to 0.69 ($\bar{x} = 0.25$), three of which (Trans #14, #15, #19) were statistically significant ($p < 0.05$). Direction of change (+/-) was largely symmetrical for both adults and sub-adults. Two of the significant values were negative and two were positive.

Discussion

These data suggest that predator-mediated cyclical selection is occurring on the lateral plates in sub-adult stickleback, with trout predation tending to produce an increase in plate means and bird predation tending to

TABLE 1. Directional selection differentials (i) on lateral plates for adult and sub-adult stickleback at Drizzle Lake

	Adult			Sub-adult		
Transect	i	t	df	i	t	df
2	0.05	0.72	802	-0.04	0.23	172
3	0.02	0.28	770	0.32	1.65	139
4	0.01	0.07	397	-0.69	1.72	44
5	-0.05	0.27*	1113	-0.05	0.15	101
6	-0.01	0.23	1978	0.26	0.95	74
7	0.03	0.56	1350	0.22	1.29	122
8	0.08	0.96	678	-0.08	0.45	123
9	-0.06	0.68	587	0.02	0.13	130
10	-0.01	0.11	425	-0.39	1.28	87
11	0.08	0.71*	424	0.43	1.13	24
12	-0.06	0.80*	884	0.27	0.69	24
13	0.03	0.51	1053	-0.32	0.97	88
14	-0.04	0.57	678	-0.33	2.43	217
15	0.08	1.08	1142	0.31	2.30*	264
16	0.07	1.24	1294	0.10	0.77	239
17	0.03	0.39	630	-0.51	1.59	129
18	-0.25	2.73*	790	0.20	0.73	86
19	0.01	0.22	1934	0.30	2.03	210

T-values show statistical significance of i. *: $p < 0.05$.

reduce plate means. Although there were slight spatial associations in the lake between plate frequencies and predator group, it was primarily temporal variability in the relative proportion of each predator group that produced the recurrent changes within and between seasons. Such a bi-directional selective regime will contribute to the maintenance of lateral plate variability within the population as there are multiple optima. These data are consistent with hypotheses of adaptive variation for continuously distributed traits (VAN VALEN, 1965; GRANT & PRICE, 1981; WILSON, 1989) and extend the conclusions from previous studies of gasterosteids that have demonstrated close associations between morphology and predation among different geographical localities (*e.g.* HAGEN & GILBERTSON, 1972; MOODIE, 1972; MOODIE & REIMCHEN, 1976; GROSS, 1978; REIMCHEN, 1980; review in REIMCHEN, 1994). If the changes observed are in fact the result of differential predation, then the data also suggest that population differentiation could occur very quickly when predation regimes change.

Lake populations from the northern hemisphere exhibit regular seasonal differences in a diversity of biophysical parameters each of which might influence strength and direction of selection. Predatory fish forage throughout the year but their consumption rate drops during winter while avian piscivores will forage throughout the year unless excluded by winter ice cover. In coastal British Columbia, large flocks of non-breeding common loons move from marine waters onto some freshwater lakes only during July and forage extensively on stickleback during a three week period (REIMCHEN & DOUGLAS, 1980). Red-necked grebe occur on coastal lakes in autumn, winter and early spring but are absent during summer. Doubled-crested cormorant (*Phalacrocorax auritus*) occur on lakes during January and February but not during the rest of the year. Common merganser (*Mergus merganser*) occur in large numbers during early spring and late autumn and occasionally forage extensively on threespine stickleback at the lake surface during twilight (REIMCHEN & DOUGLAS, 1984). In some streams of northern Europe, mergansers forage extensively on *Gasterosteus* only in spring (RAD, 1980). Anadromous cutthroat trout enter rivers during high runoff in October and November and migrate into lakes where they overwinter (REIMCHEN, 1990). When these seasonal differences are combined with yearly fluctuations in the proportion of each species, there is ample opportunity for temporal variability in selective regime and therefore, it seems reasonable that short term temporal variation in selection would be common in these populations. Such cyclical selection has been observed in other taxa (*e.g.* SHEPPARD, 1951; FORD, 1964; SAMOLLOW, 1980; review in ENDLER, 1986).

Gene flow between populations can also be expected to contribute to the persistence of variability in a strongly selected population (FALCONER, 1967). Investigations of the large-bodied stickleback from Mayer Lake show evidence for hybridization near the stream mouths between the lake and stream form (MOODIE, 1970). The large-bodied stickleback in Drizzle Lake spatially overlap with a small- bodied stickleback which breeds in streams that flow into and out of the lake. These forms differ in multiple morphological traits and there is no evidence for hybridization (REIMCHEN *et al.*, 1985). Recent mitochondrial DNA analyses of numerous freshwater and marine stickleback from the Queen Charlotte Islands demonstrate that most populations have multiple haplotypes while Drizzle Lake stickleback have only a single haplotype (O'REILLY *et al.*, 1993) which further suggests that gene flow is not a substantive cause for plate variability in this population.

Some of the phenotypic variability within populations may be associated with habitat differences among the sexes (SELANDER, 1966; SLATKIN, 1984; BOROWSKY, 1990). In stickleback populations on the Queen Charlotte Islands, males have slightly higher plate counts than females (MOODIE, 1972; MOODIE & REIMCHEN, 1976; REIMCHEN *et al.*, 1985) and therefore, sex ratio differences within the lake could shift plate distribution. I compared plate distribution of 812 males and 1289 females from Drizzle Lake; mean plate counts differed by 1.9% (male = 4.78, female = 4.69). If all of the adults near the shore were males, this would account for less than one-quarter of the 10% change in plate frequencies in the littoral habitat. In actuality, there was only a marginal excess of males near the shore (57%, N = 310) which shows that habitat differences between the sexes only contribute a small proportion to the spatial differences in plate frequencies.

It was primarily variation in sub-adults rather than adults that tracked the divergent predation regime. Presumably, this is because sub-adults are a more defined age cohort than those of the adults and are subject to much higher levels of predation. Such a trend has also been observed in studies on polymorphic traits in several diverse taxa (*i.e.* GIESEL, 1970; REIMCHEN, 1979, 1980; SAMALLOW, 1980). Seasonal movement of adults over multiple seasons would obscure site-specific or time-specific differentiation that could have developed during earlier periods in the life history.

ENDLER (1986) summarized published data on selection differentials in natural populations and calculated an average value of 0.59 among 36 species (maximum 2.0). In the current study, average values tend to be low (sub-adults = 0.27, adults = 0.05) and in most cases statistically non-significant. Because there was such close temporal tracking of the plates to predation regime, these non-significant differentials would appear to be biologically relevant. Small differentials such as these are probably the norm and distinguishing these from random sampling effects requires either exceptionally large samples or alternatively, sequential sampling over temporally varying selective pressures.

I have currently assumed that the changes in plate frequencies are the immediate result of differential predation but there are several issues which suggest that additional factors must be operating. To establish whether the observed predation rate will produce the observed changes in plate frequencies, I modelled the tracking between predation regime and lateral plates for period between June 1982 and July 1982 when plate means dropped about 6%, from 4.83 to 4.53. Total population size estimates for adult stickleback (which comprise at least five separate year classes, REIMCHEN, 1992b) ranged from 30,000 to 120,000 (REIMCHEN, 1990); for sub-adults (which comprise a single year class), I estimate a population of 80,000. Birds consumed approximately 8000 more stickleback than trout during this period, most of which were sub-adults. Initial lateral plate frequencies were 0.41, 0.35 and 0.24 for LP4, LP5 and LP6 respectively for an average of 4.83 (see Fig. 1). Let us initially assume that plate phenotypes have a capture probability by birds of 1, 1.1 and 1.2 for LP4, LP5 and LP6 respectively. After one month, mean plates for the population would decline from 4.830 to 4.824, which is about 2% of the observed decline. If we assume that LP6 had twice the capture rate of LP4 (LP4 = 1, LP5 = 1.5, LP6 = 2), mean plate number would decline from 4.830 to 4.806, only 8% of the observed decline. To produce the observed change would require either a substantially lower population estimate or a greater differential among the phenotypes.

Temporal changes in phenotype frequencies might also result from behavioural differences among the phenotypes. For example, if phenotypes with increased plates became less attractive during periods of increased bird predation, they would be underrepresented in the trapping grid. There is precedence for this suggestion. At Mayer Lake, stickleback with seven plates differ in their nesting habitat and responses to disturbance compared with non-sevens (MOODIE, 1970). Stickleback with seven plates from other populations in coastal British Columbia are less likely to approach a predatory fish than non-sevens (MOODIE *et al.*, 1973). In a stickleback population from western Europe, those with four or five plates are more aggressive than those with ten or twelve plates (HUNTINGFORD, 1981). Spatial variability in plate morph frequencies in a large quarry has also been detected with respect to depths in the water column and horizontal positions along the shore (MACLEAN, 1980). Stream channels have sharp clines in plate expression (HEUTS, 1947; BELL & RICHKIND, 1981; BAUMGARTNER, 1992). Dorsal and pelvic spine expression in stickleback from Boulton Lake, Queen Charlotte Islands, differed between littoral and limnetic habitats (REIMCHEN, 1980). Apart from the defensive complex of stickleback, other characters also exhibit within-locality variability including trophic structures (SCHLUTER & MCPHAIL, 1993) and vertebral phenotypes (REIMCHEN & NELSON, 1987). If some phenotypes have higher fitness in one habitat than another, then a behaviour that produces a preference for the appropriate habitat will be strongly favoured (POWELL & TAYLOR, 1979, GARCIA-DORADO, 1986).

Morphological and behavioral differences within threespine stickleback populations have been described at different geographical and temporal scales. Currently, the differences at each scale broadly appear to be functionally associated with ecological differences (HAGEN & GILBERTSON, 1972; MOODIE & REIMCHEN, 1976; GROSS, 1978; BELL & RICHKIND, 1981; SCHLUTER & MCPHAIL, 1993; BELL, 1994; this study) indicating that microevolutionary forces are important in shaping population differentiation. Such a conclusion reinforces the traditional paradigm of adaptive radiation of form (MAYR, 1966; CAIN, 1971) and does not support developing views (*i.e.* GOULD, 1984) that much of the geographical variation within species is a non-adaptive product of history.

References

BAŃBURA, J., PRZYBYLSKI, M. & FRANIEWICZ, P. (1989). Selective predation of the pike *Esox lucius*: comparison of lateral plates and some metric features of the three-spined stickleback *Gasterosteus aculeatus*. — Zool. Scripta 18, p. 303-309.

BAUMGARTNER, J.V. (1992). Spatial variation of morphology in a freshwater population of the threespine stickleback, *Gasterosteus aculeatus*. — Can. J. Zool. 70, p. 1140-1148.

BELL, M.A. (1994). Paleobiology and evolution of threespine stickleback. — In: The evolutionary biology of the threespine stickleback (M.A. BELL & S.A. FOSTER, eds). Oxford University Press, Oxford, p. 438-471.

—— & RICHKIND, K. A. (1981). Clinal variation of lateral plates in threespine stickleback fish. — Am. Nat. 117, p. 113-132.

BOROWSKY, R. (1990). Habitat choice by allelic variants in *Xiphophorus variatus* (Pisces: Poeciliidae) and implications for maintenance of genetic polymorphism. — Evolution 44, p.1338-1345.

CAIN, A.J. (1971). Animal species and their evolution. — Hutchinson University Library, London, 192 pp.

ENDLER, J.A. (1977). Geographic variation, speciation and clines. — Princeton University Press, Princeton, New Jersey, 246 pp.

—— (1986). Natural selection in the wild. — Princeton University Press, Princeton, New Jersey, 337 pp.

FALCONER, D.S. (1967). Introduction to quantitative genetics. — The Ronald Press Company, New York, 365 pp.

FORD, E.B. (1964). Ecological genetics. — Methuen, London.

FUTUYMA, D.J. (1979). Evolutionary biology. — Sinauer Associates, Sunderland, Mass.

—— & MORENO, G. (1988). The evolution of ecological specialization. — Ann. Rev. Ecol. Syst. 19, p. 207-233.

GACH, M.H. & REIMCHEN, T.E. (1989). Mitochondrial DNA patterns among endemic stickleback from the Queen Charlotte Islands: a preliminary survey. — Can. J. Zool. 67, p. 1324-1328.

PREDATOR HANDLING FAILURES OF LATERAL PLATE MORPHS IN GASTEROSTEUS ACULEATUS: FUNCTIONAL IMPLICATIONS FOR THE ANCESTRAL PLATE CONDITION

by

T.E. REIMCHEN [1,2]
(Department of Biology, University of Victoria, P.O. Box 3020, Victoria, BC, V8W 2Y2, Canada)

(Acc. 25-II-2000)

Summary

The completely plated morph in the threespine stickleback (*Gasterosteus aculeatus*) is considered to be the ancestral plate condition and is the predominant morph inhabiting marine waters and numerous northern freshwaters. Evolutionary aspects of this distribution have been widely addressed yet functional mechanisms remain obscure. Experiments described here using the common piscivore (*Oncorhynchus clarki*) show that the posterior plates of *G. aculeatus* interfere with the swallowing abilities of the piscivore, possibly by disrupting pharyngeal jaw retraction, and this leads to increased escape opportunities of the stickleback. The advantage of the completely plated condition is most expressed at higher ratios of prey diameter to predator mouth diameter and appears to have a defensive effect comparable to that of dorsal and pelvic spines. This attribute, combined with the physical protection that plates offer against puncturing, would be particularly beneficial where there is high probability of capture by toothed predators. Such a selection regime appears to characterize the predominantly limnetic and pelagic habitats where marine stickleback are found and may account for the wide geographical distribution of the completely plated morph and its persistence from the Miocene.

Keywords: functional morphology, *Gasterosteus*, lateral plates, stasis, predator-prey.

[1] E-mail address: reimchen@uvic.ca

[2] I am grateful to M.A. Bell, C. Bergstrom, S. Douglas and G.E. Moodie for discussion. This work was supported by the Natural Sciences and Engineering Council of Canada (A2354).

 Behaviour 137, 1081-1096

Introduction

Persistence and stasis of ancestral traits occur in the fossil record of a diversity of marine taxa (Stanley & Yang, 1987). The constancy of form, occasionally across major biophysical boundaries in the geological record, has led to an emerging view that morphological and epigenetic constraints fundamentally limit the capacity for change (Gould & Eldridge, 1977; Oster & Alberch, 1982; review in Schlichting & Pigliucci, 1998). The persistence of species with ancestral traits may be facilitated in peripheral habitats such as the oceanic depths with reduced species composition but it is not clear how ancestral traits will persist in highly competitive surface or shallow water habitats. Studies of predator-defences in marine gastropods show evidence for continuity of predator-prey interactions in the fossil record (Vermeij, 1986) which would suggest that stasis in these traits may reflect long-term stability of the adaptive landscape rather than constraints (Charlesworth *et al.*, 1982; Hecht & Hoffman, 1986; Levinton, 1988).

The threespine stickleback (*Gasterosteus aculeatus*) has been a useful model for evaluating a broad range of evolutionary processes (Bell, 1984) and may offer insight into factors influencing stasis. This species is widespread in northern marine and freshwater waters, including littoral and pelagic habitats. Bony lateral plates on these fish, which are highly variable among freshwater populations, are largely monomorphic for the completely plated morph throughout their distribution in marine waters of the Atlantic and Pacific oceans (Heuts, 1947; Münzing, 1963; reviews in Wootton, 1976, 1984; Bell, 1984). Analyses of the fossil record indicates that the completely plated condition has continuity back to the Miocene and probably represents the ancestral condition in this taxon (Bell, 1977, 1981).

Several hypotheses have been proposed to account for the occurrence of the complete morph in *G. aculeatus*. In marine populations, elevated salinity may favour this morph, perhaps due to genetic linkage between plate number and ion regulation (Bertin, 1925; Heuts, 1947). Among freshwater populations, however, there is no correlation between relative salinity and incidence of the complete morph (Hagen & Gilbertson, 1972). Low water temperatures are correlated with presence of the complete morph in marine waters (Bertin, 1925; Heuts, 1947) as well as freshwaters (Munzing, 1972; Hagen & Moodie, 1982) but it is unclear how this relates functionally to number of bony plates. Elevated frequencies of the complete morph in large

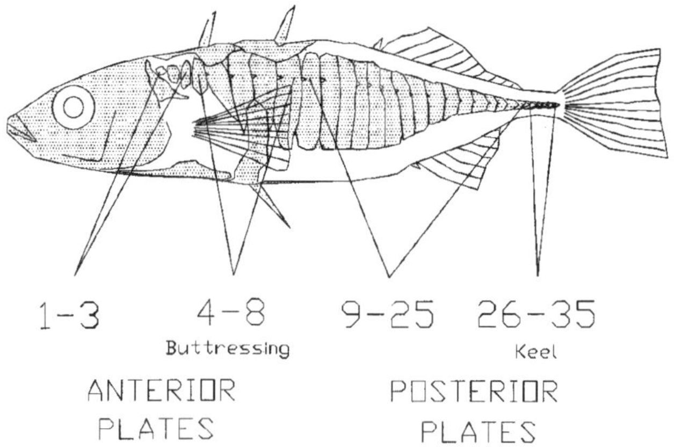

Fig. 1. Schematic of lateral plates on the complete morph of threespine stickleback. Unshaded areas show unprotected integument.

rivers of western Europe could be historical markers of late Pleistocene distribution (Münzing, 1963, 1972) and signify levels of introgression and hybridization between marine and freshwater populations (Miller & Hubbs, 1969; however, see Hagen & McPhail, 1970). These hypotheses each have a common assumption that the presence of plates along the length of the trunk is not directly functional but rather selectively neutral and/or correlated to other functional attributes. If this is correct, then the wide distribution and long term persistence of the completely plated morph would largely represent developmental constraints.

Plates on *Gasterosteus* are important during predator defence and potentially these traits extend from long-term associations with predators. The plates can be partitioned into functional-specific zones including anterior and posterior regions of the trunk (Fig. 1). Plates on both regions reduce the amount of epidermal puncturing by toothed predators and this increases probability of survival among stickleback which have escaped during handling by the predator (Reimchen, 1992). Biomechanical studies demonstrate that anterior plates from positions 4 through 8 buttress the dorsal and pelvic spines from deflection during predator handling (Reimchen, 1983). Combination of these two functional mechanisms will contribute to the wide distribution of the complete morph.

During a recent series of experiments using large-spined, low plated stickleback and a common predator (*Oncorhynchus clarki*), I observed that increased cross-sectional diameter of the stickleback (with erect spines) was directly correlated to increased escape rate of the stickleback and increased handling time by the trout (Reimchen, 1991a, b). Yet there was considerable residual variance in handling time for any given prey size and it appearred that number of lateral plates might influence predator success as the fish with the highest number of plates required the longest handling time (Reimchen, unpubl. obs.). To examine this possibility more thoroughly, I obtained stickleback from a population which had both low plated and completely plated individuals. An additional characteristic of these fish was their small dorsal and pelvic spines relative to those present in the first locality. Because spines are very effective in reducing the handling success of predatory fish and large spines are more effective than small spines (Hoogland *et al.*, 1957), I anticipated that the stickleback with small spines would be more readily swallowed compared with long-spined stickleback, independent of the number of posterior bony plates. In this study, I test predator handling success primarily in relation to number of lateral plates and secondarily in relation to the spine length.

Methods

General experimental procedures follow those described in Reimchen (1991a). Stickleback were obtained from two localities, Drizzle Lake and Delkatla Pond, both located on the northern regions of the Queen Charlotte Islands, western Canada (morphometrics in: Reimchen *et al.*, 1985; Gach & Reimchen, 1989). Drizzle Lake stickleback are all of the low morph and usually have from two to seven lateral plates (mode 5) and are subject to predation by a broad diversity of avian and salmonid piscivores (Reimchen, 1994). Delkatla stickleback are polymorphic but generally have from seven to ten plates or more than 30 plates. This population, found on an estuary, appears to be recently derived from completely plated marine ancestors and is currently subject to predation by predatory fish and birds (Reimchen, unpubl. data). In the first series of experiments, I used stickleback from Drizzle Lake and Delkatla Pond in order to assess the combined effects of spine size and number of plates on handling success. In the second series of experiments, only Delkatla fish were presented in order to compare prey which differed predominantly in plate number. Before each trial for both series of experiments, I scored plate morph and measured body length (standard length, SL). In each trial where I used fish from two populations or of different morphs, I chose fish of similar body size. Stickleback were added individually to 120 l glass tanks each containing one cutthroat trout. The sticklebacks were usually attacked within several seconds and the outcome recorded. For all trials, I noted total handling time (s) and number of escapes following initial capture. Handling time represented the interval from first capture to completion of swallowing or to final rejection of the prey when the trout showed no additional interest in the stickleback. In some trials, I was able to observe and record the

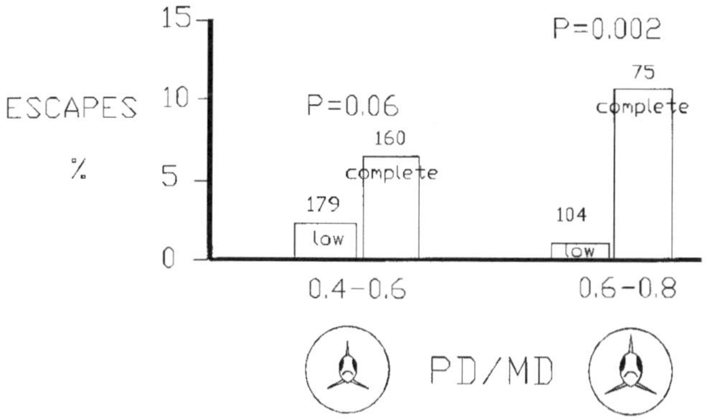

Fig. 5. Frequency of escapes for PD/MD grouping for plate morphs from Delkatla (experiment 2). Numbers show sample size and statistical probabilities for log-likelihood χ^2.

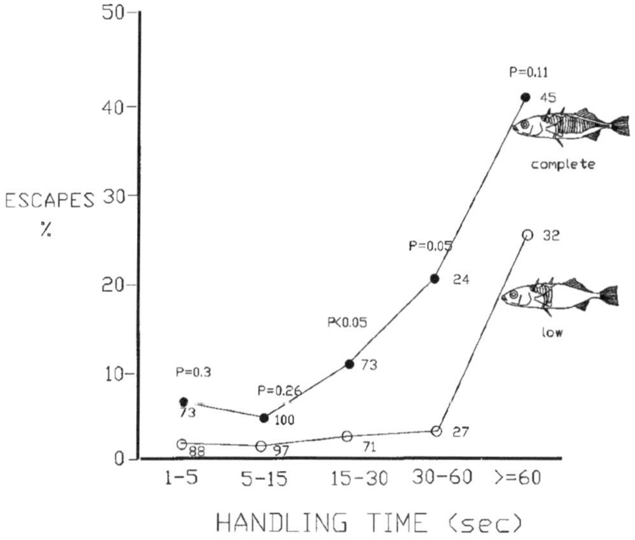

Fig. 6. Frequency of escapes against handling time for plate morphs from Delkatla (experiment 2). Numbers show sample size and statistical probabilities for log-likelihood χ^2.

Discussion

Previous biomechanical studies on *G. aculeatus* have demonstrated two functional attributes to the lateral plates: (1) four of the anterior plates provide the major structural buttressing for the dorsal and pelvic spine (Reimchen, 1983) and (2) each plate reduces the extent of epidermal lacerations during capture by toothed piscivores and this results in higher survival rate of stickleback with greater numbers of plates (Reimchen, 1992). Results from the present study demonstrate a third function. Posterior plates are associated with increased handling failures of the predator and this effect is accentuated as stickleback size reaches the maximum gape of the predator.

The actual mechanism for the differential escape rate of the plate morphs has not yet been thoroughly established but it appears to be associated with pharyngeal jaw activity. Escape of the stickleback occurred both during the initial capture phase when the trout attempts to fracture or depress the erect dorsal and pelvic spines and during the second phase when the trout attempts to swallow the stickleback. In many bony fishes, including trout and other Salmonidae, the prey are raked into the esophagus from the oral cavity with repeated protraction and retraction of the pharyngeal jaws which occur at the posterior region of the oral cavity (Liem & Kaufman, 1984; Lauder, 1985; Galis, 1993). The activity of these internal jaws is not easily monitored as their motion is not coupled with mandibular jaw action and can occur when the mouth is closed. I suspected that pharyngeal jaw action might produce a weak audible stimulus and lowered a hydrophone in the tank during a feeding event. Rhythmic scratching was detectable during the swallowing phase and presumably reflected pharyngeal raking on the bony plates of the stickleback for it abruptly ceased when the stickleback was swallowed or expelled from the mouth. Based on the prolonged pharyngeal jaw activity when a completely plated morph had been captured, I suggest that posterior lateral plates directly interfere with pharyngeal jaw action and that increased number of lateral plates increase the extent of this interference. This can result in longer handling time and increased escape rate of the stickleback. That swallowing difficulty and elevated escape rate was greater when stickleback cross-sectional diameter increased relative to the maximum gape of the trout is consistent with this hypothesis.

Alignment of the prey in the mouth is an important factor in predator handling efficiency (Reimchen, 1991b) and may influence the selective differential between the completely plated and low plated morph of stickleback. From 50 to 75% of the strikes of the trout were on the mid- and posterior trunk of the stickleback and this can result in tailfirst alignment in the mouth which in turn leads to prolonged handling time. As the pharyngeal jaws rake the fish tailfirst into the esophagus, the erect but posteriorly angled dorsal and pelvic spines can lodge against the palate or branchial arches blocking further movement of the stickleback into the esophagus. In this alignment, pharyngeal raking occurs on the trunk posterior to the spines. On a low plated fish, this is likely to lead to epidermal lacerations on the trunk and increased probability of mortality if the stickleback escapes (*e.g.* Reimchen, 1992). Increased number of posterior plates would minimize epidermal lacerations during pharyngeal raking and increase the post-capture survival.

One of the most prominent features of *G. aculeatus* are the large dorsal and pelvic spines and these probably constitute the first line of defence against predators. In one of the seminal papers on prey defences in fishes, Hoogland *et al.* (1957) demonstrated that spines increase the effective diameter of the stickleback and injure the soft parts of the predator's mouth, both of which lead to increased escape opportunities of the stickleback. In these experiments, pike (*Esox*) and perch (*Perca*) were presented with a range of prey that differed in the extent of spine expression. *G. aculeatus*, with their large dorsal and pelvic spines, were eaten at lower frequency than the nine-spined stickleback (*Pungitius*), which have comparatively small spines, and these were eaten less frequently than non-spiny prey including minnow (*Phoxinus*), roach (*Rutilus*) and rudd (*Scardinius*). Removal of dorsal and pelvic spines on *G. aculeatus* greatly increased the palatability of the stickleback although not to the equivalent level of the species lacking spines. Given this outcome, it was surprising that in my experiments, Delkatla stickleback with short dorsal and pelvic spines and with low PD/MD ratios had higher post-capture escape rate than Drizzle Lake stickleback of equivalent body size with large and robust dorsal and pelvic spines and high PD/MD ratios. One of the additional distinctions between these two populations was the presence of posterior plates on Delkatla stickleback and these plates appear to increase handling failures of the predator. If so, then some of the differences in escape rate observed by Hoogland *et al.* (1957) among the five species which were attributed to differences in spines could have been influenced by lateral plates as among the prey offered to the predators, the least preferred (*i.e. G. aculeatus*), was the only prey species with bony plates on the trunk.

The completely plated morph of *G. aculeatus* appears to exhibit a geographical association with water temperature. Among freshwater populations, this morph tends to be more prevalent in northern latitudes (Bertin, 1925; Munzing, 1963, 1972; Wootton, 1976 for review) and on the eastern edges of the continents where winter temperatures are lower (Hagen

& Moodie, 1982). Among southern populations as in California, the complete morph tends to be more common in colder streams but thermal factors are less predictive than stream gradient (Baumgartner & Bell, 1984). One of the potential selective mechanisms coupling plates and temperature is the ratio of pursuit to capture success by the predator, as this can lead to alternate defence adaptations (Reimchen, 1994). Prey defense following capture by a toothed predator favours increased complement of plates (Reimchen, 1992, 1997) and therefore any ecological variable or behavioural trait which leads to increased likelihood of capture should produce an adaptive regime favouring the completely plated morph. One such ecological variable could be low water temperature. Metabolic rates in an ectotherm will be reduced at lower temperatures and this will lead to reduced burst swimming performance (Webb, 1978). Any such reduction which results in an increased capture rate by a larger-bodied predator with greater thermal inertia, is likely to generate a selective advantage to the complete morph.

Variability in the complete morph may also result from for differential predation on plate number. Bańbura *et al.* (1989) examined pike (*Esox lucius*) stomachs and found that stickleback with 33-35 plates were eaten less frequently than those with 30-31 plates compared with population frequencies. These results are consistent with predictions from my study as they demonstrate the potential for increased escape rate for higher plate numbers. Yet the results could also be explained by differential capture of these phenotypes rather than differential escape after capture if there were behavioural differences among morphs (Huntingford, 1993). Lateral plates comprise a meristic series corresponding to the underlying body segmentation and, as such, on a completely plated morph, fish with fewer segments will have correspondingly fewer plates. Because differences in number of body segments can alter burst acceleration (Swain & Lindsey, 1984; Swain, 1992), the differential predation observed by Bańbura *et al.* (1989) could be a consequence of differential swimming performance rather than differential plate number.

One of the major trends to emerge from large-scale collections of *G. aculeatus* was the prevalence of the completely plated morph in marine waters (Wootton, 1976, for review). Elevated salinity may be one of the selective mechanisms for the occurrence of the complete morph in these habitats (Bertin, 1925; Heuts, 1947). An additional feature of marine habitats other than salinity, is the overwhelming predominance of the limnetic zone and great depth of the water column relative to that found in even the largest freshwater lakes. Threespine stickleback occur on the continental shelves but also in open surface waters of the Atlantic and the Pacific Oceans where they are exposed to a diverse range of toothed predators (Gross, 1978; Reimchen, 1994). Salmonids (*Oncorhynchus* spp.) are known to forage on stickleback 1000 km from the nearest land (Quinn & Light, 1989; Deagle *et al.*, 1996). There is little shelter available in open water habitats and a small fish would be at high risk of capture compared with littoral or benthic habitats where shelter is often available and where escape during pursuit is possible. Under open water conditions, the completely plated morph should have higher fitness than the low morph given the increased likelihood of escape during capture (this study) and the increased likelihood of survival following escape due to reduced risk of epidermal puncturing (Reimchen, 1992). Such a selective regime would contribute to the prevalence of the completely plated morph in marine waters. It would also account for some of the freshwater distribution of this morph. In recent geographical surveys of 180 freshwater lakes on the Queen Charlotte Islands (Reimchen *et al.*, 1985; Reimchen, 1989, unpubl. obs.), the complete morph was common in only three localities and these were characterized by their large size, steep shore profile, lack of littoral zone, high water clarity and prevalence of open water piscivores. These are the same general ecological features as those observed in open water marine habitats.

This study suggests a potential functional mechanism for the stasis of the ancestral plate condition in marine *G. aculeatus*. Most modern bony fishes (Euteleostei) use the pharyngeal jaw apparatus for prey handling (Lauder, 1983) and these fish have been common in marine waters from the Upper Cretaceous (Carroll, 1988), well before the first Miocene appearance of *Gasterosteus* (Bell, 1977). As stickleback comprise a common element in the diet of numerous marine piscivores, it is reasonable to hypothesize that the persistence of the complete morph from the Miocene, and its presence in some cold freshwater lakes extends from long-term continuity of an adaptive landscape mediated through post-capture handling failures in bony fishes.

Physiology & Behaviour

Frank A. von Hippel

Early European ethologists looked for mechanistic explanations for behaviour in the animal's physiology, with the view that behaviour and physiology were different expressions of the same phenomena (Barlow, 1989). The internal and external causes of behaviour, argued Tinbergen (1942) in his landmark paper, "An objectivistic study of the innate behaviour of animals," relate to one another in a hierarchy. Later, in the most important paper of his career, "On aims and methods of Ethology," Tinbergen (1963) sharpened the theoretical underpinnings of ethology with his four why's: "Huxley likes to speak of 'the three major problems of biology': that of *causation*, that of *survival* value, and that of *evolution* — to which I should like to add a fourth, that of *ontogeny*." In his discussion of ontogeny (and therefore the nature-nurture controversy), Tinbergen wrote, "If I were to elaborate this further I should have to cross swords with my friend Konrad Lorenz himself — both a pleasure and a serious task requiring the most thorough preparation — but this is not the occasion to indulge in swordplay." So were born Tinbergen's four why's: the development of behaviour ("Ontogeny"), its evolutionary history reconstructed through comparative biology ("Evolution"), its adaptive qualities ("Survival Value"), as well as its physiological basis and correlates ("Causation"). "I shall devote the next pages to some remarks on each of these four problems as they apply to behaviour," wrote Tinbergen. "If these remarks appear to some readers unsophisticated, I should like to remind them of the fact that Ethology is a science in its infancy, where even a little plain common sense can help."

Because early ethology and physiology were so tightly linked, threespine stickleback developed as a model for the study of physiology in tandem with its development as a model for behavioural studies. The first paper of this section, by A.J.H. De Ruiter and S.E. Wendelaar Bonga (1985), links together the large ethological literature on the stickleback reproductive cycle with a fascinating physiological process: "The nestbuilding behaviour displayed by male sticklebacks during the reproductive period requires osmoregulatory adaptations as drastic as seaward migration. However, whereas migration is shown by many other fish, the nestbuilding behaviour of sticklebacks has consequences for osmoregulation that are quite unique among the more than 20,000 species of teleosts." This is because the male kidney is completely converted from an osmoregulatory organ into a glue-producing organ for the purpose of nest construction. In the process it may increase to seven times its original size. This transformation process is mediated by androgens. De Ruiter and Bonga investigated how the stickleback compensates for lost kidney functions, such as the regulation of osmotically accumulated water, with other organs, thereby freeing up the kidneys to produce glue. Importantly, they noted that kidney transformation can be induced with exogenous testosterone in castrated males and in females.

Although many details of the kidney transformation had been discovered by the time of De Ruiter and Bonga's (1985) paper, the glycoprotein that comprises the glue had not yet been isolated. This occurred 14 years later, when the name "spiggin" was coined by Steffan Jakobsson (Jakobsson et al., 1999) after the Swedish word for stickleback, 'spigg.' The fact that the stickleback has this unique glue protein is part of the reason that it has been used so much as a biomonitoring tool in contaminants research, as described below.

In the second paper of this section, Bertil Borg and Ian Mayer (1995) summarized the state of knowledge of stickleback endocrinology, with a focus on effects of androgens on both behavioural and non-behavioural traits, including effects of photoperiod on reproduction. In the third and final paper of this section, Borg et al. (2004) reviewed studies that demonstrate, using experiments that have an elegance reminiscent of Tinbergen's, that the primary cells involved in acquiring information on photoperiod in sticklebacks are not in the eyes. Rather, photoreceptors in the pineal organ and perhaps elsewhere are key to regulation of reproductive physiology and behaviour. This work involved stitching black plastic covers over stickleback heads in such a way that extra-retinal photoreceptors were blocked from the light while the eyes were not.

The early ethological literature focused a great deal of energy on various competing "drives," which led to many interesting behavioural patterns, such as displacement activities. The various drive theories lost their gumption over time, however, because a firm basis in physiology was absent. The studies in this section demonstrate how far that understanding of physiology has come, with, for example, stickleback manipulated through castration and hormone replacement capsules experimentally matched with the use of sham-operated controls. This sophistication in the employment of physiological tools has been matched by the rapid march of genetic technologies.

The long history of studies on stickleback physiology has led to a decidedly practical application, that of the development of the threespine stickleback into a model organism for studying environmental contaminants in both laboratory and field settings (Handy et al., 2002; Katsiadaki et al., 2002a, 2002b, 2007; Pottinger et al., 2002; Hahlbeck et al., 2004a, 2004b; Katsiadaki et al., 2007; Sanchez et al., 2008). The development of the threespine stickleback as a model in ecotoxicology is due to a number of useful features:

Distribution and abundance: Aquatic environments are a sink for contaminants, so it is critical to have aquatic models in ecotoxicology. Threespine stickleback have a Holarctic distribution in marine, brackish and fresh waters throughout much of North America, Europe and Asia, where they are often the most abundant fish species. Therefore, they are often found in contaminated sites, whether that contamination occurs in near-shore marine, brackish or freshwater habitats, from the temperate zone to the Arctic. Because stickleback populations are typically vast in size, conservation issues with their use in research and monitoring are seldom encountered (for exceptions, see Foster et al., 2003; von Hippel, 2008). Additionally, because stickleback are not taken in commercial or sport fisheries and are not stocked, their use even in large numbers in research does not encounter resistance from permitting agencies (Katsiadaki, 2007).

Hardiness: Stickleback are common and hardy laboratory animals that are easy and inexpensive to maintain and can be easily collected from natural populations (Fig. 1). Because they are hardy animals and relatively tolerant of contaminants (though they often show measurable physiological responses at very low contaminant levels), physiological and genetic effects of even high levels of contaminants can be studied in living organisms (e.g., Katsiadaki et al., 2002b; Bernhardt et al., 2006; Bernhardt & von Hippel, 2008). More sensitive species can serve as indicators that a habitat is becoming polluted, but their early mortality prevents their use for mechanistic studies on the effects of the contaminant(s) on the living system (genetic, hormonal, or developmental). Therefore, more sensitive species are less useful than stickleback for understanding how contaminants can damage human and nonhuman health and development.

Figure 1. "Transport of sticklebacks in a 10 cm layer of fish. The survival rate after a 20 minutes transport from the sea to the polder, was 100%." From Kemper (1995).

Genetic tools: Numerous sophisticated genetic tools are available for stickleback studies (Cresko et al., 2007; see section Behavioural Genetics, Phylogenetics & Speciation). The stickleback has a well-developed genetic map (Peichel et al., 2001); it has a genome that is small (half the size of zebrafish), fully sequenced, and well-assembled. The stickleback is amenable to gene knockdown, over-expression and mutagenesis studies of the genetic mechanisms underlying developmental disruption. Furthermore, the content of the teleost genome is broadly shared with humans (Postlethwait, 2004), which allows discoveries of effects of contaminant exposure on stickleback genes potentially to be translated to human health. The power of these genetic tools can now be coupled with technologies in physiology for use in biomonitoring (Wirzinger et al., 2007), and to discover the mechanistic basis of developmental disruption caused by contaminants (e.g., Nagae et al., 2007; Brown et al., 2008). Fish embryos are transparent, so developmental disruption can often be seen in young fish. Because sticklebacks live in such a wide variety of aquatic environments, from as far south as Mexico and the Mediterranean to as far north as the Alaska and Norway, they have tremendous potential for elucidating the genetic basis of detoxification and repair abilities, if populations can be found that have adapted to contaminant exposure (Katsiadaki, 2007).

Sex determination system: Threespine stickleback have a genotypic XY sex determination system (Griffiths et al., 2000; Peichel et al., 2004) which, like the human system, is difficult to perturb. Since many contaminants, such as many pesticides and PCBs, disrupt sexual development and induce developmental abnormalities in humans by disrupting sex steroid axes, it is important to have a model organism that is similarly disrupted and that has a similar sex determination system. Stickleback fit this bill. Additionally, if environmental contaminants induce hermaphroditism in stickleback (which so far has only been demonstrated for perchlorate; Bernhardt et al., 2006), investigators can genetically test whether they are genotypic females that were masculinized or genotypic males that were feminized. Or if a contaminant has caused complete sex reversal, such that gonadal histology does not reveal intersex gonads, this too can be determined with genotyping (Katsiadaki, 2007). Such genotypic tests are not available for most animal models used in the field.

Biomarkers: Stickleback have both female-specific (vitellogenin — egg yolk protein precursor) and male-specific (spiggin) biomarkers for xenobiotic chemicals (Katsiadaki et al., 2002b; Hahlbeck et al., 2004b). In fact, the stickleback is one of only a handful of animals for which both male and female biomarkers of endocrine disruption have been found. This makes the stickleback a powerful model for endocrine disruption studies, since modulation of both male and female sex steroids can be studied in the same model. Furthermore, disruption of the thyroid axis can be tested through expression of thyroid hormones or via thyroid histology. These biomarkers can be used to study multiple avenues of endocrine disruption. For example, spiggin or kidney hypertrophy can be used as a marker for chemicals with either androgenic or anti-androgenic effects (Jolly et al., 2006, 2009; Katsiadaki et al., 2006); the latter of these categories is the more common of the two, particularly in wastewaters (Katsiadaki, 2007). The spiggin biomarker is the most sensitive known for investigating environmental chemicals that modulate androgens, and it is the only fish protein induced by androgens that has been discovered (Katsiadaki, 2007). Therefore, the stickleback is the only fish that currently has a quantifiable androgen and anti-androgen biomarker (Katsiadaki, 2007).

The different parts of the stickleback can be efficiently used in a comprehensive monitoring programme. For example, from a single fish, the kidney can be tested for spiggin, the blood plasma can be tested for vitellogenin and various hormones, the thyroid tissue can be fixed for histology to visualize hypo- or hyper-thyroidism, the gonads can be fixed for histology to visualize an intersex condition, a fin clip can be tested for genotypic sex, and the remainder of the body can be subjected to analytical chemistry for contaminant composition.

Conservation of endocrine systems: The stickleback, because it is a teleost fish, is also an excellent model for studies of endocrine disruption in humans because the teleost and mammalian (including human) endocrine systems are so similar. For example, the hypothalamus-pituitary-thyroid axis has remained highly conserved throughout the evolution of bony vertebrates (Brown et al., 2004; Carr & Norris, 2006). All teleosts and tetrapods respond to mammalian thyroid stimulating hormone (TSH), and teleost thyroid hormones are structurally identical to mammalian thyroid hormones (Eales & Brown, 1993; Carr & Norris, 2006). The same types of broad and specific similarities between teleost fishes and humans are found in the other hormone axes as well, with some minor differences such as the presence of the androgen 11-ketotestosterone (11-KT) in fish but not in people. The evolutionary homologies of the hormonal axes allow insights from investigation of contaminant exposures in stickleback to have direct implications for human health.

Additional advantages: The stickleback's relatively short life cycle allows studies to be carried out efficiently, even multi-generational studies. Their biology is perhaps the best documented of any fish, allowing considerable cross-talk of techniques and tools to be applied to complex toxicological problems. Because the stickleback is small, large numbers can be maintained in small laboratories, and thousands of embryos from *in vitro* crosses can be grown in a refrigerator-sized incubator.

Physiology & Behaviour revisited

The development of the stickleback into a potent model for ecotoxicology studies has, in a way, come full circle. Early ethologists saw behaviour and physiology as inextricably linked. Decades of stickleback research revealed

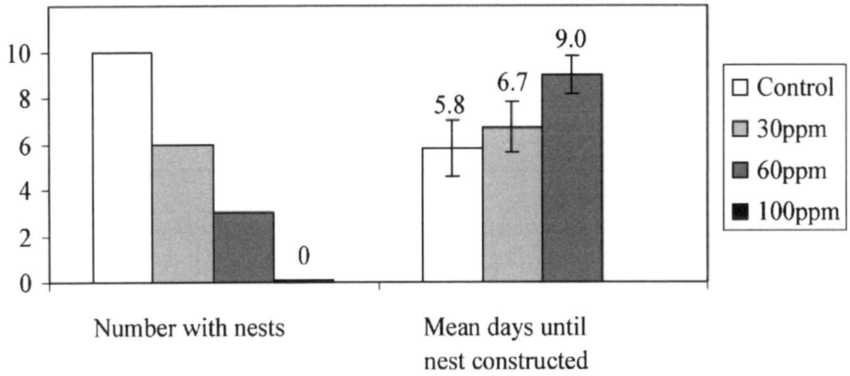

Figure 2. An example of a widespread endocrine-disrupting contaminant (perchlorate) that disrupts male reproductive behaviour. Fewer treated males built nests in a dose-dependent manner, and those that did took longer to do so. Shown are means (± 1 SE). From Bernhardt and von Hippel (2008).

the endocrine controls of specific reproductive behaviours. Now the stickleback model is used to understand how contaminants disrupt behaviour as well as physiology. For example, Bernhardt and von Hippel (2008) used the stickleback to show that the widespread contaminant perchlorate disrupts reproductive behaviour and swimming performance, while it also induces hermaphroditism (Bernhardt et al., 2006), and Sebire et al. (2009) showed that a widely used organophosphorous pesticide has anti-androgenic effects and disrupts the male's zigzag dance. Given the tight endocrine controls of the male's breeding behaviours, from establishing a territory and building a nest to courtship and parental care, sensitive and inexpensive behavioural biomarkers of specific endocrine disrupting chemicals are likely to be developed (Katsiadaki, 2007; Fig. 2). Biomarkers related to nuptial colouration are also on the horizon (e.g., Bernhardt et al., 2006).

Tinbergen despaired at the environmental destruction he witnessed in his native Holland and elsewhere in Europe (Kruuk, 2003). He died before scientists discovered that indigenous people of the Arctic, like the Inuit with whom he lived so happily in Greenland, are among the most contaminated people on Earth. These Arctic inhabitants are subject to perhaps the worst environmental injustice anywhere. Industrial pollutants, used elsewhere to fire up local economies, settle out in the Arctic where they biomagnify in the food web and contaminate animals harvested by subsistence hunters. The research programme initiated by Tinbergen on the stickleback reproductive cycle has evolved into a powerful conservation tool. This tool is now routinely employed to understand the nature of contamination in arctic and temperate environments.

References

Barlow, G.W. (1989). Has sociobiology killed ethology or revitalized it? — In: Whither ethology? Perspectives in ethology (Bateson, P.P.G. & Klopfer, P.H., eds). Plenum Press, New York, p. 1-45.

Bernhardt, R.R. & von Hippel, F.A. (2008). Chronic perchlorate exposure impairs stickleback reproductive behaviour and swimming performance. — Behaviour 145: 527-559.

Bernhardt, R.R., von Hippel, F.A. & Cresko, W.A. (2006). Perchlorate induces hermaphroditism in threespine sticklebacks. — Environ. Toxicol. Chem. 25: 2087-2096.

Borg, B., Bornestaf, C., Hellqvist, A., Schmitz, M. & Mayer, I. (2004). Mechanisms in the photoperiodic control of reproduction in the stickleback. — Behaviour 141: 1521-1530.

Borg, B. & Mayer, I. (1995). Androgens and behaviour in the three-spined stickleback. — Behaviour 132: 1025-1035.

Brown, M.M., Williams, T.D., Chipman, J.K., Katsiadaki, I., Sanders, M. & Craft, J.A. (2008). Construction of subtracted EST and normalised cDNA libraries from liver of chemical-exposed three-spined stickleback (*Gasterosteus aculeatus*) containing pollutant-responsive genes as a resource for transcriptome analysis. — Mar. Environ. Res. 66: 127-130.

Brown, S.B., Adams, B.A., Cyr, D.G. & Eales, J.G. (2004). Contaminant effects on the teleost fish thyroid. — Environ. Toxicol. Chem. 23: 1680-1701.

Carr, J.A. & Norris, D.O. (2006). The thyroid gland. — In: Endocrine disruption: biological bases for health effects in wildlife and humans (Norris, D.O. & Carr, J.A., eds). Oxford University Press, New York, NY, p. 87-110.

Cresko, W.A., McGuigan, K.L., Phillips, P.C. & Postlethwait, J.H. (2007). Studies of threespine stickleback developmental evolution: progress and promise. — Genetica 129: 105-126.

De Ruiter, A.J.H. & Wendelaar Bonga, S.E. (1985). Consequences of nestbuilding behaviour for osmoregulation in male three-spined sticklebacks. — Behaviour 93: 8-20.

Eales, J.G. & Brown, S.B. (1993). Measurement and regulation of thyroidal status in teleost fish. — Rev. Fish Biol. Fish. 3: 299-347.

Foster, S.A., Baker, J.A. & Bell, M.A. (2003). The case for conserving threespine stickleback populations: protecting an adaptive radiation. — Fisheries 28: 10-18.

Griffiths, R., Orr, K.J., Adam, A. & Barber, I. (2000). DNA sex identification in the three-spined stickleback. — J. Fish Biol. 57: 1331-1334.

Hahlbeck, E., Griffiths, R. & Bengtsson, B. (2004a). The juvenile three-spined stickleback (*Gasterosteus aculeatus* L.) as a model organism for endocrine disruption I. Sexual differentiation. — Aquat. Toxicol. 70: 287-310.

Hahlbeck, E., Katsiadaki, I., Mayer, I., Adolfsson-Erici, M., James, J. & Bengtsson, B.E. (2004b). The juvenile three-spined stickleback (*Gasterosteus aculeatus* L.) as a model organism for endocrine disruption — II — Kidney hypertrophy, vitellogenin and spiggin induction. — Aquat. Toxicol. 70: 311-326.

Handy, R.D., Runnalls, T. & Russell, P.M. (2002). Histopathologic biomarkers in three-spined sticklebacks, *Gasterosteus aculeatus*, from several rivers in southern England that meet the freshwater fisheries directive. — Ecotoxicology 11: 467-479.

von Hippel, F.A. (2008). Conservation of threespine and ninespine stickleback radiations in the Cook Inlet Basin, Alaska. — Behaviour 145: 693-724.

Jakobsson, S., Borg, B., Haux, C. & Hyllner, S.J. (1999). An 11-ketotestosterone induced kidney-secreted protein: the nest-building glue from male three-spined stickleback, *Gasterosteus aculeatus*. — Fish Physiol. Biochem. 20: 79-85.

Jolly, C., Katsiadaki, I., Le Belle, N., Mayer, I. & Dufour, S. (2006). Development of a stickleback kidney cell culture assay for the screening of androgenic and anti-androgenic endocrine disruptors. — Aquat. Toxicol. 79: 158-166.

Jolly, C., Katsiadaki, I., Morris, S., Le Belle, N., Dufour, S., Mayer, I., Pottinger, T.G. & Scott, A.P. (2009). Detection of the anti-androgenic effect of endocrine disrupting environmental contaminants using *in vivo* and *in vitro* assays in the three-spined stickleback. — Aquat. Toxicol. 92: 228-239.

Katsiadaki, I. (2007). The use of the stickleback as a sentinel and model species in ecotoxicology. — In: Biology of the three-spined stickleback (Östlund-Nilsson, S., Mayer, I. & Huntingford, F.A., eds). CRC Press, Boca Raton, p. 319-351.

Katsiadaki, I., Morris, S., Squires, C., Hurst, M.R., James, J.D. & Scott, A.P. (2006). Use of the three-spined stickleback (*Gasterosteus aculeatus*) as a sensitive *in vivo* test for detection of environmental antiandrogens. — Environ. Health Perspect. 114, Supplement 1: 115-121.

Katsiadaki, I., Sanders, M., Sebire, M., Nagae, M., Soyano, K. & Scott, A.P. (2007). Three-spined stickleback: an emerging model in environmental endocrine disruption. — Environmental Sciences 14: 263-283.

Katsiadaki, I., Scott, A.P., Hurst, M.R., Matthiessen, P. & Mayer, I. (2002a). Detection of environmental androgens: a novel method based on enzyme-linked immunosorbent assay of spiggin, the stickleback (*Gasterosteus aculeatus*) glue protein. — Environ. Toxicol. Chem. 21: 1946-1954.

Katsiadaki, I., Scott, A.P. & Mayer, I. (2002b). The potential of the three-spined stickleback (*Gasterosteus aculeatus* L.) as a combined biomarker for oestrogens and androgens in European waters. — Mar. Environ. Res. 54: 725-728.

Kemper, J.H. (1995). Role of the three-spined stickleback *Gasterosteus aculeatus* L. in the food ecology of the spoonbill *Platalea leucorodia*. — Behaviour 132: 1285-1299.

Kruuk, H. (2003). Niko's nature; a life of Niko Tinbergen and his science of animal behaviour. — Oxford University Press, Oxford.

Nagae, M., Kawasaki, F., Tanaka, Y., Ohkubo, N., Matsubara, T., Soyano, K., Hara, A., Arizono, K., Scott, A.P. & Katsiadaki, I. (2007). Detection and assessment of androgenic potency of endocrine-disrupting chemicals using three-spined stickleback, *Gasterosteus aculeatus*. — Environ. Sci. 14: 255-261.

Peichel, C.L., Nereng, K.S., Ohgi, K.A., Cole, B.L.E., Colosimo, P.F., Buerkle, C.A., Schluter, D. & Kingsley, D.M. (2001). The genetic architecture of divergence between threespine stickleback species. — Nature 414: 901-905.

Peichel, C.L., Ross, J.A., Matson, C.K., Dickson, M., Grimwood, J., Schmutz, J., Myers, R.M., Mori, S., Schluter, D. & Kingsley, D.M. (2004). The master sex-determination locus in threespine stickleback is on a nascent Y chromosome. — Curr. Biol. 14: 1416-1424.

Postlethwait, J.H. (2004). Evolution of the zebrafish genome. — In: Fish development and genetics: the zebrafish and medaka models (Zhiyuan, G. & Korzh, V., eds). World Scientific Publishing Company, Hackensack, NJ, p. 581-611.
Pottinger, T.G., Carrick, T.R. & Yeomans, W.E. (2002). The three-spined stickleback as an environmental sentinel: effects of stressors on whole-body physiological indices. — J. Fish Biol. 61: 207-229.
Sanchez, W., Katsiadaki, I., Piccini, B., Ditche, J.M. & Porcher, J.M. (2008). Biomarker responses in wild three-spined stickleback (*Gasterosteus aculeatus* L.) as a useful tool for freshwater biomonitoring: a multiparametric approach. — Environ. Int. 34: 490-498.
Sebire, M., Scott, A.P., Tyler, C.R., Cresswell, J., Hodgson, D.J., Morris, S., Sanders, M.B., Stebbing, P.D. & Katsiadaki, I. (2009). The organophosphorous pesticide, fenitrothion, acts as an anti-androgen and alters reproductive behavior of the male three-spined stickleback, *Gasterosteus aculeatus*. — Ecotoxicology 18: 122-133.
Tinbergen, N. (1942). An objectivistic study of the innate behaviour of animals. — Bibliotheca Biotheoretica D 1: 39-98.
Tinbergen, N. (1963). On aims and methods of ethology. — Z. Tierpsychol. 20: 410-433.
Wirzinger, G., Weltje, L., Gercken, J. & Sordyl, H. (2007). Genotoxic damage in field-collected three-spined stickleback (*Gasterosteus aculeatus* L.): a suitable biomonitoring tool? — Mutat. Res. 628: 19-30.

CONSEQUENCES OF NESTBUILDING BEHAVIOUR FOR OSMOREGULATION IN MALE THREE-SPINED STICKLEBACKS

by

A. J. H. DE RUITER and S. E. WENDELAAR BONGA[1]

(Department of Animal Physiology, University of Groningen,
P.O. Box 14, 9750 AA Haren, The Netherlands)

(With 6 Figures)

(Acc. 7-IX-1984)

During the last decades the complicated behaviour of three-spined sticklebacks has not only attracted the attention of ethologists, but also of many students of osmoregulation. The migratory behaviour, as shown especially by the *trachurus* form of the three-spined stickleback, makes high demands upon the osmoregulatory capacities of the fish during its journey from freshwater to seawater and vice versa. Many investigators have studied the physiological mechanisms of adaptation that enable these fish to live in environments that are totally different in osmolality and ion composition (KOCH & HEUTS, 1943; LAM, 1968; WENDELAAR BONGA, 1973, 1976; WENDELAAR BONGA & VEENHUIS, 1974a, b; see also WOOTTON, 1976).

The nestbuilding behaviour displayed by male sticklebacks during the reproductive period requires osmoregulatory adaptations as drastic as seaward migration. However, whereas migration is shown by many other fish, the nestbuilding behaviour of sticklebacks has consequences for osmoregulation that are quite unique among the more than 20,000 species of teleosts. The materials selected for nestbuilding by male sticklebacks are glued together with a mucous substance secreted by the kidneys. To this end the kidneys are transformed, structurally and functionally, from an important osmoregulatory organ into a mucus secreting gland. This phenomenon offers the opportunity of studying the osmotic and ionic homeostasis in a freshwater fish that lacks normal kidney functions. Our studies have been concerned with the glandular transformation of the kidneys, the endocrine control of this process, and its consequences for hydromineral control.

[1]) Present address: Department of Zoology, University of Nijmegen, The Netherlands.

Some principles of hydromineral regulation in freshwater fish

Ion concentrations in the blood and other body fluids are maintained at rather constant levels, which results in an osmolality of about 330 mOsmol/l in freshwater sticklebacks. Freshwater is strongly hypotonic compared to the body fluids (less than 5 mOsmol/l), and this results in osmotic water influx and loss of ions via the gills (CONTE, 1969). The kidneys normally play an important role in the elimination of the water surplus, via excretion of large volumes of urine (HICKMAN & TRUMP, 1969). Due to the high ion-reabsorptive capacity of the kidney tubules, the eliminaton of the water can take place without substantial loss of ions (HICKMAN & TRUMP, 1969). Any loss of ions via urine or via diffusion through the gills are compensated by active uptake of ions from the food in the intestine and from the ambient water by the chloride cells in the gills (CONTE, 1969; BIERTHER, 1970). Thus, in normal freshwater fish the kidneys, intestine and gills are vital organs for maintaining water and ion homeostasis. It is therefore surprising that male sticklebacks can live in freshwater with their kidneys transformed into mucus glands.

Structure and function of the normal stickleback kidney

The kidneys are located in the body cavity, basolaterally to the spinal column, and consist of numerous nephrons—coiled tubule-like structures —embedded in haematopoietic tissue. The nephrons are connected by short branches to the two ureters, which run laterally of the kidneys and terminate in the urinary bladder. A nephron consists of a renal corpuscle, and the attached nephronic tubule which consists of four segments.

The renal corpuscle consists of a Bowman's capsule with inside a tuft of bloodcapillaries, the glomerulus. In the three-spined stickleback the renal corpuscles are situated in small groups, mostly at the periphery of the kidney.

The nephronic tubules consist of a short neck-segment, which connects the Bowman's capsule to the remaining part of the tubule, and three main segments, namely the first and second proximal segments and the remaining collecting tubule, which opens into the ureters (Figs 1 and 2).

A glomerular filtrate is formed by means of ultrafiltration of the blood. This is processed into urine in the nephronic tubule segments, via reabsorption and secretion of substances such as ions, small proteins, sugars and amino acids. Reabsorption of ions, water and macromolecules takes place in the first proximal segment, while in the other segments mainly ions and water are reabsorbed.

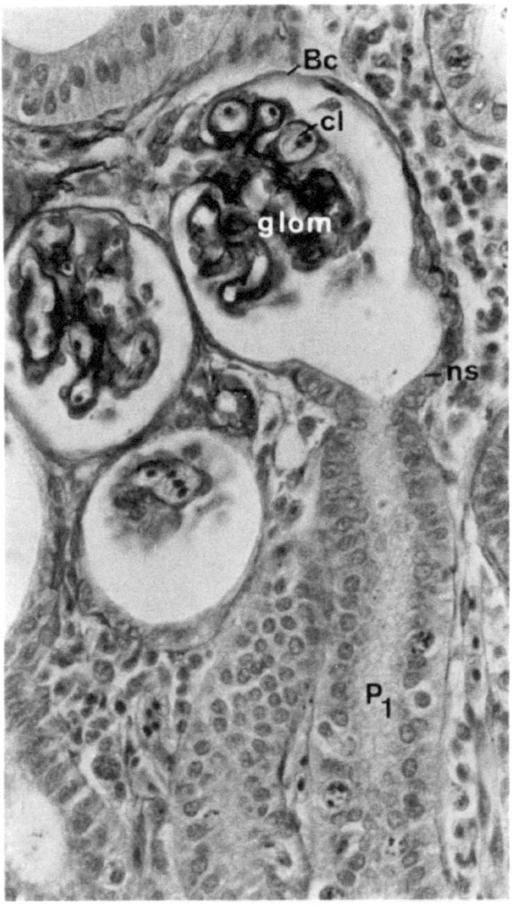

Fig. 1. Photomicrograph of a part of an immature freshwater male stickleback renal section showing a group of renal corpuscles. One of them is showing the attachment of the neck segment (*ns*) and the first proximal segment (*P1*) of the nephronic tubule with Bowman's capsule (*Bc*). (*glom*) glomerulus; (*cl*) capillary lumen. Silvermethenamine staining. × 475.

A so called basal labyrinth is present in the basal parts of the epithelial cells of both the proximal segments and the collecting tubule (Fig. 4A). The basal labyrinth is an intracellular membrane system, consisting of membrane folds connected with the basal and lateral cell membranes (Wendelaar Bonga & Veenhuis, 1974a, b; Figs 2 and 4A). The presence

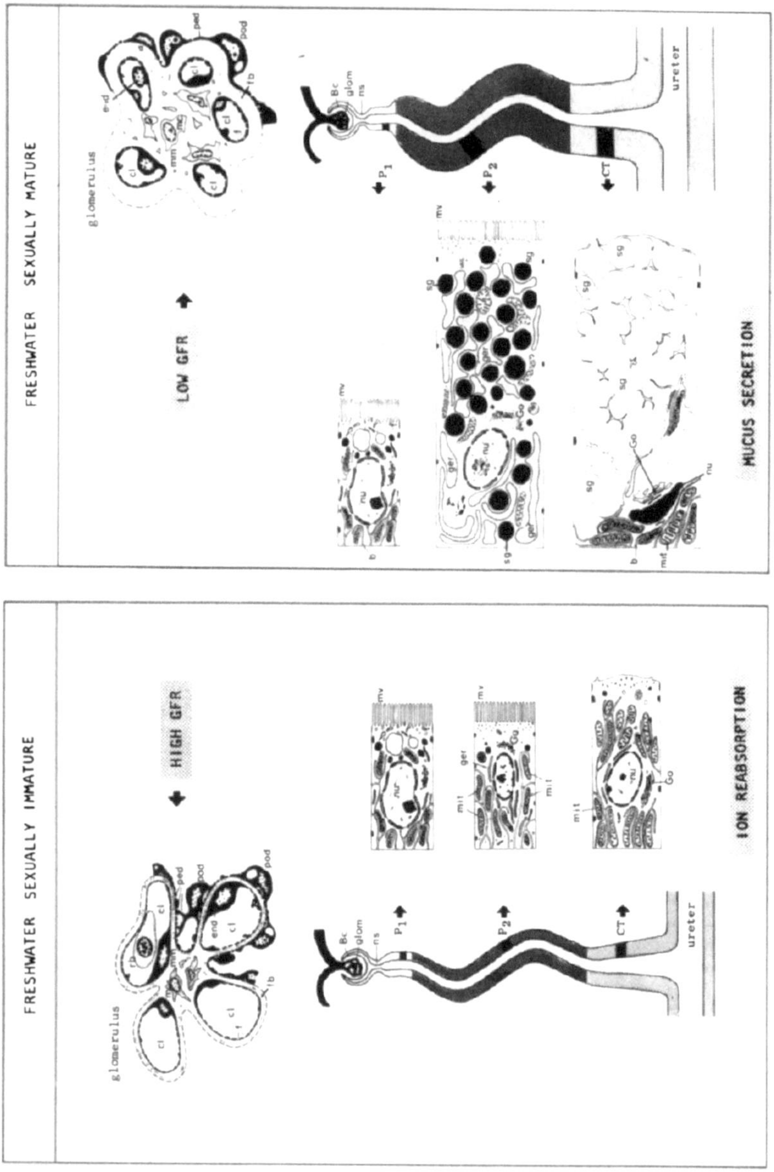
FRESHWATER SEXUALLY MATURE
LOW GFR
glomerulus
ureter
MUCUS SECRETION
FRESHWATER SEXUALLY IMMATURE
HIGH GFR
glomerulus
ureter
ION REABSORPTION

Fig. 2

of a basal labyrinth is characteristic for ion and water transporting cells (BERRIDGE & OSCHMAN, 1972). It is generally accepted that ion transporting enzyme systems, *e.g.* $Na+/K+$ activated ATPase, are located in the lateral cell membranes and especially in the membranes of the basal labyrinth (DIBONA & MILLS, 1979; FLIK *et al.*, 1980). These enzyme systems utilize ATP produced by mitochondria, which are closely associated with the membranes of the basal labyrinth (Figs 2 and 4A). Thus in summary, the kidneys of freshwater fish are characterised by a high glomerular filtration rate and a high ion-reabsorptive activity in comparison to seawater fish.

Effects of testosterone on structure and function of the stickleback kidney

In spring, rising water temperature and increasing daylength stimulate the production of testosterone in the gonads of the males (BAGGERMAN, 1957, 1966, 1968; GOTTFRIED & VAN MULLEM, 1967). This androgen plays an important role in spermatogenesis and is essential for the development of the secondary sex characteristics, such as nuptial coloration of the skin, as well as reproductive and nestbuilding behaviour (HOAR, 1962; WAI & HOAR, 1963; BAGGERMAN, 1966, 1968; see also WOOTTON, 1976). As soon as the nuptial coloration of the skin becomes noticeable, the first signs of glandular activity appear in the kidneys. The most important structural changes in the kidneys concern the transformation of the second proximal tubule and the collecting tubule, which comprise about 90% of the nephronic tubule cells (Fig. 2). The basal labyrinth almost completely disappears in these segments, probably as a

Fig. 2. Diagram of a nephron of a sexually immature and a sexually mature male stickleback. In the glomeruli of mature males the capillary lumina (*cl*) and number of fenestrations (*f*) in the endothelium (*end*) are smaller. In these males the filtration barrier (*fb*; = capillary wall) has thickened, while the number of filtration-slit-membranes between the pedicels (*ped*) of the podocytes (*pod*) also has decreased. These phenomena point to a reduced glomerular filtration rate. In contrast to immature males in the mature male nephronic tubule cells of the second proximal segment (*P2*) and the collecting tubule (*CT*) the basal labyrinth (*b*) and closely associated mitochondria (*mit*), characteristic structures for ion reabsorptive cells, almost have completely disappeared. The cytoplasm of these cells is almost completely occupied by stored secretory granules (*sg*) and granular endoplasmic reticulum (*ger*; mainly in the P2), which is very characteristic for mucus secreting cells. The single epithelial cells of the three main segments are drawn to scale to demonstrate the considerable differences in cell size between immature and mature males. (*nu*) nucleus; (*Go*) Golgi apparatus; (*mv*) microvilli.

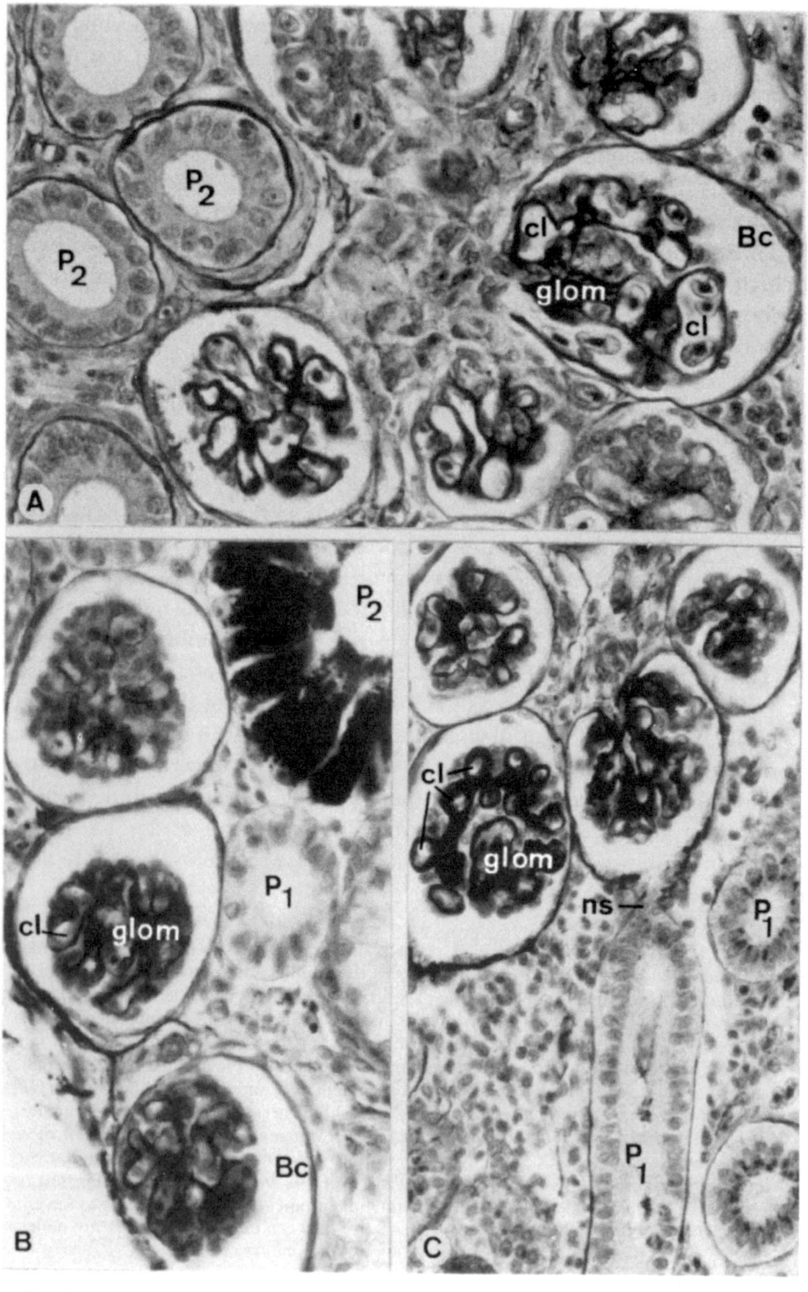

Fig. 3

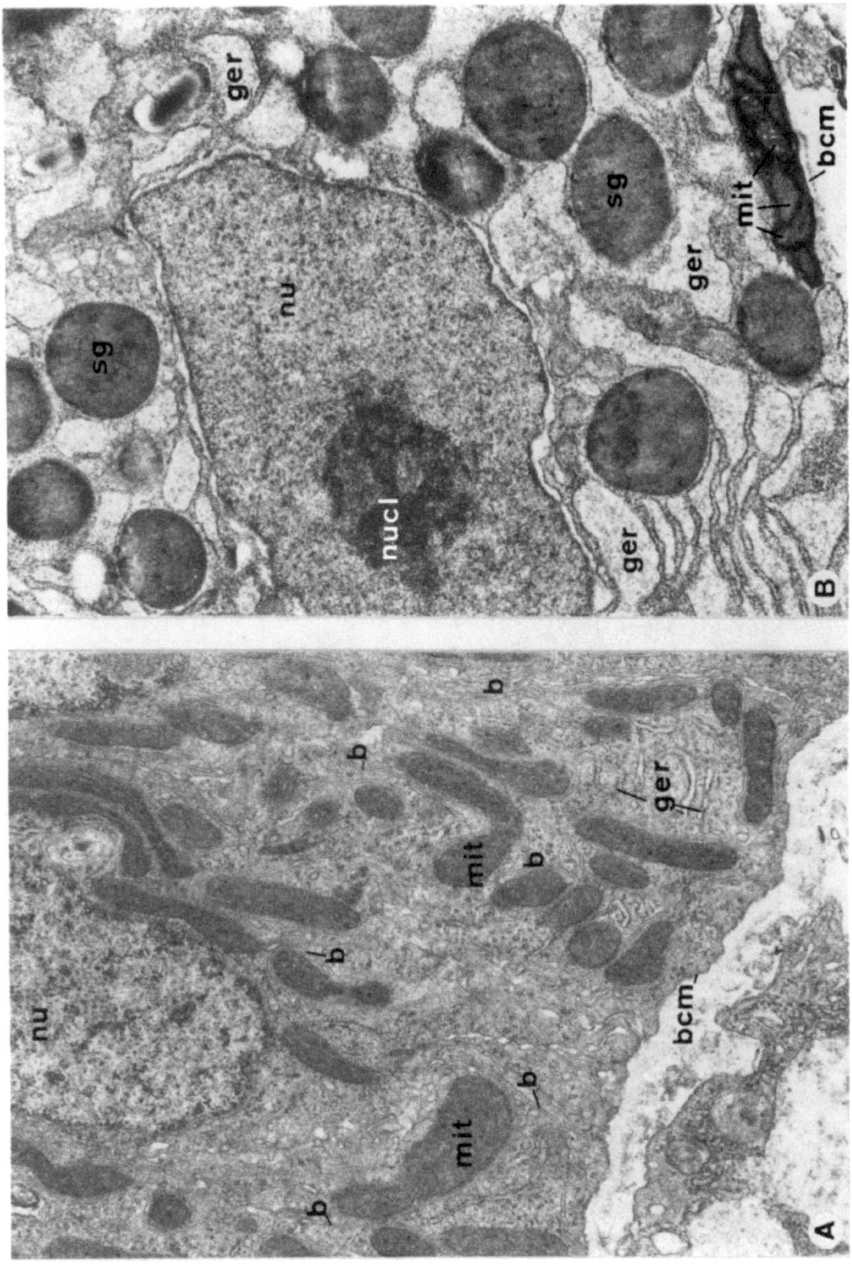
ger
sg
nu
nucl
ger
ger
sg
ger
mit
bcm
B
nu
b
mit
b
b
b
mit
b
ger
b
bcm
A

result of autophagous digestion. This implies that the ion reabsorbing capacity of the kidneys, and thus the capacity to produce hypotonic urine, is greatly reduced. Simultaneously these cells have grown considerably and an extensive granular endoplasmic reticulum developes, especially in the second proximal segment. The Golgi apparatus also proliferates and starts forming secretory granules. After this transformation, the cells of the second proximal tubule show all the structural characteristics of serous gland cells and those of the collecting tubule attain the appearance typical of mucous cells (DE RUITER & MEIN, 1982; Fig. 2). The transformed kidney of a sexually mature stickleback resembles a salivary gland more than an excretory organ and has increased up to sevenfold in size.

This glandular transformation of the kidney can also be induced experimentally by administration of methyltestosterone to castrated males or (intact) females (WAI & HOAR, 1963; MOURIER, 1972; DE RUITER & MEIN, 1982). Furthermore, our *in vitro* studies have shown that the effect of testosterone on the kidneys is direct, and not mediated by other hormones, since the glandular transformation can also be induced in small kidney pieces cultured for 10 days in the presence of testosterone (DE RUITER & MEIN, 1982).

The changes in the nephronic tubules are accompanied by changes in the glomeruli. At light microscopic level, measurements have shown that in sexually mature males the diameter of the renal corpuscles and of glomerular bloodcapillaries decreases markedly, which points to a reduced glomerular bloodflow (DE RUITER, 1980a; Fig. 3). In addition, electron microscopical observations have clearly shown that thickening and partial obstruction of the filtration barrier takes place. This barrier consists

---

← ← Fig. 3A, B, C. Photomicrograph of renal corpuscles as they are present in: A) immature freshwater males, B) mature freshwater males and C) immature seawater males. The diameters of both glomeruli and capillary lumina of the mature freshwater (B) and immature seawater (C) males are smaller than those of immature freshwater males (A). (P2) second proximal segment. See Fig. 1 for other abbreviations. Silvermethenamine staining. ×875.

← Fig. 4A, B. Electron micrograph of the basal part of a second proximal tubule cell of an immature (A) and of a mature (B) male. The basal labyrinth (*b*) in mature males, in contrast to that in immatures, has almost completely disappeared and is replaced by dilated cisterns of the granular endoplasmic reticulum (*ger*) and numerous secretory granules (*sg*). This together with a well developed nucleolus (*nucl*) in the nucleus (*nu*) points to a high mucus secretory activity. (*bcm*) basal cell membrane. 4A × 34500; 4B × 13000.

of capillary endothelium, mesangium, basal lamina and podocytes (Fig. 2). *In vivo* and *in vitro* experiments have shown that these glomerular structural changes also are induced directly by testosterone (DE RUITER, 1981). The changes that occur in the glomeruli of sexually mature males are similar to those that take place in the glomeruli of freshwater fish that migrate to seawater. Since it is known that in the latter these changes are connected with a firm reduction of the glomerular filtration rate (GFR) and low urine production (HICKMAN & TRUMP, 1969), we have concluded that the GFR and urine production in sexually mature males also is reduced (DE RUITER, 1980a, 1981). This conclusion has been confirmed experimentally for urine production (DE RUITER, 1980b). Thus the effects of testosterone on the kidneys of male sticklebacks lead to reduction of glomerular filtration (DE RUITER, 1981), the loss of most of the ion-reabsorptive capacity of the nephronic tubules, and a firm reduction of urine production. The reduction of urine flow is understandable, since the loss of ion reabsorptive capacity of the nephronic tubules would lead to substantial loss of ions and other valuable substances via the urine. On the other hand, the loss of the kidneys as the main pathway for the elimination of osmotically accumulated water will greatly disturb water and ion balance unless the loss of normal kidney function is compensated elsewhere in the body.

Compensatory mechanisms for water elimination

In sexually mature male sticklebacks living in freshwater there are two alternative ways to drain off the surplus of water; via the secretion of mucus by the kidneys, and the excretion of intestinal fluid. The loss of water via renal mucus secretion may partially compensate the reduction of urine excretion. However, we have demonstrated that the main compensatory pathway for the elimination of water in mature males is the intestine (DE RUITER, 1980b, DE RUITER *et al.*, 1985). As has been observed for some other teleost species, sticklebacks normally excrete through the anal opening some fluid, produced in the intestine. This fluid is more or less iso-osmotic with the body fluids. The collection of this intestinal fluid is complicated by the circumstance that, in sticklebacks, the urinary bladder opens into the posterior part of the intestine, and thus urine is mixed with intestinal fluid. Using micro-surgery and cannulation procedures, we were able to collect separately the fluids produced in the intestine and the kidneys (Fig. 5). We found that the rate of intestinal fluid production in mature males increased up to five-fold after glandular transformation

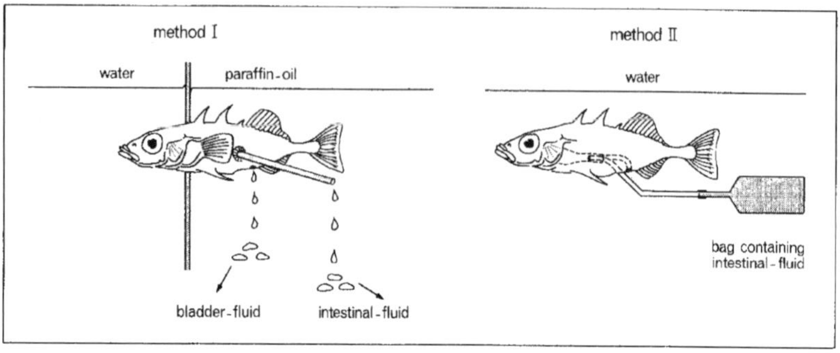

Fig. 5. Schematic representation of the methods used for the collection of urine or bladder fluid (= urine + renal mucus) and intestinal fluid samples. Method I; After microsurgical displacement of the anal opening to the flank, the abdomen of the fish was submerged in paraffin oil, which made it possible to collect bladder fluid and intestinal fluid separately, both from the bottom of the paraffin bath. Method II; A permanent intestinal cannula connected with an exchangeable small plastic bag made it possible to collect daily intestinal fluid samples from free swimming fish. For details see DE RUITER, 1980b.

of the kidneys. Furthermore, the osmolality of the fluid is reduced significantly, possibly by increased reabsorption of ions from this fluid in the posterior intestine (DE RUITER, 1980b, DE RUITER *et al.*, 1985).

Electron microscopical observations on intestinal and gall-bladder showed that the epithelial cells lining these organs, are provided with a well developed basal labyrinth, similar to that of the nephronic tubule cells (DE RUITER *et al.*, 1985; Fig. 6). The presence of such a labyrinth provides structural evidence that these epithelia are capable of transcellular transport of water and ions (DIAMOND & BOSSERT, 1967, 1968; BERRIDGE & OSCHMAN, 1972).

By means of morphometrical techniques we established that the extent of basal labyrinth in the epithelium of the posterior intestine increases markedly in sexually mature males and in testosterone-treated castrates or female fish (DE RUITER *et al.*, 1985). Thus in mature males the testosterone induced disappearance of most of the basal labyrinth in the nephronic tubule cells is accompanied by a considerable increase in extent of this membrane system in the intestine.

Concluding remarks

Nestbuilding behaviour in male sticklebacks strongly affects hydromineral regulation. The transformation of the kidneys into mucus

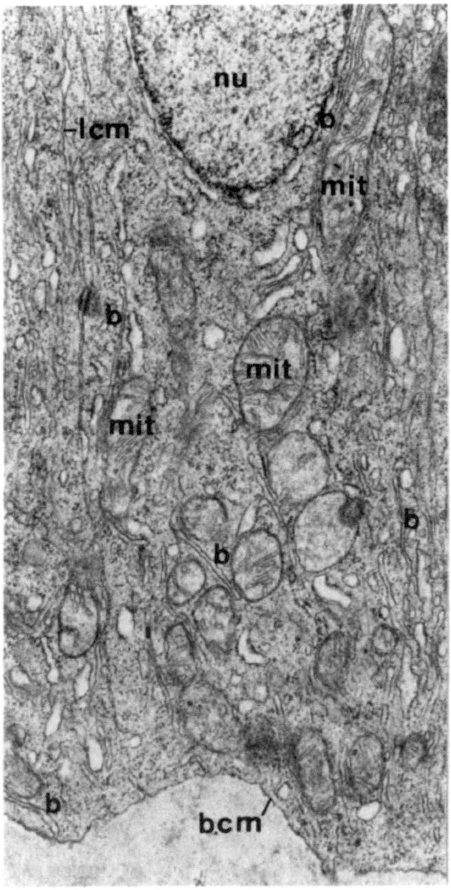

Fig. 6. Electron micrograph of the basal part of a stickleback intestinal epithelial cell provided with a well developed basal labyrinth (*b*). For abbreviations see legends of other figures. × 22000.

glands leads to complete loss of normal kidney functions. The loss of the water excreting capacity of the kidneys is partially counterbalanced by the secretion of mucus and, more importantly, by enhanced release of intestinal fluid. Other kidney functions, such as formation and removal of nitrogenous waste products, have not yet been studied, but may be taken over by the gills.

All the complicated structural and functional changes that follow sexual maturation in males can be induced by administration of testosterone in castrated males or even females. Our *in vitro* studies have shown that the glandular transformation of the nephronic tubules is a direct effect of testosterone on the tubular cells. The processes accompanying this transformation, such as the reduction of both the extent of renal tubule basal labyrinth and glomerular filtration rate, the increase of intestinal water permeability, growth of the basal labyrinth in the intestinal cells, and enhanced secretion of intestinal fluid, are also induced by testosterone, either directly or mediated by other hormones. Thus, in sticklebacks testosterone is a genuine multifunctional hormone: in addition to the well known control of spermatogenesis, nuptial coloration and reproductive behaviour it has pronounced effects on hydromineral regulation.

Summary

Nestbuilding behaviour, which takes place in sexually mature males as a consequence of an enhanced level testosterone in the blood, considerably influences hydromineral balance. As shown by *in vivo* and *in vitro* experiments testosterone is directly involved in the glandular transformation of the nephronic tubule cells into gland cells, that secrete mucus for nestbuilding. Light and electron microscopical observations as well as physiological data provide strong evidence that, in sexually mature males both ion reabsorptive capacity of most of the nephronic tubule cells and glomerular filtration rate *c.q.* urine production has decreased considerably. As a consequence of this the water balance of these fish, living in freshwater, would be greatly disturbed. However in these males the intestine and apparently also the gall-bladder, compensate for the loss in renal function in maintaining hydromineral balance. Again, this is influenced by testosterone.

References

Baggerman, B. (1957). An experimental study on the timing of breeding and migration in the three-spined stickleback (*Gasterosteus aculeatus* L.). — Archs néerl. Zool. 12, p. 105-318.

—— (1966). On the endocrine control of reproductive behaviour in the male three-spined stickleback (*Gasterosteus aculeatus* L.). — Symp. Soc. Exp. Biol. 20, p. 427-456.

—— (1968). Hormonal control of reproductive and parental behaviour in fishes. — In: Perspectives in endocrinology: Hormones in the lives of lower vertebrates (E. J. W. Barrington & C. Barker Jørgensen, eds). pp. 351-404. Acad. Press, London, New York.

Bierther, M. (1970). Die Chloridzellen des Stichlings. — Z. Zellforsch 107, p. 421-446.

Berridge, M. J. & Oschman, J. L. (1972). Transporting epithelia. — Acad. Press, London, New York.

Conte, F. P. (1969). Salt secretion. — In: Fish physiology, Vol. 1 (W. S. Hoar & J. D. Randall, eds). pp. 241-292. Acad. Press, London, New York.

Diamond, J. M. & Bossert, W. H. (1967). Standing gradient osmotic flow: a mechanism for coupling of water and solute transport in epithelia. — J. Gen. Physiol. 50, p. 2061-2083.

—— & —— (1968). Functional consequences of ultrastructural geometry in 'backwards' fluid-transporting epithelia. — J. Cell Biol. 37, p. 694-702.

DIBONA, D. R. & MILLS, J. W. (1979). Distribution of Na^{+}-pump sites in transporting epithelia. — Fed. Proc. 38, p. 134-143.

FLIK, G., RUITER, A. J. H. DE, VEENHUIS, M. & WENDELAAR BONGA, S. E. (1980). Cytochemical demonstration of K-NNPase activity in teleost tissue. — In: Eurem 80, Electron microscopy 1980, Vol. 2, Biology. (P. BREDEROO & W. DE PRIESTER, eds). pp. 298-299. Proc. VII Europ. Congr. on Electron Microsc., The Hague.

GOTTFRIED, H. & MULLEM, P. J. VAN (1967). On the histology of the interstitium and the occurrence of steroids in the stickleback (*Gasterosteus aculeatus* L.) testis. — Acta Endocrinol. 56, p. 1-15.

HOAR, W. S. (1962). Hormones and the reproductive behaviour of the male three-spined stickleback (*Gasterosteus aculeatus*). — Anim. Behav. 10, p. 247-266.

HICKMAN, C. P. & TRUMP, B. F. (1969). The kidney. — In: Fish Physiology, Vol. 1 (W. S. HOAR & J. D. RANDALL, eds). pp. 91-239. Acad. Press, London, New York.

KOCH, H. J. & HEUTS, M. J. (1943). Régulation osmotique, cycle sexual et migration de reproduction chez les Epinoches. — Archs int. Physiol. 53, p. 253-266.

LAM, T. J. (1968). Effect of prolactin on plasma electrolytes of early-winter marine threespine stickleback, *Gasterosteus aculeatus*, form *trachurus*, following the transfer from sea- to freshwater. — Can. J. Zool. 46, p. 1095-1097.

MOURIER, J. P. (1972). Étude de la cytodifférenciation du rein de l'épinoche femelle après traitement par la méthyltestostérone. — Z. Zellforsch. 123, p. 96-111.

RUITER, A. J. H. DE (1980a). Changes in glomerular structure after sexual maturation and seawater adaptation in males of the euryhaline teleost *Gasterosteus aculeatus* L. — Cell Tissue Res. 206, p. 1-20.

—— (1980b). Effects of testosterone on kidney structure and hydromineral regulation in the teleost *Gasterosteus aculeatus* L. — Thesis. University of Groningen.

—— (1981). Testosterone-dependent changes *in vivo* and *in vitro* in the structure of renal glomeruli of the teleost *Gasterosteus aculeatus* L. — Cell Tissue Res. 219, p. 253-266.

—— & MEIN, C. G. (1982). Testosterone-dependent transformation of nephronic tubule cells into serous and mucous gland cells in stickleback kidneys *in vivo* and *in vitro*. — Gen. Comp. Endocrinol 47, p. 70-83.

——, HOOGEVEEN, Y. L. & WENDELAAR BONGA, S. E. (1985). Ultrastructure of intestinal and gall-bladder epithelium in the teleost *Gasterosteus aculeatus* L., as related to their osmoregulatory function. — Cell Tissue Res. 240, p. 191-198.

WAI, E. H. & HOAR, W. S. (1963). The secundary sex characters and reproductive behaviour of gonadectomized sticklebacks treated with methyltestosterone. — Can. J. Zool. 41, p. 611-628.

WENDELAAR BONGA, S. E. (1973). Morphometrical analysis with the light and electron microscope of the kidney of the anadromous three-spined stickleback *Gasterosteus aculeatus*, form *trachurus*, from fresh water and from sea water. — Z. Zellforsch. 137, p. 563-588.

—— (1976). The effect of prolactin on kidney structure of the euryhaline teleost *Gasterosteus aculeatus* during adaptation to fresh water. — Cell Tissue Res. 166, p. 319-338.

—— & VEENHUIS, M. (1974a). The membranes of the basal labyrinth in the kidney cells of the stickleback, *Gasterosteus aculeatus*, studied in ultrathin sections and freeze-etch replicas. — J. Cell Sci. 14, p. 319-338.

—— & —— (1974b). The effect of prolactin on the number of membrane-associated particles in the kidney cells of the euryhaline teleost *Gasterosteus aculeatus* during transfer from seawater to fresh water: a freeze-etch study. — J. Cell Sci. 16, p. 687-701.

WOOTTON, R. J. (1976). The biology of the sticklebacks. — Acad. Press, London, New York.

ANDROGENS AND BEHAVIOUR IN THE THREE-SPINED STICKLEBACK

by

BERTIL BORG and IAN MAYER[1])

(Department of Zoology, University of Stockholm, S-106 91 Stockholm, Sweden)

(Acc. 22-V-1995)

Summary

In vitro incubations showed 11-ketoandrostenedione (11KA) to be the main androgen produced by the testes of the breeding stickleback. In breeding males the plasma levels of 11-ketotestosterone (11KT) are higher than those of other measured androgens, and display a clear peak in the breeding season. 11KA can be converted extra-testicularily to 11KT. In breeding females the plasma levels of 11KT are lower than in breeding males, whereas the opposite is the case for testosterone. Castration leads to the disappearence of male reproductive behaviour and secondary sexual characters. 11-androgens are more effective than other androgens in stimulating secondary sexual characters, notably kidney hypertrophy. All aspects of male reproductive behaviour can be restored in castrated males with 11KA capsules. 11KA was more effective than other tested androgens in this respect. The effectiveness of 11KA on nesting is not influenced by photoperiod.

Introduction

The three-spined stickleback, *Gasterosteus aculeatus*, is one of the most extensiveley studied fishes as concerns the effects of gonadectomy and androgens on behaviour (for reviews, see LILEY, 1969; LILEY & STACEY, 1983). The suitability of the stickleback for these types of studies is due not only to the fact that this species readily displays all aspects of reproductive behaviour in captivity, but also to the fact that the stickleback, unlike most other fishes, can be relatively easily castrated without testicular remnants regenerating. It has been found in several studies (HOAR, 1962a, b; WAI & HOAR, 1963; BAGGERMAN, 1966; SMITH & HOAR, 1967) that castration reduces/abolishes all types of reproductive behaviour in male sticklebacks, with the partial exception of aggressiveness under certain conditions (BAGGERMAN, 1966). Male reproductive behaviour can

[1]) Supported by Swedish Natural Science Research Council. We also want to thank Dr. Rüdiger SCHULZ, Utrecht for helping us with the radioimmunoassay.

be stimulated by androgen administration in intact and castrated males (Hoar, 1962a, b; Wai & Hoar, 1963; Smith & Hoar, 1967) and to some extent also in females (Hoar, 1962a, b; Wai & Hoar, 1963). The androgen-treatments in the early studies were all performed using methyltestosterone (MT), administered via the water in large doses. MT is a synthetic androgen not occurring naturally and its effects may not be physiological. In the present article we summarize published and previously unpublished studies concerning those gonadal steroids which occur naturally in the stickleback and what their effects are on behaviour and other reproductive parameters.

Steroid hormones occurring in the stickleback

The metabolic pathways of steroids can be studied in different organs by incubating tissue-pieces or homogenates with radiolabelled precursors followed by extraction, chromatographic separation of compounds formed, and identification combining chromatography, microchemical reactions and crystallization to constant specific activity. When testes from nesting male sticklebacks were tissue-incubated with tritiated pregnenolone, the main synthesis pathway was found to be: pregnenolone – progesterone – 17α-hydroxyprogesterone – androstenedione (A4) – 11β-hydroxyandrostenedione (OHA) – 11-ketoandrostenedione (11KA) (Borg *et al.*, 1989b), with 11KA being found in the largest amounts. Some 11-ketotestosterone (11KT) was also formed, but not testosterone (T) or 11β-hydroxytestosterone (OHT). The formation of 11KA in testes of postbreeding fish was only about 2% of that in the testes of the nesting ones.

Mayer *et al.* (1990b) measured the androgens 11KT, 11KA, OHT, OHA and T by means of specific radioimunoassays (RIA) in plasma sampled from breeding stickleback males captured in the field. Of these androgens 11KT was present in the largest amounts. Furthermore, when these androgens were studied in males sampled over an annual cycle, 11KT displayed a marked peak in the breeding season, whereas the other androgens did not. Thus, 11KT appeared to be the major androgen in the blood whereas 11KA was the main product in the testes incubations. This apparent contradiction can be resolved by the observation that extratesticular conversion of 11KA into 11KT can be carried out *in vitro* by blood cells of this species (Mayer *et al.*, 1990a). Castration alone drastically reduces plasma levels of both 11KT and T, whereas the levels of 11KA, OHT and OHA were not notably influenced (MAYER *et al.*, 1990b). When castrated males were implanted with Silastic capsules containing 11KA plasma levels of 11KT, but not of T, were increased (MAYER *et al.*, 1990b), when implanted with A4-capsules the opposite was the case.

Previously published information on androgens in sticklebacks deals with males only. In order to investigate the androgens present in females, blood samples were collected from mature females, many of them with running (*i.e.* eggs easily expelled when gently pressed), ovulated roe, from the northern Baltic in June 1987. Two samples, each with plasma from c. 20 females, were taken. The plasma was extracted with ether, the steroids present separated using thin layer chromatography, eluted and analyzed by means of RIAs. The results are shown in Table 1, where they can also be compared with the levels found in males sampled at the same time and analyzed together with the females. Although the material is small it appears that the levels of 11KT are much higher in males than in females, whereas the opposite is the case for T. The levels 11KA tended to be higher in males than in females, but the levels within the male samples were very variable and sometimes lower than in the female samples. No difference was found in the levels of OHT and OHA. RIAs for the progestin 17α, 20β-dihydroxy-4-pregnene-3-one (17,20P) and for 17β-estradiol (E2) were also run for the same samples. The levels of 17,20P were 1-2 ng/ml and E2 was non-detectable in the female samples.

Some androgens, notably T and A4, can be aromatized to estrogens, whereas others, like the 11-androgens, are non-aromatizable. Aromatiza-

TABLE 1. Plasma levels of androgens in breeding male and female sticklebacks

Group	Androgens (ng/ml) 11KT	T	OHT	OHA	11KA
Males	43.3±3.6	2.3±0.2	4.4±2.0	7.2±1.2	16.6±3.9
Females, sample 1	0.6	24.0	4.0	10.3	2.1
Females, sample 2	1.6	14.1	1.2	8.9	5.8

Data for males from MAYER *et al.* (1990b), means ± SE shown, N = 10, samples of each 10-15 males. 11KT: 11-ketotestosterone; 11KA: 11-ketoandrostenedione; OHT: 11β-hydroxytestosterone; OHA: 11β-hydroxyandrostenedione; T: testosterone.

tion can occur in the brain of vertebrates and is particularily active in teleosts, including the stickleback (Borg *et al.*, 1987).

Non-behavioral effects of androgens in sticklebacks

Male secondary sexual characters in the stickleback include breeding colours and a hypertrophied kidney. The breeding colours include light-blue eyes and a red ventral side. The red colour is produced by erythrophores containing carotenoid pigments, but also to some extent by a disappearance of the guanine-layer which makes the red colour of muscle and blood visible. The visual impression of the erythrophores depends on their content of carotenoid pigments and on what extent the pigment is concentrated or dispersed. Changes in dispersion can cause rapid changes in the intensity of colouration. The male kidney hypertrophies in the breeding season and produces a glue which is used in the building of the nest. These characters disappear after castration and can be restored with MT (*e.g.* Wai & Hoar, 1963). While the female kidney does not undergo hypertrophy during the natural breeding season, it too can be stimulated following MT administration (Mourier, 1972). The 11-androgens have been found to be more effective than the other androgens in stimulating the kidney: de Ruiter & Mein (1982) found that 11KT was more effective than MT in stimulating the stickleback kidney *in vitro*. Further, Andersson *et al.* (1988) observed that 11KA was more effective than A4 in this respect in intact, post-breeding males and also in inducing breeding colours. Recently, Borg *et al.* (1993) carried out a dose-response experiment on the effectiveness of several androgens in stimulating the kidney in castrated stickleback males. 11KT was found to more effective than any other tested androgen. Injections of 11KT was only slightly more effective than 11KA and OHA, but *c.* 300 times more effective as T. In the natural reproductive cycle of the stickleback, spermatogenesis is quiescent during the breeding season when the testes contains only spermatozoa and a few spermatogonia. Spermatogenesis commences after the end of breeding and is completed in late autumn, early winter (*e.g.* Borg, 1982). MT-treatment can prevent the post-breeding onset of spermatogenesis (Borg, 1981). Administation of 11KA in Silastic capsules was more effective than capsules containing A4 in inhibiting the onset of spermatogenesis in post-breeding males (Andersson *et al.*, 1988). There is also a marked negative individual correlation between spermatogenesis and kidney hypertrophy in un- and control-treated sticklebacks (Borg, 1981, 1982; Andersson *et al.*, 1988). This supports the view that androgens are physiologically inhibiting spermatogenesis in the stickleback. The relationship between spermatogenesis and androgens in other experimentally studied fishes is often very different from that found in the stickleback, but so are the natural seasonal pattern (for review, see Borg, 1994). Both 11KA and A4 were effective in stimulating aromatase activity in a hypothalamic area in the brain of castrated sticklebacks (Borg *et al.*, 1989a).

Behavioral effects of androgens in sticklebacks

Borg (1987) implanted male sticklebacks, castrated when already in breeding condition, with Silastic capsules containing 11KA, T, A4 or 5α-dihydrotestosterone or empty capsules. While at least some fish in all androgen-treated groups established territories and built nests, no castrated fish implanted with empty capsules did. The proportion of males building nests was highest in the 11KA treated group. Besides nest-building behaviour, some of these males were also tested for courtship and parental behaviour. These males exhibited the complete natural repertoire of courtship behaviours; including zig-zag swimming, leading, dorsal pricking, fanning, glueing, creeping through, showing nest enstrance, culminating in inducing the females to spawn in their nests and the final creeping through the nest 'fertilizing' the eggs. Further, these males also showed parental behaviour, including fanning, and were able to rear adopted roe through to hatching (Borg, 1987). Similar results were also attained in later experiments using implants of 5α-reduced 11-ketoandrostenedione (5α-androstane-3,11, 17-trione), although this androgen was less effective than 11KA, *i.e.* nest-building only occurring in a few specimens where leaking capsules had led to increased amounts of the androgen.

We have on a number of occasions implanted intact adult female sticklebacks with Silastic capsules containing 11KA. While this resulted in a clear stimulation of kidney hypertrophy, it did not induce any of the typical male behaviour patterns.

The reproductive cycle in the stickleback is strongly influenced by the photoperiod. When sticklebacks are caught in winter and exposed to a combination of long photoperiod and high temperature they rapidly attain full maturation, whereas this does not occur in short photoperiod (*e.g.* BAGGERMAN, 1957). A photoperiodic effect was also observed in castrated males treated with MT (HOAR, 1962b); under long photoperiod a higher proportion of the males built nests than under short photoperiod. More recently, an experiment was also performed to determine whether a similar effect could be obtained in castrated fish treated with 11-androgens. Stickleback males were caught in the Öresund, Sweden in January 1989, and castrated or sham-operated and implanted with Silastic capsules (outer diameter 1.2 mm, length 5 mm, sealed with silicon glue) filled with 11KA (only castrated) or empty capsules on the 27-28 January. They were kept under a photoperiod of Light:Dark (LD) 8:16 h and c. 5°C until the 29 January when the fish were divided between two 1200 l aquaria containing brackish water (0.5% salinity), sand, plants and algae. One aquarium was exposed to a photoperiod of LD 10:14 and the other to LD 16:8. The water was aerated and filtered and kept at 20°C. The fish were fed daily with frozen mysids. On the 11 February earthenware pots and tubes were put into the aquaria in order to provide better shelter for nest-building. The aquaria were inspected daily for nests, and if present the owners were caught out and the nests destroyed. Fish were counted as non-nesting only if they were in apparent good health at the termination of the experiment on the 29 March, otherwise they were excluded. Castrated fish were dissected in order to ascertain the completeness of castration (in this experiment all were complete) and to confirm the presence of the capsules. The results are shown in Table 2. As expected, castrated controls implanted with empty capsules did not build

TABLE 2. Nest-building in sham-operated stickleback males and in castrated fish implanted with empty capsules or with capsules containing 11-ketoandrostenedione (11KA), under different photoperiods (Light:Dark 16:8 and 10:14 h) in winter

Group	% nestbuilders (N)	
	LD 16:8	LD 10:14
sham	95.6% (46)	19.1% (47)***
castrated controls	0% (8)	0% (13)
castrated 11KA	51.4% (37)	52.9% (34)

N refers to nest-builders + non-nesting fish. *** Significantly different from sham-operated fish in LD 16:8, $p < 0.001$, Chi-square test.

nests. The proportion of nest-builders was higher in sham-operated fish kept under LD 16:8 than under LD 10:14. However, there was no effect of photoperiod in the 11KA fish in this respect.

General discussion

In the stickleback plasma levels of 11KT were higher in males than in females, whereas the opposite was the case for T. This is in agreement with the findings in most other studied teleosts (for review, see FOSTIER *et al.*, 1983; BORG, 1994). So far, the functional reason why 11-androgens are important androgens in male teleosts, rather than T as in tetrapods, is completely unknown. Since teleosts have T as well as T-receptors (for review, see BORG, 1994), 11KT seems to be an additional hormone rather than a replacement for T. Phylogenetically it appears more likely that the 11KT system has appeared among the actinopterygian fishes (it also occurs in sturgeons) rather than disappeared in tetrapods, since 11KT is also largely missing in elasmobranchs. Although 11KT was the dominating androgen in the plasma of breeding stickleback males, 11KA was the main steroid formed in testes incubations, 11KT being formed by extra-testicular conversion of 11KA. It was surprising that the levels of 11KA were not influenced by castration and 11KA treatment, whereas the levels of 11KT were (MAYER *et al.*, 1990b). Similar results have also been obtained with larger materials in the African catfish, *Clarias gariepinus* (SCHULZ *et al.*, 1993). It can be speculated that part of the circulating 11KA is rapidly converted to 11KT whereas the rest is unaccessible, perhaps being strongly protein-bound. The marked sex difference in 11KT levels, the seasonal cycle of the different androgens as well as their relative effectiveness on secondary sexual characters, and to some extent also on behaviour, suggests that 11KT is physiologically the most important androgen in the male stickleback. In contrast, the low levels of T in the male, the high levels in the female, together with the low effectiveness of T on male secondary sexual characters, and partly on behaviour suggests that T is not a physiologically important androgen in the stickleback. The effectiveness of MT on behaviour and secondary sexual characters could be due to the high doses given and probably does not reflect the physiological action of T in the stickleback, especially not the effects on the female. The high effectiveness of 11KT on secondary sexual characters found in the stickleback is in general agreement with what has been found in other teleost fishes (for review, see BORG, 1994), although, unlike in the stickleback, dose-response experiments in castrated fish have not been carried out in other species. Presently, there is only limited information on the effects of the administration of 11-androgens on behaviour in fishes other than the stickleback; KINDLER *et al.* (1991) observed an increase in reproductive behaviour in intact male bluegill sunfish, Lepomis macrochirus, after adminstration of 11KT, but not of T. More recently, BERGLUND *et al.* (1994) found that implantation of 11KA-capsules (no other androgen tested) blocked downstream migratory behaviour in immature Atlantic salmon (*Salmo salar*) at the time of smoltification; mature parr males generally show a lesser degree of downstream migration compared to immature smolts. However, 11-androgens do not seem to be effective in stimulating reproductive behaviour in all fishes. Castration of male rainbow trout, *Oncorhynchus mykiss*, abolished reproductive behaviour (MAYER *et al.*, 1994). Implantation of 11KA capsules had no effect, whereas 17,20P capsules were stimulatory. This is interesting when comparing the natural seasonal cycle of steroid hormones in different fishes. In the stickleback, 11KT peaks in the breeding season while 17,20P appears to be largely absent in the male (MAYER *et al.*, 1990b). In salmonids and many other fishes, on the other hand, androgens decline prior to the attainment of full maturation, whereas levels of 17,20P rise (for review, see BORG, 1994). BORG (1987) found that all aspects of reproductive behaviour in castrated stickleback could be restored by implantation of 11KA capsules. This indicates that no other gonadal hormones are needed for expression of reproductive behaviour in sticklebacks. It does not, however, show that 11-androgens are directly stimulating all aspects of reproductive behaviour. It is possible that some behaviours appear when certain key stimuli, such as nest, roe etc. are present without being directly controlled by androgens. Parental behaviour is of particular interest in this respect. In a number of fishes with male parental care plasma levels of 11KT decline from the spawning to the parental phase (bluegill sunfish, KINDLER *et al.*, 1989; demoiselle, *Chromis dispilus*, PANKHURST, 1990; garibaldi, *Hypsypops rubicundus*, SIKKEL, 1993). Sticklebacks have not been studied in this respect. Displacement fanning declines drastically in nesting sticklebacks after castration, disappearing after a few days (*e.g.* BAGGERMAN, 1966) and can be stimulated at least to some extent with androgens (*e.g.* MT, SMITH & HOAR, 1967).

BAGGERMAN (1968) castrated sticklebacks on the third day after they had fertilized their eggs. In the 4-5 days following this operation, there was hardly any difference in parental fanning or in agonistic, sexual and nest-building behaviour compared to controls. These results should, however, be treated with caution, since androgen stimulation on most characters could be expected to persist for some time after castration, BAGGERMAN (1968) reported that castration 3-4 days post-fertilization drastically lowered parental fanning, sexual and nest-building behaviours in the post-hatching phase from about a week after the operation. WAI & HOAR (1963) and SMITH & HOAR (1967) found that MT-treated males that had been castrated at least 7 days before being allowed to build nests, to perform parental fanning, though to a lesser extent than in intact fish, whereas the behaviour (naturally) did not occur in fish castrated that had not built nests, and gradually disapperared in fish castrated after they had received eggs in the nest. However, since the roe was unfertilized it probably did not provide the same stimuli as the healthy roe of the intact fish. BORG (1987) did not quantify parental behaviour in long-term castrated, 11KA-treated males – it was, however, sufficient to raise the roe through to hatching. Androgens are not not the only hormones involved in controlling fanning in sticklebacks. Parental behaviour in the stickleback in addition appears to be stimulated by prolactin (MOLENDA & FIEDLER, 1971; SLIJKHUIS *et al.*, 1984).

11-androgens (predominantly in males) and 17,20P (which occurs in both sexes) are important reproductive hormones in teleost fishes, whereas there are little data that suggest that they should be important hormones in any non-actinopterygian vertebrate. The studies above suggest that both these types of hormones play an important role in stimulating male reproductive behaviour in fish. The effectiveness of 11-androgens on behaviour in fish is in contrast to the situation in mammals also in another respect, in that in most studies on mammalian males (*e.g.* BEYER *et al.*, 1973) aromatization is essential for the effects of androgens on reproductive behaviour. To summarize, the regulation of male reproductive behaviour in teleosts appears to have little in common with what is found in mammals. The variation observed among different fishes in this respect is marked and might further increase when more species have been studied.

References

ANDERSSON, E., MAYER, I. & BORG, B. (1988). Inhibitory effect of 11-ketoandrostenedione and androstenedione on spermatogenesis in the three-spined stickleback, *Gasterosteus aculeatus* L. — J. Fish Biol. 33, p. 835-840.

BAGGERMAN, B. (1957). An experimental study on the timing of breeding and migration in the three-spined stickleback (*Gasterosteus aculeatus* L.). — Arch. Néerl. Zool. 12, p. 105-317.

—— (1966). On the endocrine control of reproductive behaviour in the male three-spined stickleback (*Gasterosteus aculeatus* L.). — Symp. Soc. Exp. Biol. 20, p. 427-456.

—— (1968). Hormonal control of reproduction and parental behaviour in fishes. — In: Perspectives in endocrinology: hormones in the lives of lower vertebrates (E.J.W. BARRINGTON & C. BARKER JØRGENSEN, eds). Academic Press, London, p. 351-404.

BERGLUND, I., LUNDQVIST, H. & FÄNGSTAM, H. (1994). Downstream migration of immature salmon (*Salmo salar*) smolts blocked by implantation of the androgen 11-keto-androstenedione. — Aquaculture 121, p. 269-276.

BEYER, C., LARSSON, K., PEREZ-PALACIOS, G. & MORALI, G. (1973). Androgen structure and male sexual behaviour in the castrated rat. — Horm. Behav. 4, p. 99-108.

BORG, B. (1981). Effects of methyltestosterone on spermatogenesis and secondary sexual characters in the three-spined stickleback (*Gasterosteus aculeatus* L.). — Gen. Comp. Endocrinol. 44, p. 177-180.

—— (1982). Seasonal effects of photoperiod and temperature on spermatogenesis and male secondary sexual characters in the three-spined stickleback (*Gasterosteus aculeatus* L.). — Can. J. Zool. 60, p. 3377-386.

—— (1987). Stimulation of reproductive behaviour by aromatizable and non-aromatizable androgens in the male three-spined stickleback, *Gasterosteus aculeatus* L. — In: Proceedings of V. Congr. Europ. Ichthyol. Stockholm, p. 269-271.

—— (1994). Androgens in teleost fishes. — Comp. Biochem. Physiol. 109C, p. 219-245.

——, ANDERSSON, E., MAYER, I. & LAMBERT, J.G.D. (1989a). Aromatase activity in the brain of the three-spined stickleback, *Gasterosteus aculeatus*. III. Effects of castration under different conditions and of replacement with different androgens. — Exp. Biol. 48, p. 149-152.

——, ANTONOPOULOU, E., ANDERSSON, E., CARLBERG, T. & MAYER, I. (1993). Effectiveness of several androgens in stimulating kidney hypertrophy, a secondary sexual character, in castrated male three-spined stickleback, *Gasterosteus aculeatus*. — Can. J. Zool. 71, p. 2327-2329.

——, SCHOONEN, W.G.E.J. & LAMBERT, J.G.D. (1989b). Steroid metabolism in the testes of the breeding and nonbreeding three-spined stickleback, *Gasterosteus aculeatus*. — Gen. Comp. Endocrinol. 73, p. 40-45.

——, TIMMERS, R.J.M. & LAMBERT, J.G.D. (1987). Aromatase activity in the brain of the three-spined stickleback, *Gasterosteus aculeatus*. I. Distribution and effects of season and photoperiod. — Exp. Biol. 47, p. 63-68.

FOSTIER, A., JALABERT, B., BILLARD, R., BRETON, B. & ZOHAR, Y. (1983). The gonadal steroids. — In: Fish physiology, Vol. 9A (W.S. HOAR, D.J. RANDALL & E.M. DONALDSON, eds). Academic Press, New York, p. 277-372.

HOAR, W.S. (1962a). Reproductive behavior of fish. — Gen. Comp. Endocrinol., Suppl. 1, p. 206-216.

—— (1962b). Hormones and the reproductive behaviour of the male three-spined stickleback (*Gasterosteus aculeatus*). — Anim. Behav. 10, p. 247-266.

KINDLER, P.M., PHILIPP, D.P., GROSS, M.R. & BAHR, J.M. (1989). Serum 11-keto-testosterone and testosterone concentrations associated with reproduction in male bluegill (*Lepomis macrochirus*: Centrarchidae). — Gen. Comp. Endocrinol. 75, p. 446-453.

——, Bahr, J.M. & Philipp, D.P. (1991). The effect of exogenous 11-ketotestosterone, testosterone, and cyproterone acetate on prespawning and parental care behaviors of male bluegill. — Horm. Behav. 25, p. 410-423.

Liley, N.R. (1969). Hormones and reproductive behaviour. — In: Fish physiology, Vol. 3. (W.S. Hoar & D.J. Randall eds). Academic Press, New York, p. 73-116.

—— & Stacey, N.E. (1983). Hormones, pheromones and reproductive behaviour. — In: Fish physiology, Vol. 9B. (W.S. Hoar, D.J. Randall & E.M. Donaldson, eds). Academic Press, New York, p. 1-63.

Mayer, I., Borg, B. & Schulz, R. (1990a). Conversion of 11-ketoandrostenedione to 11-ketotestosterone by blood cells of six fish species. — Gen. Comp. Endocrinol. 77, p. 70-74.

——, —— & —— (1990b). Seasonal changes in and effect of castration/androgen replacement on the plasma levels of five androgens in the male three-spined stickleback, *Gasterosteus aculeatus* L. — Gen. Comp. Endocrinol. 79, p. 23-30.

——, Liley, N.R. & Borg, B. (1994). Stimulation of spawning behaviour in castrated rainbow trout (*Oncorhynchus mykiss*) by 17α, 20β-dihydroxy-4-pregnene-3-one, but not by 11-ketoandrostenedione. — Horm. Behav. 28, p. 181-190.

Molenda, E. & Fiedler, K. (1971). Die Wirkung von Prolaktin auf das Verhalten von Stichlings-Männchen (*Gasterosteus aculeatus* L.). — Z. Tierpsychol. 28, p. 463-474.

Mourier, J.-P. (1972). Etude de la cytodifferenciation du rein de l'Epinoche femelle après traitement par la methyltestosterone. — Z. Zellforsch. 123, p. 96-111.

Pankhurst, N.W. (1990). Changes in plasma levels of gonadal steroids during spawning behaviour in territorial demoiselles *Chromis dispilus* (Pisces, Pomacentridae) sampled underwater. — Gen. Comp. Endocrinol. 79, p. 215-225.

de Ruiter, A.J.H. & Mein, C.G. (1982). Testosterone-dependent transformation of nephronic tubule cells into serous and mucous gland cells in stickleback kidneys *in vivo* and *in vitro* — Gen. Comp. Endocrinol. 47, p. 70-83.

Schulz, R.W., Paczoska-Eliasiewicz, H., Satijn, D.G.P.E. & Goos, H.J.Th. (1993). The feedback regulation of pituitary GTH-II secretion in male African catfish (*Clarias gariepinus*): Participation of 11-ketotestosterone. — Fish. Physiol. Biochem. 11, p. 107-115.

Sikkel, P.C. (1993). Changes in plasma androgen levels associated with changes in male reproductive behavior in a brood cycling marine fish. — Gen. Comp. Endocrinol. 89, p. 229-237.

Slijkhuis, H., de Ruiter, A.J.H., Baggerman, B. & Wendelaar Bonga, S.E. (1984). Parental fanning behavior and prolactin cells activity in the male three-spined stickleback *Gasterosteus aculeatus* L. — Gen. Comp. Endocrinol. 54, p. 297-307.

Smith, R.J.F. & Hoar, W.S. (1967). The effect of prolactin and testosterone on the parental behaviour of the male stickleback *Gasterosteus aculeatus*. — Anim. Behav. 15, p. 342-352.

Wai, E.H. & Hoar, W.S. (1963). The secondary sex characters and reproductive behaviour of gonadectomized sticklebacks treated with methyltestosterone. — Can. J. Zool. 41, p. 611-628.

MECHANISMS IN THE PHOTOPERIODIC CONTROL OF REPRODUCTION IN THE STICKLEBACK

by

BERTIL BORG[1,2], **CECILIA BORNESTAF**[1], **ANNA HELLQVIST**[1], **MONIKA SCHMITZ**[3] and **IAN MAYER**[1,4]

([1] Department of Zoology, Stockholm University, SE-106 91 Stockholm, Sweden; [3] Department of Aquaculture, Swedish University of Agricultural Sciences, SE-901 83 Umeå, Sweden)

(Acc. 13-IV-2004)

Summary

In sticklebacks, sexual maturation is stimulated by long photoperiods but not by short photoperiods, even at high temperatures. Extra-retinal photoreception can mediate this response, and appears to be more important than retinal photoreception. Although plasma melatonin levels are high at night and low during the day, experiments using melatonin administration via the water indicate that melatonin is of no or little importance for the photoperiodic response. Androgens can be aromatised to estrogens in the stickleback brain. Treatment with aromatase inhibitors stimulates maturation of males also under short photoperiod, suggesting that aromatase is involved in the suppressive actions of short photoperiod. Expression of both follicle stimulating hormone (FSH)-β and luteinizing hormone (LH)-β is higher under long than under short photoperiod. FSH-β is controlled by a negative steroid feedback on the brain-pituitary-gonad axis under short photoperiod and by a positive steroid feedback under long photoperiod. It is suggested that the former can suppress reproduction under short photoperiod and the latter can stimulate breeding under long photoperiod.

Keywords: Photoperiod, stickleback, reproduction, photoreceptors, melatonin, aromatase.

[2] Corresponding author's e-mail address: bertil.borg@zoologi.su.se

[4] We thank the Swedish Research Council for Environment, Agricultural Sciences and Spatial Planning (M.S.), the Swedish Natural Science Research Council (BB) and the Swedish Forestry and Agriculture Research Council (IM)

Behaviour 141, 1521-1530
Also available online -

Introduction

In almost all animals living in temperate regions, reproduction is a seasonal event. In order to produce offspring at a time when the chance of survival is highest, animals need to sense environmental cues telling them what season it is. The most important external cue is the photoperiod, although other environmental factors such as temperature are also involved in the control of seasonal reproduction. Some fishes like cyprinids and sticklebacks spawn when days are lengthening in spring-summer (long-day breeders), whereas others, such as most salmonids, spawn in autumn-winter when days are shortening (short-day breeders). In both cases, numerous experimental studies have demonstrated effects of photoperiod on the timing of reproduction (*e.g.* Duston & Bromage, 1987; Awaji & Hanyu, 1988). It is now well established that reproduction in the three-spined stickleback, *Gasterosteus aculeatus,* can be controlled by photoperiod manipulation (*e.g.* Baggerman, 1957; Borg, 1982a), with sexual maturation being stimulated by long photoperiods, and inhibited by short photoperiods. The present article reviews the mechanisms by which photoperiodic effects on stickleback reproduction are mediated.

Photoreceptor organs

In order to influence reproduction, photic information needs to be detected by photoreceptors. In mammals, the presence of photoreceptors is only clearly established in the eyes. In contrast, in most non-mammalian vertebrates extra-retinal photoreceptors are also present, the most recognised of these being in the pineal organ (Ekström & Meissl, 1997). The pineal organ is not directly photosensitive in mammals, but in fishes, including the stickleback (van Veen *et al.*, 1980), it contains well-defined photoreceptor cells, structurally similar to those found in the retina. In addition, deep encephalic photoreception is known from many vertebrates, and photoreceptors can also be present in the parapineal organ and in the skin. Benoit (1964) observed that extra-retinal photoreceptors can mediate photoperiodic effects on reproduction in birds, as blinded ducks can be stimulated to mature by long photoperiods. Mediation of photoperiodic effects by extra-retinal photoreceptors has been observed in many animals, including sticklebacks (Borg, 1982b). Menaker *et al.* (1970) demonstrated that when the roof of the skull of sparrows is shielded so that light is prevented from reaching the encephalic photoreceptors, sexual maturation is not stimulated even if the eyes are functioning normally, indicating a clear functional separation between the different photoreceptors.

Although it is more difficult to cover the head of fishes compared to birds, it was possible to stitch black plastic covers or transparent covers for controls, to the top of stickleback heads, so as to shield encephalic receptors while leaving the eyes unaffected. Using this approach, it was possible to study the effects of head-covering on reproductive parameters in both male and female sticklebacks. Kidney hypertrophy is a pronounced male secondary sexual characteristic in sticklebacks, where the hypertrophied kidney produces a glue-protein, spiggin (Jakobsson *et al.*, 1999; Jones *et al.*, 2001), which is used in the construction of the nest. Compared to controls, males with covered heads displayed less kidney hypertrophy after exposure to a stimulatory long photoperiod in winter (Bornestaf & Borg, 2000a, b). Similarly, female sticklebacks with covered heads displayed lower ovarian weights than controls after exposure to long photoperiods, and further the proportion of females reaching full maturation (ovulation) was lower in the covered group compared to the controls (Borg & Bornestaf, unpubl.). Bornestaf & Borg (2000b) found a significant inhibitory effect of black head-covering compared to transparent head-covering on kidney hypertrophy in male sticklebacks kept under long photoperiods of intermediate light intensities (17 and 2 Lux). In contrast, under high intensity (320 Lux) both groups matured fully, while under darkness both groups exhibited low kidney epithelium height (KEH) values indicating lack of maturation. It is possible that under high intensity sufficient light was entering the skull from the sides and from below to stimulate extra-retinal receptors to signal that days were long, whereas under intermediate intensities, this is not the case. It is, however, possible that the eyes are also able to mediate photoperiodic effects if intensities are sufficiently high and that this is the reason for the small difference in maturation between black and transparent covered sticklebacks under the high light intensity. Either way, these results suggest that extra-retinal photoreceptors are more important, or at least more sensitive, than retinal photoreceptors in mediating photic information in the control of reproduction in the stickleback.

Melatonin

Circulating levels of the hormone melatonin is generally present at high levels during the night and at low levels during the day in most organisms, including teleost fishes (for review, see Mayer *et al.*, 1997a). Plasma melatonin levels in the stickleback conform to this pattern (Mayer *et al.*, 1997b). It is now well established that melatonin mediates photoperiodic information in the control of seasonal reproductive cycles in mammals, being stimulatory in short-day breeders (*e.g.* Bittman *et al.*, 1983) and inhibitory in long-day breeders (*e.g.* Carter & Goldman, 1983). There is, however, little evidence for a similar physiological role of melatonin in the control of non-mammalian reproduction. Similarily as in mammals, pinealectomy drastically diminishes circulating night-time melatonin levels in fishes (for review, see Mayer *et al.*, 1997a), demonstrating that the pineal organ is the major source of melatonin in the blood. It has also been shown in several studies that pinealectomy can influence reproduction in fishes. This does not, however, demonstrate that melatonin *per se* is controlling reproduction since the pineal organ may secrete substances other than melatonin. Further, photic information received by the light-sensitive pineal organ is also likely to influence other systems through the extensive pinealofugal innervation of many brain areas that occurs in fishes, including the stickleback (Ekström & van Veen, 1983).

The administration of constant-release melatonin implants in fishes has generally failed to affect fish reproduction (Mayer *et al.*, 1997a). It seems probable that these treatments, giving constant high plasma melatonin levels, may act differently in the fish compared to the normal daily cycle, due possibly to receptor down-regulation. Injections of melatonin, on the other hand, have in several studies suppressed maturation in fishes, including the stickleback (Borg & Ekström, 1981). However, the physiological relevance of these experiments can be questioned since injections probably give initially high then rapidly declining plasma melatonin levels, very different from the natural diurnal cycle. In mammals, adminstration of melatonin via cannulation has overcome these problems, but this is presently not feasable in fishes. However, fish can take up melatonin administrated via the water and it is possible to closely mimick, both in terms of temporal pattern and plasma levels, the normal daily cycle in circulating melatonin levels by adding melatonin and changing water at specific times. Using this approach, it was found that melatonin adminstrated for 16 hours a day (mimicking a non-stimulatory short day, LD 8:16 photoperiod) over extensive periods to stickleback males (Mayer *et al.*, 1997b) and females (Bornestaf *et al.*, 2001) kept under a stimulatory long photoperiod had no or only a marginal effect on sexual maturation compared to controls treated with solvent alone. In stark constrast, sexual maturation was strongly suppressed in those fish kept under the control non-stimulatory LD 8:16 photoperiod. Similarily, attempts to mimick natural daily melatonin cycles by administration of melatonin via the food to salmonids had only minor effects on reproduction (Amano *et al.*, 2000). To summarize, to date there is little experimental evidence that melatonin plays an important role in the physiological control of seasonal reproduction in fish.

Brain-pituitary-gonadal axis

The gonads of vertebrates produce sex hormones and gametes under the stimulation of gonadotropic hormones (GTHs) from the pituitary. Teleost fishes, including the stickleback (Hellqvist *et al.*, 2004a) have two GTHs, follicle-stimulating hormone (FSH) and luteinizing hormone (LH). Both hormones consist of an α-chain, which is common for both GTHs, and a unique β-chain. In the stickleback, the expression of both LH-β and FSH-β is higher under long photoperiod than under short photoperiod in winter (Hellqvist *et al.*, 2004a, b) and both GTHs display the highest seasonal expression during the breeding season (Hellqvist *et al.*, 2004c). The pituitary is in turn controlled by the brain, in particular the hypothalamus. The brain (hypothalamus)-pituitary-gonad axis is under the control of gonadal feedback mechanisms. In most cases these feedbacks are negative, with gonadal hormones suppressing the release of GTHs. However, cases of positive steroid feedback effects, when gonadal hormones instead stimulate GTH secretion, are known from fishes (*e.g.* Crim *et al.*, 1981; Borg *et al.*, 1998).

It has been observed in mammals (*e.g.* Turek, 1977) that the suppressive effects of steroid hormones on GTH secretion are much stronger under non-stimulatory photoperiods than under stimulatory photoperiods. This is likely to be part of the mechanism(s) by which reproduction is suppressed under the former. Hellqvist *et al.* (2004b) investigated the feedback effects of gonadal hormones on the expression of both GTHs in male sticklebacks kept

1526 BORG, BORNESTAF, HELLQVIST, SCHMITZ & MAYER

under long (LD 16:8) and short photoperiod (LD 8:16) in winter by experiments involving castration alone or combined with replacement with the androgens testosterone (T) or 11-ketoandrostenedione (11KA). The results indicated that LH-β expression was under the control of a positive steroid feedback under both photoperiods. In contrast, FSH-β was controlled by a positive gonadal feedback under long photoperiod when expression was suppressed by castration and increased by androgen administration. Under short photoperiod, the pattern was the opposite, indicating the presence of a negative steroid feedback. In the seasonal cycle, the expression of FSH-β peaks earlier than that of LH-β (Hellqvist *et al.*, 2004c). Thus, it is likely that FSH rather than LH controls the onset of maturation. A positive steroid feedback acting stimulatory on FSH expression may lead to sexual maturation under long photoperiod, whereas a negative steroid feedback under short photoperiod might suppress reproduction. Sexual maturation in the stickleback is an all-or-nothing response. Under a long photoperiod the fish mature while under a short photoperiod maturation is inhibited, whereas under intermediate photoperiods some fish will mature as rapidly as under long photoperiod and others do not mature at all (Baggerman, 1972, 1980). There is a gradual effect of photoperiod (and of season), but each fish either matures fully or not. It is tempting to speculate that differences in steroid feedback mechanisms may control the 'switch' regulating whether a fish will mature or not.

Aromatase

Some androgens, such as T, can be aromatised to estrogens, whereas others, like 5α-dihydrotestosterone (DHT) and 11KA, cannot. Aromatase, the enzyme responsible for this conversion, is present in the brain of vertebrates, although especially high in teleost fishes (Callard *et al.*, 1981). In the stickleback, levels of aromatase activity are particularily high in the hypothalamus and pituitary (Borg *et al.*, 1987), suggesting a role of aromatase in the brain-pituitary-gonad axis. This is known from other fishes; in salmonids, aromatisation is important for positive steroid feedback control of the production of LH (*e.g.* Crim *et al.*, 1981). An inhibitory role of aromatase on sexual maturation is indicated in the Atlantic salmon, where treatment with aromatase inhibitors leads to a higher proportion of males maturing as parr (Antonopoulou *et al.*, 1995). In order to investigate whether aromatisation is involved also in the photoperiodic control of sexual maturation in the stickleback, Bornestaf *et al.* (1997) studied the effects of administering aromatase inhibitors to males kept under either stimulatory long (LD 16:8) or inhibitory short (LD 8:16) photoperiods. Males were implanted with Silastic capsules containing different aromatase inhibitors, the steroidal 1,4,6-androstatrien-3,17-dione (ATD) and the non-steroidal fadrozol (CGS), or empty capsules as controls. Under long photoperiod all groups displayed marked kidney hypertrophy, indicating maturation, while under the short photoperiod, both aromatase inhibitors, especially CGS, stimulated kidney hypertrophy compared to the non-maturing controls. Hellqvist *et al.* (2004b) obtained similar results, with CGS again being more effective than ATD in overcoming the inhibitory effects of short photoperiod on male maturation. In the same study it was also observed that the high expression of LH-β and FSH-β under long photoperiod was not influenced by CGS, whereas the otherwise low expressions under short photoperiod were increased. These studies suggest that aromatisation is involved in the mechanisms by which short photoperiod inhibits maturation in the stickleback.

Conclusions and future perspectives

While more is known on the mechanisms of photoperiodic control of reproduction in the stickleback than in perhaps any other fish, many important questions still remain unsolved. For example, while it is known that extra-retinal photoreception can mediate photoperiodic effects in the control of reproduction in the stickleback, the controlling mechanism(s) is still unknown. However, it is unlikely that circulating melatonin levels play an important role. It is an intriguing possibility that photoperiodic effects are instead mediated via pinealofugal innervation of the brain.

The steroid feedback regulation of FSH-β-expression is markedly different under long than short photoperiods, and may be part of the mechanism(s) by which photoperiod controls maturation. However, further experimental studies are needed in order to test if this is the case. An especially interesting possibility is that differences in steroid feedback mechanisms on the brain-pituitary-gonadal axis may play a role in the all-or-nothing response in sexual maturation in sticklebacks under different photoperiods, as well as in sexual maturation in fishes in general.

Aromatisation of androgens to estrogens appears to play a role in the short photoperiod inhibition of sexual maturation, though this effect is separate from the negative feedback effect of gonadal steroids on FSH-β-expression, as that effect could be exerted also by the non-aromatisable androgen 11KA.

References

Amano, M., Iigo, M., Ikuta, K., Kitamura, S., Yamada, H. & Yamamori, K. (2000). Roles of melatonin in gonadal maturation of underyearling precocious male masu salmon. — Gen. Comp. Endocrinol. 120, p. 190-197.

Antonopoulou, E., Mayer, I., Berglund, I. & Borg, B. (1995). Effects of aromatase inhibitors on sexual maturation in Atlantic salmon, *Salmo salar*, male parr. — Fish Physiol. Biochem. 14, p. 15-24.

Awaji, M. & Hanyu, I. (1988). Effects of water temperature and photoperiod on the beginning of spawning season in the orange-red type medaka. — Zool. Sci. 5, p. 1059-1064.

Baggerman, B. (1957). An experimental study on the timing of breeding and migration in the three-spined stickleback (*Gasterosteus aculeatus* L.). — Arch. néerl. Zool. 12, p. 105-317.

— — (1972). Photoperiodic responses in the stickleback and their control by a daily rhythm of photosensitivity. — Gen. Comp. Endocrinol. Suppl. 3: 466-476.

— — (1980). Photoperiodic and endogenous control of the annual reproductive cycle in teleost fishes. — In: Environmental physiology of fishes (M.A. Ali, ed.). Plenum, New York, p. 533-567.

Benoit, J. (1964). The role of the eye and of the hypothalamus in the photostimulation of gonads in the duck. — Ann. N. Y. Acad. Sci. 117, p. 204-215.

Bittman, E.L., Demsey, R.J. & Karsch, F.J. (1983). Pineal melatonin secretion drives the reproductive response to daylength in the ewe. — Endocrinology 113, p. 2276-2283.

Borg, B. (1982a). Seasonal effects of photoperiod and temperature on spermatogenesis and male secondary sexual characters in the three-spined stickleback, *Gasterosteus aculeatus* L. — Can. J. Zool. 60, p. 3377-3386.

— — (1982b). Extraretinal photoreception involved in photoperiodic effects on reproduction in male three-spined sticklebacks, *Gasterosteus aculeatus*. — Gen. Comp. Endocrinol. 47, p. 84-87.

— — & Ekström, P. (1981). Gonadal effects of melatonin in the three-spined stickleback, *Gasterosteus aculeatus* L., during different seasons and photoperiods. — Repr. Nutr. Develop. 21, p. 919-927.

— —, Timmers, R.J.M. & Lambert, J.G.D. (1987). Aromatase activity in the brain of the three-spined stickleback, *Gasterosteus aculeatus*. I. Distribution and effects of season and photoperiod. — Exp. Biol. 47, p. 63-68.

— —, Antonopoulou, E., Mayer, I., Andersson, E., Berglund, I. & Swanson, P. (1998). Effects of gonadectomy and androgen treatments on pituitary and plasma levels of gonadotropins in mature male Atlantic salmon, *Salmo salar*, parr — Positive feedback control of both gonadotropins. — Biol. Reprod. 58, p. 814-820.

Bornestaf, C., Antonopoulou, E., Mayer, I. & Borg, B. (1997). Effects of aromatase inhibitors on reproduction in male three-spined sticklebacks, *Gasterosteus aculeatus*, exposed to long and short photoperiods. — Fish Physiol. Biochem. 16, p. 419-423.

— — & Borg, B. (2000a). Extra-retinal photoreceptors are more important than retinal photoreceptors for sexual maturation in the three-spined stickleback (*Gasterosteus aculeatus*). — In: Reproductive physiology of fish. Institute of Marine Research and University of Bergen, p. 138 (abstract).

— — & — — (2000b). Extraretinal photoreception is more important than retinal photoreception for sexual maturation in the three-spined stickleback (*Gasterosteus aculeatus*). — In: Mechanisms in the photoperiodic control of reproduction in the three-spined stickleback, *Gasterosteus aculeatus* (C. Bornestaf). Doctoral Thesis, Stockholm, 2000.

— —, Mayer, I. & Borg, B. (2001). Melatonin and maturation pace in female three-spined sticklebacks, *Gasterosteus aculeatus*. — Gen. Comp. Endocrinol. 122, p. 341-348.

Callard, G.V., Petro, Z., Ryan, K.J. & Claiborne, J.B. (1981). Estrogen synthesis *in vitro* and *in vivo* in the brain of a marine teleost (*Myoxocephalus*). — Gen. Comp. Endocrinol. 43, p. 243-255.

Carter, D.S. & Goldman, B.D. (1983). Antigonadal effects of timed melatonin infusion in pinealectomized male Djungarian hamsters (*Phodopus sungorus sungorus*): Duration is the critical parameter. — Endocrinology 113, p. 1261-1267.

Crim, L.W., Peter, R.E. & Billard, R. (1981). Onset of gonadotropic hormone accumulation in the immature trout pituitary gland in response to estrogen or aromatizable androgen steroid hormones. — Gen. Comp. Endocrinol. 44, p. 374-381.

Duston, J. & Bromage, N. (1987). Constant photoperiod regimes and the entrainment of the annual cycle of reproduction in the female rainbow trout (*Salmo gairdneri*). — Gen. Comp. Endocrinol. 65, p. 373-384.

Ekström, P. & Meissl, H. (1997). The pineal organ of teleost fishes. — Rev. Fish Biol. Fisheries 7, p. 199-284.

— — & van Veen, Th. (1983). Central connections of the pineal organ in the three-spined stickleback, *Gasterosteus aculeatus* L. (Teleostei). — Cell Tissue Res. 232, p. 141-155.

Hellqvist, A., Bornestaf, C., Borg, B. & Schmitz, M. (2004a). Cloning and sequencing of the FSH-β and LH-β-subunit in the three-spined stickleback, *Gasterosteus aculeatus*, and effects of photoperiod and temperature on LH-β and FSH-β mRNA expression. — Gen. Comp. Endocrinol. 135, p. 167-174.

— —, Schmitz, M. & Borg, B. (2004b). Effects of castration and androgen replacement and effects of aromatase inhibitors on FSH-β and LH-β mRNA expression under different photoperiods in the three-spined stickleback, *Gasterosteus aculeatus*. — In: The brain-pituitary-gonadal axis and gonadotropic hormones in the three-spined stickleback, *Gasterosteus aculeatus* (A. Hellqvist). Doctoral Thesis, Stockholm, 2003.

— —, — — & — — (2004c). Seasonal changes in expression of LH-β and FSH-β mRNA in male and female three-spined stickleback, *Gasterosteus aculeatus*. — In: The brain-pituitary-gonadal axis and gonadotropic hormones in the three-spined stickleback, *Gasterosteus aculeatus* (A. Hellqvist). Doctoral Thesis, Stockholm, 2003.

Jakobsson, S., Borg, B., Haux, C. & Hyllner, S.J. (1999). An 11-ketotestosterone induced kidney-secreted protein: the nest building glue from male three-spined stickleback, *Gasterosteus aculeatus*. — Fish Physiol. Biochem. 20, p. 79-85.

Jones, I., Lindberg, C., Jakobsson, S., Hellman, U., Hellqvist, A., Borg, B. & Olsson, P.-E. (2001). Molecular cloning and characterization of spiggin: an androgen regulated extraorganismal adhesive with structural similarities to von Willebrand Factor-related proteins. — J. Biol. Chem. 276, p. 17857-17863.

Mayer, I., Bornestaf, C. & Borg, B. (1997a). Melatonin in non-mammalian vertebrates: Physiological role in reproduction? (review). — Comp. Biochem. Physiol. 118C, p. 515-531.

— —, — —, Wetterberg, L. & Borg, B. (1997b). Melatonin does not prevent long photoperiod stimulation of secondary sexual characters in the male three-spined stickleback *Gasterosteus aculeatus*. — Gen. Comp. Endocrinol. 108, p. 386-394.

Menaker, M., Roberts, R., Elliott, J. & Underwood, H. (1970). Extraretinal light perception in the sparrow, III: The eyes do not participate in photoperiodic photoreception. — Proc. Natl. Acad. Sci. USA 67, p. 320-325.

Turek, F. (1977). The interaction of photoperiod and testosterone in regulating serum gonadotropin levels in castrated male hamsters. — Endocrinology 101, p. 1210-1215.

van Veen, Th., Ekström, P., Borg, B. & Møller, M. (1980). The pineal complex of the three-spined stickleback, *Gasterosteus aculeatus* L. A light-, electron microscopic and fluorescence histochemical investigation. — Cell Tissue Res. 209, p. 11-28.

Behavioural Genetics, Phylogenetics & Speciation

Frank A. von Hippel & William A. Cresko

The inheritance of traits proved to be a major problem for Charles Darwin, both because he was unaware of the work of his contemporary Gregor Mendel, and because DNA would not be discovered as the genetic material for another century after the publication of *The Origin of Species*. Darwin struggled with how best to present inheritance of traits without possessing the vocabulary of genetics, and within the context of the widely accepted views of inherited habit proposed by Jean-Baptiste Lamarck. Darwin did not state that Lamarck was wrong, but rather that natural selection provided a more powerful, though not mutually exclusive, explanation. Darwin (1859) wrote, "I believe that the effects of habit are in many cases of subordinate importance to the effects of the natural selection of what may be called spontaneous variations of instincts; — that is of variations produced by the same unknown causes which produce slight deviations of bodily structure." Those spontaneous variations would later be called mutations, and those unknown causes would later be revealed as genes and alleles.

A century later, James Watson and Francis Crick (1953) ushered in the modern field of molecular genetics through the publication of their classic paper, "Molecular structure of nucleic acids." Just one page long, this paper must be the shortest to ever win the Nobel Prize (1962) for its authors. And in the most understated sentence of the paper, Watson and Crick wrote, "It has not escaped our notice that the specific pairing we have postulated immediately suggests a possible copying mechanism for the genetic material." Although it would be a few years before Meselson and Stahl (1958) showed definitively that DNA is replicated semi-conservatively, Watson and Crick clearly presaged how genetic variation might be formed through mutations when DNA was copied. The discovery of DNA as the genetic material led to dizzying advancement of the field of genetics, which only accelerates as time progresses. This work has culminated in the examination of genetic patterns and processes across entire genomes. Within the span of a century and a half, we have progressed from not understanding the inheritance of traits to being able to attack biological problems with complete genetic information from numerous individuals.

Work on stickleback has advanced our understanding of genetics in many ways. Given that the stickleback developed as a model in ethology before doing so in other fields, and during approximately the same time frame as the discovery of DNA as the genetic material, it seems natural that the genetics underlying stickleback behaviour attracted the interest of scientists before behavioural genetics became widespread. The first paper of this section is one of the early studies of the genetics of aggression. In excerpts republished here, Theo Bakker (1986) placed his behavioural genetics research in the context of the sociobiology debate that raged throughout the 1970's and 1980's.

More recently, the threespine stickleback system has developed into a model for understanding the evolutionary, genetic and developmental mechanisms underlying rapid adaptation. A particular strength of the system is the large number of freshwater populations independently derived from a common extant oceanic ancestor. Threespine stickleback are found in marine, brackish and freshwater habitats of much of the northern hemisphere (Wootton, 1976). The morphology of the oceanic form is similar throughout its range and has changed little in the last ten million years (Bell & Foster, 1994), allowing the use of extant oceanic populations as ancestor-surrogates for those in fresh water (Taylor et al., 1997; Cresko, 2000; Taylor & McPhail, 2000; Hendry et al., 2001, 2002; Hendry & Taylor, 2004; McKinnon et al., 2004).

Freshwater populations often retain a majority of the ancestral oceanic neutral genetic diversity (Cresko, 2000), but also typically undergo rapid phenotypic divergence. As a result, thousands of stickleback populations throughout the world vary in numerous traits, such as cartilage and dermal bones, craniofacial morphology, and response to thermal stress (Wootton, 1976; Bell & Foster, 1994; Peichel, 2005; Cresko et al., 2007; e.g., see Fig. 1). This divergence has occurred across a variety of geographic and temporal scales, from derived freshwater populations on different continents that are tens of thousands of years old to populations founded only a few decades ago. The existence of similar forms in similar habitats that are widely disjunct throughout their range has firmly established that these forms evolved independently in parallel due to similar selection pressures (Barrett et al.,

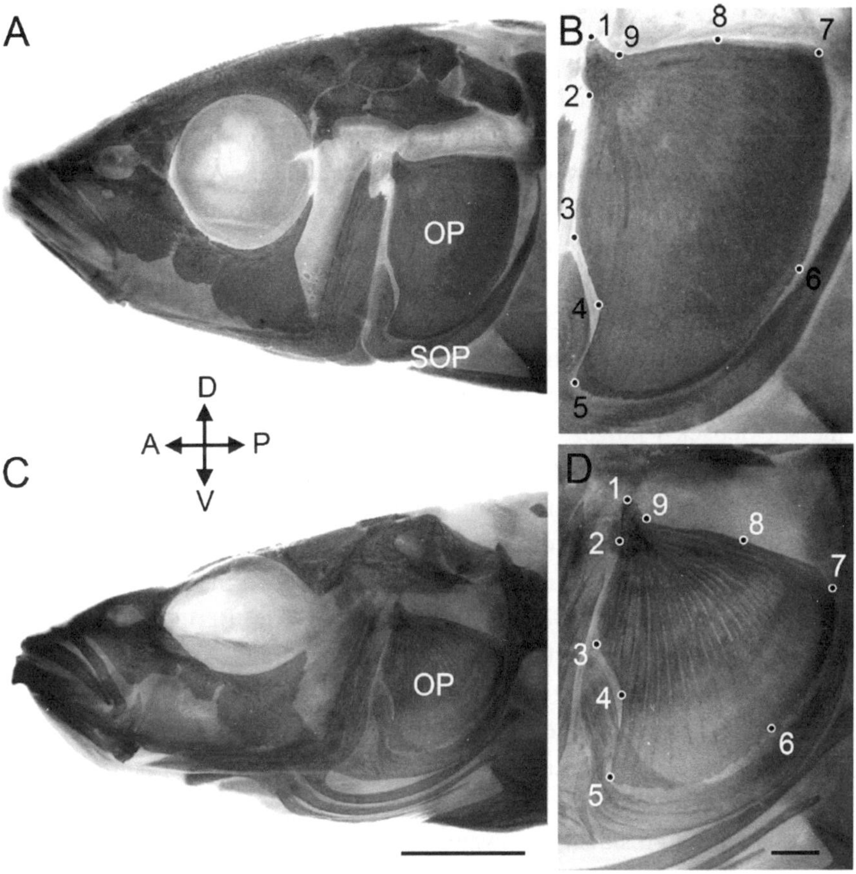

Figure 1. OP shape evolves in AK sticklebacks. Left side facial views of Alizarin Red S stained fish (A, C) and enlargements of the opercular region (B, D). (A, B): Alaskan anadromous fish captured in Mud Lake (SL = 70.5 mm). (C, D): Alaskan resident lacustrine fish captured in Visnaw Lake (SL = 61.9 mm). Landmarks used in the TPS analyses are shown in B and D. Scale bars: 5 mm (A, C), 1 mm (B, D). From Kimmel et al. (2008).

2008). This temporal and spatial variation provides an ideal situation for studying parallel evolution and the genetic mechanisms underlying it (Peichel, 2005; Cresko et al., 2007).

In the second paper of this section, Michael Bell (1995) discussed phylogenetic inferences and their application to the study of morphological or behavioural evolution. Common ancestry must be ruled out to ascribe parallel phenotypic divergence to natural selection. In an accompanying retrospective, Bell also discusses the historical context of the debate between those who argued for the prominence of dispersal versus those who argued for the prominence of selection in the radiation of stickleback populations.

Stickleback evolution in freshwater habitats has given rise not only to a diverse array of phenotypes, but also to new species (Rundle & Schluter, 1998; Schluter, 2000; McKinnon & Rundle, 2002; McKinnon et al., 2004). Numerous threespine stickleback species pairs exist sympatrically or parapatrically, that provide extraordinary opportunities to study speciation in nature (Schluter & Nagel, 1995; Foster et al., 1998; Schluter, 2000; McKinnon & Rundle, 2002; Bell et al., 2004; McKinnon et al., 2004). One of the categories of species pairs is that of the ancestral oceanic type (marine and anadromous) and the resident freshwater derived type, which coexist throughout most of the stickleback's range. In the third paper of this section, von Hippel and Weigner (2004) presented the first discovery of stickleback species pairs in a dynamic glacial environment, including in a pro-glacial (ice contact) lake. von Hippel and Weigner discovered both lake and stream pairs of anadromous and resident freshwater species in the newly-formed (<100 year old) drainage system of the Bering Glacier, Alaska. The Bering Glacier is the largest glacier in the world outside of Greenland and Antarctica, and it is the largest temperate surging glacier on Earth. These glacial surges continuously reset the clock on ecological succession in freshwater environments, and yet even in these extremely young habitats the species pairs are relatively common. Similarly, Gelmond et al. (in press) found species pairs of anadromous and resident freshwater threespine stickleback in ponds formed after the 1964 Great Alaska Earthquake on Middleton Island, Alaska. Distinguishing between different hypotheses for the formation and maintenance of these species pairs requires the use of genetic techniques. Thus, the threespine stickleback represents an elegant natural experiment on rapid adaptation that has been waiting for the development of suitable molecular tools to be fully exploited.

A fundamental problem in biology is to understand the genetic basis of evolving traits. However, most work has focused on a small number of model organisms for which little ecological or evolutionary information is available. Recently, the physical and monetary constraints that have historically limited genetic analysis to inbred organisms in the laboratory have dissipated, opening a huge array of phenotypes in natural populations to genetic analysis. Similar to the need for model developmental genetic systems, understanding the genetic and genomic basis of phenotypic variation in natural populations requires model systems (Cresko et al., 2007). Threespine stickleback is such an emerging model organism for the developmental genetic and genomic analysis of quantitative traits in the wild.

Significant progress has been made on this front due to the development of genomic and developmental tools for stickleback (Peichel et al., 2001; Hosemann et al., 2004; Kingsley et al., 2004; Cresko et al., 2007; M. Miller et al., 2007b; Baird et al., 2008), including the first draft of the genome sequence (2007). Developmental genetic work on stickleback has focused on a small number of traits. Researchers have isolated these traits in mapping crosses, identified associated genomic regions, and then examined the effects of alleles on cellular and developmental processes (reviewed by Peichel, 2005; Cresko et al., 2007). A handful of stickleback characters have been successfully dissected in this way (e.g., Peichel et al., 2001; Colosimo et al., 2004; Cresko et al., 2004; Shapiro et al., 2004; Colosimo et al., 2005; Kimmel et al., 2005; C. Miller et al., 2007a; Albert et al., 2008).

Stickleback developmental genetic research has focused on the use of experimental laboratory crosses for genetic mapping of trophic morphology, armour variation, and sex determination loci, among other traits (reviewed by Cresko et al., 2007). In one of the first stickleback mapping studies, Peichel et al. (2001, 2004) localized the sex determination locus to the distal end of linkage group XIX. Other experimental mapping examined variation in the extent of lateral plate and pelvic armour formation between ancestral and descendant populations. Crosses between complete- and low-plated populations revealed that a single major QTL or Mendelian factor governs the number of all but the most anterior lateral plates, and that the completely plated phenotype is specified by a dominant allele (Colosimo et al., 2004; Cresko et al., 2004). Analysis showed that the major lateral plate loci mapped to the same region of the genome (linkage group IV) in three independent Alaskan populations and populations in British Columbia and California (Colosimo et al., 2004; Cresko et al., 2004). Subsequent work by Colosimo et al. (2005) identified the gene *Ectodysplasin-A* (*Eda*) as segregating alleles that are associated with reduced lateral plates. An amazing aspect of the *Eda* low allele is that the clade to which it belongs is around two million years old (Colosimo et al., 2005), and most low plate alleles appear to be identical by descent. Thus the *Eda* alleles, and presumably other regions of the stickleback genome, have interacted in alternative oceanic and freshwater environments over extensive periods. A single Mendelian factor or major QTL was also found to exert substantial influence over pelvic expression, and both mapping and complementation crosses suggested that the same locus at the distal end of linkage group VII may be involved in pelvic reduction in populations in Alaska and British Columbia (Cresko et al., 2004; Shapiro et al., 2004, 2006). In addition to this work on armour variation, several other traits have been mapped, from head bone shape variation (Kimmel et al., 2005), to pigmentation (C. Miller et al., 2007a), to body shape (Albert et al., 2008). The general picture that emerges from this work leads to the surprising conclusion that parallel phenotypic divergence in these characters often has a parallel genetic basis, involving the same genes and maybe even alleles in independently derived populations due to the reuse of standing genetic variation in oceanic populations (reviewed by Cresko et al., 2007).

An exciting area of recent research is the genetic analysis of stickleback behavioural variation among populations and incipient species, a research approach that brings stickleback studies full circle to their ethological foundations. Threespine stickleback populations display various levels of divergence, from no differentiation to complete reproductive isolation (Foster et al., 1998). The use of genetic tools to discover the underlying genetic variation responsible for reproductive isolation likely has the greatest potential with the most highly differentiated populations that are already reproductively isolated in the wild but amenable to laboratory crosses. In the final paper of this section, Jun Kitano, Seiichi Mori and Catherine Peichel examined the divergence of male courtship displays in the Japan Sea/Pacific Ocean species pair. They discovered that the Pacific Ocean males perform a fast zig-zag dance while the Japan Sea males engage in a slower "rolling dance." The two members of the species pair differ in both the type and tempo of their courtship movements, in addition to body size, nuptial colouration and nest architecture. This system therefore provides an excellent opportunity to explore genetic variation underlying the behaviour responsible for reproductive isolation.

The powerful genetic tools now available have finally enabled the field of behavioural genetics to realize its potential, and research on stickleback will be integral in moving the field forward. Perhaps, with the sociobiology debate now well behind us, progress in behavioural genetics can proceed without the vitriol of the past. This progress also enables a far greater integration of Tinbergen's (1963) four why's into modern research programmes. Study of the genetics underlying behaviour promises to link insights in "ontogeny," "evolution," "survival value" and "causation." Central in this linkage is the courtship behaviour of the stickleback, the very place where it all began (ter Pelkwijk & Tinbergen, 1937).

References

Albert, A.Y., Sawaya, S., Vines, T.H., Knecht, A.K., Miller, C.T., Summers, B.R., Balabhadra, S., Kingsley, D.M. & Schluter, D. (2008). The genetics of adaptive shape shift in stickleback: pleiotropy and effect size. — Evolution 62: 76-85.

Baird, N.A., Etter, P.D., Atwood, T.S., Currey, M.C., Shiver, A.L., Lewis, Z.A., Selker, E.U., Cresko, W.A. & Johnson, E.A. (2008). Rapid SNP discovery and genetic mapping using sequenced RAD markers. — PLoS ONE 3: e3376.

Bakker, T.C.M. (1986). Aggressiveness in sticklebacks (*Gasterosteus aculeatus* L.): a behaviour-genetic study. — Behaviour 98: 1-144.

Barrett, R.D., Rogers, S.M. & Schluter, D. (2008). Natural selection on a major armor gene in threespine stickleback. — Science 322: 255-257.

Bell, M.A. (1995). Intraspecific systematics of *Gasterosteus aculeatus* populations: implications for behavioral ecology. — Behaviour 132: 1131-1152.

Bell, M.A., Aguirre, W.E. & Buck, N.J. (2004). Twelve years of contemporary armor evolution in a threespine stickleback population. — Evolution 48: 814-824.

Bell, M.A. & Foster, S.A. (eds) (1994). The evolutionary biology of the threespine stickleback. — Oxford University Press, Oxford.

Colosimo, P.F., Hosemann, K.E., Balabhadra, S., Villarreal, G.J., Dickson, M., Grimwood, J., Schmutz, J., Myers, R.M., Schluter, D. & Kingsley, D.M. (2005). Widespread parallel evolution in sticklebacks by repeated fixation of ectodysplasin alleles. — Science 307: 1928-1933.

Colosimo, P.F., Peichel, C.L., Nereng, K.S., Blackman, B.K., Shapiro, M.D., Schluter, D. & Kingsley, D.M. (2004). The genetic architecture of parallel armor plate reduction in threespine sticklebacks. — PLoS Biol. 2: 0635-0641.

Cresko, W.A. (2000). The ecology and geography of speciation: a case study using an adaptive radiation of threespine stickleback in Alaska. — PhD thesis, Clark University.

Cresko, W.A., Amores, A., Wilson, C., Murphy, J., Currey, M., Phillips, P.C., Bell, M.A., Kimmel, C.B. & Postlethwait, J.H. (2004). Parallel genetic basis for repeated evolution of armor loss in Alaska threespine stickleback populations. — P. Natl. Acad. Sci. USA 101: 6050-6055.

Cresko, W.A., McGuigan, K.L., Phillips, P.C. & Postlethwait, J.H. (2007). Studies of threespine stickleback developmental evolution: progress and promise. — Genetica 129: 105-126.

Darwin, C. (1859). On the origin of species by means of natural selection. — Murray, London.

Foster, S.A., Scott, R.J. & Cresko, W.A. (1998). Nested biological variation and speciation. — Philos. T. Roy. Soc. B 353: 207-218.

Gelmond, O., von Hippel, F.A. & Christy, M.S. (in press). Rapid ecological speciation in three-spined stickleback, Middleton Island, Alaska: the roles of selection and geographic isolation. — J. Fish Biol.

Hendry, A.P., Day, T. & Taylor, E.B. (2001). Population mixing and the adaptive divergence of quantitative traits in discrete populations: a theoretical framework for empirical tests. — Evolution 55: 459-466.

Hendry, A.P. & Taylor, E.B. (2004). How much of the variation in adaptive divergence can be explained by gene flow? An evaluation using lake-stream stickleback pairs. — Evolution 58: 2319-2331.

Hendry, A.P., Taylor, E.B. & McPhail, J.D. (2002). Adaptive divergence and the balance between selection and gene flow: lake and stream stickleback in the Misty system. — Evolution 56: 1199-1216.

von Hippel, F.A. & Weigner, H. (2004). Sympatric anadromous-resident pairs of threespine stickleback species in young lakes and streams at Bering Glacier, Alaska. — Behaviour 141: 1441-1464.

Hosemann, K.E., Colosimo, P.E., Summers, B.R. & Kingsley, D.M. (2004). A simple and efficient microinjection protocol for making transgenic sticklebacks. — Behaviour 141: 1345-1355.

Kimmel, C.B., Aguirre, W.E., Ullmann, B., Currey, M. & Cresko, W.A. (2008). Allometric change accompanies opercular shape evolution in Alaskan threespine sticklebacks. — Behaviour 145: 669-691.

Kimmel, C.B., Ullmann, B., Walker, C., Wilson, C., Currey, M., Phillips, P.C., Bell, M.A., Postlethwait, J.H. & Cresko, W.A. (2005). Evolution and development of facial bone morphology in threespine sticklebacks. — P. Natl. Acad. Sci. USA 102: 5791-5796.

Kingsley, D.M., Zhu, B.L., Osoegawa, K., De Jong, P.J., Schein, J., Marra, M., Peichel, C., Amamiya, C., Schluter, D., Balabhadra, S., Friedlander, B., Cha, Y.M., Dickson, M., Grimwood, J., Schmutz, J., Talbot, W.S. & Myers, R. (2004). New genomic tools for molecular studies of evolutionary change in threespine sticklebacks. — Behaviour 141: 1331-1344.

McKinnon, J.S., Mori, S., Blackman, B.K., David, L. & Kingsley, D.M. (2004). Evidence for ecology's role in speciation. — Nature 429: 294-298.

McKinnon, J.S. & Rundle, H.D. (2002). Speciation in nature: the threespine stickleback model systems. — Trends Ecol. Evol. 17: 480-488.

Meselson, M. & Stahl, F.W. (1958). The replication of DNA in *Escherichia coli*. — P. Natl. Acad. Sci. USA 44: 671-682.

Miller, C.T., Beleza, S., Pollen, A.A., Schluter, D., Kittles, R.A., Shriver, M.D. & Kingsley, D.M. (2007a). cis-regulatory changes in Kit ligand expression and parallel evolution of pigmentation in sticklebacks and humans. — Cell 131: 1179-1189.

Miller, M.R., Dunham, J.P., Amores, A., Cresko, W.A. & Johnson, E.A. (2007b). Rapid and cost-effective polymorphism identification and genotyping using restriction site associated DNA (RAD) markers. — Genome Res. 17: 240-248.

Peichel, C.L. (2005). Fishing for the secrets of vertebrate evolution in threespine sticklebacks. — Dev. Dyn. 234: 815-823.

Peichel, C.L., Nereng, K.S., Ohgi, K.A., Cole, B.L.E., Colosimo, P.F., Buerkle, C.A., Schluter, D. & Kingsley, D.M. (2001). The genetic architecture of divergence between threespine stickleback species. — Nature 414: 901-905.

Peichel, C.L., Ross, J.A., Matson, C.K., Dickson, M., Grimwood, J., Schmutz, J., Myers, R.M., Mori, S., Schluter, D. & Kingsley, D.M. (2004). The master sex-determination locus in threespine stickleback is on a nascent Y chromosome. — Curr. Biol. 14: 1416-1424.

ter Pelkwijk, J.J. & Tinbergen, N. (1937). Eine reizbiologische analyse einiger verhaltensweisen von *Gasterosteus aculeatus* L. — Z. f. Tierpsychol. 1: 193-200.
Rundle, H.D. & Schluter, D. (1998). Reinforcement of stickleback mate preferences: sympatry breeds contempt. — Evolution 52: 200-208.
Schluter, D. (2000). The ecology of adaptive radiations. — Oxford University Press, Oxford.
Schluter, D. & Nagel, L.M. (1995). Parallel speciation by natural selection. — Am. Nat. 146: 292-301.
Shapiro, M.D., Bell, M.A. & Kingsley, D.M. (2006). Parallel genetic origins of pelvic reduction in vertebrates. — P. Natl. Acad. Sci. USA 103: 13753-13758.
Shapiro, M.D., Marks, M.E., Peichel, C.L., Blackman, B.K., Nereng, K.S., Jónsson, B., Schluter, D. & Kingsley, D.M. (2004). Genetic and developmental basis of evolutionary pelvic reduction in threespine sticklebacks. — Nature 428: 717-723.
Taylor, E.B. & McPhail, J.D. (2000). Historical contingency and ecological determinism interact to prime speciation in sticklebacks, *Gasterosteus*. — P. Roy. Soc. Lond. B Bio. 267: 2375-2384.
Taylor, E.B., McPhail, J.D. & Schluter, D. (1997). History of ecological selection in sticklebacks: uniting experimental and phylogenetic approaches. — In: Molecular evolution and adaptive radiation (Givnish, T.J. & Sytsma, K.J., eds). Cambridge University Press, New York, p. 511-534.
Tinbergen, N. (1963). On aims and methods of ethology. — Z. Tierpsychol. 20: 410-433.
Watson, J.D. & Crick, F.H.C. (1953). Molecular structure of nucleic acids; a structure for deoxyribose nucleic acid. — Nature 171: 737-738.
Wootton, R.J. (1976). The biology of the sticklebacks. — Academic Press, London.

AGGRESSIVENESS IN STICKLEBACKS (GASTEROSTEUS ACULEATUS L.): A BEHAVIOUR-GENETIC STUDY

by

THEO C. M. BAKKER[1])
(Ethology Department, Zoological Laboratory, University of Leiden, P.O.B. 9516, 2300 RA Leiden, The Netherlands)

(With 50 Figures)
(Acc. 1-IX-1985)

Contents

[1]) I feel greatly indebted to Prof. Dr P. SEVENSTER for leaving me the subject of this study, for his stimulating interest, his helpful criticism and advice, for the enlightening discussions with him, and for his many valuable suggestions to improve the text of this manuscript. I am very grateful to Mrs E. J. DUINEVELD-DEKKER for her excellent participation in the experimental work. Without her continuous dedication it would have been impossible to carry out this study. I further wish to express my gratitude to Dr M. 't HART for his help in the initial stages of the study and to Mrs E. FEUTH-DE BRUIJN who was always uncomplicatedly willing to assist. I have greatly benefitted from discussions with Dr G. A. van OORTMERSSEN and Prof. Dr J. J. A. van IERSEL on an earlier draft of the manuscript. Many thanks go to Dr W. J. ROWLAND who made valuable comments and considerably improved my English. Dr F. R. van DIJKEN, Prof. Dr W. SCHARLOO and Prof. Dr K. BAKKER are acknowledged for their critical reading the manuscript. Finally I want to thank Mr G. VELDHUYZEN for technical assistence and Mr M. L. BRITTIJN for drawing the figures.

The greater part of this research was supported by the Foundation of Fundamental Biological Research (BION), which is subsidized by the Netherlands Organization for the Advancement of Pure Research (ZWO). Financial aid to publish this paper was also provided by ZWO.

1. General introduction

1.1. Genes and aggression.

Aggression has been a favourite topic of scientific research since the second world-war. Since aggressive behaviour is nearly universal among animals, including human beings, there has been much speculation about its cause. Because the resulting possible explanations often conflict with the different images of man as conceived by the various religious, political and cultural groups, they are bound to lead to heated debate. In essence these debates are usually centred around the question of genetic determinism, which maintains that aggression is inherited and inevitable. Questions like "Is aggressiveness controlled by genes?" had led to the well-known "innate *versus* learned" dichotomy in the study of behaviour. This dichotomy was fed by extreme explanations of the causation of aggression such as the frustration-aggression model (DOLLARD *et al.*, 1939) or the learning theory model (BANDURA & WALTERS, 1963) on one side, and those such as the Lorenzian-Freudian drive model (LORENZ, 1966) on the other. In the sixties the evidence accumulated allowing for a synthesis of both views. It was recognized that genes are as indispensable to behaviour as they are to *e.g.* morphological structure. This becomes evident from, for example, selection experiments with non-inbred populations of *Drosophila*: "There appears to be no character—morphogenetic, behavioral, physiological, or cytological—that cannot be selected in *Drosophila*." (LEWONTIN, 1974). However, a genetic basis for behaviour does not mean that behaviour is solely genetically determined; there is always an interaction between the genotype and the environment in the development of any biological character. So differences between individuals may be caused by either genetic differences or environmental differences or both. The question "Is aggressiveness controlled by genes?" is therefore false and should be replaced by "To what extent can the variation of aggressiveness be attributed to genotypic variation?".

The false innate-learned dichotomy of behaviour gradually seemed to disappear in the 7th decade of this century. However, with the recent interest in sociobiology (WILSON, 1975) the debate has revived. In sociobiology social behaviour is approached from an evolutionary viewpoint, and thus the genetic factor in behaviour is stongly emphasized. Moreover, this emphasis is reinforced by some of the more outrageous claims of many sociobiologists (*e.g.* WILSON, 1975, 1978) and the misleading terminology that is used to indicate genetic involvement in the expression of any character. A primary issue brought about by sociobiology revolves around the question of genetic determination of behaviour, especially where it concerns human beings. There exists a general aversion against the idea of genetic determination of behaviour, because people incorrectly think that genetic determination means something like fate, which cannot be avoided. The English sociobiologists indeed are generally more cautious when applying sociobiology to humans (*e.g.* DAWKINS, 1976; MAYNARD SMITH, 1972). But as sociobiologists they also stongly stress the role of genes in the social behaviour of animals. In the context of aggression this renewed interest in the genetic programming of behaviour has led to the "Evolutionarily Stable Strategy"-concept (*e.g.* MAYNARD SMITH & PRICE, 1973). Like all concepts in sociobiology, the ESS-concept starts from the principle that at least part of the variation in aggressiveness between individuals must be genetic in origin.

Although aggression has been a favourite topic in behavioural studies, fundamental behaviour genetic research on aggression is rather scarce. Most such studies have been concerned with the comparison of the aggressiveness of inbred strains of mice and rats. LAGERSPETZ & LAGERSPETZ (1974) and MICHARD & CARLIER (1985) summarize the results of 14, resp.

20 of such studies with mouse strains. Although the mutual comparison of these studies is difficult, due to various methodological differences, they indicate that the differences in aggression levels between mouse strains are genetically based. Similar investigations with similar results have been made with rat strains (*e.g.* HALL & KLEIN, 1942), breeds of chickens (*e.g.* FENNELL, 1945), and breeds of dogs (SCOTT, 1958; SCOTT & FULLER, 1965).

Behavior genetic methods which yield more useful information about the genetic architecture of a trait, namely assessing resemblance between relatives, crosses, and directional selection, have been scarcely applied to aggressive behaviour. The few examples that exist with regard to aggressiveness in fish (see introduction of chapter 3) confirm the picture that part of the variation of aggressive behaviour is attributable to genotypic variation. In section 3.6. directional selection experiments with aggressiveness as the criterion of selection will be reviewed and discussed in the light of the results of the present study. In all, nine selection studies have been published, five with house mice, three with domestic chickens and one with paradise fish. All but one revealed a reasonable amount of genetic variation for aggressiveness. Because of the high intensities of selection applied in those investigations, significant changes in the level of aggression are gained even after only a very limited number of generations of selection. One has to realize that such strong artificial selection pressures do often not obtain in nature. Furthermore, under natural circumstances the relevant environmental variables are far less constant than in the selection studies. Nevertheless, the few selection studies show that there is considerable genetic variation in aggressiveness present in the populations studied.

Hitherto I have used aggression in a broad, functional sense, such as "a physical act or threat of action by one individual that reduces the freedom or genetic fitness of another" (WILSON, 1975). The usual definitions of aggression encompass aggressive behaviour, even in the most divergent situations. Several authors have attempted to divide aggression into discrete categories. LORENZ (1966) categorizes aggression in terms of the three main situations in which it may occur: social (or intra-specific) aggression, predatory aggression, and anti-predator aggression. MOYER (1968) has classified aggressive behaviour in more detail and distinguishes eight categories. I will not occupy myself with the criticism of this classification, but see *e.g.* HUNTINGFORD (1976c). Although MOYER states that the categories are not mutually exclusive, he nevertheless ascribes the causation of aggression in the different categories to different neural and endocrine factors. HUNTINGFORD (1976c) has reviewed literature concerning the relationship between inter- and intra-specific aggression and she concludes that: "The most realistic assessment of this collection of facts would seem to be that social, anti-predator and predatory aggression are neither invariably linked nor inevitably distinct motivationally.". In those species where intra- and inter-specific aggression do covary, as *e.g.* in the three-spined stickleback (HUNTINGFORD, 1976a, b, 1982), a possible common physiological basis for intra- and inter-specific aggression may have evolved under the influence of convergent selection pressures (HUNTINGFORD, 1976c).

The same indistinct situation with respect to the identity of causation of intra- and inter-specific aggression arises if only the causation of social aggression is considered. According to LORENZ's view (1966) a unitary drive underlies the social aggression shown in various contexts. The classification of MOYER (1968) further divides the "social aggression" category of LORENZ, and with that, its causation. Although data have been published supporting both views, behaviour genetic research that could help to elucidate the genetic architecture of intra-specific aggression in different situations is unfortunately almost lacking. In this respect, the present study attempts to obtain some understanding of the genetics of aggression in the three-spined stickleback. It concentrates on intra-specific aggressiveness in a number of different situations, which is quantified in most of these situations by the duration of overt aggression, *i.e.* biting and bumping. In this study, which aims at the genetic causation of aggression, no further definition of aggression (in a functional sense) is needed.

1.2. Aim of the study.

The aim of the study is two-fold:

1. To assess the extent to which the inter-individual variation of aggressiveness, as measured in each of a number of different situations, can be ascribed to genetic variation.
2. To assess the extent to which variation of aggressiveness in these different situations is influenced by common genetic factors.

In this study the intra-specific aggressiveness of three-spined sticklebacks has been measured in different situations. In each situation individual aggression scores have been obtained. To realize the first aim three of these aggression scores have been used as a criterion of selection in two-way selection experiments. Thus a total of six different selection lines and one, unselected control line have been bred and maintained. In order to realize the second aim of this multiple selection research, individuals selected for the aggression score in a particular situation were also scored for aggressiveness in other situations.

At first sight the three-spined stickleback seems a poor choice for an experimental animal for genetic research. Its generation time is rather long (5 to 7 months), its chromosome number approaches that of man (diploid number: 42, MURAMOTO *et al.*, 1969), and inbred strains are lacking. Contrary to these restrictions, which hinder the kind of refined genetic analyses that have been carried out with *Drosphila*, there are some advantages which make the stickleback very suited for behaviour genetic research, as has been shown by a study of SEVENSTER & 't HART (1974). Since the publication of TER PELKWIJK & TINBERGEN (1937) the three-spined stickleback has been a favourite subject in ethological research and hence its behaviour has become well-known. It has an extensive behavioural repertoire, which can be relatively simply and objectively recorded. It is easily kept and bred in the laboratory under semi-natural conditions. Its way of life permits a high degree of standardization of environmental variables. Isolation, which can be carried out even from the egg stage, does not lead to any apparent behavioural abnormalities (CULLEN, 1960; SEVENSTER, 1968). Further advantages of the three-spined stickleback for behaviour genetic research are its great reproductive capacity and its well-studied ecology and endocrinology.

1.3. Aggressive behaviour of sticklebacks.

Aggressive behaviour is shown by sticklebacks in a wide variety of situations. I will confine myself here to a description of the situations in which intraspecific aggression has been measured in the present study. For a more detailed survey of the life-cycle of the three-spined stickleback, the interested reader is referred to publications of VAN IERSEL (1953), SEVENSTER (1961), and WOOTTON (1976, 1984).

In schools of juvenile sticklebacks consisting of both sexes the first signs of aggression (*juvenile aggression*) become visible at an early age, which in the laboratory is about four weeks after hatching. During the first weeks after the onset of aggression the attacks are "unilateral". The attacked juvenile flees, sometimes before a bite or bump can be delivered. The fleeing juvenile swims rapidly in a straight line away from the aggressor and stops abruptly at some distance from the latter. Most attacks are restricted to one bite or bump, unless the attacked fish fails to flee. Chas-

Retrospective: Intraspecific systematics of *Gasterosteus aculeatus* populations: Implications for Behavioral Ecology

Article published in *Behaviour* volume 132 (1995), p. 1131–1152

Michael A. Bell
(Department of Ecology and Evolution, Stony Brook University, Stony Brook, NY 11794-5245, USA)

Unlike most fields of science, evolutionary biology comprises two distinct approaches that have traditionally been pursued in isolation. One approach concerns reconstruction of the history of life, the domain of palaeontology and systematics, and the other concerns the mechanisms that have influenced this history, which are studied by population biologists. However, in the 1980s, a consensus emerged that inference of evolutionary mechanisms using comparative data requires a phylogenetic context (Brooks & McLennan, 1991; Harvey & Pagel, 1991). Associations between phenotypes and environments could reflect common ancestry, the independent action of similar evolutionary mechanisms, or a combination of both. While associations between phenotypes and environments could illuminate the action of natural selection (Endler, 1986), they could also indicate that closely related, phenotypically similar populations had colonized similar habitats. Thus, before similarities among populations could be used to infer natural selection, the influence of common ancestry had to be factored out. Felsenstein (1985) proposed the method of "independent contrasts" to do this (but see Rohlf, 2006 for a technical criticism), and it soon became common practice to take phylogeny into account in comparative studies (Brooks & McLennan, 1991; Harvey & Pagel, 1991). However, robust phylogenies based on characters that are functionally and genetically independent from the similarities under investigation are needed, and they can be difficult to obtain.

The main goals of my paper (Bell, 1995) were to emphasize the importance of taking phylogeny into account for comparative studies of behaviour and to present a simple way to do it for comparative studies of threespine stickleback populations. I started by discussing behaviours as homologous properties that vary among groups (i.e., characters). Next, I discussed the history of interpretation of geographical variation among populations of threespine stickleback. I sketched extreme alternative explanations (i.e., common ancestry vs. adaptation) for occurrence of similar phenotypes in geographically distant populations and illustrated these alternatives using pelvic girdle reduction in the threespine stickleback. Pelvic reduction is hardly a behavioural character, but it is one for which there already was considerable information (Bell, 1974, 1987, 1988; Reimchen, 1980; Giles, 1983; Bell et al., 1993; Bell & Ortí, 1994). I closed the paper by suggesting criteria for selection of behavioural phenotypes and stickleback populations for comparative studies. It has become common practice to take phylogeny into account to distinguish the effects of common ancestry and local adaptation in comparative studies of threespine stickleback phenotypes (e.g., Taylor & McPhail, 1999, 2000; Reusch et al., 2001; Colosimo et al., 2005; Mäkinen et al., 2006; Aguirre, 2007).

By 1994 when I wrote this paper, adaptive interpretation of stickleback phenotypic variation had become well established, and comparisons among populations were important in this research programme (reviewed by Bell & Foster, 1994; Reimchen, 1994). Phylogeographic analyses (Avise, 1994) of threespine stickleback populations were also starting to appear (Haglund et al., 1992; Ortí et al., 1994; Deagle et al., 1996), but their utility to interpret comparative phenotypic data was limited. Adaptive interpretation of threespine stickleback variation originated with Hagen's (1967) and McPhail's (1969) seminal studies, which established the importance of threespine stickleback in evolutionary biology (see Foster & Baker, 2004; Pennisi, 2004; Coulata & Pennisi, 2005; Gibson, 2005). Their studies focused on the interactions of a few populations that represented separate biological species within the threespine stickleback species complex. They emphasized divergence of morphological and pigmentation traits between parapatric biological species, the effects of these traits on fitness, and evidence for reproductive isolation. The elaborate stickleback courtship behaviour was a logical place to look for isolating mechanisms because it had already been described in detail by Tinbergen (1951) and other ethologists. Since then, positive assortative mating between biological species of threespine stickleback has proven to be important for reproductive isolation (McPhail, 1994; McKinnon & Rundle, 2002; Boughman, 2007).

However, some of the most interesting phenotypic variation within the threespine stickleback species complex occurs among allopatric populations that may or may not represent separate biological species. Hagen's (1967) misadventure in stickleback taxonomy, which was a peripheral issue for him, violated the rules of zoological nomenclature and elicited strong criticism from Miller and Hubbs (1969). They used lateral plate counts from several populations, which they misinterpreted from the perspective of a "dispersalist-neutralist paradigm." Lateral plates are enlarged scales that form a single row along the sides of the body. They argued that western North American threespine stickleback represent three subspecies that can be diagnosed using lateral plate phenotypes. Marine, anadromous (collectively oceanic) and freshwater populations with modal plate counts of about 33 per side formed the northern subspecies, *Gasterosteus aculeatus aculeatus*. The central freshwater subspecies, *G. a. macrocephalus*, had no more than ten anterior plates per side, while *G. a. williamsoni* was the southern freshwater subspecies in which most individuals lacked plates. They interpreted samples that were both phenotypically and geographically intermediate to these subspecies to indicate introgressive hybridization between subspecies. Their analysis implicitly assumed that lateral plate phenotypes were functionally equivalent, the "neutralist" component of their paradigm, and attributed local variation to dispersal between subspecies, the "dispersalist" component. While Hagen's (1967) comments on stickleback taxonomy violated the rules of zoological nomenclature, Miller and Hubbs's (1969) paradigm had no biological basis.

Treatment of geographical variants of threespine stickleback as separate species or subspecies had a long history. It is reflected by inclusion of five nominal species of *Gasterosteus* in the figure on the cover of this volume, which was originally published in *British Fresh-Water Fishes* (Houghton, 1879). However, even as long ago as 1879, the author questioned whether these nominal species were more than just geographical variants of a single species.

By 1969, Hagen and McPhail (1970) had already realized that lateral plate phenotypes are adaptive, so Hagen set to work to disprove Miller and Hubbs's (1969) assumption that the plates are non-adaptive indicators of phylogeny. In a terrific burst of scientific creativity, Hagen's lab produced several papers on the adaptive significance of lateral plates. Hagen and Gilbertson's (1972) comparative study of lateral-plate and other phenotypes using mostly low-plated, freshwater populations in relation to geographic location and environmental variables indicated strongly that local adaptation provided a robust explanation for variation among stickleback populations. Plate number did not decline with distance from the coast, which would be expected if gene flow from oceanic, completely plated *G. a. aculeatus* to lower-plated *G. a. microcephalus* caused high plate counts in the latter (Hagen & Gilbertson, 1972). However, plate number of low-plated stickleback populations (i.e., *G. a. microcephalus* of Miller & Hubbs, 1969) was lower in habitats lacking predatory fishes and higher (mode = seven per side) where predatory fishes were abundant, indicating that fish predation favoured higher plate counts. A year later, Hagen and Gilbertson (1973) published evidence that seven-plated stickleback were underrepresented in trout stomachs and that mean plate counts in low-plated populations increased rapidly after a population became exposed to trout predation. Subsequent research primarily by Reimchen (e.g., 1994; Spoljaric & Reimchen, 2007; see also Bell et al., 2004; Marchenko & Schluter, 2007; Barrett et al., 2008) has refined but generally supported Hagen's findings that geographical variation of lateral plate phenotypes reflects adaptation to predation and other environmental factors.

Taken together, the research of Hagen and his collaborators (Hagen, 1973; Hagen & Gilbertson, 1972, 1973; Moodie et al., 1973) made a compelling case that local natural selection exerts a powerful influence over plate phenotypes. However, the contribution of common ancestry to plate variation, the factor emphasized to the exclusion of all other causes by Miller and Hubbs (1969), received only limited attention in Hagen's research programme. Population comparisons continue to be a rich source of insight into phenotypic evolution in threespine stickleback (e.g., Reimchen, 1989; Bell et al., 1994; Vamosi & Schluter, 2004; Spoljaric & Reimchen, 2007), but the use of formal phylogenetic analyses to factor out the effects of common ancestry has remained limited (but see Thompson et al., 1997; Taylor & McPhail, 1999, 2000; Colosimo et al., 2005).

Independent derivation of freshwater threespine stickleback populations from local oceanic populations is still the cornerstone for the assumption that geographically distant, freshwater populations can be treated as phylogenetically independent cases. Lindsey (1962) was probably the first to propose this model for the origin of freshwater threespine stickleback. McPhail and Lindsey (1970) developed it more explicitly, and I (Bell, 1976) elaborated it in light of Hagen and Gilbertson's (1972) evidence for local adaptation of armor and trophic traits in freshwater populations. I (Bell, 1976) proposed hypotheses for molecular genetic variation within and among threespine stickleback populations to test this model, and these hypotheses generally have not been falsified (Withler & McPhail, 1985; Taylor & McPhail, 1999, 2000; Colosimo et al., 2005; Mäkinen et al., 2006; but see Reusch et al., 2001). The emergence of specialized molecular tools for threespine stickleback (Cresko et al., 2007; Kingsley & Peichel, 2007; Baird et al., 2008) has made it relatively easy to devise phylogenetic analyses to test the hypothesis that similar phenotypes in separate populations evolved independently.

The threespine stickleback has important attributes for comparative studies, which I emphasized in my paper: (1) Numerous morphological properties of oceanic threespine stickleback have varied little during the last 13 million years (Bell et al., 2009) and throughout their geographic range (Bell & Foster, 1994; Walker & Bell, 2000). Thus, the ancestral conditions for many traits in freshwater populations are relatively uniform worldwide in the oceanic ancestor. (2) Freshwater stickleback have evolved repeatedly from oceanic populations. (3) Freshwater populations diverge in characteristic ways from oceanic stickleback (Bell & Foster, 1994). Thus (from 1 plus 2), shared traits in geographically distant, freshwater populations that contrast with the state in oceanic ancestors evolved independently of each other. (4) Numerous threespine stickleback populations are distributed throughout Holarctic fresh waters. Therefore, *numerous geographically distant, freshwater populations have evolved freshwater traits independently and can be assembled to perform statistically valid tests for associations between phenotypes and environments without correcting for common ancestry*. This was the major point of my paper; even without an explicit phylogenetic analysis to account for the influence of common ancestry, geographically distant, freshwater threespine stickleback populations can be used in comparative studies to infer the adaptive significance of derived behaviours or other phenotypic characters that are absent in oceanic ancestors.

Of course, this does not eliminate the value of phylogeographic analysis to validate independent evolution of similar phenotypes in separate freshwater populations. Examples of this approach include studies by Thompson et al. (1997) and Taylor and McPhail (1999, 2000), which tested the hypothesis that phenotypically similar members of pairs of biological species within the species complex had evolved independently, and by Colosimo et al. (2005), which tested the hypothesis that *Ectodysplasin* alleles responsible for the low-plate morph in almost all freshwater populations throughout the Holarctic evolved from a single mutational transition from a complete-morph allele. The recent development of extensive molecular resources (Cresko et al., 2007; Kingsley & Peichel, 2007; Baird et al., 2008) has made it relatively simple to justify the use of separate freshwater populations as independent replicates in the iterative process of diversification by natural selection. Furthermore, these tools allow researchers to test whether parallel diversification in geographically isolated populations is due to parallel changes in genomic regions, genes and alleles.

Acknowledgements

I thank D.W. Hagen and J.D. McPhail for developing an adaptive perspective for stickleback evolution, T.E. Reimchen for clever insights and tireless dedication that sharpened their perspective, D.M. Kingsley and C.M. Peichel for turning the threespine stickleback into a genomic model, and my departmental colleagues for the stimulating intellectual environment within which I am fortunate to work. Contribution 1182 from Ecology and Evolution at Stony Brook University.

References

Aguirre, W.E. (2007). The pattern and process of evolutionary diversification. Lessons from a threespine stickleback adaptive radiation. — PhD dissertation, Stony Brook University, Stony Brook, New York.

Avise, J.C. (1994). Molecular markers, natural history and evolution. — Chapman & Hall, New York, NY.

Baird, N.A., Etter, P.D., Atwood, T.S., Currey, M.C., Shiver, A.L., Lewis, Z.A., Selker, E.U., Cresko, W.A. & Johnson, E.A. (2008). Rapid SNP discovery and genetic mapping using sequenced RAD markers. — PLoS ONE 3: e3376.

Barrett, R.D.H., Rogers, S.M. & Schluter, D. (2008). Natural selection on a major armor gene in threespine stickleback. — Science 322: 255-257.

Bell, M.A. (1974). Reduction and loss of the pelvic girdle in *Gasterosteus* (Pisces): a case of parallel evolution. — Nat. Hist. Mus. Los Angeles Co. Contrib. Sci. 257: 1-36.

Bell, M.A. (1976). Evolution of phenotypic diversity in *Gasterosteus aculeatus* superspecies on the Pacific coast of North America. — Syst. Zool. 25: 211-227.

Bell, M.A. (1987). Interacting evolutionary constraints in pelvic reduction of threespine sticklebacks, *Gasterosteus aculeatus* (Pisces: Gasterosteidae). — Biol. J. Linnean Soc. 31: 347-382.

Bell, M.A. (1988). Stickleback fishes: bridging the gap between population biology and paleobiology. — Trends Ecol. Evol. 2: 320-325.
Bell, M.A. (1995). Intraspecific systematics of *Gasterosteus aculeatus* populations: implications for behavioral ecology. — Behaviour 132: 1131-1152.
Bell, M.A., Aguirre, W.E. & Buck, N.J. (2004). Twelve years of contemporary armor evolution in a threespine stickleback population. — Evolution 58: 814-824.
Bell, M.A. & Foster, S.A. (1994). Introduction to the evolutionary biology of the threespine stickleback. — In: The evolutionary biology of the threespine stickleback (Bell, M.A. & Foster, S.A., eds). Oxford University Press, Oxford, p. 1-27.
Bell, M.A. & Ortí, G. (1994). Pelvic reduction in threespine stickleback from Cook Inlet lakes: geographic distribution and intrapopulation variation. — Copeia 1994: 314-325.
Bell, M.A., Ortí, G., Walker, J.A. & Koenings, J.P. (1993). Evolution of pelvic reduction in threespine stickleback fish: a test of competing hypotheses. — Evolution 47: 906-914.
Bell, M.A., Stewart, J.D. & Park, P.J. (2009). The world's oldest fossil threespine stickleback fish. — Copeia 2009: 256-265.
Boughman, J.W. (2007). Speciation in sticklebacks. — In: Biology of the three-spined stickleback (Östlund-Nilsson, S., Mayer, I. & Huntingford, F.A., eds). CRC Press, Boca Raton, FL, p. 83-126.
Brooks, D.R. & McLennan, D.A. (1991). Phylogeny, ecology and behavior: a research program in comparative biology. — University of Chicago Press, Chicago, IL.
Colosimo, P.F., Hoseman, K.E., Balabhadra, S., Villareal Jr., G., Dickson, M., Grimwood, J., Schmutz, J., Myers, R.M., Schluter, D. & Kingsley, D.M. (2005). Widespread parallel evolution in sticklebacks by repeated fixation of *Ectodysplasin* alleles. — Science 307: 1928-1933.
Coulata, E. & Pennisi, E. (2005). Breakthrough of the year: evolution in action. — Science 310: 1878-1880.
Cresko, W.A., McGuigan, K.L., Phillips, P.C. & Postlethwait, J.H. (2007). Studies of threespine stickleback developmental evolution: progress and promise. — Genetica 29: 105-206.
Deagle, B.E., Reimchen, T.E. & Levin, D.B. (1996). Origins of endemic stickleback from the Queen Charlotte Islands: mitochondrial and morphological evidence. — Can. J. Zool. 74. 1045-1056.
Endler, J.A. (1986). Natural selection in the wild. — Monographs in Population Biology, No 21. Princeton University Press, Princeton, NJ.
Felsenstein, J. (1985). Phylogenies and the comparative method. — Am. Nat. 125: 1-15.
Foster, S.A. & Baker, J.A. (2004). Evolution in parallel: new insights from a classic system. — Trends Ecol. Evol. 19: 456-459.
Gibson, G. (2005). The synthesis and evolution of a supermodel. — Science 307: 1890-1891.
Giles, N. (1983). The possible role of environmental calcium levels during the evolution of phenotypic diversity in Outer Hebridean populations of the three-spined stickleback. — J. Zool. London 199: 535-544.
Hagen, D.W. (1967). Isolating mechanisms in threespine sticklebacks (*Gasterosteus*). — J. Fish. Res. Bd. Can. 30: 845-894.
Hagen, D.W. (1973). Inheritance of numbers of lateral plates and gill rakers in *Gasterosteus aculeatus*. — Heredity 30: 303-312.
Hagen, D.W. & Gilbertson, L.G. (1972). Geographic variation and environmental selection in *Gasterosteus aculeatus* L. in the Pacific northwest, America. — Evolution 26: 32-51.

Hagen, D.W. & Gilbertson, L.G. (1973). Selective predation and the intensity of selection acting upon the lateral plates of threespine sticklebacks. — Heredity 30: 273-287.

Hagen, D.W. & McPhail, J.D. (1970). The species problem within *Gasterosteus aculeatus* on the Pacific coast of North America. — J. Fish. Res. Bd. Can. 27: 147-155.

Haglund, T.R., Buth, D.G. & Lawson, R. (1992). Allozyme variation and phylogenetic relationships of Asian, North American, and European populations of the threespine stickleback. — Copeia 1992: 432-443.

Harvey, P.H. & Pagel, D.M. (1991). The comparative method in evolutionary biology. — Oxford University Press, Oxford.

Houghton, W. (1879). British fresh-water fishes. — Webb & Bowler Limited, Southernhay East (1981 reprint).

Kingsley, D.M. & Peichel, C.L. (2007). The molecular genetics of evolutionary change in sticklebacks. — In: Biology of the three-spined stickleback (Östlund-Nilsson, S., Mayer, I. & Huntingford, F.A., eds). CRC Press, Boca Raton, FL, p. 41-81.

Lindsey, C.C. (1962). Experimental study of meristic variation in a population of threespine stickleback, *Gasterosteus aculeatus*. — Can. J. Zool. 40: 271-312.

Mäkinen, H.S., Cano, J.M. & Merilä, J. (2006). Genetic relationships among marine and freshwater populations of the European three-spined stickleback (*Gasterosteus aculeatus*) revealed by microsatellites. — Mol. Ecol. 15: 1519-1534.

Marchenko, K.B. & Schluter, D. (2007). Parallel evolution by correlated response: lateral pate reduction in threespine stickleback. — Evolution 61: 1084-1090.

McKinnon, J.S. & Rundle, H.D. (2002). Speciation in nature: the threespine stickleback model systems. — Trends Ecol. Evol. 17: 480-488.

McPhail, J.D. (1969). Predation and the evolution of a stickleback (*Gasterosteus*). — J. Fish. Res. Bd. Can. 26: 3183-3208.

McPhail, J.D. (1994). Speciation and the evolution of reproductive isolation in the sticklebacks (*Gasterosteus*) of south-western British Columbia. — In: The evolutionary biology of the threespine stickleback (Bell, M.A. & Foster, S.A., eds). Oxford University Press, Oxford, p. 399-437.

McPhail, J.D. & Lindsey, C.C. (1970). Freshwater fishes of northwestern Canada and Alaska. — Bull. Fish. Research Board of Canada 173: 1-381.

Miller, R.R. & Hubbs, C.L. (1969). Systematics of *Gasterosteus aculeatus*, with particular reference to intergradations and introgression along the Pacific coast of North America: a commentary on a recent contribution. — Copeia 1969: 52-69.

Moodie, G.E.E., McPhail, J.D. & Hagen, D.W. (1973). Experimental demonstration of selective predation in *Gasterosteus aculeatus*. — Behaviour 47: 95-105.

Ortí, G., Bell, M.A., Reimchen, T.E. & Meyer, A. (1994). Global survey of mitochondrial DNA sequences in the threespine stickleback: evidence for recent migrations. — Evolution 48: 608-622.

Pennissi, E. (2004). Changing a fish's bony armor in the wink of a gene. — Science 304: 1736-1739.

Reimchen, T.E. (1980). Spine deficiency and polymorphism in a population of *Gasterosteus aculeatus*: an adaptation to predators? — Can. J. Zool. 58: 1232-1244.

Reimchen, T.E. (1989). Loss of nuptial color in threespine sticklebacks (*Gasterosteus aculeatus*). — Evolution 43: 450-460.

Reimchen, T.E. (1994). Predators and morphological evolution in threespine stickleback. — In: The evolutionary biology of the threespine stickleback (Bell, M.A. & Foster, S.A., eds). Oxford University Press, Oxford, p. 240-276.

Reusch, T.B.H., Wegner, K.M. & Kalbe, M. (2001). Rapid genetic divergence in postglacial populations of threespine stickleback (*Gasterosteus aculeatus*): the role of habitat type, drainage and geographical proximity. — Mol. Ecol. 10: 2435-2434.

Rohlf, F.J. (2006). A comment on "phylogenetic correction". — Evolution 60: 1509-1515.

Spoljaric, M. & Reimchen, T.E. (2007). 10,000 years later: evolution of body shape in Haida Gwaii threespine stickleback. — J. Fish Biol. 70: 1484-1503.

Taylor, E.B. & McPhail, J.D. (1999). Evolutionary history of an adaptive radiation in species pairs of threespine sticklebacks (*Gasterosteus*): insights from mitochondrial DNA. — Biol. J. Linnean Soc. 66: 271-291.

Taylor, E.B. & McPhail, J.D. (2000). Historical contingency and ecological determinism interact to prime speciation in sticklebacks, *Gasterosteus*. — Proc. Roy. Soc. Lond. B Biol. 267: 2375-2384.

Thompson, C.E., Taylor, E.B. & McPhail, J.D. (1997). Parallel evolution of lake-stream pairs of threespine sticklebacks (*Gasterosteus*) inferred from mitochondrial DNA variation. — Evolution 51: 1955-1965.

Tinbergen, N. (1951). The study of instinct. — Oxford University Press, Oxford.

Vamosi, S.M. & Schluter, D. (2004). Character shifts in the defensive armor of sympatric sticklebacks. — Evolution 58: 376-385.

Walker, J.A. & Bell, M.A. (2000). Net evolutionary trajectories of body shape evolution within a microgeographic radiation of threespine sticklebacks (*Gasterosteus aculeatus*). — J. Zool. London 52: 293-302.

Withler, R.E. & McPhail, J.D. (1985). Genetic variability in freshwater and anadromous sticklebacks (*Gasterosteus aculeatus*) of southern British Columbia. — Can. J. Zool. 63: 528-533.

INTRASPECIFIC SYSTEMATICS OF GASTEROSTEUS ACULEATUS POPULATIONS: IMPLICATIONS FOR BEHAVIORAL ECOLOGY

by

MICHAEL A. BELL[1,2])

(Department of Ecology and Evolution, State University of New York, Stony Brook, New York 11794-5245, USA)

(With 5 Figures)
(Acc. 2-VI-1995)

Summary

The threespine stickleback is widespread in diverse coastal marine and freshwater habitats of the Holarctic. Extensive information on its behavior and evolution provide an exceptional background for behavioral ecology. The behavioral phenotype links studies of proximate and ultimate (evolutionary) causes of behavior. Analyses of the ultimate causes of behavior should be placed in a phylogenetic context. There have been four phases of interpretation of phenotypic variation in threespine stickleback: (1) taxonomic, (2) dispersalist-neutralist, (3) adaptationist, and (4) adaptationist-phylogenetic. *Gasterosteus aculeatus* should be treated as a highly variable species complex. Phylogenetic inferences based on geographical location and allozyme and nucleotide variation facilitate discrimination between ancestry and local adaptation as ultimate causes of phenotypic variation among populations. The phylogeny of threespine stickleback populations and the use of phylogenetic information to analyze causation of pelvic girdle variation among populations are reviewed. Criteria for selection of populations and behavioral traits for comparative studies are discussed.

Introduction

The threespine stickleback, *Gasterosteus aculeatus*, has been studied extensively by behaviorists and evolutionary biologists. It is a species complex

[1]) e-mail address: MABELL@LIFE.BIO.SUNYSB.EDU

[2]) I thank G. Ortí for providing Fig. 3 and its caption and for generating much of the original DNA data cited in this paper. I thank publishers of works in which Fig. 3, 4 (Evolution) and 5 (Oxford University Press) were originally published for permission to use them here. My research was supported by NSF grant BSR-8905758. G. Ortí and J.A. Walker provided constructive criticism of this paper. I thank T. Klepaker for his unpublished record of pelvic reduction in a Norwegian stickleback population. RFLP data of Bell *et al.* are from an unpublished manuscript by M.A. Bell, G. Ortí, T. Städler, L. Nevermann, and A. Meyer. This paper is contribution 937 from the Graduate Program in Ecology and Evolution at the State University of New York at Stony Brook.

consisting of countless, phenotypically divergent, allopatric populations and biological species that are distributed widely in coastal marine and fresh waters of the northern Holarctic (*e.g.* BELL, 1984a; BELL & FOSTER, 1994; MCPHAIL, 1994; NELSON, 1994). Placement of phenotypic diversity within this complex into a phylogenetic context will improve interpretation of experimental results using different populations (reviewed in BROOKS & MCLENNAN, 1991; HARVEY & PAGEL, 1991).

G. aculeatus was described by LINNAEUS (1758), but several invalid nominal species with obscure phylogenetic relationships were added by 1910 (HUBBS, 1929; WOOTTON, 1976). In 1934 TINBERGEN introduced *G. aculeatus* as a model system in ethology (VAN DEN ASSEM & SEVENSTER, 1985), and in the late 1960's, HAGEN (1967) and MCPHAIL (1969) initiated its use in population biology to study speciation and adaptation (reviewed in, *e.g.* BELL, 1976, 1984a; BELL & FOSTER, 1994; MCPHAIL, 1994; REIMCHEN, 1994). Population biologists exploited phenotypic differentiation, ethological methodology (reviewed by ROWLAND, 1994), and the usual methods in evolutionary biology to investigate stickleback evolution.

Wide-spread armor variation proved to be habitat specific and responsive to local natural selection (*e.g.* HAGEN & GILBERTSON, 1972, 1973; reviewed by REIMCHEN, 1994). Moreover, population biologists discovered extensive interpopulation behavioral variation, which has been confirmed for foraging (reviewed in HART & GILL, 1994), antipredator (HUNTINGFORD, 1982; GILES & HUNTINGFORD, 1984; HUNTINGFORD & WRIGHT, 1992; HUNTINGFORD *et al.*, 1994), and reproductive behavior (WHORISKEY & FITZGERALD, 1985; FOSTER, 1988, 1994, this volume; RIDGWAY & MCPHAIL, 1988).

During the past decade, it has become more widely appreciated that common ancestry and adaptation both contribute to phenotypic similarity among related groups (reviewed in BROOKS & MCLENNAN, 1991; HARVEY & PAGEL, 1991). Phylogeny must be factored out before similarities among groups can be used to infer adaptive causes for variation among groups. TEMRIN & SILLEN-TULLBERG (1994), for example, recently used phylogenetic information to infer the number of evolutionary transitions between monogamy and polygamy in birds and to show that most such transitions are to higher states of polygamy. The purposes of this paper are to review the phylogeny of threespine stickleback populations and methods to factor out common ancestry to facilitate inference of the role of natural selection in behavioral divergence.

Levels of causation and causal links between behavior and evolution

Biological research tends to emphasize one of two levels of causation. 'Proximate causation' concerns the ontogeny and dynamics of individuals and the 'ultimate causation' concerns causes for phenomena manifested by evolutionary lineages (Mayr, 1982; Francis, 1995). This dichotomy is reflected in behavioral research (Fig. 1; *e.g.* Gould, 1982; Real, 1994; Francis, 1995).

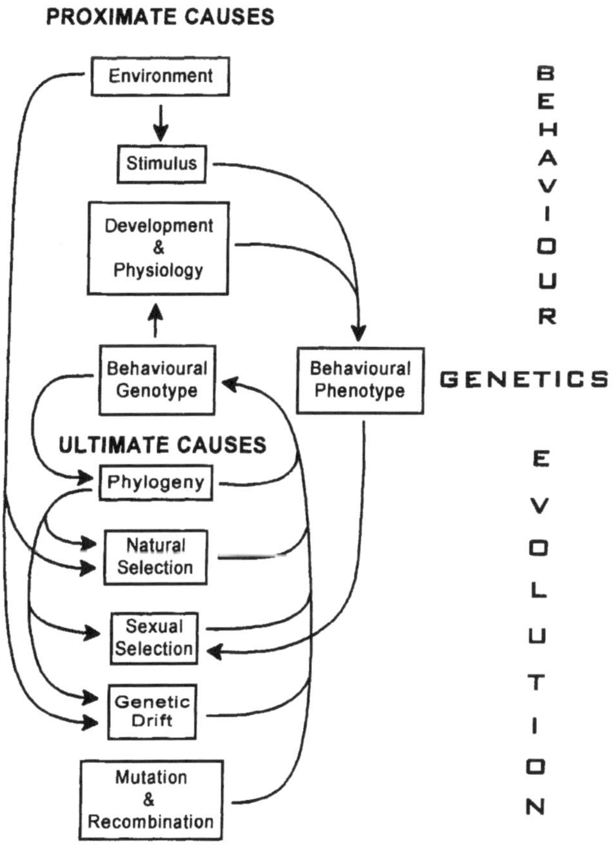

Fig. 1. Relationships of major causes and effects in behavior, genetics, and evolution. The behavioral phenotype is the nexus of all three fields. Behavior and genetics deal principally with proximate causes, and evolution deals principally with ultimate causes.

Studies of proximate behavioral causes divide them into two general classes: intrinsic properties and external stimuli. Intrinsic properties have some genetic basis (*e.g.* Boake, 1994) and include developmental and physiological mechanisms. Recent and current environments are the source of stimuli for behaviors. Together, intrinsic properties and external stimuli produce the 'behavioral phenotype'.

Studies of behavior evolution, generally referred to as behavioral ecology, emphasize the ultimate causes for behavioral phenotypes. From an evolutionary perspective, the environment imposes both ultimate and proximate causes of variation for behavioral phenotypes. The present environment indicates the conditions under which behavioral phenotypes evolved and provides stimuli, which in evolutionary studies may be identifiable causes of phenotypic plasticity, or at worst constitute unexplained variation (Boake, 1994; Wilson *et al.*, 1994; Francis, 1995). Ultimate causes include (1) natural selection, which produces or increases the adaptive fit between a phenotype and the environment; (2) sexual selection, which enhances traits that favor mate acquisition, (3) phylogeny, which accounts for the phenotype that existed prior to adaptation to current conditions; and (4) stochastic processes (*i.e.* mutation, genetic recombination, genetic drift). These ultimate causes produce a behavioral genotype that interacts with proximate causes of behavior to yield the behavioral phenotype.

The nexus for proximate and ultimate causation is the behavioral phenotype. Studies of behavioral evolution must separate proximate and ultimate causes and then evaluate the importance of phylogeny, natural selection, and stochastic processes. Comparison of phenotypes among groups in relation to environmental differences is a valuable source of hypotheses for ultimate causes (reviewed by Endler, 1986). In recent years, however, the importance and difficulty of separating common ancestry and adaptation as causes for phenotypic similarity have become more fully appreciated (reviewed in Brooks & McLennan, 1991; Harvey & Pagel, 1991). The methods for data acquisition and analysis for phylogenetic inference have also improved and become applicable to more closely related groups than before (reviewed by Avise *et al.*, 1987; Avise, 1994).

History of interpretation of variation in *Gasterosteus aculeatus*

Interpretation of phenotypic variation within the *G. aculeatus* species complex can be divided into four phases, (1) taxonomic, (2) dispersalist-neutralist, (3) adaptationist, and (4) adaptationist-phylogenetic. The taxonomic phase began with description of the species by LINNAEUS (1758). Between 1792 and 1910, 41 junior synonyms of *G. aculeatus* were described (WOOTTON, 1976) based mostly on armor characters, particularly the bony lateral plates.

Because this phase was purely descriptive, it ignored polyphyletic origin of similar phenotypes. In some respects, it was counterproductive because description sometimes monopolized study, excluding explanation. For example, REGAN (1909), treated *G. aculeatus* as a wide-spread variable species but recognized three weakly-armored nominal species of *Gasterosteus* from Italy, North Africa, and southern California, USA. Rather than addressing the causes for their disjunct geographical distribution or of convergent evolution, he focused on diagnostic characters and nomenclature.

Next, BERTIN (1925) demonstrated continuous morphological variation among European populations of *G. aculeatus*, which he attributed entirely to phenotypic plasticity. Although there is now abundant evidence that these phenotypic differences have a genetic basis (*e.g.* HEUTS, 1947; MÜNZING, 1959; LINDSEY, 1962; HAGEN, 1973; MCPHAIL, 1994; BAŃBURA & BAKKER, this volume; reviewed by BELL, 1984a), even the most divergent extant populations continue to be lumped with the original Linnean species, *G. aculeatus* (see below). The only exception to this generalization is recognition of *Gasterosteus wheatlandi*, which is phenotypically (HUBBS, 1929; SARGENT *et al.*, 1984; BELL, 1994; BOWNE, 1994; NELSON, 1994) and genetically (HAGLUND *et al.*, 1992; BUTH & HAGLUND 1994; ORTÍ *et al.*, 1994) divergent from *G. aculeatus* but always breeds in sympatry with it. BERTIN's monograph closed the taxonomic phase but went too far in denying genetic differentiation within *G. aculeatus.*

During the neutralist-dispersalist phase, armor characters were recognized as genetic differentiation, but they were treated as neutral markers of populations that had diverged during glacial isolation and intergraded where contact was re-established by postglacial dispersal. MÜNZING

(1963) interpreted lateral plate variation in European populations to indicate occurrence of a high-plated isolate northwest of the Scandinavian Ice Sheet and of a low-plated isolate in the Mediterranean during the last glacial advance. MILLER & HUBBS (1969) gave a similar interpretation for lateral plate variation in western North American populations. Although MÜNZING (1963) postulated local evolution of reduced armor in two populations near the Black Sea, he attributed similar phenotypes elsewhere to postglacial dispersal. The neutralist-dispersalist phase perpetuated typological thinking about stickleback variation and non-adaptive interpretation of variable stickleback phenotypes.

HAGEN & MCPHAIL (1970) criticized of the neutralist-dispersalist interpretation, heralding the deluge of adaptationist research that followed. HAGEN's research program set the standard for this approach: HAGEN & GILBERTSON (1972) demonstrated associations of armor and trophic phenotypes with habitat type, that some lateral plate phenotypes experience selective predation (HAGEN & GILBERTSON, 1973), and that some phenotypes are heritable (HAGEN, 1973). Numerous comparative studies have confirmed and refined HAGEN & GILBERTSON's conclusions (*e.g.* MOODIE & REIMCHEN, 1976; LAVIN & MCPHAIL, 1985, 1986; REIMCHEN *et al.*, 1985; BOURGEOIS *et al.*, 1994; reviewed by BELL, 1984a; REIMCHEN, 1994) and extended them to other characters, including body size (MCPHAIL, 1977; BELL, 1984b), nuptial coloration (MCPHAIL, 1969; HAGEN, *et al.*, 1980; REIMCHEN, 1989), pelvic structure (GILES, 1983; BELL *et al.*, 1985, 1993; BELL, 1987, 1988), body form (REIMCHEN *et al.*, 1985; BAUMGARTNER *et al.*, 1988; WALKER, 1995), trophic structures (MOODIE & REIMCHEN, 1976; GROSS & ANDERSON, 1984; LAVIN & MCPHAIL, 1985, 1986; SCHLUTER & MCPHAIL, 1992; MCPHAIL, 1994), and predator avoidance behavior (reviewed by HUNTINGFORD *et al.*, 1994). Although large-scale comparative studies have not been published, there is also evidence for ecotypic differentiation of reproductive (reviewed by FOSTER, 1994, this volume) and foraging behaviors (reviewed by HART & GILL, 1994, MCPHAIL, 1994).

The geographical isolation of most populations used in these comparative studies makes it improbable that common ancestry caused resemblances among populations from similar habitats, but this issue generally was not addressed explicitly. Rather, it was assumed incorrectly that the populations were free to evolve independently and that the number of populations in the analysis determines the number of degrees of freedom (HARVEY & PAGEL, 1991). Thus, these papers are not part of the phylogenetic-adaptationist phase. An exception to this generalization is exclusion of common ancestry as the cause for similar pelvic reduction in two postglacial insular populations in British Columbia and a Pliocene (actually Miocene, BELL, 1994) form in Nevada, USA (BELL, 1974) by reference to the geographical distances and geological time separating them. The wide distribution of pelvic reduction subsequently documented in freshwater populations of *G. aculeatus* requires that it evolved repeatedly (BELL, 1987). The same argument could be applied to the other phenotypic characters noted in the previous paragraph (BELL, 1976, 1984a; BELL & FOSTER, 1994).

Two methods can be used to infer that phenotype-environment correlations among freshwater populations of *G. aculeatus* reflect local adaptation and not common ancestry. Geographical separation, limited time since establishment of freshwater populations, and absence of widespread freshwater phenotypes from anadromous and marine populations can be used to infer their independent evolution (BELL, 1987, 1988; BELL & FOSTER, 1994; MCPHAIL, 1994). The second method requires cladistic analysis of populations and mapping of phenotypic states onto the cladogram (reviewed in BROOKS & MCLENNAN, 1991; HARVEY & PAGEL, 1991).

Nomenclature of biological species within the *Gasterosteus aculeatus* species complex

Except for *G. wheatlandi* (see above) and the dubious fossil species, *G. doryssus* (BELL, 1994), all nominal species of *Gasterosteus* have been placed in synonymy with *G. aculeatus* (WOOTTON, 1976). Yet, several studies document existence within the *G. aculeatus* complex of pairs of sympatric biological species (reviewed by MCPHAIL, 1994), none of which has been described as a taxonomic species. Except for the 'white stickleback', a widespread, undescribed, marine, form in eastern Canada (BLOUW & HAGEN, 1990; MACDONALD *et al.*, this volume a, b), the well documented biological species of *G. aculeatus* are endemic to geographically restricted habitats (reviewed by MCPHAIL, 1994). Although some endemic biological species are highly divergent (*e.g.* MCPHAIL, 1969; HAGEN & MOODIE, 1979; SCHLUTER & MCPHAIL, 1992), phenotypic divergence of comparable magnitude commonly occurs in allopatric populations (reviewed in

Hagen & McPhail, 1970; Bell, 1976, 1984a; Bell & Foster, 1994). It would be very difficult to diagnose these endemics properly without reference to their geographical location, and equally divergent allopatric populations could not easily be shown to be biological species. Thus, I concur with J.D. McPhail and D.M. Blouw (pers. comm., 1994) that it is neither practical nor useful to describe the biological species within the *G. aculeatus* species complex as separate taxonomic species.

Implications of phylogeny for comparative behavioral studies using *Gasterosteus aculeatus*

A common method to infer evolutionary mechanisms for population differentiation is to analyze interpopulation associations between mean phenotypes and environments (Endler, 1986). Typically each population is implicitly treated as a statistically independent observation (Harvey & Pagel, 1991). However, phenotype-environment associations may have evolved only once and spread by dispersal from the point of origin to many similar habitat patches. Thus, all populations that contribute to this phenotype-environment association may belong to a single clade, and the phenotype-environment association may be a characteristic of the clade, not of selection in that habitat type for that phenotype. The observed phenotype-environment association may be independent of the phenotype's ecological function, and a strong phenotype-environment association would not indicate that the phenotype evolved in response to conditions in that environment. Thus, it is important to infer whether phenotypic similarities of populations from a particular habitat type are due to common ancestry or to repeated evolution (Fig. 2).

Inference of phylogenetic independence of similar phenotypes in separate populations of *Gasterosteus aculeatus*

Two methods can be used to infer phylogenetic independence of similar phenotypes in different *G. aculeatus* populations. The first employs geographical separation of populations, and resembles the traditional practice of comparing allopatric radiations in distantly-related clades (*e.g.* placental mammals and Australian [Keast, 1972] or South American marsupial mammals [Simpson, 1980]) or of parallel differentiation of introduced conspecific populations on separate continents (*e.g.* house

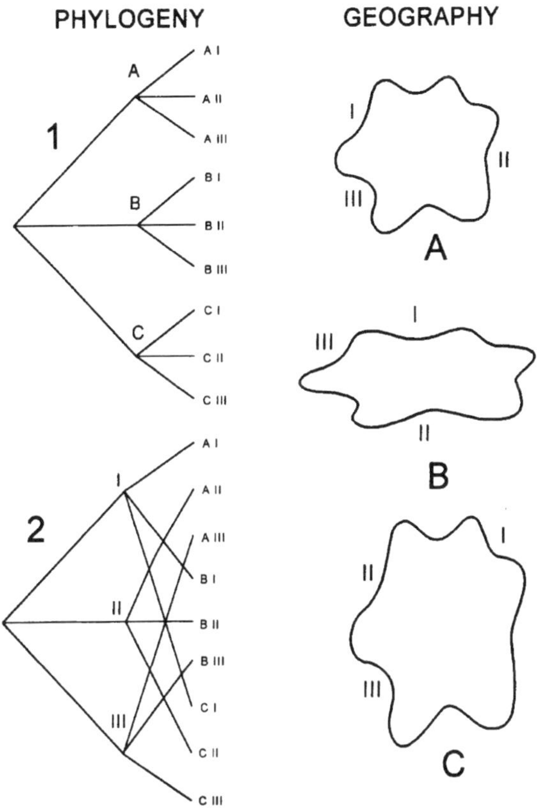

Fig. 2. Hypothetical distribution of three phenotypes (I, II & III) in three regions (A, B & C) on right, and the two extreme alternative phylogenetic hypotheses to explain the distribution of the three phenotypes in the three regions. The upper cladogram (1) depicts colonization of each region only once before divergence of the three phenotypes within each region. The lower cladogram (2) depicts evolution of each phenotype only once, followed by colonization of each region by each phenotype. Each population can be used as a statistically independent observation when cladogram 1 depicts the true phylogeny.

sparrows [Johnston & Selander, 1973], *Drosophila melanogaster* [Oakshott *et al.*, 1982]). Geographical separation, relatively recent colonization, and absence of a phenotype in the putative common ancestor together indicate iterative evolution.

These three requirements are easily met by many freshwater populations of *G. aculeatus*. They have been derived locally from marine or anadromous populations innumerable times (reviewed by Bell, 1987,

1988; Bell & Foster, 1994; McPhail, 1994). They are ubiquitous in lowland habitats of the boreal Holarctic, providing ample opportunity for selection of appropriate samples for comparative studies. Many freshwater populations occur in recently (*i.e.* ≤ 25,000 years ago) deglaciated regions, limiting the time for dispersal through fresh water (reviewed in *e.g.* Bell, 1987, 1988; Bell & Foster, 1994; McPhail, 1994). Many widespread freshwater phenotypes, including for example, armor reduction (reviewed by Bell, 1987, 1988; Reimchen, 1994), low gill raker counts (Hagen & Gilbertson, 1972; Moodie & Reimchen, 1976; Gross & Anderson, 1984), gigantism (McPhail, 1994), reduced red nuptial coloration (Reimchen, 1989), increased burst swimming speed (Taylor & McPhail, 1986), and reduced courtship behavior and diversionary displays (Foster, 1988, 1994, this volume), appear to be absent in marine and anadromous populations. These derived (apomorphic) states must have evolved many times.

A second approach (reviewed by Avise *et al.*, 1987; Avise, 1994), which is in its infancy in *G. aculeatus* (Lauder, 1995), is to construct a cladogram for populations and show that the distribution of apomorphic states on the cladogram requires their repeated evolution. Populations in which apomorphic states must have evolved independently represent statistically independent cases and can be employed in comparative studies. Phenotypic states of *G. aculeatus* are prone to homoplasy (*i.e.* evolution of derived states repeatedly in separate lineages) and are unreliable for phylogenetic inference. However, molecular characters, including allozyme variants and nucleotide sequences, should more closely reflect phylogeny than local adaptation (Avise, *et al.*, 1987; Avise, 1994), and can be used to demonstrate that some phenotypes have evolved repeatedly in *G. aculeatus*.

Phylogeny of *Gasterosteus aculeatus* populations

Cladistic analyses using allozyme (Haglund *et al.*, 1992; Buth & Haglund, 1994) and DNA sequence variation (Ortí *et al.*, 1994) among populations of *G. aculeatus* produced comparable results. Inclusion of specimens from populations used in two other previous analyses (Haglund *et al.*, 1992; O'Reilly *et al.*, 1993) enabled Ortí *et al.* (1994) to produce a relatively comprehensive cladogram (Fig. 3). Two major groups are referred to as 'Japanese' and 'Euro-American' clades because

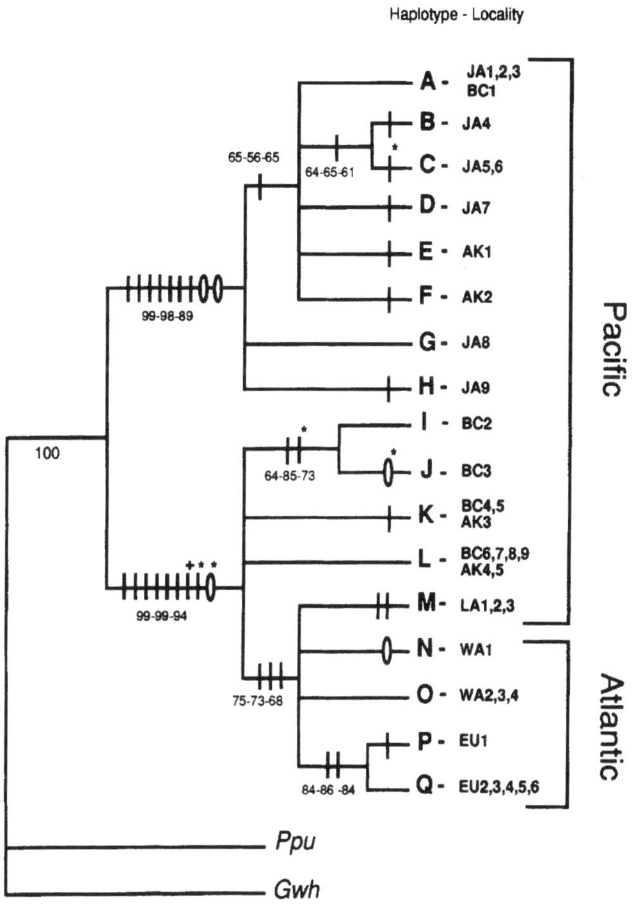

Fig. 3. Majority rule bootstrap consensus tree summarizing geographic locations and relationships among 36 specimens of *Gasterosteus aculeatus* and two outgroups, *Pungitius pungitius* (Ppu) and *Gasterosteus wheatlandi* (Gwh). Each terminal represents a mitochondrial DNA haplotype (A-Q) characterized by 747 base pairs of the cytochrome *b* sequence. Geographic locations are JA, Japan; AK, Alaska, USA; BC, British Columbia, Canada; LA, Los Angeles, California, USA; WA, western Atlantic; (Quebec, Nova Scotia, New York); EU, Europe (Scotland, England, Sweden, France). Three numbers along internodal segments are bootstrap values derived using parsimony, neighbor-joining, and maximum-likelihood methods, respectively. Lines across branches indicate transition substitutions and ovals transversions. Asterisks indicate amino-acid replacements. (See Ortí *et al.* [1994] for additional information on sources of samples and methods.)

of their geographical distribution among samples analyzed by HAGLUND *et al.* (1992). However, ORTÍ *et al.* (1994) detected the Japanese clade in Alaska and showed that it and the Argonaut form characterized by O'REILLY *et al.* (1993) in British Columbia are equivalent. The Japanese clade is absent from the Atlantic basin, but the Euro-American clade entered the Atlantic basin via the Arctic Ocean, when the Bering Strait opened (HERMAN & HOPKINS, 1980; MAGGS, 1990; VERMEIJ, 1991). The Atlantic clade subsequently split into eastern North American and European clades.

Using this cladogram, one can select freshwater populations in each clade that must have evolved similar apomorphic phenotypes independently. Since the major clades are associated with geographic regions between which dispersal through the ocean must have occurred, it is safe to conclude that any phenotypic state that is present in freshwater populations but absent from marine ones evolved independently in each of these major clades.

Use of geographical distance and phylogeny to infer independent evolution of reduced pelvic phenotypes in *Gasterosteus aculeatus*

Use of geographical distance and cladistic analysis to infer phylogenetic independence of the evolution of a phenotypic state in *G. aculeatus* may be illustrated using pelvic reduction. Pelvic reduction forms a stereotyped sequence of structural loss (Fig. 4), and the mean and distribution of reduced pelvic phenotypes varies greatly among populations (Table 1; see also BELL *et al.*, 1985; BELL, 1987; BELL & ORTÍ, 1994). Although pelvic reduction occurs in only a small fraction of *G. aculeatus* populations (BELL, 1987; BELL *et al.*, 1993), more than 100 cases have been observed. It is absent from marine populations and occurs mainly in lakes that lack discharge streams and predatory fishes and have low calcium ion concentrations (BELL *et al.*, 1993). The geographical dispersion and isolation of populations with pelvic reduction indicate strongly that it evolved independently within at least six Holarctic regions since the last glacial retreat (Fig. 5). The Nevada occurrence is from the Miocene (BELL, 1994) and the California case is the only extant occurrence of pelvic reduction south of glacial ice during the last glacial advance. All regions are separated by large expanses of land that lack stickleback populations or by the ocean,

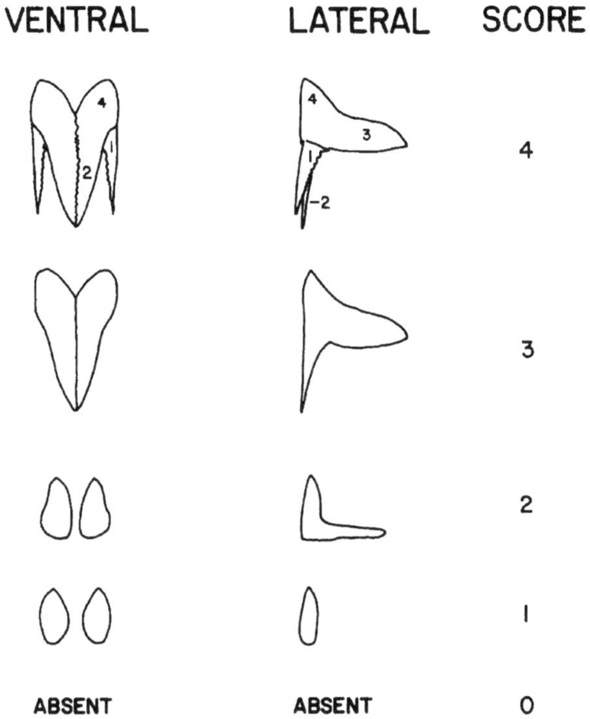

Fig. 4. Usual sequence of pelvic phenotypes (anterior up) in *Gasterosteus aculeatus* with pelvic reduction. The numbers on the top structures indicate the order in which pelvic elements are lost, and these labelled elements are (1) pelvic spine, (2) posterior process, (3) ascending branch, and (4) anterior process. Pelvic scores for individuals (Table 1) are the sum of the scores (right) of both halves of the pelvic skeleton. (From Bell *et al.*, 1993)

where pelvic reduction has not been observed. These far-flung populations could not have been derived from a common ancestor with pelvic reduction that dispersed from a single point of origin to establish the existing distribution within the last 22,000 years. Its almost universal association with unusual environmental circumstances provides an adaptive alternative to common ancestry as the explanation of its distribution.

Addition of cladistic information strengthens this interpretation. There is no compelling evidence for pelvic reduction in any population with exclusively Japanese-clade ancestry (see below). However, it occurs in western North America, eastern North America, and western Europe,

TABLE 1. Geographical location, clade membership, and relative frequencies (rounded to nearest %) of pelvic girdle phenotypes of *Gasterosteus aculeatus* populations with pelvic reduction from Cook Inlet, Alaska

Lake	Area[1]	Haplotype[2]		Pelvic score distribution (%)[3]									
		J	EA	0	1	2	3	4	5	6	7	8	N
Bear Paw	MS	15	9	8	15	65	4	6	2	0	0	0	210
Boot	MS	0	8	8	9	66	5	8	1	1	0	0	2087
Bruce	MS	0	15	4	4	10	7	29	13	28	3	2	272
Coyote	K	2	7	33	25	40	1	1	0	0	0	0	361
Echo[4]	K	0	5	1	0	0	0	0	1	85	4	10	783
Kalmback	K	0	21	8	4	10	3	6	3	11	5	50	536
Morvoro	MS	0	7	5	8	15	2	13	5	8	5	40	62
Prator	MS	4	0	2	3	41	11	19	10	9	3	2	283
Toad	MS	0	2	9	15	73	2	2	0	0	0	0	257

Pelvic phenotype frequencies from BELL & ORTí 1994; haplotype frequencies from BELL *et al.*, unpubl. data.)
[1] MS: Mat-Su Valley; K: Kenai Peninsula.
[2] J: Japanese clade; EA: Euro-American clade.
[3] 0-8, scores defined in Fig. 4, N = sample size.
[4] Unweighted mean for three samples (BELL & ORTí 1994).

each of which is occupied principally or exclusively by different subclades within the Euro-American clade. As the plesiomorphic state for the pelvic skeleton is full expression (BELL, 1987), which is monomorphic in marine populations of the Euro-American clade, pelvic reduction must have evolved independently in each of its three major subclades. Although, the intraspecific cladogram for *G. aculeatus* (Fig. 3) by itself is still too crude to imply more than a few independent cases of pelvic reduction, its occurrence in freshwater populations within each subclade of the Euro-American clade is very important in principle because it firmly establishes that pelvic reduction must have evolved more than once within *G. aculeatus*. Pelvic reduction also occurs in ninespine (*Pungitius pungitius*; NELSON, 1971; ZIUGANOV & ZOTIN, this volume) and brook sticklebacks (*Culaea inconstans*; NELSON & ATTON, 1971), showing that pelvic reduction is easily derived in sticklebacks (NELSON, 1993).

Geographical separation, isolation, and recent origin are the only sources of evidence for repeated evolution of pelvic reduction in Cook Inlet, Alaska. However, a strong case for repeated evolution can be made even without phylogenetic support. Pelvic reduction is widely distributed in lake populations of *G. aculeatus* in the Mat-Su Valley and on the Kenai

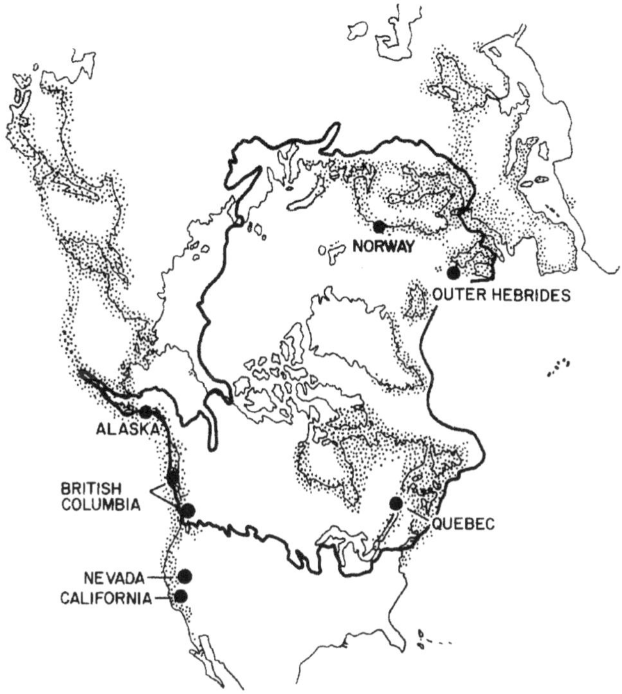

Fig. 5. North polar projection (with North America down) showing the distribution of *Gasterosteus aculeatus* (stippled), regions in which pelvic reduction has been reported (black dots), and the approximate southern limit of contiguous continental glaciation and sea ice during the last glacial advance (heavy line). The Nevada case is from the Miocene (modified from BELL & FOSTER, 1994.)

Peninsula, which are isolated by about 50 km of Cook Inlet sea water (BELL & ORTÍ, 1994). Cook Inlet was largely filled with glacial ice until 22,000 years ago (REGER & UPDIKE, 1983; R.D. REGER, pers. comm. 1993), and most (possibly all) freshwater populations of *G. aculeatus* in lakes surrounding it must have evolved from marine or anadromous populations since deglaciation (BELL *et al.*, 1985, 1993; BELL & ORTÍ, 1994). About a quarter of 179 samples (with $N \geq 20$) distributed equally between the Mat-Su Valley and the Kenai Peninsula had substantial ($\geq$ 5%) frequencies of reduced pelvic phenotypes (BELL & ORTÍ, 1994). In both regions, populations with substantial pelvic reduction are widely distributed and interspersed with populations lacking pelvic reduction

(Bell & Ortí, 1994). Most populations with substantial pelvic reduction are also isolated in lakes with no discharge streams, and conditions associated with pelvic reduction include absence of native predatory fishes and low dissolved calcium concentration (Bell & Ortí, 1994). The isolation and distribution of populations with pelvic reduction and limited time since deglaciation all indicate that evolution of pelvic reduction has evolved many times in Cook Inlet populations. Thus, it can be assumed that natural selection controlled pelvic evolution independently within most lakes, and it is legitimate to treat them as an independent cases (Bell *et al.*, 1993).

A preliminary study of the distribution of the Euro-American and Japanese clades in Cook Inlet (Bell *et al.*, unpubl. data) is consistent with treatment of lake populations as phylogenetically independent observations. Although the Japanese clade is infrequent, both it and the Euro-American clade are widely distributed and interspersed among populations. Substantial pelvic reduction occurs in samples in which both clades occur or either occurs alone (Table 1). These preliminary results add evidence concerning two important issues. The wide distribution and interspersion of the clades indicates that fresh water colonization occurs easily. Presence of pelvic reduction in populations with either or both clades does not suggest that common ancestry is important in pelvic reduction.

Choosing threespine stickleback populations and behavioral phenotypes for comparative studies

Identification of plesiomorphic and apomorphic states of a phenotype and of populations that must have evolved the apomorphic state independently (homoplasy) are keys to applying the comparative method (reviewed in Brooks & McLennan, 1991; Harvey & Pagel, 1991). There are statistical methods to adjust for phylogenetic distance among groups (*e.g.* Felsenstein, 1985; Garland & Carter, 1994), but they cannot be applied to fine-scale divergence (*i.e.* within regions) among freshwater populations of *G. aculeatus* because it is not yet possible to produce well resolved cladograms for them. Appropriate phenotypes and populations can be selected in three steps. Initially one must (A) *select behavioral phenotypes* for which it is appropriate to treat freshwater populations as independent observations:

A.1. The ancestral states of behavioral phenotypes must be inferred using marine and anadromous populations, and possibly other stickleback genera (see McLennan, 1993 for proposed behavioral character states of stickleback genera).

A.2. Apomorphic states of behavioral phenotypes that are consistently absent from marine populations must be identified in freshwater populations.

Behavioral phenotypes that are absent from marine populations and present in freshwater ones are suitable for comparative studies. In order to increase the probability of inferring the plesiomorphic state of behavioral phenotypes, one should (B) *select marine populations* using one or more of the following criteria, listed in order of their value:

B.1. Include populations that can be shown using molecular markers (Haglund *et al.*, 1992; O'Reilly *et al.*, 1993; Buth & Haglund, 1994; Ortí *et al.*, 1994) to belong to as many of the major clades as possible.

B.2. Compare marine populations from throughout the Holarctic range.

B.3. Within a region dominated by one clade, make observations from as large a geographical area as possible.

Finally, one should select freshwater populations (C) to *infer adaptive significance* or *functional associations* between behavioral phenotypes using the following criteria:

C.1. Include populations that can be shown using molecular markers to belong to as many of the major clades as possible (see B. 1., above).

C.2. Compare populations from throughout the Holarctic range and otherwise maximize geographical distances among populations compared (B.2).

C.3. Minimize the probability that dispersal through freshwater could cause close common ancestry of geographically separated populations within a region by using populations from: (a) different drainages that open separately to the ocean and are separated by high mountain divides across which dispersal by *G. aculeatus* is unlikely; (b) recently deglaciated regions in which the time for dispersal through fresh water is limited; and (c) isolated lakes without discharge streams.

Although comparative studies of behavior using highly corroborated cladograms to select phylogenetically independent cases of character evolution offer a powerful tool to infer causes for adaptive divergence of behavior, inferred causes must be treated as hypotheses. HAGEN & GILBERTSON's (1993) analysis of selective predation and temporal change of lateral plate phenotypes that vary in relation to environmental conditions was an early experimental test for an adaptive interpretation of geographical variation *G. aculeatus* (HAGEN & GILBERTSON, 1972). Similarly, MCPHAIL (1984) inferred the ecological cause for trophic and body-form divergence between members of a lacustrine species pair, for which he presented an explicitly adaptive interpretation (MCPHAIL, 1994). SCHLUTER & MCPHAIL (1992) compared several species pairs and solitary populations, which strengthened and generalized MCPHAIL's adaptive interpretation. SCHLUTER's (1994) experimental analysis of another lacustrine species pair supports MCPHAIL's (1984) adaptive interpretation.

References

VAN DEN ASSEM, J. & SEVENSTER, P. (1985). Preface. In: Fifty years of behaviour study in sticklebacks. Papers read at the First International Conference on Stickleback Behaviour (J. VAN DEN ASSEM & P. SEVENSTER, eds). Behaviour 93, p. v-vi.

AVISE, J.C. (1994). Molecular markers, natural history and evolution. — Chapman and Hall, New York.

——, ARNOLD, J., BALL, R.M., BERMINGHAM, E., LAMB, T., NEIGEL, J.E. & SAUNDERS, N.C. (1987). Intraspecific phylogeography: the mitochondrial DNA bridge between population genetics and systematics. — Ann. Rev. Ecol. Syst. 18, p. 489-522.

BAŃBURA, J. & BAKKER, Th.C.M. (1995). Lateral plate morph genetics revisited: evidence for a fourth morph in three-spined sticklebacks. — Behaviour 132, p. 1153-1171.

BAUMGARTNER, J.V., BELL, M.A. & WEINBERG, P.H. (1988). Body form differences between the Enos Lake species pair of threespine sticklebacks (*Gasterosteus aculeatus* complex). — Can. J. Zool 66, p. 467-474.

BELL, M.A. (1974). Reduction and loss of the pelvic girdle in *Gasterosteus* (Pisces): a case of parallel evolution. — Natur. Hist. Mus. Los Angeles Co. Contrib. Sci. 257, p. 1-36.

—— (1976). Evolution of phenotypic diversity in *Gasterosteus aculeatus* superspecies on the Pacific coast of North America. — Syst. Zool. 25, p. 211-227.

—— (1984a). Evolutionary phenetics and genetics: the threespine stickleback, *Gasterosteus aculeatus*, and related species. — In: Evolutionary genetics of fishes (B.J. TURNER, ed.). Plenum Press, New York, p. 431-528.

—— (1984b). Gigantism in threespine sticklebacks: implications for causation of body size evolution. — Copeia 1984, p. 530-534.

—— (1987). Interacting evolutionary constraints in pelvic reduction of threespine sticklebacks, *Gasterosteus aculeatus* (Pisces, Gasterosteidae). — Biol. J. Lin. Soc. 31, p. 347-382.

—— (1988). Stickleback fishes: bridging the gap between population biology and paleobiology. — Trends Ecol. Evol. 3, p. 320-325.

—— (1994). Paleobiology and evolution of threespine stickleback. — In: The evolutionary biology of the threespine stickleback (M.A. BELL & S.A. FOSTER, eds). Oxford University Press, Oxford, p. 438-471.

—— & FOSTER, S.A. (1994). Introduction to the evolutionary biology of the threespine stickleback. — In: The evolutionary biology of the threespine stickleback (M.A. BELL & S.A. FOSTER, eds). Oxford University Press, Oxford, p. 1-27.

——, Francis, R.C. & Havens, A.C. (1985). Pelvic reduction and its directional asymmetry in threespine sticklebacks from the Cook Inlet region, Alaska. — Copeia 1985, p. 437-444.

—— & Ortí, G. (1994). Pelvic reduction in threespine stickleback from Cook Inlet lakes: geographical distribution and intrapopulation variation. — Copeia 1994, p. 314-325.

——, ——, Walker, J.A. & Koenings, J.P. (1993). Evolution of pelvic reduction in threespine stickleback fish: a test of competing hypotheses. — Evolution 47, p. 906-914.

Bertin, L. (1925). Recherches bionomiques, biométriques et systématiques sur les Epinoches (Gastérostéides). — Ann. L'Inst. Océanogr., Monaco 2, p. 1-204.

Blouw, D.M. & Hagen, D.W. (1990). Breeding ecology and evidence of reproductive isolation of a widespread stickleback fish (Gasterosteidae) from Nova Scotia, Canada. — Biol. J. Linn. Soc. 39, p. 195-217.

Boake, C.R.B. (ed.)(1994). Quantative genetic studies of behavioral evolution. — University of Chicago Press, Chicago.

Bourgeois, J.F., Blouw, D.M., Koenings, J.P. & Bell, M.A. (1994). Multivariate analysis of geographic covariance between phenotypes and environments in the threespine stickleback, *Gasterosteus aculeatus*, from the Cook Inlet area, Alaska. — Can. J. Zool. 72, p. 1497-1509.

Bowne, P.S. (1994). Systematics and morphology of the Gasterosteiformes. — In: The evolutionary biology of the threespine stickleback (M.A. Bell & S.A. Foster, eds). Oxford University Press, Oxford, p. 28-60.

Brooks, D.R. & McLennan, D.A. (1991). Phylogeny, ecology, and behavior. University of Chicago Press, Chicago.

Buth, D.G. & Haglund, T.R. (1994). Allozyme variation in the *Gasterosteus aculeatus* complex. — In: The evolutionary biology of the threespine stickleback (M.A. Bell & S.A. Foster, eds). Oxford University Press, Oxford, p.61-84.

Endler, J.A. (1986). Natural selection in the wild. — Monographs in population biology, No. 21. Princeton University Press, Princeton, New Jersey.

Felsenstein, J. (1985). Phylogenies and the comparative method. — Am. Nat. 126, p. 1-25

Foster, S.A. (1988). Diversionary displays of paternal stickleback: defenses against cannibalistic groups. — Behav. Ecol. Sociobiol. 22, p. 335-340.

—— (1994). Evolution of reproductive behaviour of threespine stickleback. — In: The evolutionary biology of the threespine stickleback (M.A. Bell & S.A. Foster, eds). Oxford University Press, Oxford, p. 381-398.

—— (1995). Understanding the evolution of behaviour in threespine stickleback: the value of geographic variation. Behaviour 132, p. 1107-1129.

Francis, R.C. (1995). Evolutionary neurobiology. — Trends Ecol. Evol. 10, p. 276-281.

Garland, T. Jr. & Carter, P.A. (1994). Evolutionary physiology. — Ann. Rev. Physiol. 56, p. 579-621.

Giles, N. (1983). The possible role of environmental calcium levels during the evolution of phenotypic diversity in Outer Hebridean populations of the three-spined stickleback, *Gasterosteus aculeatus*. — J. Zool., London 199, p. 535-544.

—— & Huntingford, F.A. (1984). Predation risk and inter-population variation in antipredator behaviour in the three-spined stickleback, *Gasterosteus aculeatus* L. — Anim. Behav. 32, p. 264-275.

Gould, J.L. (1982). Ethology, the mechanisms and evolution of behavior. — W.W. Norton & Co., New York.

Gross, H.P. & Anderson, J.M. (1984). Geographic variation in the gill rakers and diet of European threespine sticklebacks. — Copeia 1984, p. 87-97.

Hagen, D.W. (1967). Isolating mechanisms in threespine sticklebacks (*Gasterosteus*). — J. Fish. Res. Bd. Can. 24, p. 1637-1692.

—— (1973). Inheritance of numbers of lateral plates and gill rakers in *Gasterosteus aculeatus*. — Heredity 30, p. 303-312.

—— & Gilbertson, L.G. (1972). Geographic variation and environmental selection in *Gasterosteus aculeatus* L. in the Pacific northwest, America. — Evolution 26, p. 32-51.
—— & —— (1973). Selective predation and the intensity of selection acting upon the lateral plates of threespine sticklebacks. — Heredity 30, p 273-287.
—— & McPhail, J.D. (1970). The species problem within *Gasterosteus aculeatus* on the Pacific coast of North America. — J. Fish. Res. Bd. Canada 27, p. 147-155.
—— & Moodie, G.E.E. (1979). Polymorphism for breeding colors in *Gasterosteus aculeatus.* I. Their genetics and geographic distribution. — Evolution 33, p. 641-648.
——, —— & Moodie, P.F. (1980). Polymorphism for breeding colors in *Gasterosteus aculeatus.* II. Reproductive success as a result of convergence for threat display. — Evolution 34, p. 1050-1059.
Haglund, T.R., Buth, D.G. & Lawson, R. (1992). Allozyme variation and phylogenetic relationships of Asian, North American, and European populations of the threespine stickleback. — Copeia 1992, p. 432-443.
Hart, P.J.B. & Gill, A.B. (1994). Evolution of foraging behaviour in the threespine stickleback. — In: The evolutionary biology of the threespine stickleback (M.A. Bell & S.A. Foster, eds). Oxford University Press, Oxford, p. 207-239.
Harvey, P.H. & Pagel, M.D. (1991). The comparative method in evolutionary biology. — Oxford University Press, Oxford.
Herman, Y. & Hopkins, D.M. (1980). Arctic oceanic climate in Late Cenozoic time. — Science 209, p. 557-562.
Heuts, M.J. (1947). Experimental studies on adaptive evolution in *Gasterosteus aculeatus* L. — Evolution 1, p. 89-102.
Hubbs, C.L. (1929). The Atlantic American species of the fish genus *Gasterosteus.* — Occ. Pap. Mus. Zool., Univ. Michigan 200, p. 1-9.
Huntingford, F.A. (1982). Do inter- and intraspecific aggression vary in relation to predation pressure in sticklebacks? — Anim. Behav. 30, p. 909-916.
—— & Wright, P.J. (1992). Inherited population differences in avoidance conditioning in three-spined sticklebacks, *Gasterosteus aculeatus.* — Behaviour 122, p. 264-273.
——, ——, & Tierney, J.F. (1994). Adaptive variation in antipredator behaviour in threespine stickleback. — In: The evolutionary biology of the threespine stickleback (M.A. Bell & S.A. Foster, eds). Oxford University Press, Oxford, p. 277-296.
Johnston, R.F. & Selander, R.K. (1973). Evolution in the house sparrow. III. Variation in size and sexual dimorphism in Europe and North and South America. — Am. Nat. 107, p. 373-390.
Keast, A. (1972). Australian mammals: zoogeography and evolution. — In: Evolution, mammals and southern continents (A. Keast, F.C. Erk & B. Glaes, eds). State University of New York Press, Albany, p. 195-246.
Lauder, G.V. (1995). A model of variability. — Science 267, p. 1192.
Lavin, P.A. & McPhail, J.D. (1985). The evolution of freshwater diversity in threespine stickleback (*Gasterosteus aculeatus*): site-specific differentiation of trophic morphology. — Can. J. Zool. 63, p. 2632-2638.
—— & —— (1986). Adaptive divergence of trophic phenotype among freshwater populations of threespine stickleback (*Gasterosteus aculeatus*). — Can. J. Fish. Aquat. Sci. 43, p. 2455-2463.
Lindsey, C.C. (1962). Experimental study of meristic variation in a population of threespine stickleback, *Gasterosteus aculeatus.* — Can. J. Zool. 40, p. 271-312.
Linnaeus, C. (1758). Systema naturae per regna tria naturae, secundum classes, ordines, genera, species com characteribus, differentiis, synonymis, locis. Editio decima, reformata, Tom. I. — Laurentii Salvii, Holmiae, Stockholm.
MacDonald, J.F., Bekkers, J., MacIsaac, S.M. & Blouw, D.M. (1995a). Intertidal breeding and aerial development of embryos of a stickleback fish (*Gasterosteus*). — Behaviour 132, p. 1183-1206.

——, MacIsaac, S.M., Bekkers, J. & Blouw, D.M. (1995b). Experiments on embryo survivorship, habitat selection, and competitive ability of a stickleback fish (*Gasterosteus*) which nests in the rocky intertidal zone. — Behaviour 132, p. 1207-1221.

MacLennan, D.A. (1993). Phylogenetic relationships in the Gasterosteidae: an updated tree based on behavioral characters with a discussion of homoplasy. — Copeia 1993, p. 318-325.

Maggs, C.A. (1990). Distribution and evolution of noncoraline crustose red algae in the north Atlantic. — In: Evolutionary biogeography of the marine algae of the North Atlantic (D. J. Garbary & G.R. South, eds). NATO ASI Series, Berlin, p. 241-264.

Mayr, E. (1982). The growth of biological thought. — Harvard University Press, Cambridge, Mass.

McPhail, J.D. (1969). Predation and the evolution of a stickleback (*Gasterosteus*). — J. Fish. Res. Bd. Canada 26, p. 3183-3208.

—— (1977). Inherited interpopulation difference in size at first reproduction in threespine stickleback, *Gasterosteus aculeatus* L. — Heredity 38, p. 53-60.

—— (1984). Ecology and evolution of sympatric sticklebacks (*Gasterosteus*): morphological and genetic evidence for a species pair in Enos Lake, British Columbia. — Can. J. Zool. 62, p. 1402-1408.

—— (1994). Speciation and the evolution of reproductive isolation in the sticklebacks (*Gasterosteus*) of south-western British Columbia. — In: The evolutionary biology of the threespine stickleback (M.A. Bell & S.A. Foster, eds). Oxford University Press, Oxford, p. 399-437.

Miller, R.R. & Hubbs, C.L. (1969). Systematics of *Gasterosteus aculeatus*, with particular reference to intergradation and introgression along the Pacific coast of North America: a commentary on a recent contribution. — Copeia 1969, p. 52-69.

Moodie, G.E.E. & Reimchen, T.E. (1976). Phenetic variation and habitat differences in *Gasterosteus* populations of the Queen Charlotte Islands. — Syst. Zool. 25, p. 49-61.

Münzing, J. (1959). Biologie, Variabilität und Genetik von *Gasterosteus aculeatus* L. (Pisces). Untersuchungen im Elbegebeit. — Int. Revue ges. Hydrobiol. 44, p. 317-382.

—— (1963). The evolution of variation and distributional patterns in European populations of the three-spined stickleback, *Gasterosteus aculeatus*. — Evolution 17, p. 320-332.

Nelson, J.S. (1971). Absence of the pelvic complex in ninespine sticklebacks, *Pungitius pungitius*, collected in Ireland and Wood Buffalo National Park region, Canada, with notes on meristic variation. — Copeia 1971, p. 707-717.

—— (1993). Analysis of the multiple occurrence of pelvic fin absence in extant fishes. — Matsya 15/16, p. 21-38.

—— (1994). Fishes of the world, 3rd ed. — John Wiley & Sons, New York.

—— & Atton, F.M. (1971). Geographic and morphological variation in the presence and absence of the pelvic skeleton in the brook stickleback, *Culaea inconstans* (Kirtland), in Alberta and Saskatchewan. — Can. J. Zool. 49, p. 343-352.

Oakshott, J.G., Gibson, J.B., Anderson, P.R., Knibb, W.R., Anderson, D.G. & Chambers, G.K. (1982). Alcohol dehydrogenase and glycerol-3-phosphate dehydrogenase clines in *Drosophila melanogaster* on different continents. — Evolution 36, p. 86-96.

O'Reilly, P., Reimchen, T.E., Beech, R. & Strobeck, C. (1993). Mitochondrial DNA in *Gasterosteus aculeatus* and Pleistocene glacial refugium on the Queen Charlotte Islands, British Columbia. — Evolution 47, p. 678-684.

Ortí, G., Bell, M.A., Reimchen, T.E. & Meyer, A. (1994). Global survey of mitochondrial DNA sequences in the threespine stickleback: evidence for recent migrations. — Evolution 48, p. 608-622.

Real, L.A. (ed.) (1994). Behavioral mechanisms in evolutionary ecology. — University of Chicago Press, Chicago.

Regan, C.T. (1909). The species of threespined sticklebacks (*Gasterosteus*). — Ann. Mag. Nat. Hist. 4, p. 435-437.

Reger, R.D. & Updike, R.G. (1983). Upper Cook Inlet region and the Matanuska Valley. — In: Guidebook to permafrost and Quaternary geology along the Richardson and Glenn Highways between Fairbanks and Anchorage, Alaska (T. Péwé & R.D. Reger, eds). Alaska Division of Geological and Geophysical Surveys, Anchorage, p. 185-259.

Reimchen, T.E. (1984). Status of un-armored and spine-deficient populations (Charlotte unarmored stickleback) of threespine stickleback, *Gasterosteus* sp., on the Queen Charlotte Islands, British Columbia. — Can. Field-Natur. 98, p. 120-126.

—— (1989). Loss of nuptial color in threespine sticklebacks (*Gasterosteus aculeatus*). — Evolution 43, p. 450-460.

—— (1994). Predators and morphological evolution in threespine stickleback. — In: The evolutionary biology of the threespine stickleback (M.A. Bell & S.A. Foster, eds). Oxford University Press, Oxford, p. 240-276.

——, Stinson, E.M. & Nelson, J.S. (1985). Multivariate differentiation of parapatric and allopatric populations of threespine stickleback in the Sangan River watershed, Queen Charlotte Islands. — Can. J. Zool. 63, p. 2944-2951.

Ridgway, M.S. & McPhail, J.D. (1988). Raiding shoal size and a distraction display in male sticklebacks (*Gasterosteus*). — Can. J. Zool. 66, p. 201-205.

Rowland, W.J. (1994). Proximate determinants of stickleback behaviour: an evolutionary perspective. — In: The evolutionary biology of the threespine stickleback (M.A. Bell & S.A. Foster, eds). Oxford University Press, Oxford, p. 297-344.

Sargent, R.C., Bell, M.A., Krueger, W.H. & Baumgartner, J.V. (1984). A lateral plate cline, sexual dimorphism, and phenotypic variation in the blackspotted stickleback, *Gasterosteus wheatlandi*. — Can. J. Zool. 62, p. 368-376.

Schluter, D. (1994). Experimental evidence that competition promotes divergence in adaptive radiation. — Science 266, p. 798-801.

—— & McPhail, J.D. (1992). Ecological character displacement and speciation in sticklebacks. — Am. Nat. 140, p. 85-108.

Simpson, G.G. (1980). Splendid isolation, the curious history of mammals in South America. — Yale University Press, New Haven, Connecticut.

Taylor, E.B. & McPhail, J.D. (1986). Prolonged and burst swimming in anadromous and freshwater threespine stickleback, *Gasterosteus aculeatus*. — Can. J. Zool. 64, p. 416-420.

Temrin, H. & Sillen-Tullberg, B. (1994). The evolution of avian mating systems: a phylogenetic analysis of male and female polygamy and length of pair bond. — Biol. J. Linn. Soc. 52, p. 121-149.

Vermeij, G. (1991). Anatomy of an invasion: the trans-Arctic interchange. — Paleobiology 17, p. 281-307.

Walker, J.A. (in press). Principal components of shape variation within an endemic radiation of threespine stickleback. — In: Advances in morphometrics (L.F. Marcus, M. Corti, D. Slice & G. Naylor, eds). Plenum, New York.

Whoriskey, F.G. & FitzGerald, G.J. (1985). Sex, cannibalism and sticklebacks. — Behav. Ecol. Sociobiol. 18, p. 15-18.

Wilson, D.S, Clark, A.B., Coleman, K. & Dearstyne, T. (1994). Shyness and boldness in humans and other animals. — Trends Ecol. Evol. 9, p. 442-446.

Wootton, R.J. (1976). The biology of the sticklebacks. — Academic Press, London.

Ziuganov, V.V. & Zotin, A.A. (1995). Pelvic girdle polymorphism and reproductive barriers in the ninespine stickleback *Pungitius pungitius* (L.) from northwest Russia. — Behaviour 132, p. 1095-1105.

SYMPATRIC ANADROMOUS-RESIDENT PAIRS OF THREESPINE STICKLEBACK SPECIES IN YOUNG LAKES AND STREAMS AT BERING GLACIER, ALASKA

by

FRANK A. VON HIPPEL[1] **and HEIDI WEIGNER**[2]
(Department of Biological Sciences, University of Alaska Anchorage, 3211 Providence Dr., Anchorage, AK 99508-4614, USA)

(Acc. 24-X-2004)

Summary

1. Resident freshwater and anadromous pairs of threespine stickleback species (*Gasterosteus aculeatus* species complex) have been reported from only a few lakes.
2. We discovered such pairs in four lakes and four streams, including recently deglaciated terrain, near the Bering Glacier, Alaska.
3. Bering Glacier surges are accompanied and ended by massive glacial outburst floods, which create unmatched dynamics in the surrounding lakes and streams.
4. This is the first report of a stickleback species pair in a proglacial lake (Vitus), which did not exist 100 years ago.
5. These species pairs may have arisen through sympatric speciation from an anadromous ancestor, double invasion of anadromous stickleback, or independent colonization by resident and anadromous forms.

Keywords: threespine stickleback, species pair, rapid speciation, Bering Glacier.

---

[1] Corresponding author's e-mail address: affvh@uaa.alaska.edu

[2] The U.S. Bureau of Land Management provided significant funding and logistical support for this project. John Payne, Scott Guyer, and Chris Noyles of the BLM made our field work possible. Dedicated field assistance was provided by Xavier Sotelo and Eric Sjoden, and John Tucker ably flew us to remote lakes and streams. Numerous helpful comments on the manuscript were provided by Mike Bell, Bill von Hippel, Bruce Molnia, Matthew Travis, and an anonymous reviewer.

Behaviour 141, 1441-1464
Also available online -

Introduction

Species pairs of threespine stickleback (*Gasterosteus aculeatus* species complex) with parapatric or sympatric ranges occur in six varieties (Table 1; reviewed by McPhail, 1994; McKinnon & Rundle, 2002). Of these, only the resident freshwater and anadromous pair has a broad geographical distribution. All other types have restricted ranges, suggesting that peculiarities of local geology and ecology and possibly stochastic events gave rise to them. For example, the benthic-limnetic pairs are thought to have arisen 11,000 years ago as the result of marine stickleback being trapped in freshwater due to isostatic rebound following the melting of Pleistocene ice (McPhail, 1993, 1994). These fish would then be subject to strong directional selection, probably towards a benthic ecology. It is thought that when the lakes were reopened to colonization, possibly 9,000 years ago, a second invasion of marine stickleback evolved to exploit the limnetic niche. Similarly, local geological history may explain the origin of the Sea of Japan species pair (Higuchi & Goto, 1996; Yamada *et al.*, 2001). Conversely, the resident-anadromous pairs have originated independently and often, throughout the stickleback's range, constituting numerous replicates with different histories and local ecologies. These pairs therefore provide a unique opportunity to study the forces that drive parallel evolution and speciation.

TABLE 1. *Six varieties of species pairs of threespine stickleback*

Species pair type	Location
White marine/typical marine	Nova Scotia
Japan Pacific/Sea of Japan	Coast of Hokkaido Island[1]
Lake resident/stream resident	Vancouver & Graham Islands, British Columbia
Stream dwelling nuptially-red & nuptially-black	Olympic Peninsula, Washington
Lake dwelling limnetics & benthics	Six lakes on three islands in the Straight of Georgia, British Columbia
Resident freshwater/anadromous	Throughout much of Holarctic range of species complex

[1] In Lakes Harutori, Akkeshi and Hyotan these two marine forms are also reproductively sympatric with a resident form (Mori, 1990; Higuchi *et al.*, 1996; Kitano *et al.*, 2003), comprising the only known case of a species triplet, except possibly three morphs that specialize on different substrates and temperatures in an Icelandic geothermal pond (Jónsson & Ingólfsdóttir, 2003).

The ecology of lakes differs greatly from that of streams, and therefore the lake-dwelling resident-anadromous pairs should be considered separately from the stream-dwelling pairs. Stream pairs are common in lowland streams throughout much of the stickleback's range, including coastline areas of Europe, Asia and North America (McPhail, 1994). Lake pairs, on the other hand, have been reported only from the Azabachije Lake basin in Kamchatka (Ziuganov *et al.*, 1987), Lake Sana on Iturup Island in the Kuril Islands (discussed in Mori, 1990), in Lakes Harutori, Akkeshi and Hyotan on Hokkaido Island (Mori, 1990; Higuchi *et al.*, 1996; Kitano *et al.*, 2003), and in Mud Lake in southcentral Alaska (Karve, 2004). In this paper, we report eight additional resident-anadromous pairs of threespine stickleback species, including four lake pairs, from the drainage of the Bering Glacier, Alaska.

We report morphological results for four pairs (two lake and two stream) from samples collected in 2002 and 2003. One pair is in proglacial (ice-contact) Vitus Lake. This is the first discovery of a stickleback species pair in a proglacial lake. We also report the location of four additional species pairs discovered in 2003 and 2004 (Fig. 1, Table 2). All of these species pairs exist in areas just above sea level. Vitus Lake was completely covered with ice during the Little Ice Age local maximum around 1900 (Fig. 1). Given that Vitus Lake did not exist 100 years ago, and that glacial surges have caused massive disturbances including glacial outburst floods (*jökulhlaups*) as recently as August 1994 and September 1995 (Fatland & Lingle, 1998), the species pair has existed in this location for less than 100 years, and possibly only since the 1995 flood.

The youth of the Vitus Lake species pair raises interesting questions about the speed and mechanisms by which sympatric fish pairs can form as new aquatic habitat is created in the wake of receding glaciers. Research on the speciation process is hindered by the fact that speciation is typically identified only after it has occurred (McPhail, 1994). Incipient species, therefore, are the most useful for understanding genetic, ecological and behavioral processes leading to new species (McPhail, 1994; Johannesson, 2001). Previous work on stickleback species pairs has sought to explain the origin of these pairs many thousands of years ago. At Bering Glacier we have the opportunity to study the formation of species pairs in real time or shortly after their origin. In this paper we demonstrate that the species pairs are real, and we hypothesize pathways for their origin and maintenance.

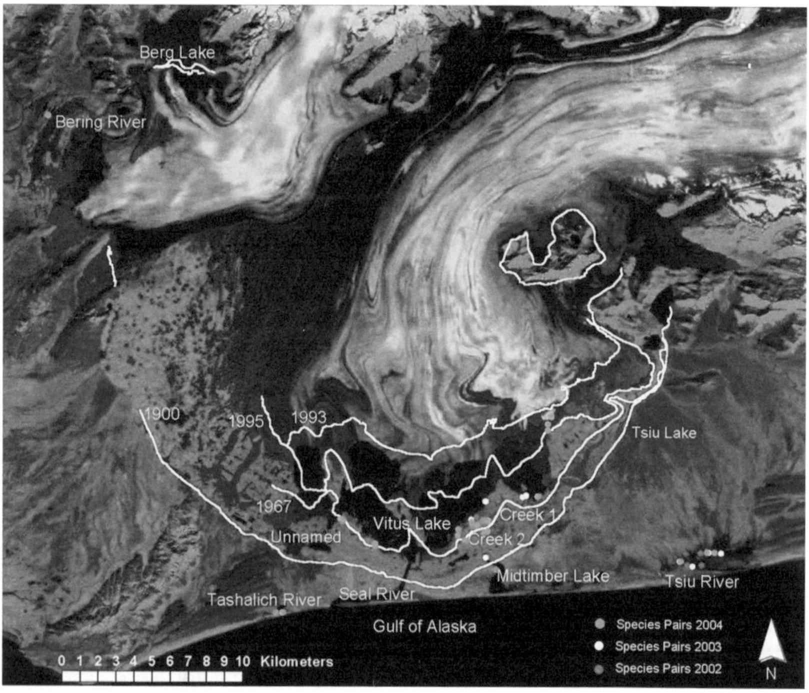

Fig. 1. Landsat imagery of the Bering Glacier region showing the location of eight stickleback species pairs (Vitus Lake, Midtimber Lake, Tsiu River, creeks 1 and 2, unnamed lake, Tashalich River, and Bering River side-lake) and other landscape features discussed in text. The location of sampling points with a pair of stickleback species is shown for samples collected in 2002, 2003 and 2004. Image was captured September 10, 2001. The position of the ice is shown at Little Ice Age maximum of 1900 as well as at maximum 1967 surge position, before the most recent surge in 1993, and at maximum recent surge position in 1995.

Ecological instability of aquatic habitats below Bering Glacier

Bering Glacier is the largest and longest glacier outside of Greenland and Antarctica (Molnia & Post, 1995; Molnia *et al.*, 1996; Jaeger & Nittrouer, 1999). It began to recede from near its late Holocene maximum at the start of the 20th century, creating several large proglacial lakes. The largest of these is Vitus Lake (25 km long, 10 km wide, to 180 m deep, 160 km^2 in area), which varies in size depending on glacial dynamics (Brouwers & Forester, 1993; Muller & Fleisher, 1995; Molnia *et al.*, 1996). With the lake surface at ~2 m elevation (Brouwers & Forester, 1993), most of the lake's water lies below sea level. Vitus Lake has changed markedly during its brief life from a

TABLE 2. *Resident-anadromous species pairs in the Bering Glacier drainage*

Location	GPS coordinates	Habitat type
Vitus Lake	60.1°N, 143.3°W	proglacial lake
Midtimber Lake	60.08°N, 143.35°W	lake
Tsiu River/wetlands	60.08°N, 143.06°W	river-wetland complex
Creek 1	60.10482°N, 143.35008°W	creek draining into Vitus Lake
Creek 2	60.09770°N, 143.35117°W	creek draining into Vitus Lake
Unnamed lake	60.10195°N, 143.55113°W	small lake upstream of Vitus Lake
Tashalich River	60.03905°N, 143.62566°W	river in the process of becoming a lake
Bering River side-lake	60.35771°N, 143.97263°W	small lake adjacent to large river

The specificity of the GPS coordinates varies depending on the size of the lake or stream.

small freshwater lake to a large system with tidal exchange through the Seal River. It now has a thin freshwater to brackish layer maintained by ice melt on top of a saline extension of the Gulf of Alaska (Brouwers & Forester, 1993).

Bering Glacier surges every 20-30 years at the expense of upstream ice thickness (~1900, ~1920, ~1938-1940, 1957-1960, 1965-1967, 1993-1995) and is the Earth's largest surging glacier (Meier & Post, 1969; Molnia & Post, 1995). The 1938-1940 surge occupied nearly all of Vitus Lake. By 1948 the lake had become 10 km^2 in area, but was mostly reoccupied with ice in the 1957-1960 surge, resulting in a 3 km^2 lake (Molnia & Post, 1995). Most of the existing lake has formed since 1967 (Fig. 1) by glacial recession of >2 km some years (Molnia & Post, 1995; Molnia *et al.*, 1996). The latest surge (summer 1993 - summer 1995), which included glacial advances of 100 m per day (Roush, 1996) and a 1,500 m advance in 17 days (Herzfeld & Mayer, 1997), consumed about 2/3 of Vitus Lake (90% of the remaining lake was covered by icebergs) and reoccupied 16,734 ha of non-ice landcover before the glacier retreated again (Gray *et al.*, 1994; Payne *et al.*, 1997; Fig. 1). Part of the glacier's terminus was displaced downstream by about 9 km, partially or completely covering every island in Vitus Lake (Molnia, 1995).

The surges are accompanied and ended by massive glacial outburst floods originating from a subglacial conduit system, which, together with the rapid advances and retreats of the glacier, create unmatched dynamics in the surrounding aquatic habitats (Fleisher *et al.*, 1998; Jaeger & Nittrouer, 1999). The outburst flood of July 27-August 9, 1994 drastically altered proglacial topography (Merrand & Hallet, 1996; Fleisher *et al.*, 1998) and caused the loss of red (sockeye) salmon (*Oncorhynchus nerka*) and silver (coho) salmon (*Oncorhynchus kisutch*) runs in the Seal River. Surges dramatically alter lake bathymetry, turbidity, and water temperature (Fleisher *et al.*, 1998), as well as suspended sediment discharge through Seal River (Gray *et al.*, 1994; Merrand *et al.*, 1994).

The dynamic nature of the proglacial lakes is not limited to episodes of surging. For example, a quarter century of glacial retreat before the latest surge expanded the width of Vitus Lake more than threefold. Glacial retreat also allowed water to break out of Tsiu Lake on August 10-11, 1989 and draw down both Tsiu Lake and Tsivat Lake 17 m when the water discharged into Vitus Lake and abandoned its former outlet over the course of a single week (Fleisher *et al.*, 1993, 1998; see Fig. 1). This created a massive turbid plume in Vitus Lake that was visible for years. The 1993 surge closed the new outlet, quickly raised the Tsiu lake level 14 m, and forced drainage once again through the former outlet. The sedimentation rate from glacial meltwater and icebergs dropping their load is in the range of meters per year (Brouwers & Forester, 1993; Molnia *et al.*, 1996). Additionally, subglacial supercooled water (-0.15 to -1.0°C) is discharged through basal conduits into the proglacial lakes (Fleisher *et al.*, 1993, 1998).

Methods

Field sites and morphological variables

Bering Glacier is accessible only by air. The U.S. Bureau of Land Management maintains a seasonal field station on the south shore of Vitus Lake. Sampling sites on Vitus Lake, creeks 1 and 2, and unnamed lake were reached from the field station by foot, all terrain vehicle, inflatable power boat and helicopter. Midtimber Lake was reached by foot and helicopter, and the Tsiu River and wetlands, the Bering River, Tashalich River and numerous other sites were reached by helicopter.

Fish were trapped July 13-30, 2002, May 5-8, June 2-5, and July 16-August 15, 2003, and May 10-12, June 1-4, June 20-July 5, July 17-August 9, 2004 using 1/8 inch mesh unbaited minnow traps and unbaited semi-oval traps. Stickleback were killed with an overdose of MS-222 anesthetic, fixed in 10% buffered formalin, preserved in 70% ethanol, and stained with Alizarin red S.

Morphological traits were analyzed for the Vitus, Midtimber, Tsiu and creek 1 species pairs collected in 2002 and 2003. Only fish >32 mm standard length (SL; anterior tip of upper jaw to posterior end of hyperal plate) were scored because lateral plate development is complete in most threespine stickleback populations by this size (Bell, 2001). The following traits were measured using digital calipers: SL, lengths of left and right pelvic spine and

TABLE 5. *Morphological variables of 2002 samples of resident and anadromous stickleback in the Tsiu River and wetlands, Alaska*

Variable name	Resident FW mean, $N = 201$ (SD)	Anadromous mean, $N = 86$ (SD)	Univariate F	p
SL	45.3 (3.7)	62.4 (4.7)	1137.0	0.000
Left pelvic spine length	4.3 (0.5)	8.8 (0.9)	1625.81	0.000
Right pelvic spine length	4.5 (0.5)	8.8 (0.9)	–	–
Second dorsal spine length	3.2 (0.4)	6.2 (0.7)	935.03	0.000
Distance dorsal to hyperal	7.0 (0.9)	7.9 (0.9)	148.97	0.000
Dorsal fin base length	10.5 (1.1)	15.7 (1.6)	80.62	0.000
Body depth	10.1 (1.0)	12.9 (1.2)	15.71	0.000
Snout length	3.6 (0.4)	5.6 (0.7)	145.17	0.000
Orbit width	4.2 (0.4)	5.2 (0.4)	0.05	NS
Left pectoral fin length	5.8 (0.7)	12.1 (1.3)	1323.43	0.000
Lateral plate morph	1.0 (0)	3.0 (0)	–	–
Number left lateral plates	5.3 (1.0)	33.1 (0.8)	–	–
Number right lateral plates	5.3 (1.8)	33.4 (0.8)	–	–
Number gaps in left plates	0 (0.1)	0 (0)	–	–
Number gaps in right plates	0 (0.2)	0 (0)	–	–
Pelvic score left	4 (0.1)	4 (0)	–	–
Pelvic score right	4 (0.1)	4 (0)	–	–
Number gill rakers	18.7 (0.9)	23.4 (1.2)	1143.54	0.000

Mean lengths and standard deviations are in mm. Reported here are the univariate F's from a MANOVA based on z-transformed meristic traits and adjusted residuals of z-transformed morphometric traits (see Methods). The univariate F for SL is the result of a one-way ANOVA of raw values. Fish were assigned as resident or anadromous by cluster analysis. Resident sample size for left and right pelvic spine length was 200. Anadromous sample size for left and right pelvic spine length was 85 and left pectoral fin length 84.

Discussion

Lacustrine pairs of resident and anadromous threespine stickleback species have been found only in a few lakes in the north Pacific (Ziuganov *et al.*, 1987; Mori, 1990; Higuchi *et al.*, 1996; Kitano *et al.*, 2003; Karve, 2004). The results of this study demonstrate that such pairs also exist in relatively young lakes at Bering Glacier (Table 2), also part of the north Pacific basin. Most of Alaska, Russia, and islands in the north Pacific and north Atlantic are roadless and remote and therefore have not been sampled for stickleback. Pairs of resident and anadromous stickleback may be reasonably common in lakes throughout the northern part of their range, particularly in newly

TABLE 6. *Morphological variables of 2003 samples of resident and anadromous stickleback in the Tsiu River and wetlands, Alaska*

Variable name	Resident FW mean, $N = 208$ (SD)	Anadromous mean, $N = 200$ (SD)	Univariate F	p
SL	46.4 (3.4)	65.3 (4.4)	2408.76	0.000
Left pelvic spine length	4.7 (0.5)	9.3 (0.8)	2815.08	0.000
Right pelvic spine length	4.7 (0.5)	9.4 (0.8)	–	–
Second dorsal spine length	3.3 (0.4)	6.8 (0.7)	1661.53	0.000
Distance dorsal to hyperal	7.3 (0.9)	8.1 (0.9)	421.58	0.000
Dorsal fin base length	10.7 (1.1)	16.7 (1.5)	121.36	0.000
Body depth	10.6 (0.9)	14.2 (1.2)	0.22	NS
Snout length	3.6 (0.4)	5.4 (0.5)	539.31	0.000
Orbit width	4.1 (0.4)	5.2 (0.3)	36.10	0.000
Left pectoral fin length	6.2 (0.6)	12.9 (1.0)	4027.15	0.000
Lateral plate morph	1.0 (0.2)	3.0 (0)	–	–
Number left lateral plates	5.3 (2.5)	33.5 (0.7)	–	–
Number right lateral plates	5.3 (2.5)	33.5 (0.7)	–	–
Number gaps in left plates	0 (0.2)	0 (0)	–	–
Number gaps in right plates	0 (0.2)	0 (0)	–	–
Pelvic score left	4.0 (0)	4.0 (0)	–	–
Pelvic score right	4.0 (0)	4.0 (0)	–	–
Number gill rakers	18.9 (0.9)	23.3 (1.0)	1785.73	0.000
Number of dorsal fin rays	10.3 (0.7)	11.7 (0.7)	300.50	0.000
Operculum width	3.8 (0.5)	7.6 (0.7)	1817.19	0.000

Mean lengths and standard deviations are in mm. Reported here are the univariate F's from a MANOVA based on z-transformed meristic traits and adjusted residuals of z-transformed morphometric traits (see Methods). The univariate F for SL is the result of a one-way ANOVA of raw values. Fish were assigned as resident or anadromous by cluster analysis. Resident sample size for left and right pelvic spine length was 204 and pectoral fin length 207. Anadromous sample size for left pelvic spine length was 197, right pelvic spine length 195, dorsal spine length 198, and pectoral fin length and number of gill rakers 199.

formed lakes left behind by receding glaciers. We found these pairs in four out of 44 lakes sampled. This is probably an underestimate of the true frequency because anadromous runs in southcentral Alaska peak in late May and early June (Karve, 2004; von Hippel, unpubl. data) and only 15 of our 104 sampling days (14%) took place at this time (see Methods).

The presence of a species pair in proglacial Vitus Lake demonstrates that substantial ecological and evolutionary interactions, such as interspecific competition for territories and hybridization, may be occurring in low pro-

TABLE 7. *Morphological variables of 2003 samples of resident and anadromous stickleback in creek 1, near the Bering Glacier, Alaska*

Variable name	Resident FW mean, $N = 14$ (SD)	Anadromous mean, $N = 34$ (SD)	Mann-Whitney U	p
SL	44.2 (5.6)	64.5 (5.7)	8.0	0.000
Left pelvic spine length	5.3 (0.7)	9.3 (1.0)	0	0.000
Right pelvic spine length	5.5 (0.8)	9.3 (1.1)	–	–
Second dorsal spine length	3.5 (0.5)	6.4 (0.9)	0	0.000
Distance dorsal to hyperal	6.3 (0.8)	8.2 (1.0)	147.0	0.039
Dorsal fin base length	10.5 (1.7)	16.1 (1.7)	203.0	NS
Body depth	9.8 (1.1)	13.9 (1.4)	183.0	NS
Snout length	3.6 (0.4)	5.3 (0.7)	98.0	0.001
Orbit width	3.8 (0.4)	5.1 (0.5)	146.0	0.037
Left pectoral fin length	6.2 (0.8)	12.3 (1.3)	0	0.000
Lateral plate morph	1.9 (0.9)	3.0 (0)	85.0	0.000
Number left lateral plates	16.5 (12.6)	33.9 (0.7)	23.0	0.000
Number right lateral plates	16.9 (12.8)	33.6 (0.8)	–	–
Number gaps in left plates	0.4 (1.1)	0 (0)	–	–
Number gaps in right plates	0.2 (0.6)	0 (0)	–	–
Pelvic score left	4.0 (0)	4.0 (0)	–	–
Pelvic score right	4.0 (0)	4.0 (0)	–	–
Number gill rakers	19.2 (1.1)	22.7 (1.2)	5.5	0.000
Number of dorsal fin rays	11.3 (0.8)	11.5 (0.7)	200.0	NS
Operculum width	3.6 (0.8)	7.5 (0.7)	0	0.000

Mean lengths and standard deviations are in mm. Reported here are the Mann-Whitney U statistics based on z-transformed meristic traits and adjusted residuals of z-transformed morphometric traits, with the exception of SL, which was a direct comparison of raw SL values (see Methods). The significance of analyses remains the same when analyzed by one-way ANOVAs. Anadromous sample size for left and right pelvic spine and dorsal spine was 33.

ductivity proglacial lakes. Robinson & Wilson (1994) found that divergence in morphological traits is more likely to occur when fish are in species-poor communities. Such communities should have reduced predatory and competitive pressures allowing more variants to survive and diverge (McPhail, 1994). This might explain the observation that species pairs of fishes occur most often in recently glaciated landscapes, which have depauperate fish faunas (McPhail, 1994). In the lakes around Bering Glacier it may have more to do with a low density of fish, leading to reduced competition and predation, rather than low species richness (Weigner & von Hippel, unpubl. data), although predators may play a role in promoting species pairs (Vamosi, 2002; Vamosi & Schluter, 2002, 2004; Rundle *et al.*, 2003, 2004). The low density, in turn, is probably due to the low productivity in such new habitats and the massive abiotic disturbances that repeatedly 'reset' the ecological clock. The colonization and evolution of fish in proglacial lakes has seldom been studied, but these lakes have the very conditions, such as reduced competition and predation, that may promote the formation of species pairs. Understanding of fish evolution in these ice-contact systems awaits their exploration.

Morphological differences between anadromous and resident forms at the Bering Glacier are typical of these species pairs. Anadromous stickleback are large, completely-plated and have numerous (about 23) gill rakers per arch, and resident fish are small, have variable plate morphology and fewer (about 19-20) gill rakers per arch (Tables 3-7). Anadromous and resident forms differ in body size, shape, foraging traits, armor, nuptial coloration and courtship behavior, a suite of traits subject to both natural and sexual selection (reviewed by McPhail, 1994; McKinnon & Rundle, 2002). The greater number of gill rakers and larger eyes of the anadromous form in the Bering Glacier area are consistent with a more planktivorous diet, as expected (McPhail, 1994).

It is common for resident populations to differ morphologically from each other (Bell & Foster, 1994), and this was also shown to be the case in the comparisons between the Vitus, Midtimber and Tsiu resident forms (Table 8). These resident forms appear to represent independently evolving populations. Our results of significant morphological differences between the Vitus and Tsiu anadromous forms suggest that interpopulation variation and homing behavior in anadromous stickleback is well worth further investigation.

While we do not know how long it has been since the resident populations were derived from anadromous ancestors, the Vitus Lake species pair, as well as the nearby pairs in creek 1, creek 2 and unnamed lake, are quite young. At the start of the 20th Century, Bering Glacier terminated in an outwash plain, and Vitus Lake did not exist (Molnia & Post, 1995; Fig. 1). Vitus Lake experienced massive disturbance by the surging Bering Glacier and glacial outburst floods as recently as 1965-1967 and 1993-1995. Therefore, the Vitus Lake species pair is younger than 100 years old. Although they may well have persisted through the floods, they may have formed after 1967 or 1995. The other four species pairs (Table 2) may also be quite young. The youth of these species pairs and their formation in a proglacial lake (Vitus) and nearby, formerly proglacial lakes and streams in constant states of flux raise the question of their origin and maintenance.

Origin of species pairs

Explanations for the origin of species pairs in British Columbia and elsewhere often rely on geological events, such as isostatic rebound, that take hundreds or thousands of years to complete. Glacial retreat in Alaska suggests, however, that the retreat of the ice sheets elsewhere about 11,000 years ago was not a simple process, and that catastrophic events such as surges and glacial outburst floods may have played an integral role in the (rapid?) formation of these better known species pairs.

At least three pathways may have given rise to the Bering Glacier species pairs: sympatric speciation from an anadromous ancestor, double invasion of anadromous stickleback, or colonization by resident stickleback from adjacent areas either before or after colonization by anadromous stickleback from the sea. Sympatric speciation is an unlikely explanation for all sites, and double invasion may only be feasible in Midtimber Lake. For all other sites, independent colonization appears to be the most likely explanation for the origin of species pairs.

Sympatric speciation hypothesis

For sympatric speciation to occur, some offspring of anadromous fish must fail to return to the sea, and mate assortatively with each other, while anadromy is retained by other fish from the same population. This might occur through major selection differentials acting on opposite sides of an ecotone (McPhail, 1994), but such disruptive selection seems unlikely in the Bering Glacier area where the locations of ecotones constantly shift due to abiotic disturbance. Sympatric speciation has also been discounted in other species pairs (McPhail, 1994).

Double invasion hypothesis

Double invasion could be mediated by the local geomorphology of Midtimber Lake. It was formerly a high velocity ice-marginal river that drained Bering Glacier (Molnia & Post, 1995). Between 1920-1938, Midtimber River separated from the glacier as Vitus Lake captured the discharge that had previously flowed into Midtimber (Molnia & Post, 1995). By the early 1950s, Vitus Lake drained entirely through the Seal River, and Midtimber became tidally dominated and collected meltwater from stagnant ice remaining from the 1967 surge (Molnia & Post, 1995; B. Molnia, pers. comm.). The river became a lake that has continually shrunk. Since the early 1970s winter wave action in the Gulf of Alaska has often closed the outlet with sand, and Midtimber's outlet opens only occasionally after heavy rains (B. Molnia, pers. comm.). We found the outlet opened from heavy rains in the latter half of August, 2003 and May, 2004, but otherwise we did not see it open. Due to the typical outlet blockages, Midtimber Lake has had opportunities for a double invasion. Freshwater forms of fish often evolve after sea-run fish become land-locked by an extrinsic barrier (McDowall, 1988).

The typical outlet blockage could explain why the proportion of anadromous fish seems to be so low and the frequency of intermediate phenotypes high in Midtimber Lake. For example, partial lateral plate morphs accounted for one-quarter of the 2003 sample and gill raker counts ranged from 17-23. If anadromous fish only occasionally enter Midtimber, and then become trapped, directional selection would continually force their evolution towards a resident freshwater phenotype. Additionally, the low frequency of anadromous fish in Midtimber may lead to gene flow because it would be difficult for anadromous fish to find conspecific mates. In particular, anadromous females may be likely to mate with resident males (Furin & von Hippel, unpubl. data). The regular (perhaps frequent) influx of anadromous fish during years when the outlet is open would lead to the variable distribution that we see in the morphology of Midtimber fish, either through selection and/or gene flow. Midtimber Lake may have a leaky species pair, with gene flow from anadromous fish continually eroding divergence towards a resident phenotype.

Outlet blockages leading to double invasion may only persist for some years until excessive meltwater and precipitation breach the dunes, so this double-invasion model requires that trapped fish evolve quickly enough to favor assortative mating after, perhaps, less than a decade of isolation. Results from several populations of threespine stickleback indicate that the rate of evolution in freshwater isolates can be rapid enough to cause appreciable morphological change in just one decade (Bell, 2001). For example, Loberg Lake stickleback evolved from a nearly monomorphic (96%) complete morph population in 1990 to a low-morph dominated population in

2001 (75% low, 11% complete), and during this time gill raker number and body form also evolved rapidly (Aguirre, 2003; Bell *et al.*, 2004). Moreover, reproductive isolation evolved between low and complete morphs in the White Sea basin within eight generations of introduction to a pond (Ziuganov, 1995). Since body size is an important cue for assortative mating (McPhail & Hay, 1983; Borland, 1986; Nagel & Schluter, 1998; McKinnon *et al.*, 2004), and the F_1 of anadromous stickleback raised in freshwater are smaller than their parents (Mori, 1990), nearly instant isolation may occur by the first generation after establishment of anadromous stickleback in fresh water.

It would be advantageous for individuals of the two forms to mate assortatively because hybrids would likely have reduced fitness relative to the two parental forms. Hybrids are ecologically inferior and have reduced mating success and survival in pairs of benthic-limnetic species of stickleback due to the sharp division of habitats (Schluter, 1995; Vamosi & Schluter, 1999; Vamosi *et al.*, 2000; Vamosi & Schluter, 2002). Hagen (1967) showed that anadromous and resident populations within the Little Campbell River produced a narrow hybrid zone, but substantial gene flow back into either parental species was lacking. In contrast, lake resident-anadromous species pairs in this study may experience extensive introgression. The life history switch from anadromous to resident may present the greatest obstacle for hybrid fitness in the resident-anadromous pairs, given the large differences in migration behavior and physiological regulation (Johannesson, 2001). Gene flow would be reduced if the two forms differed in the timing or place of reproduction or if courtship was disrupted by differences in morphology and behavior (McPhail, 1994; McKinnon & Rundle, 2002).

Imprinting of fry on their father's phenotype and choice of nest microhabitat might enable the rapid evolution of reproductive isolation in stickleback without necessitating a genetic correlation between female preference and male phenotype (McPhail, 1994). If double invasion facilitated by temporary barriers leads to species pairs in a couple of decades or less, 'rapid speciation' may be a term that needs a temporal recalibration by at least two orders of magnitude. The time scale for speciation due to genetic change in diploid organisms is typically viewed as 100,000-1,000,000 years, while a potential case of several allopatric speciation events in cichlids within the last 200-300 years 'represents an almost unbelievable rate of speciation' (Coyne & Orr, 2004). Yet, perhaps the bulk of divergence in all types of stickleback species pairs (and species pairs of other fishes) occurred at the point of their formation, after which relative stasis dominated.

Independent colonization hypothesis

The species pair in Vitus Lake probably formed by independent colonization by resident stickleback from adjacent areas and by anadromous fish from the sea. Before Vitus Lake formed around 1900, nearby ponds, lakes and streams that must have contained resident and/or anadromous stickleback existed. As the drainage was rearranged with the retreating ice, these forms may have independently invaded the newly created Vitus Lake from these pre-existing waters. Similar independent colonizations could have given rise to the other species pairs in the area.

The Tsiu River and wetlands has the greatest differentiation between the forms (Tables 5 & 6), and may be the oldest of the pairs, given its greater distance from the glacier (Fig. 1). All anadromous fish were complete lateral plate morphs, and nearly all resident fish were low plate morphs. This suggests that while gene flow between anadromous and resident forms may occur, it is sufficiently limited to consider them to be separate biological species. However, resident fish in Vitus Lake, Midtimber Lake and creek 1 include low, partial and complete lateral plate morphs.

Phenotypic variation in the resident populations may reflect introgression by anadromous stickleback (leaky species pairs), stable polymorphism, or transient polymorphism during evolutionary transition. In the presence of gene flow, independent colonization allows reticulate evolution. Lateral plate polymorphism is controlled by a major locus and a series of modifiers (Colosimo *et al.*, 2004). This polymorphism could be due to gene flow from the anadromous form preventing fixation at the low morph, which dominates in Alaskan lakes (Bell, 2001). Alternatively, partial and complete morph phenotypes may be present in resident fish that colonize lakes during the independent colonization process. However, these results are also consistent with a recent origin from an anadromous ancestor, with the resident form in the process of evolving to a low-plated fish.

Differentiating between origin hypotheses

Unfortunately, results to date do not allow us to reject any of the alternative hypotheses for origin of stickleback species pairs. Divergence in the midst of gene flow is a pattern in the numerous independent cases of resident-anadromous species pairs (McKinnon & Rundle, 2002). However, distinguishing between the double-invasion and independent colonization hypotheses may be possible by testing for genetic distance between members of a pair and adjacent populations that might have been the source of colonizers (*e.g.* Taylor & McPhail, 2000). If the species pair is the result of a double invasion separated by a short-term block of the outlet, then two members of a pair should be derived in quick succession from the same anadromous ancestor and might be nearly identical genetically but genetically more distant from adjacent populations with the same life history. Conversely, if the pair is the result of independent colonization by pre-existing resident freshwater and anadromous forms, then this should be reflected in a greater genetic distance between members of a pair than between adjacent populations with the same life history.

Persistence of species pairs

Regardless of their origin, the persistence of these species pairs begs explanation. Whether these new pairs arose in allopatry, parapatry or sympatry, their maintenance or continued divergence takes place in sympatry, making speciation in this system a complex process. Their maintenance may be due to reduced competition and predation in these lakes and streams due to the low density of fishes, allowing multiple variants to survive. The ecology of this region is dominated by abiotic factors such as temperature, suspended sediments, salinity, glacial surges and floods. There is no steady-state in these systems; disturbance is the norm.

Given that lake-dwelling species pairs of resident and anadromous stickleback are currently known from only a few locations, these species pairs may typically not persist for long after lakes are formed. For example, it is possible that such species pairs formed numerous times along the periphery of the melting Laurentide and Cordilleran ice sheets, but that within any particular lake gene flow between the forms eventually melded two into one (as may be happening in Midtimber Lake) or that competitive exclusion eventually caused the extinction of one. The Bering Glacier is a remnant of the Cordilleran ice sheet and an excellent analog to conditions at the end of the Pleistocene. The ecology of new lakes changes markedly as they evolve from ice-contact lakes with incredible sediment loads and periodic outburst floods to ice-free lakes with relatively stable hydrology. The dynamic nature and reduced competition and predation of ice-contact and nearby lakes may lead to the rapid formation of species pairs, but subsequent quiescence of the geomorphology and ecological stability may then lead to their dissolution.

References

Aguirre, W.E. (2003). Body shape evolution of threespine stickleback in Loberg Lake, Alaska. — American Society of Ichthyologists and Herpetologists Meeting, Manaus, Brazil, 26 June-1 July, 2003.

Bell, M.A. (2001). Lateral plate evolution in the threespine stickleback: getting nowhere fast. — Genetica 112-113, p. 445-461.

— —, Aguirre, W.E. & Buck, N.J. (2004). Twelve years of contemporary armor evolution in a threespine stickleback population. — Evolution 58, p. 814-824.

— — & Foster, S.A. (1994). Introduction to the evolutionary biology of the threespine stickleback. — In: The evolutionary biology of the threespine stickleback (M.A. Bell & S.A. Foster, eds). Oxford University Press, Oxford, p. 1-27.

— — & Ortí, G. (1994). Pelvic reduction in threespine stickleback from Cook Inlet Lakes: Geographical distribution and intrapopulation variation. — Copeia 1994, p. 314-325.

— —, — —, Walker, J.A. & Koenings, J.P. (1993). Evolution of pelvic reduction in threespine stickleback fish: a test of competing hypotheses. — Evolution 47, p. 906-914.

Borland, M.A. (1986). Size-assortative mating in threespine stickleback from two sites on the Salmon River, British Columbia. — M.Sc. thesis, University of British Columbia, Vancouver.

Brouwers, E.M. & Forester, R.M. (1993). Ostracode assemblages from modern bottom sediments of Vitus Lake, Bering Piedmont Glacier, Southeast Alaska. — In: Geologic studies in Alaska by the U.S. Geological Survey, 1992 (C. Dusel-Bacon & A.B. Till, eds). U.S. Geological Survey, Reston, p. 228-235.

Colosimo, P.F., Peichel, C.L., Nereng, K., Blackman, B.K., Shapiro, M.D., Schluter, D. & Kingsley, D.M. (2004). The genetic architecture of parallel armor plate reduction in threespine sticklebacks. — PLoS Biology 2, p. 0635-0641.

Coyne, J.A. & Orr, H.A. (2004). Speciation. — Sinauer Associates, Inc., Sunderland.

Fatland, D.R. & Lingle, C.S. (1998). Analysis of the 1993-95 Bering Glacier (Alaska) surge using differential SAR interferometry. — J. Glaciol. 44, p. 532-546.

Fleisher, P.J., Cadwell, D.H. & Muler, E.H. (1998). Tsivat Basin Conduit System persists through two surges, Bering Piedmont Glacier, Alaska. — Geol. Soc. Am. Bull. 110, p. 877-887.

— —, Franz, J.M. & Gardner, J.A. (1993). Bathymetry and sedimentary environments in proglacial lakes at the Eastern Bering Piedmont Glacier of Alaska. — J. Geol. Educ. 41, p. 267-274.

Gray, J.R., Hart, R.J. & Molnia, B.F. (1994). 1994 changes in physical and sedimentary characteristics of proglacial Vitus Lake resulting from the surge of Bering Glacier, Alaska. — Eos 75(44), p. 63.

Divergence of male courtship displays between sympatric forms of anadromous threespine stickleback

Jun Kitano[1)], **Seiichi Mori**[2)] **& Catherine L. Peichel**[1,3)]

([1] Division of Human Biology, Fred Hutchinson Cancer Research Center, Seattle, WA 98109, USA; [2] Biological Laboratory, Gifu-keizai University, Ogaki, Gifu 503-8550, Japan)

(Accepted: 30 December 2007)

Summary

Courtship displays are an important component of animal reproduction, and divergence in these displays can play an integral role in promoting or maintaining reproductive isolation between species. The courtship behaviour of the threespine stickleback (*Gasterosteus aculeatus*) has been extensively characterized. In Japan, there are two lineages of threespine stickleback that are genetically differentiated and reproductively isolated when found in regions of sympatry. We have previously shown that males of the Pacific Ocean lineage perform the zig-zag dance, while sympatric males of the Japan Sea lineage perform a different dance, which we have termed the rolling dance (Kitano et al., 2007, Biol. J. Linn. Soc. 91: 671-685). As a first step towards understanding the role of the courtship display in female mate choice and sexual isolation between this sympatric pair, we conducted a more detailed analysis of divergence between the courtship displays of the Pacific Ocean and Japan Sea males. Kinematic analysis revealed that the tempo of the zig-zag dance is two times faster than the tempo of the rolling dance. A detailed analysis of head movements during the courtship dance revealed that the Japan Sea males erect their gill covers, extend their red throats, and open their mouths more frequently during the rolling dance than the Pacific Ocean males do during the zig-zag dance. These results demonstrate that there is extensive divergence between sympatric Pacific Ocean and Japan Sea males in both the motor patterns and the tempo of their courtship displays. Thus, the Japanese sympatric stickleback pair provides a good system to study the role of courtship behaviour in speciation, as well as the genetic mechanisms that underlie the divergence of kinematics and motor patterns in courtship behaviour.

Keywords: *Gasterosteus aculeatus*, zig-zag dance, courtship behaviour, motor pattern, kinematics, gill cover erection, high-speed video, tempo, nest.

[3)] Corresponding author's e-mail address: cpeichel@fhcrc.org

Behaviour 145, 443-461
Also available online - www.brill.nl/beh

Introduction

Movement is an important component of animal communication in a variety of organisms (Tinbergen, 1951; Lorenz, 1958; von Frisch, 1967; Brown et al., 2005). Movement can directly transmit information, such as the location of food resources in honeybee (von Frisch, 1967) or asymmetry of body shape in human (Brown et al., 2005). In addition, movement can enhance the detection or the stimulatory effects of other aspects of a signal, such as color, shape, sound, and posture (Hailman, 1977; Fleishman, 1992; Partan & Marler, 1999; Künzler & Bakker, 2001; Cooper & Goller, 2004; Rosenthal et al., 2004). For example, vocalization in túngara frogs is usually accompanied by the inflation of a conspicuous vocal sac, which enhances the attractiveness of males (Rosenthal et al., 2004). The kinematic variables of a movement will determine the efficiency of its use in animal communication. For example, movements should be performed at an optimum velocity in order to effectively elicit the receiver's response (Fleishman, 1992; Rowland, 1995) and should be designed to enhance the signal over moving background noise (Fleishman, 1992; Peters & Evans, 2003). Thus, kinematic studies of movements are necessary to understand the functions of these displays in animal communication. Furthermore, divergence in courtship displays can play an important role in sexual isolation between closely related species (Mayr, 1942; Tinbergen, 1951); thus, studies on divergence in the kinematics of courtship displays can contribute to an understanding of the speciation process.

The courtship rituals of the male threespine stickleback (*Gasterosteus aculeatus*) have been extensively studied in the context of animal communication, behavioural ecology, behavioural evolution, and speciation (Tinbergen, 1951; Wootton, 1984; McLennan et al., 1988; Bell & Foster, 1994; Foster, 1995, 1999; Foster et al., 1996; McKinnon & Rundle, 2002). In a majority of threespine stickleback populations, the appearance of a gravid female induces a territorial male to perform a stereotyped courtship dance known as a zig-zag dance (Tinbergen, 1951, 1953; Wootton, 1984). The male zig-zag dance plays an important role in attracting female attention, as well as in female mate choice and sexual isolation (Tinbergen, 1951; McPhail & Hay, 1983; Jamieson & Colgan, 1989; Rowland, 1995; Künzler & Bakker, 2001; but see Milinski & Bakker, 1990; Boughman et al., 2005). Despite the fact that one kinematic variable (the tempo of the male courtship dance) has been shown to influence female responses in the threespine stickleback (Rowland, 1995), there have been very few studies of variation in the kinematics of the male courtship dance across threespine stickleback populations.

In Japan, there are two genetically, morphologically and behaviourally distinct lineages of anadromous threespine stickleback, the Japan Sea and Pacific Ocean lineages (Haglund et al., 1992; Higuchi & Goto, 1996; Ishikawa & Mori, 2000; Ishikawa et al., 2006; Kitano et al., 2007). Although the distributions of these two lineages are overlapping in eastern Hokkaido, Japan, they are reproductively isolated even in regions of sympatry (Higuchi & Goto, 1996; Kitano et al., 2007). Our previous mate choice experiment revealed asymmetric sexual isolation between the sympatric Pacific Ocean anadromous (PA) and Japan Sea anadromous (JA) forms: the PA females strongly prefer the PA males, while the JA females show no preference for JA males over PA males (Kitano et al., 2007). In order to understand which male mating signals contribute to sexual isolation between the Japanese sympatric pair, it is necessary to make a detailed comparison of the male mating signals in this sympatric pair. We have previously shown that these two sympatric forms have divergent courtship behaviours. Pacific Ocean males perform the zig-zag dance, while Japan Sea males do not perform the zig-zag dance and instead perform a behaviour that we have called the rolling dance (Kitano et al., 2007), which has previously been called the lateral display (Ishikawa & Mori, 2000; Ishikawa et al., 2006). In addition to using this sympatric pair to study the potential role of courtship behaviour in sexual isolation, the striking differences in courtship behaviour provide an opportunity to study the genetic basis for the evolution of courtship behaviour, using the recently established genomic tools in stickleback (Peichel et al., 2001; Kingsley & Peichel, 2007). To this end, we conducted a more detailed analysis of divergence in the kinematics and display behaviours of male courtship dances between Pacific Ocean and Japan Sea males.

Material and methods

Stickleback samples

Anadromous threespine sticklebacks of the sympatric Pacific Ocean (PA) and Japan Sea (JA) forms were collected from Akkeshi Bay (Kume et al., 2005), Japan, in May 2003, 2005 and 2006. First, the PA and JA forms erect their gill covers, the head width became 1.30 ± 0.09 times wider than the head width before the display ($N = 5$).

Divergence of nest weight

We did not find any significant correlation between male standard length and nest weight within the populations ($p = 0.070$, $N = 11$ for PA males; $p = 0.194$, $N = 16$ for JA males). However, we did find that PA males constructed significantly heavier nests (4.5 ± 0.5 g) than JA males (1.7 ± 0.3 g; $F_{1,25} = 27.659$, $p < 0.001$).

Discussion

The zig-zag dance of PA males is about two times faster than the rolling dance of JA males. The tempo of the zig-zag dance is important for the effective induction of a female response (Rowland, 1995). For example, in an Atlantic population, females responded to videos of the zig-zag dance that were played back at normal speed or videos that were played back 1.5–2 times faster than normal, but did not respond very well to videos that were played back at either half speed or at a much faster speed (Rowland, 1995). Thus, there may be an optimum speed at which the male dance attracts females, and this optimum speed may differ for females of the PA and JA forms. It is hypothesized that movement plays a more important role in dense habitats than in open habitats, because motion can effectively draw the receiver's attention in the presence of moving backgrounds of dense vegetation (McKenzie & Keenleyside, 1970; Fleishman, 1992; Peters & Evans, 2003). In a region where the two forms are breeding in sympatry, the JA males tend to breed in the more open habitat, while the PA males tend to breed in a more complex vegetated habitat (Kume et al., 2005). Additional studies on the breeding habitats of multiple sympatric populations of these two forms will be required to understand the relationship between breeding ecology and courtship divergence in the Japanese sympatric pair.

In many stickleback populations, males roll their bodies during both nest-showing behaviour and dorsal pricking behaviour (McLennan et al., 1988). Therefore, it is possible that the Japan Sea form has incorporated one or both of these behavioural components into their courtship dance. Although nest-showing behaviour did not appear to be different between the two forms (Kitano, unpubl. obs.), the Japan Sea forms consistently perform more vigorous

Divergence of male displays in stickleback 455

dorsal pricking than the PA forms (Ishikawa & Mori, 2000; Ishikawa et al., 2006; Kitano et al., 2007). Investigation of the genetic correlation between different behavioural components (e.g., Bell, 2005) may help to understand the evolutionary origin of the rolling dance.

Mouth opening, throat extension and gill cover erection were more frequently observed in the rolling dance of JA males than in the zig-zag dance of PA males. However, the functions of these behaviours during stickleback courtship remain elusive. Previously, a positive association between the zig-zag dance and mouth opening was reported from an Atlantic Ocean population of stickleback (Tinbergen, 1953; Symons, 1965). However, in the Japanese sympatric pair, the rolling JA males open their mouths more frequently than the zig-zagging PA males. Mouth opening has also been observed in allopatric Japan Sea populations (Ishikawa et al., 2006), suggesting that the associations between behavioural components differ across populations. Because the red throat coloration of male threespine stickleback is an important ornament for attracting females (Rowland, 1994; Rowe et al., 2004), throat extension and mouth opening in the JA males may help to enhance the presentation of red coloration, which is particularly intense inside the mouth and on the ventral side of the throat, but not very intense on the lateral side of the body in JA males (Kitano et al., unpubl. data; Figure 3). Expansion of conspicuous throat ornaments is widely seen in the animal kingdom (Darwin, 1874); specific examples include extension of a pigmented dewlap in anoline lizards (Williams & Rand, 1977; Fleishman, 1992), inflation of a red throat pouch in the great frigatebird (Dearborn & Ryan, 2002), and vocal sac inflation in frog (Rosenthal et al., 2004). Displays of the col orful interior of the mouth have also been observed in birds, such as the tooth-billed bowerbird (Frith & Frith, 1993).

Gill cover erection is prevalent in male–male agonistic interactions in several teleost fish species, such as Siamese fighting fish (Robertson & Sale, 1974; Polnau & Ma, 2001), cichlids (Baerends & Baerends, 1950) and river

Figure 3. (A) Throat extension and mouth opening in a JA male during the rolling dance. (B) Mouth opening and gill cover erection in a JA male during the rolling dance. (C) Comparison of the frequency of mouth opening and throat extension between PA and JA males. Means ± SE are shown. $N = 5$ for each population. (D) Dorsal view of gill cover erection during the rolling dance of a JA male. This figure is published in colour at http://www.ingenta.com

bullhead (Morris, 1954), and is thought to help the fish look bigger in male–male combat. Although we have not observed gill cover erection during male–male agonistic interactions in the JA males (Kitano et al., unpubl. data), female threespine stickleback of several populations, including the JA population, have a preference for larger males (Moodie, 1982; Rowland, 1989a; Kitano et al., unpubl. data). Thus, it is possible that gill cover erection in JA males may help the males look bigger during courtship to compensate for their smaller body size, which might result from adaptation of the JA forms to different ecological resources than those of the PA forms (Kitano et al., 2007).

Our previous mate choice experiments revealed asymmetric sexual isolation: PA females strongly prefer PA males, while JA females show no preference for JA males over PA males (Kitano et al., 2007). These data suggest that the PA and JA females have different mating preferences. Because the sympatric PA and JA males have diverged in both the tempo and the motor patterns of their courtship displays, as well as in several other mating signals, including body size and nuptial color (Kitano et al., 2007), females could use any or all of these signals in making mate choice decisions. In addition, female sticklebacks can use the male nest as a signal in mate choice (Östlund-Nilsson, 2000, 2001; Barber et al., 2001; Östlund-Nilsson & Holmlund, 2003), and we found that nest sizes were significantly different between the sympatric forms, with PA males building larger nests. Nest area and shape also differ between the Japanese sympatric forms (M. Kume, pers. commun.), suggesting that nest divergence might also contribute to reproductive isolation between the sympatric Japanese sticklebacks. Further mate choice experiments will allow us to dissect the roles of different male mating signals in female mate choice and sexual isolation between the Japanese sympatric pair.

Although variation in courtship behaviour has been observed between many stickleback populations (Wilz, 1973; McPhail & Hay, 1983; Ridgway & McPhail, 1984; Foster, 1994, 1995, 1999; Foster et al., 1996; Hay & McPhail, 2000; Boughman et al., 2005), we know very little about the relative contributions of genetic and environmental factors to behavioural variation. A recent study has shown that there is a great deal of phenotypic plasticity in male courtship behaviours within a population, although there may also be a genetic component to differences between populations (Shaw et al.,

2007). Male sticklebacks can change their reproductive behaviour depending on the stage of breeding season (Symons, 1965; Kynard, 1978; Mori, 1993), female receptivity (Rowland, 2002), or female morphology (Rowland, 1982, 1989b). In this study, we have only investigated the first dance that male performed and, thus, did not examine temporal changes in behaviour throughout courtship. However, the PA and JA males perform divergent courtship behaviours in the laboratory throughout the breeding season of May–July (Ishikawa & Mori, 2000; Ishikawa et al., 2006; Kitano et al., 2007; this study). Furthermore, most components of the divergent courtship behaviours that have been found in wild-caught fish also differ between laboratory-reared PA and JA forms (Kitano & Peichel, unpubl. data), suggesting that there is a genetic component to courtship divergence in the Japanese stickleback sympatric pair. Together with the feasibility of making hybrids between these forms (Kitano et al., 2007) and the availability of genomic tools in stickleback (Peichel et al., 2001; Kingsley & Peichel, 2007), it should be possible to use the Japanese sympatric pair as a system to study the genetic architecture of evolutionary changes in the kinematics and motor patterns of courtship behaviour.

Acknowledgements

We thank M. Kume for fish collection and discussion, J. Kitajima and M. Nishida for technical assistance, and all members of the Peichel Lab, especially A. Bruner, for fish care. We thank A. Greenwood and three anonymous reviewers for constructive comments that greatly improved the manuscript. J.K. was a Uehara Memorial Foundation Fellow and Instructor of Kyoto University. This research was supported in part by a Career Award in the Biomedical Sciences from the Burroughs Wellcome Fund to C.L.P. and by Grant-in-Aid for Scientific Research from the Ministry of Education, Culture, Sports, Science and Technology of Japan, and Japan Fund for Global Environment from Environment Restoration and Conservation of Japan to S.M.

References

Barber, I., Nairn, D. & Huntingford, F.A. (2001). Nests as ornaments: revealing construction by male sticklebacks. — Behav. Ecol. 12: 390-396.

Baerends, G.P. & Baerends-van Roon, J.M. (1950). An introduction to the study of the ethology of cichlid fishes. — Brill, Leiden.

Bell, A.M. (2005). Behavioural differences between individuals and two populations of stickleback (*Gasterosteus aculeatus*). — J. Evol. Biol. 18: 464-473.

Bibliography of stickleback papers published in *Behaviour*, 1948–2008

Tinbergen, N. & van Iersel, J.J.A. (1948). "Displacement reactions" in the three-spined stickleback. — Behaviour 1: 56-63.

Morris, D. (1952). Homosexuality in the ten-spined stickleback (*Pygosteus pungitius* L.). — Behaviour 4: 233-261.

Hinde, R.A. (1953). Appetitive behaviour, consummatory act, and the hierarchical organisation of behaviour — with special reference to the great tit (*Parus major*). — Behaviour 5: 189-224.

van Iersel, J.J.A. (1953). An analysis of the parental behaviour of the male three-spined stickleback (*Gasterosteus aculeatus* L.). — Behaviour, Supplement III: 1-159.

Morris, D. (1955). The causation of pseudofemale and pseudomale behaviour: a further comment. — Behaviour 8: 46-56.

Hoogland, R., Morris, D. & Tinbergen, N. (1957). The spines of sticklebacks (*Gasterosteus* and *Pygosteus*) as means of defence against predators (*Perca* and *Esox*). — Behaviour 10: 205-236.

Morris, D. (1957). "Typical Intensity" and its relation to the problem of ritualisation. — Behaviour 11: 1-12.

Morris, D. (1958). The reproductive behaviour of the ten-spined stickleback (*Pygosteus pungitius* L.). — Behaviour, Supplement VI: 1-154.

Guiton, P. (1960). On the control of behaviour during the reproductive cycle of *Gasterosteus aculeatus*. — Behaviour 15: 163-184.

Tugendhat, B. (1960a). The normal feeding behavior of the three-spined stickleback (*Gasterosteus aculeatus* L.). — Behaviour 15: 284-318.

Tugendhat, B. (1960b). The disturbed feeding behavior of the three-spined stickleback: I. Electric shock is administered in the food area. — Behaviour 16: 159-187.

Segaar, J. (1961). Telencephalon and behaviour in *Gasterosteus aculeatus*. — Behaviour 18: 256-287.

Sevenster, P. (1961). A causal study of a displacement activity (fanning in *Gasterosteus aculeatus* L.). — Behaviour, Supplement IX: 1-170.

van den Assem, J. (1967). Territory in the three-spined stickleback, *Gasterosteus aculeatus* L. An experimental study in intra-specific competition. — Behaviour, Supplement XVI: 1-164.

Beukema, J.J. (1968). Predation by the three-spined stickleback (*Gasterosteus aculeatus* L.): the influence of hunger and experience. — Behaviour 31: 1-125.

Symons, P.E.K. (1968). Analysis of spine-raising in the male three-spined stickleback. — Behaviour 26: 1-75.
van den Assem, J. & van der Molen, J.N. (1969). Waning of the aggressive response in the three-spined stickleback upon constant exposure to a conspecific. I, A preliminary analysis of the phenomenon. — Behaviour 34: 286-323.
Jenni, D.A., van Iersel, J.J.A. & van Den Assem, J. (1969). Effects of pre-experimental conditions on nest site selection and aggression in *Gasterosteus aculeatus* L. — Behaviour 35: 61-75.
Peeke, H.V.S. (1969). Habituation of conspecific aggression in the three-spined stickleback (*Gasterosteus aculeatus* L.). — Behaviour 35: 137-156.
Wootton, R.J. (1971). Measures of the aggression of parental male three-spined sticklebacks. — Behaviour 40: 228-261.
Jenni, D.A. (1972). Effects of conspecifics and vegetation on nest site selection in *Gasterosteus aculeatus* L. — Behaviour 42: 97-117.
Wootton, R.J. (1972). The behaviour of the male three-spined stickleback in a natural situation: a quantitative description. — Behaviour 41: 232-241.
Moodie, G.E.E., McPhail, J.D. & Hagen, D.W. (1973). Experimental demonstration of selective predation on *Gasterosteus aculeatus*. — Behaviour 47: 95-105.
't Hart, M. (1978). A study of a short term behaviour cycle creeping through in the three-spined stickleback (*Gasterosteus aculeatus* L.). Part 2. — Behaviour 67: 1-65.
Kynard, B.E. (1978). Breeding behavior of a lacustrine population of threespine sticklebacks (*Gasterosteus aculeatus* L.). — Behaviour 67: 178-206.
Li, S.K. & Owings, D.H. (1978). Sexual selection in the three-spined stickleback: II. Nest raiding during the courtship phase. — Behaviour 64: 298-304.
Peeke, H.V.S., Figler, M.H. & Blankenship, N. (1979). Retention and recovery of habituated territorial aggressive behavior in the three-spined stickleback (*Gasterosteus aculeatus* L.): the roles of time and nest reconstruction. — Behaviour 69: 171-182.
Ridley, M. & Rechten, C. (1981). Female sticklebacks prefer to spawn with males whose nests contain eggs. — Behaviour 76: 152-161.
Rowland, W.J. (1982). The effects of male nuptial coloration on stickleback aggression: a reexamination. — Behaviour 80: 118-126.
Bakker, T.C.M. & Sevenster, P. (1983). Determinants of dominance in male sticklebacks (*Gasterosteus aculeatus* L.). — Behaviour 86: 55-71.
Feuth-de Bruijn, E. & Sevenster, P. (1983). Parental reactions to young in sticklebacks (*Gasterosteus aculeatus* L.). — Behaviour 83: 186-203.
Segaar, J., De Bruin, J.P.C., van der Meche, A.P. & van der Meche-Jacobi, M.E. (1983). Influence of chemical receptivity on reproductive behaviour of the male three-spined stickleback (*Gasterosteus aculeatus* L.). — Behaviour 86: 100-165.
Audet, C., FitzGerald, G.J. & Guderley, H. (1985). Prolactin and cortisol control of salinity preferences in *Gasterosteus aculeatus* and *Apeltes quadracus*. — Behaviour 93: 36-55.
Baerends, G.P. (1985). Do the dummy experiments with sticklebacks support the IRM-concept? — Behaviour 93: 258-277.
Baggerman, B. (1985). The roles of daily and annual biological rhythms in the photoperiodic regulation of the breeding season in the stickleback *Gasterosteus aculeatus* L. — Behaviour 93: 1-7.
Bakker, T.C.M. (1985). Two-way selection for aggression in juvenile, female and male sticklebacks (*Gasterosteus aculeatus* L.), with some notes on hormonal factors. — Behaviour 93: 69-81.

Borg, B. (1985). Field studies on three-spined sticklebacks in the Baltic. — Behaviour 93: 153-157.

Campbell, R.N. (1985). Morphological variation in the three-spined stickleback (*Gasterosteus aculeatus*) in Scotland. — Behaviour 93: 161-168.

Chauvin-Muckensturm, B. (1985). Complex manipulatory abilities in sticklebacks (*Gasterosteus aculeatus*). — Behaviour 93: 150-152.

De Ruiter, A.J.H. & Wendelaar Bonga, S.E. (1985). Consequences of nestbuilding behaviour for osmoregulation in male three-spined sticklebacks. — Behaviour 93: 8-20.

FitzGerald, G.J. & Whoriskey, F.G. (1985). The effects of interspecific interactions upon male reproductive success in two sympatric sticklebacks, *Gasterosteus aculeatus* and *G. wheatlandi*. — Behaviour 93: 112-126.

Gaudreault, A. & FitzGerald, G.J. (1985). Field observations of intraspecific and interspecific aggression among sticklebacks (Gasterosteidae). — Behaviour 94: 203-211.

Giles, N. & Huntingford, F.A. (1985). Variability in breeding biology of three-spined sticklebacks (*Gasterosteus aculeatus*): problems with measuring population differences in aggression. — Behaviour 93: 57-68.

't Hart, M. (1985). The influence of fertilization on the creeping through cycle of the three-spined stickleback. — Behaviour 93: 194-202.

Ketele, A.G.L. & Verheyen, R.F. (1985). Competition for space between the three-spined stickleback, *Gasterosteus aculeatus* L. f. *leiura* and the nine-spined stickleback, *Pungitius pungitius* (L.). — Behaviour 93: 127-138.

McPhail, J.D. (1985). Evidence for sympatric species of *Gasterosteus* in Enos Lake, Vancouver Island, British Columbia. — Behaviour 93: 158-160.

Milinski, M. (1985). Risk of predation of parasitized sticklebacks (*Gasterosteus aculeatus* L.) under competition for food. — Behaviour 93: 203-216.

Mori, S. (1985). Reproductive behaviour of the landlocked three-spined stickleback, *Gasterosteus aculeatus microcephalus*, in Japan. — Behaviour 93: 21-35.

Rowland, W.J. & Sevenster, P. (1985). Sign stimuli in the threespine stickleback (*Gasterosteus aculeatus*): a re-examination and extension of some classic experiments. — Behaviour 93: 241-257.

Russell, W.M.S. & Russell, C. (1985). Sticklebacks and ethology. — Behaviour 93: 184-193.

Ryabov, I.N. (1985). The behaviour of *Gasterosteus aculeatus* L. in the zone of action of warm waste waters. — Behaviour 93: 56-56.

Salfert, I.G. & Moodie, G.E.E. (1985). Filial egg-cannibalism in the brook stickleback, *Culaea inconstans* (Kirtland). — Behaviour 93: 82-100.

Sargent, R.C. (1985). Territoriality and reproductive tradeoffs in the threespine stickleback, *Gasterosteus aculeatus*. — Behaviour 93: 217-226.

Segaar, J. & De Bruin, J.P.C. (1985). Behavioural consequences of chemoreception during reproductive cycles of the three-spined stickleback (*Gasterosteus aculeatus* L.). — Behaviour 93: 139-149.

Sevenster, P. & van Roosmalen, M.E. (1985). Cognition in sticklebacks: some experiments on operant conditioning. — Behaviour 93: 170-183.

Thomas, G., Kacelnik, A. & van der Meulen, J. (1985). The three-spined stickleback and the two-armed bandit. — Behaviour 93: 227-240.

Wootton, R.J. (1985). Effects of food and density on the reproductive biology of the threespine stickleback with a hypothesis on population limitation in sticklebacks. — Behaviour 93: 101-111.

Wyers, E.J. (1985). Cognitive-behavior and sticklebacks. — Behaviour 95: 1-10.

Ziuganov, V.V. (1985). Ethological differences and the probability of introgressive hybridization between two species of the ninespine stickleback, *Pungitius pungitius* and *Pungitius platygaster*, in Eurasia. — Behaviour 93: 169.

Bakker, T.C.M. (1986). Aggressiveness in sticklebacks (*Gasterosteus aculeatus* L.): a behaviour-genetic study. — Behaviour 98: 1-144.

Crawford, S.S. & Colgan, P.W. (1989). Motivational models of courtship in male threespine sticklebacks (*Gasterosteus aculeatus*). — Behaviour 109: 285-302.

Ibrahim, A.A. & Huntingford, F.A. (1989). Laboratory and field studies of the effect of predation risk on foraging in three-spined sticklebacks (*Gasterosteus aculeatus*). — Behaviour 109: 46-57.

Huntingford, F.A. & Wright, P.J. (1992). Inherited population differences in avoidance-conditioning in three-spined stickleback, *Gasterosteus aculeatus*. — Behaviour 122: 264-273.

Krause, J. (1992). Ideal free distribution and the mechanism of patch profitability assessment in three-spined sticklebacks (*Gasterosteus aculeatus*). — Behaviour 123: 27-37.

Bumann, D. & Krause, J. (1993). Front individuals lead in shoals of three-spined sticklebacks (*Gasterosteus aculeatus*) and juvenile roach (*Rutilus rutilus*). — Behaviour 125: 189-198.

Gelowitz, C.M., Mathis, A. & Smith, R.J.F. (1993). Chemosensory recognition of northern pike (*Esox lucius*) by brook stickleback (*Culaea inconstans*): population differences and the influence of predator diet. — Behaviour 127: 105-118.

Krause, J. (1993). Transmission of fright reaction between different species of fish. — Behaviour 127: 37-48.

McLennan, D.A. (1993). Changes in female breeding behaviour across the ovulatory cycle in the brook stickleback, *Culaea inconstans* (Kirtland). — Behaviour 126: 191-218.

Mori, S. (1993). The breeding system of the three-spined stickleback, *Gasterosteus aculeatus* (Forma *Leiura*) with reference to spatial and temporal patterns of nesting activity. — Behaviour 126: 97-124.

Bakker, T.C.M. & Rowland, W.J. (1995). Male mating preference in sticklebacks: effects of repeated testing and own attractiveness. — Behaviour 132: 935-949.

Bakker, T.C.M. & Sevenster, P. (1995). Sticklebacks as models for animal behaviour and evolution — Preface. — Behaviour 132: 907-913.

Banbura, J. & Bakker, T.C.M. (1995). Lateral plate morph genetics revisited: evidence for a fourth morph in three-spined sticklebacks. — Behaviour 132: 1153-1171.

Barber, I. & Huntingford, F.A. (1995). The effect of *Schistocephalus solidus* (Cestoda: Pseudophyllidea) on the foraging and shoaling behaviour of three-spined sticklebacks, *Gasterosteus aculeatus*. — Behaviour 132: 1223-1240.

Baube, C.L., Rowland, W.J. & Fowler, J.B. (1995). The mechanisms of colour-based mate choice in female threespine sticklebacks: hue, contrast and configurational cues. — Behaviour 132: 979-996.

Bell, M.A. (1995). Intraspecific systematics of *Gasterosteus aculeatus* populations: implications for behavioral ecology. — Behaviour 132: 1131-1152.

Borg, B. & Mayer, I. (1995). Androgens and behaviour in the three-spined stickleback. — Behaviour 132: 1025-1035.

Foster, S.A. (1995). Understanding the evolution of behavior in threespine stickleback: the value of geographic variation. — Behaviour 132: 1107-1129.

Kemper, J.H. (1995). Role of the three-spined stickleback *Gasterosteus aculeatus* L. in the food ecology of the spoonbill *Platalea leucorodia*. — Behaviour 132: 1285-1299.

MacDonald, J.F., Bekkers, J., Macisaac, S.M. & Blouw, D.M. (1995a). Intertidal breeding and aerial development of embryos of a stickleback fish (*Gasterosteus*). — Behaviour 132: 1183-1206.

MacDonald, J.F., MacIsaac, S.M., Bekkers, H. & Blouw, D.M. (1995b). Experiments on embryo survivorship, habitat selection, and competitive ability of a stickleback fish (*Gasterosteus*) which nests in the rocky intertidal zone. — Behaviour 132: 1207-1221.

Mackney, P.A. & Hughes, R.N. (1995). Foraging behaviour and memory window in sticklebacks. — Behaviour 132: 1241-1253.

McDonald, C.G., Reimchen, T.E. & Hawryshyn, C.W. (1995). Nuptial colour loss and signal masking in *Gasterosteus*: an analysis using video imaging. — Behaviour 132: 963-977.

Mori, S. (1995). Factors associated with and fitness effects of nest-raiding in the three-spined stickleback, *Gasterosteus aculeatus*, in a natural situation. — Behaviour 132: 1011-1023.

Peeke, H.V.S. (1995). Habituation of a predatory response in the stickleback (*Gasterosteus aculeatus*). — Behaviour 132: 1255-1266.

Perrin, N. (1995). Signalling, mating success and paternal investment in sticklebacks (*Gasterosteus aculeatus*): a theoretical model. — Behaviour 132: 1037-1057.

Reimchen, T.E. (1995). Predator-induced cyclical changes in lateral plate frequencies of *Gasterosteus*. — Behaviour 132: 1079-1094.

Rowland, W.J. (1995). Do female stickleback care about male courtship vigour? Manipulation of display tempo using video playback. — Behaviour 132: 951-961.

Sevenster, P., Feuth-de Bruijn, E. & Huisman, J.J. (1995). Temporal structure in stickleback behaviour. — Behaviour 132: 1267-1284.

Willmott, H.E. & Foster, S.A. (1995). The effects of rival male interaction on courtship and parental care in the fourspine stickleback, *Apeltes quadracus*. — Behaviour 132: 997-1010.

Wootton, R.J., Fletcher, D.A., Smith, C. & Whoriskey, F.G. (1995). A review of reproductive rates in sticklebacks in relation to parental expenditure and operational sex ratios. — Behaviour 132: 915-933.

Ziuganov, V.V. (1995). Reproductive isolation among lateral plate phenotypes (low, partial, complete) of the threespine stickleback, *Gasterosteus aculeatus*, from the White Sea basin and the Kamchatka Peninsula, Russia. — Behaviour 132: 1173-1181.

Ziuganov, V.V. & Zotin, A.A. (1995). Pelvic girdle polymorphism and reproductive barriers in the ninespine stickleback *Pungitius pungitius* (L.) from northwest Russia. — Behaviour 132: 1095-1105.

Brown, G.E. & Godin, J.G.J. (1997). Anti-predator responses to conspecific and heterospecific skin extracts by threespine sticklebacks: alarm pheromones revisited. — Behaviour 134: 1123-1134.

Peeke, H.V.S. & Figler, M.H. (1997). Form and function of habituation and sensitization of male courtship in the three-spined stickleback (*Gasterosteus aculeatus* L.). — Behaviour 134: 1273-1287.

Barber, I. & Arnott, S.A. (2000). Split-clutch IVF: a technique to examine indirect fitness consequences of mate preferences in sticklebacks. — Behaviour 137: 1129-1140.

Bergstrom, C.A. & Reimchen, T.E. (2000). Functional implications of fluctuating asymmetry among endemic populations of *Gasterosteus aculeatus*. — Behaviour 137: 1097-1112.

Bolyard, K.J. & Rowland, W.J. (2000). The effects of spatial context and social experience on the territorial aggression of male threespine stickleback. — Behaviour 137: 845-864.

Bornestaf, C. & Borg, B. (2000). Endogenous breeding cycles in male threespine sticklebacks, *Gasterosteus aculeatus*. — Behaviour 137: 921-932.

Fernandez, C., Hermida, M., Amaro, R. & San Miguel, E. (2000). Lateral plate variation in Galician stickleback populations in the rivers Mino and Limia, NW Spain. — Behaviour 137: 965-979.

Girvan, J.R. & Braithwaite, V.A. (2000). Orientation behaviour in sticklebacks: modified by experience or population specific? — Behaviour 137: 833-843.

Grand, T.C. (2000). Risk-taking by threespine stickleback (*Gasterosteus aculeatus*) pelvic phenotypes: does morphology predict behaviour? — Behaviour 137: 889-906.

Hay, D.E. & McPhail, J.D. (2000). Courtship behaviour of male threespine sticklebacks (*Gasterosteus aculeatus*) from old and new hybrid zones. — Behaviour 137: 1047-1063.

Hechter, R.P., Moodie, P.F. & Moodie, G.E.E. (2000). Pectoral fin asymmetry, dimorphism and fecundity in the brook stickleback, *Culaea inconstans*. — Behaviour 137: 999-1009.

Ishikawa, M. & Mori, S. (2000). Mating success and male courtship behaviors in three populations of the threespine stickleback. — Behaviour 137: 1065-1080.

Jenkins, J.R. & Rowland, W.J. (2000). Stimulus-specific and response-specific habituation in courting stickleback: developmental and functional considerations. — Behaviour 137: 933-945.

Kraak, S.B.M., Bakker, T.C.M. & Hocevar, S. (2000). Stickleback males, especially large and red ones, are more likely to nest concealed in macrophytes. — Behaviour 137: 907-919.

Krause, J., Cheng, D.J.S., Kirkman, E. & Ruxton, G.D. (2000). Species-specific patterns of refuge use in fish: the role of metabolic expenditure and body length. — Behaviour 137: 1113-1127.

McKinnon, J.S., Demayo, R.F., Granquist, R. & Weggel, L. (2000). Female red throat coloration in two populations of threespine stickleback. — Behaviour 137: 947-963.

McLennan, D.A. (2000). The macroevolutionary diversification of female and male components of the stickleback breeding system. — Behaviour 137: 1029-1045.

Peeke, H.V.S. & Morgan, L.E. (2000). Behavioral differentiation of adjacent marine and fluvial populations of threespine stickleback in California: a laboratory study. — Behaviour 137: 1011-1027.

Reimchen, T.E. (2000). Predator handling failures of lateral plate morphs in *Gasterosteus aculeatus*: functional implications for the ancestral plate condition. — Behaviour 137: 1081-1096.

Robinson, B.W. (2000). Trade offs in habitat-specific foraging efficiency and the nascent adaptive divergence of sticklebacks in lakes. — Behaviour 137: 865-888.

Wootton, R.J. & Smith, C. (2000). A long-term study of a short-lived fish: the demography of *Gasterosteus aculeatus*. — Behaviour 137: 981-997.

Mussen, T.D. & Peeke, H.V.S. (2001). Nocturnal feeding in the marine threespine stickleback (*Gasterosteus aculeatus* L.): modulation by chemical stimulation. — Behaviour 138: 857-871.

Chivers, D.P., Mirza, R.S. & Johnston, J.G. (2002). Learned recognition of heterospecific alarm cues enhances survival during encounters with predators. — Behaviour 139: 929-938.

Mirza, R.S. & Chivers, D.P. (2003). Fathead minnows learn to recognize heterospecific alarm cues they detect in the diet of a known predator. — Behaviour 140: 1359-1369.

Rush, V.N., McKinnon, J.S., Abney, M.A. & Sargent, R.C. (2003). Reflectance spectra from free-swimming sticklebacks (*Gasterosteus*): social context and eye-jaw contrast. — Behaviour 140: 1003-1019.

Aguirre, W.E., Doherty, P.K. & Bell, M.A. (2004). Genetics of lateral plate and gillraker phenotypes in a rapidly evolving population of threespine stickleback. — Behaviour 141: 1465-1483.

Barber, I., Walker, P. & Svensson, P.A. (2004). Behavioural responses to simulated avian predation in female three-spined sticklebacks: the effect of experimental *Schistocephalus solidus* infections. — Behaviour 141: 1425-1440.

Borg, B., Bornestaf, C., Hellqvist, A., Schmitz, M. & Mayer, I. (2004). Mechanisms in the photoperiodic control of reproduction in the stickleback. — Behaviour 141: 1521-1530.

Candolin, U. (2004). Effects of algae cover on egg acquisition in male three-spined stickleback. — Behaviour 141: 1389-1399.

Frommen, J.G. & Bakker, T.C.M. (2004). Adult three-spined sticklebacks prefer to shoal with familiar kin. — Behaviour 141: 1401-1409.

von Hippel, F.A. & Weigner, H. (2004). Sympatric anadromous-resident pairs of threespine stickleback species in young lakes and streams at Bering Glacier, Alaska. — Behaviour 141: 1441-1464.

Hosemann, K.E., Colosimo, P.E., Summers, B.R. & Kingsley, D.M. (2004). A simple and efficient microinjection protocol for making transgenic sticklebacks. — Behaviour 141: 1345-1355.

Keivany, Y. & Nelson, J.S. (2004). Phylogenetic relationships of sticklebacks (Gasterosteidae), with emphasis on ninespine sticklebacks (*Pungitius* spp.). — Behaviour 141: 1485-1497.

Kingsley, D.M., Zhu, B.L., Osoegawa, K., De Jong, P.J., Schein, J., Marra, M., Peichel, C., Amamiya, C., Schluter, D., Balabhadra, S., Friedlander, B., Cha, Y.M., Dickson, M., Grimwood, J., Schmutz, J., Talbot, W.S. & Myers, R. (2004). New genomic tools for molecular studies of evolutionary change in threespine sticklebacks. — Behaviour 141: 1331-1344.

Le Comber, S.C., Faulkes, C.G., van Look, K.J.W., Holt, W.V. & Smith, C. (2004). Recovery of sperm activity after osmotic shock in the three-spined stickleback: implications for pre-oviposition ejaculation. — Behaviour 141: 1555-1569.

Mayer, I., Borg, B. & Pall, M. (2004). Hormonal control of male reproductive behaviour in fishes: a stickleback perspective. — Behaviour 141: 1499-1510.

McLennan, D.A. (2004). Male brook sticklebacks' (*Culaea inconstans*) response to olfactory cues. — Behaviour 141: 1411-1424.

Pall, M.K., Liljander, M. & Borg, B. (2004). Prolactin diminishes courtship behaviour and stimulates fanning in nesting male three-spined sticklebacks, *Gasterosteus aculeatus*. — Behaviour 141: 1511-1519.

Rick, I.P., Modarressie, R. & Bakker, T.C.M. (2004). Male three-spined sticklebacks reflect in ultraviolet light. — Behaviour 141: 1531-1541.

Rowland, W.J. (2004). Sexual arousal increases mate selectivity in the stickleback. — Behaviour 141: 1371-1387.

Starner, H., Pahlsson, C. & Linden, M. (2004). Tandem repeat polymorphism and heteroplasmy in the mitochondrial DNA control region of threespine stickleback (*Gasterosteus aculeatus*). — Behaviour 141: 1357-1369.

Wootton, R.J. (2004). A simulation study of the consequences of growth depression and compensatory growth in the threespine stickleback, *Gasterosteus aculeatus* L. — Behaviour 141: 1543-1553.

Christen, M. & Milinski, M. (2005). The optimal foraging strategy of its stickleback host constrains a parasite's complex life cycle. — Behaviour 142: 979-996.

Ward, J.L. & McLennan, D.A. (2006a). Changes in agonistic, courtship and parental displays of the male brook stickleback, *Culaea inconstans*, across the breeding cycle. — Behaviour 143: 33-56.

Ward, J.L. & McLennan, D.A. (2006b). The relative influences of sexual and natural selection upon the evolution of male nuptial colouration in the brook stickleback, *Culaea inconstans*. — Behaviour 143: 483-510.

Webster, M.M., Atton, N., Ward, A.J.W. & Hart, P.J.B. (2007a). Turbidity and foraging rate in threespine sticklebacks: the importance of visual and chemical prey cues. — Behaviour 144: 1347-1360.

Webster, M.M., Ward, A.J.W. & Hart, P.J.B. (2007b). Boldness is influenced by social context in threespine sticklebacks (*Gasterosteus aculeatus*). — Behaviour 144: 351-371.

Baker, J.A., Heins, D.C., Foster, S.A. & King, R.W. (2008). An overview of life-history variation in female threespine stickleback. — Behaviour 145: 579-602.

Barber, L., Wright, H.A., Arnott, S.A. & Wootton, R.J. (2008). Growth and energetics in the stickleback-*Schistocephalus* host-parasite system: a review of experimental infection studies. — Behaviour 145: 647-668.

Bernhardt, R.R. & von Hippel, F.A. (2008). Chronic perchlorate exposure impairs stickleback reproductive behaviour and swimming performance. — Behaviour 145: 527-559.

Cresko, W.A. (2008). Special issue on stickleback behavior and evolution — Preface. — Behaviour 145: 409-411.

Foster, S.A., Shaw, K.A., Robert, K.L. & Baker, J.A. (2008). Benthic, limnetic and oceanic threespine stickleback: profiles of reproductive behaviour. — Behaviour 145: 485-508.

Frommen, J.G., Luz, C., Mazzi, D. & Bakker, T.C.M. (2008). Inbreeding depression affects fertilization success and survival but not breeding coloration in threespine sticklebacks. — Behaviour 145: 425-441.

Heins, D.C. & Baker, J.A. (2008). The stickleback-*Schistocephalus* host-parasite system as a model for understanding the effect of a macroparasite on host reproduction. — Behaviour 145: 625-645.

von Hippel, F.A. (2008). Conservation of threespine and ninespine stickleback radiations in the Cook Inlet Basin, Alaska. — Behaviour 145: 693-724.

Hoffmann, E., Osterman, A., Mayer, I. & Borg, B. (2008). 11-ketotestosterone is not responsible for the entire testicular effect on male reproductive behaviour in the threespine stickleback. — Behaviour 145: 509-525.

Jones, F.C., Brown, C. & Braithwaite, V.A. (2008). Lack of assortative mating between incipient species of stickleback from a hybrid zone. — Behaviour 145: 463-484.

Kimmel, C.B., Aguirre, W.E., Ullmann, B., Currey, M. & Cresko, W.A. (2008). Allometric change accompanies opercular shape evolution in Alaskan threespine sticklebacks. — Behaviour 145: 669-691.

Kitano, J., Mori, S. & Peichel, C.L. (2008). Divergence of male courtship displays between sympatric forms of anadromous threespine stickleback. — Behaviour 145: 443-461.

Reimchen, T.E., Ingram, T. & Hansen, S.C. (2008). Assessing niche differences of sex, armour and asymmetry phenotypes using stable isotope analyses in Haida Gwaii sticklebacks. — Behaviour 145: 561-577.

Wisenden, B.D. (2008). Active space of chemical alarm cue in natural fish populations. — Behaviour 145: 391-407.

Wootton, R.J. (2008). Demography of the threespine stickleback, *Gasterosteus aculeatus*, in the Afon Rheidol 1972-1998: search for mechanisms, using simulation models. — Behaviour 145: 603-623.